Viruses and Interferon

Current Research

Edited by

Karen Mossman

Department of Pathology and Molecular Medicine
Institute for Infectious Disease Research
McMaster University
Hamilton, ON
Canada

Caister Academic Press

Caister Academic Press
Norfolk, UK

www.caister.com

British Library Cataloguing-in-Publication Data
A catalogue record for this book is available from the British Library

ISBN: 978-1-904455-81-3

Cover image adapted from Figure 3.4.

Printed and bound in Great Britain

Contents

Other Books of Interest

Emerging Trends in Antibacterial Discovery	2011
Epigenetics	2011
Metagenomics	2011
Nitrogen Cycling in Bacteria	2011
The Biology of Paramyxoviruses	2011
Helicobacter pylori	2011
Microbial Bioremediation of Non-metals	2011
Lactic Acid Bacteria and Bifidobacteria	2011
Essentials of Veterinary Parasitology	2011
Hepatitis C: Antiviral Drug Discovery and Development	2011
Streptomyces: Molecular Biology and Biotechnology	2011
Alphaherpesviruses: Molecular Virology	2011
Recent Advances in Plant Virology	2011
Vaccine Design: Innovative Approaches and Novel Strategies	2011
Salmonella: From Genome to Function	2011
Insect Virology	2010
Environmental Microbiology: Current Technology and Water Applications	2010
Sensory Mechanisms in Bacteria: Molecular Aspects of Signal Recognition	2010
Bifidobacteria: Genomics and Molecular Aspects	2010
Molecular Phylogeny of Microorganisms	2010
Nanotechnology in Water Treatment Applications	2010
Iron Uptake and Homeostasis in Microorganisms	2010
Caliciviruses: Molecular and Cellular Virology	2010
Epstein-Barr Virus: Latency and Transformation	2010
Lentiviruses and Macrophages: Molecular and Cellular Interactions	2010
RNA Interference and Viruses: Current Innovations and Future Trends	2010
Retroviruses: Molecular Biology, Genomics and Pathogenesis	2010
Frontiers in Dengue Virus Research	2010
Metagenomics: Theory, Methods and Applications	2010
Microbial Population Genetics	2010
Borrelia: Molecular Biology, Host Interaction and Pathogenesis	2010
Aspergillus: Molecular Biology and Genomics	2010
Environmental Molecular Microbiology	2010
Neisseria: Molecular Mechanisms of Pathogenesis	2010
Lab-on-a-Chip Technology: Biomolecular Separation and Analysis	2009
Lab-on-a-Chip Technology: Fabrication and Microfluidics	2009
Real-Time PCR: Current Technology and Applications	2009
ABC Transporters in Microorganisms	2009
Pili and Flagella: Current Research and Future Trends	2009

Contributors

Gaya K. Amarasinghe
Department of Biochemistry, Biophysics and Molecular Biology
Iowa State University
Ames, IA
USA

amarasin@iastate.edu

Christopher F. Basler
Department of Microbiology
Mount Sinai School of Medicine
New York, NY
USA

chris.basler@mssm.edu

Jessica A. Campbell
Department of Internal Medicine
Washington University School of Medicine
St. Louis, MO
USA

jcampbe4@alumni.nd.edu

Srikanth Chiliveru
Department of Medical Microbiology and Immunology
Aarhus University
Aarhus
Denmark

srikant@microbiology.au.dk

Stephanie J. DeWitte-Orr
Department of Pathology and Molecular Medicine
Institute for Infectious Disease Research
McMaster University
Hamilton, ON
Canada

dewitte@mcmaster.ca

Mariano Esteban
Department of Cellular and Molecular Biology
Centro Nacional de Biotecnología (CSIC)
Ciudad Universitaria Cantoblanco
Madrid
Spain

mesteban@cnb.csic.es

Eleanor N. Fish
Toronto General Research Institute
University Health Network;
Department of Immunology
University of Toronto
Toronto, ON
Canada

en.fish@utoronto.ca

Takashi Fujita
Department of Molecular Genetics
Institute for Virus Research;
Graduate School of Biostudies
Kyoto University
Kyoto
Japan

tfujita@virus.kyoto-u.ac.jp

Michael Gale Jr.
Department of Immunology
University of Washington School of Medicine
Seattle, WA
USA

mgale@uw.edu

Adolfo García-Sastre
Department of Microbiology;
Department of Medicine
Division of Infectious Diseases;
Global Health and Emerging Pathogens Institute
Mount Sinai School of Medicine
New York, NY
USA

Adolfo.Garcia-Sastre@mssm.edu

Bryan M. Gebhardt
Department of Ophthalmology
Louisiana State University Health Sciences Center
New Orleans, LA
USA

bgebha@isuhsc.edu

William P. Halford
Department of Microbiology and Immunology
Southern Illinois University School of Medicine
Springfield, IL
USA

halford@siumed.edu

Deborah J. Lenschow
Departments of Internal Medicine and Pathology and Immunology
Washington University School of Medicine
St. Louis, MO
USA

dlenscho@dom.wustl.edu

Helene M. Liu
Department of Immunology
University of Washington School of Medicine
Seattle, WA
USA

minyiliu@uw.edu

Karen L. Mossman
Department of Pathology and Molecular Medicine
Institute for Infectious Disease Research
McMaster University
Hamilton, ON
Canada

mossk@mcmaster.ca

Kazuhide Onoguchi
Department of Molecular Genetics
Institute for Virus Research
Kyoto University
Kyoto
Japan

konoguch@virus.kyoto-u.ac.jp

Søren R. Paludan
Department of Medical Microbiology and Immunology
University of Aarhus
Aarhus
Denmark

srp@microbiology.au.dk

Beatriz Perdiguero
Department of Cellular and Molecular Biology
Centro Nacional de Biotecnología
Ciudad Universitaria Cantoblanco
Madrid
Spain

perdigue@cnb.csic.es

Ramtin Rahbar
Toronto General Research Institute
University Health Network;
Department of Immunology
University of Toronto
Toronto, ON
Canada

r.rahbar@utoronto.ca

Marisela Rodriguez
Department of Internal Medicine
Washington University School of Medicine
St. Louis, MO
USA

mrrodrig@dom.wustl.edu

Kiyohiro Takahasi
Department of Molecular Genetics
Institute for Virus Research;
Institute for Innovative NanoBio Drug Discovery and Development
Graduate School of Pharmaceutical Sciences
Kyoto University
Kyoto
Japan

kiyohiro@pharm.kyoto-u.ac.jp

Gijs A. Versteeg
Department of Microbiology
Mount Sinai School of Medicine
New York, NY
USA

gijs.versteeg@mssm.edu

Ben X. Wang
Toronto General Research Institute
University Health Network;
Department of Immunology
University of Toronto
Toronto, ON
Canada

ben.wang@utoronto.ca

Mitsutoshi Yoneyama
Division of Molecular Immunology
Medical Mycology Research Center
Chiba University
Chiba;
PRESTO
Japan Science and Technology Agency
Saitama
Japan

myoneyam@faculty.chiba-u.jp

Preface

Since their discovery in 1957, based on their ability to 'interfere' with virus replication, interferons have been shown to play important roles in shaping innate and adaptive immune responses, regulating the cell cycle and fighting tumour cells. Human interferons are classified into three major types based on the receptor complex that they bind and signal through. Viruses are potent inducers of type I and type III interferon, predominantly through the cellular recognition of viral double-stranded RNA. Interferon rapidly induces an 'anti-viral state' in surrounding cells following the induction of hundreds of interferon-stimulated genes. To ensure their survival, viruses have evolved multiple strategies to circumvent the anti-viral effects of interferons.

In this book, a panel of international experts reviews the current status of the unique interplay between viruses and interferons. The book opens with a summary of the antiviral effects of extracellular double-stranded RNA, the 'viral toxin', followed by a review of the properties of type I and type III interferons, and the role of interferon-stimulated genes. The following five chapters are devoted to understanding the diverse strategies used by clinically relevant human viruses to subvert host interferon responses. The book will conclude with a summary of the clinical uses of interferon to fight viral diseases and cancer.

Karen Mossman

The Antiviral Effects of Extracellular dsRNA

1

Stephanie J. DeWitte-Orr and Karen L. Mossman

Abstract

Viral double-stranded RNA (dsRNA), a replication by-product of almost all viruses, has been studied for over 55 years, first as a toxin, then as a type I IFN inducer, a viral mimetic and an immunomodulator for therapeutic purposes. Not only does dsRNA function as a pathogen-associated molecular pattern (PAMP), sensed by host germline encoded pattern recognition receptors (PRRs) to stimulate innate immune responses, it also acts as a bridge to activate antiviral adaptive immune responses. DsRNA is generated intracellularly during a virus infection, but is released into the extracellular space during cell lysis. This review will focus on the structure and generation (both endogenous and viral) of extracellular dsRNA, and the host sensing mechanisms that result in type I IFN- and RNAi-mediated antiviral responses. The possible therapeutic applications of these findings will also be discussed. The goal of this review is to highlight the importance of this unique nucleic acid, with a focus on how its extracellularity influences its effects on the host and how these effects can be manipulated for our therapeutic purposes.

Introduction

Double-stranded (ds)RNA is produced by almost all viruses during their replication (Jacobs and Langland, 1996). These long, duplexed RNA molecules accumulate within the host cell, and act as a potent pathogen-associated molecular pattern (PAMP). Much is known of the effects of intracellular dsRNA, which begin when it is sensed by the intracellular pattern recognition receptors (PRRs) for dsRNA, namely the toll-like receptors (TLRs), RIG-I like receptors (RLRs) and the NOD-like receptors (NLRs). The end result of these pathways is production of type I interferons (IFNs) and pro-inflammatory cytokines and establishment of an antiviral state (Kawai and Akira, 2009). Importantly, these intracellular response pathways, which will be discussed in detail in later chapters, address only part of dsRNA's influence on the host. During cell lysis, dsRNA molecules are released into the extracellular environment. Although the effects of extracellular dsRNA has been observed for many years, very little is known of this molecule, including its form, dissemination, and mechanism of action. DsRNA appears to have specific attributes, which make it an ideal signalling molecule. As opposed to ssRNA, dsRNA is relatively nuclease resistant and is able to persist in both extracellular and intracellular environments. Also, there are a handful of direct and a much larger number of indirect studies that demonstrate both local and systemic effects of extracellular dsRNA. This chapter will discuss why dsRNA is uniquely fit to function as an extracellular signalling molecule, and how these characteristics make dsRNA-based therapeutics a promising new avenue for the future.

Double-stranded RNA: the molecule

Double-stranded RNA structure

The unique physical properties of dsRNA determine its biological function. DsRNA generated during a viral infection, termed 'native' dsRNA, is hypothesized to be associated with a number of host and/or viral proteins. The ultimate tertiary structure of a native dsRNA molecule and the composition of associated proteins is not known and has proven difficult to elucidate. The structure of naked, synthetic dsRNA has been determined using fibre diffraction, X-ray diffraction and high-resolution nuclear magnetic resonance (NMR) analysis (Nicholson, 2006). Essentially, dsRNA is composed of two antiparallel strands of RNA in a right-handed helix, which displays Watson–Crick basepairing. DsRNA demonstrates an A-helix motif with an 11-fold helical pitch, compared with the most common form of DNA in biological systems, which is a B-helix with a 10-fold helical pitch. The helical structure of dsRNA contains a narrow, deep major groove and a shallow minor groove. It is the minor groove, with its available 2′-hydroxyl groups, which provides an interactive surface for binding proteins, other nucleic acids and small ligands. The ribose conformation of dsRNA is C3′-endo, causing the A-helix to be underwound compared with a B-helix. The net effect of this conformational difference is a change in helical rise, from 2.7 Å for dsRNA compared with 3.4 Å for B-DNA (Tang and Draper, 1994). The gel electrophoretic mobility of dsRNA is slower than DNA molecules of equal length. This difference in gel migration has been attributed to a lower net charge density caused by an enhanced condensation of counter ions from the narrowly spaced phosphate groups of dsRNA (Gast and Sänger, 1994). The diameter of dsRNA is 30% larger than DNA, which may also influence its electrophoretic mobility (Van Den Hout *et al.*, 2010). DsRNA is also less flexible than DNA, and contains a natural curvature (30–40° over 80 bp; Tang and Draper, 1994).

Double-stranded RNA determined experimentally

The unique structural characteristics of dsRNA enable it to be identified experimentally from other nucleic acids. A summary of these experimental criteria can be found in Table 1.1. Currently, the two most commonly used methods for identifying dsRNA are differential nuclease resistance/sensitivity assays and immunofluorescence using an anti-dsRNA monoclonal antibody (e.g. J2 and K1). The resistance of dsRNA to degradation by many RNases is a result of its structure, as its major groove is too narrow for interactions with macromolecules such as nucleases and its high T_m makes it difficult to destabilize (Libonati *et al.*, 1980). The anti-dsRNA antibodies detect dsRNA but not ssRNA, DNA or RNA–DNA hybrids, independent of sequence. It has been suggested that these antibodies are structural probes, which recognize the A-helical configuration of dsRNA

Table 1.1 Experimental criteria used to define dsRNA compared with ssRNA

	dsRNA	ssRNA
Molecular structure	Watson–Crick basepairing, two complementary RNA strands in antiparallel formation	Single-stranded RNA strand
Molar absorbance	dsRNA < ssRNA	
Absolute hypochromicity	dsRNA > ssRNA	
T_m value	dsRNA* > ssRNA	
Cellulose CF-11 binding in EtOH containing buffers	Yes	No
Sedimentation coefficient	≥ 8S	Variable
Degraded by RNase A, B or T1	No	Yes
Degraded by RNase III	Yes	No
Detected using anti-dsRNA monoclonal antibody (J2)	Yes	No

*dsRNA has a T_m greater than that of ssRNA, as well as DNA of similar base composition.

Criteria listed are a summary of dsRNA characteristics as defined in the literature (Libonati *et al.*, 1980; Schonborn *et al.*, 1991; Nicholson, 2006).

(Schonborn *et al.*, 1991). When investigating dsRNA experimentally it is imperative to avoid using phenol in the extraction protocols. Over 40 years ago it was shown that these extraction methods promote formation of dsRNA artefacts (Borst and Weissmann, 1965; Oberg and Philipson, 1971). Unfortunately some recent publications investigating dsRNA have used phenol-based methods. Therefore, care should be taken when interpreting results from dsRNA studies using phenol-based extraction methods.

Double-stranded RNA production

Viral dsRNA

As mentioned in the introduction, it is commonly accepted that most viruses produce dsRNA sometime during their replicative cycle. Although not every virus has been tested for dsRNA production, there are only a handful of viruses which when tested have been found to be negative for dsRNA production. These viruses include influenza A virus (strain PR8, ΔNS1), La Crosse virus, Sendai virus and Newcastle disease virus (Weber *et al.*, 2006; Kato *et al.*, 2008). It should be noted that these results were based on recognition of dsRNA by the J2 antibody alone, and it is impossible to say whether all viral dsRNAs can be detected with this antibody. Indeed, there are conflicting reports regarding dsRNA production by influenza virus. DsRNA can be detected from cells infected with influenza virus, both *in vitro* and *in vivo*, using differential nuclease resistance/sensitivity assays (Majde *et al.*, 1991) and immunofluorescence (DeWitte-Orr and Mossman, unpublished). There are a number of different factors between studies, including strain, cell type, cellular compartmentalization and detection method. These differences highlight that multiple factors require consideration before making conclusions on the ability of a particular virus to produce dsRNA.

The mechanisms of dsRNA production are not well elucidated. It is generally accepted that ssRNA viruses produce dsRNA from replicative intermediates or some portion of the genome self-annealing. For dsRNA viruses, the genome is believed to be the source of dsRNA. Viruses with dsRNA genomes go to great lengths to keep their genomes 'undetectable' to the host dsRNA sensors. For example, the reovirus genome remains in the capsid during the entire replication cycle (Urbano and Urbano, 1994). With regards to retroviruses, secondary structures made by ssRNA molecules may be the source of duplexed RNA. Finally, DNA viruses are thought to produce dsRNA via bidirectional convergent transcription. Complementary mRNAs are then produced which when annealed produce long strands of dsRNA. As for localization, viral dsRNA appears to accumulate in the cellular compartment utilized for replication. Thus, dsRNA is cytoplasmic when viruses replicate in the cytoplasm. In most cases where viruses replicate in the nucleus, dsRNA can be found in both the nucleus and cytoplasm (Kumar and Carmichael, 1998). Also, the length of the dsRNA molecules generated seems to reflect viral genome length (Kato *et al.*, 2008; DeWitte-Orr *et al.*, 2009).

Host dsRNA

To state that dsRNA is only present in a cell during viral infections is an over-simplification. In the 1970s, endogenous dsRNA was found in the mammalian nucleus (Jelinek and Darnell, 1972). Currently, at least two sources for these nuclear dsRNA molecules are known: (1) natural antisense transcripts (NATs) produced from annealing of overlapping mRNA transcripts and (2) microRNA (miRNA) precursors, which contain secondary stem–loop structures. In both cases the function of these duplexed regions within RNA molecules is primarily to control gene transcription (Lehner *et al.*, 2002; Werner, 2005). In the case of miRNA, and perhaps also NATs, gene transcription control occurs via the RNA interference (RNAi) pathway (Matranga and Zamore, 2007; Werner *et al.*, 2009). Transcriptional control maintains homeostatic conditions within the cell and does not represent an antiviral response. In plants and invertebrates, long viral dsRNA is cleaved into short interfering (si)RNA duplexes to control viral gene transcription via the RNAi pathway, as a means of effective innate antiviral activity. In mammals, however, siRNA generation has yet to be an identified antiviral strategy (Umbach and Cullen, 2009). The effects of dsRNA in non-mammalian systems in general and RNAi pathways in particular will be discussed later in this chapter.

If endogenous dsRNA molecules are present in a host cell, the question arises, why do these molecules not trigger a type I IFN mediated antiviral response? The answer to this question has not been fully elucidated; however, the factors dictating a dsRNA-induced antiviral response appear to be location, length, chemical composition and abundance (Wang and Carmichael, 2004; Marques *et al.*, 2006; Nallagatla *et al.*, 2008). It has been shown that increased length and abundance of dsRNA molecules leads to a stronger antiviral response (DeWitte-Orr *et al.*, 2009), by facilitating receptor clustering (de Bouteiller *et al.*, 2005; Lemaire *et al.*, 2008). As opposed to host dsRNAs, which are commonly found in short stretches (≤20 bp), viral dsRNAs are often much longer. Sensors of viral dsRNA have a minimum length requirement for recognition, often >30 bp (de Bouteiller *et al.*, 2005; Lemaire *et al.*, 2008). Also, a 5′ triphosphate group or the lack of nucleoside modifications may also function as a flag, distinguishing a 'foreign' dsRNA molecule (Nallagatla *et al.*, 2008). With continued study of dsRNA sensors will come a better understanding of their discriminatory powers. It is unclear what the effects of endogenous dsRNA would be when extracellular, under non-homeostatic conditions, such as following cell lysis. In this situation, host dsRNAs could act as danger-associated molecular patterns (DAMPs). Recently, evidence for this hypothesis has been found in mice, where necrotic cell death stimulated TLR3 activation and subsequent inflammation, independent of viral infection (Cavassani *et al.*, 2008). Endogenous RNA molecules functioning as ligands for dsRNA sensors will be discussed further in the autoimmunity section of this chapter.

History of extracellular dsRNA as a toxin

The history of dsRNA begins with reports that a by-product of virus replication induced a strong systemic response, independent of the infectious particle. In the 1950s, this unidentified replication by-product was termed a viral toxin (Cox, 1953). It took many years to determine that this viral toxin was in fact dsRNA generated during a viral infection. Interest in dsRNA was tightly linked to the discovery of type I IFNs. When in 1957 Isaacs and Lindemann discovered the antiviral activities of type I IFNs (Isaacs and Lindemann, 1957), there began a search for a stimulus for IFNs that could be used as an antiviral therapeutic. In 1963, Isaacs *et al.* found that exogenous ssRNA could induce type I IFNs; however, this stimulus was deemed too weak for therapeutic use. In the same year, it was found that viruses produced duplexed RNA (Montagnier, 1963) and soon after it was discovered that dsRNA (either viral or synthetic) could be added exogenously, either *in vitro* or *in vivo*, to induce type I IFNs (Lampson *et al.*, 1967). Dr. M.R. Hilleman of the Merck Institute for Therapeutic Research was the first to discover that dsRNA was a potent inducer of type I IFNs and to suggest that the synthetic RNA duplex, polyinosinic:polycytidylic acid (poly-IC), could be used as an antiviral therapeutic agent. The potency of dsRNA held promise for antiviral therapies; early studies showed that a single dsRNA molecule when detected by a cell is sufficient to produce IFNs (Marcus, 1977). Soon poly-IC was being hailed as 'the promise of a drug that gives people wide-range immunity to viruses' (Anon, 1967). This optimism did not last, as it was soon discovered that systemic extracellular poly-IC treatment caused dramatic adverse effects, understandable considering the stability and potency of dsRNA, and by the 1970s the 'intrinsic toxicity of dsRNA' was being discussed. Effects of systemic dsRNA treatments were being compared with the pathological effects of the viral 'toxins' of the 1950s (Carter and De Clercq, 1974). Perhaps because of these toxic effects, the focus of dsRNA-based research shifted, from antiviral therapies to a model for inducing IFN signalling pathways experimentally. The benefit of this strategy was that dsRNA, particularly poly-IC, has given researchers a tool for studying IFN-based pathways in a reproducible manner, without the complications of a live viral infection. These experiments are still being performed, and are providing valuable information regarding dsRNA- and IFN-induced signalling pathways. These experiments, however, represent artificial conditions, necessitating a re-focus to studying dsRNA-mediated antiviral responses using 'native' dsRNA molecules. This approach is of particular importance if viable dsRNA-based therapeutics are to be developed. The most accurate understanding of dsRNA-mediated responses will be achieved when we investigate the effects of native viral dsRNA, within the context of a live, virus infection.

Extracellular dsRNA signalling

Historically, there are a number of studies that suggest viral dsRNA mediates both local and systemic effects, ultimately contributing to the pathogenesis of a viral infection. Not only does viral dsRNA mediate antiviral responses intracellularly, where it is generated, but when released from lysed cells, extracellular dsRNA can also influence antiviral responses. Reports of these effects are not limited to the early studies of viral 'toxins'. Though not always reported as such, any study that introduces dsRNA into a biological system without a liposome-based transfection reagent reports the effects of extracellular dsRNA, as these responses begin with cells taking up dsRNA from the extracellular space.

Perhaps the best-studied model for dsRNA as a modulator of viral pathogenesis is with influenza virus infections. Animals administered viral dsRNA systemically showed 'flu-like' symptoms similar to those infected with influenza (Fang *et al.*, 1999). Also, extracellular dsRNA could be detected from influenza-infected cell cultures (Majde *et al.*, 1998) and from the lungs of influenza-infected mice (Majde *et al.*, 1991). A model for this phenomenon has been suggested, where dsRNA from a virally infected cell is released into the extracellular environment, where it is taken up by infiltrating macrophages which are primed and activated by locally produced cytokines. These macrophages produce more cytokines, leading to the systemic flu-like symptoms (Majde, 2000; von Herrath and Bot, 2003).

In this proposed model, it is unclear how far dsRNA can travel from the site of infection. Though never studied directly, it has been suggested that the electronegative properties of dsRNA would limit its ability to disseminate (Majde, 2000). However, many *in vivo* experiments using synthetic dsRNA report systemic effects. For example, both intraperitoneal (Kumar *et al.*, 2008) and intravenous (McCartney *et al.*, 2009) injections of 'naked' poly-IC have shown to affect cells in organs such as the spleen and liver of mice. In plants and nematodes, dsRNA molecules can travel between cells and to distant tissues (Obbard *et al.*, 2009). Until recently it was thought that only plants and nematodes supported systemic RNAi responses mediated by these dsRNA molecules; however, more recently in Drosophila it was found that viral dsRNA could travel some distance from the site of active infection and induce antiviral RNAi responses in uninfected cells (Saleh *et al.*, 2009). Though the extent is unclear, these studies suggest that extracellular dsRNA does have the capacity to disseminate. Clearly, more research needs to be done to examine the fate of extracellular dsRNA, and its effects on systemic host responses; however, its influence during viral infections cannot be denied and should be taken into account when studying and treating viral infections.

Double-stranded RNA sensors

Host cells express a number of different proteins which are able to sense viral dsRNA, with a representative in almost every cellular compartment (Fig. 1.1). DsRNA recognition is generally thought to be sequence independent and instead based on its unique A-type helical structure and/or its net negative charge (Bobst *et al.*, 1976). A number of host proteins, including the dsRNA dependent protein kinase (PKR), utilize a well-conserved dsRNA-binding domain (DRBD) to recognize viral dsRNA (Saunders and Barber, 2003). Other sensors use DRBD-independent mechanisms for sensing dsRNA. For example, toll-like receptor 3 (TLR3) can bind both A and B helical nucleic acids; however, only A-type nucleic acids mediate dimerization, which is required for activation (Pirher *et al.*, 2008). It is believed that the dsRNA backbone and the position of the grooves are major determinants for binding (Liu *et al.*, 2008). Also, RIG-I-like receptors (RLRs) bind RNA ligands via a conserved large positively charged surface in the RNA binding loop of their

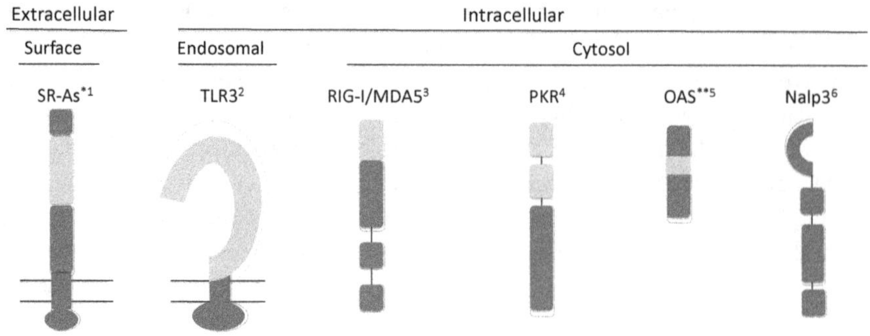

	SR-As	TLR3	RIG-I/MDA5	PKR	OAS	Nalp3
Binding domain	collagenous	ECD	CTD	DRBD	active site	N/E
Binding Mechanism	charge	structure	charge	structure	charge/ sequence	N/E
Minimum Binding length (bp)	N/E	40-50	N/E	30	30	N/E
Activation	N/E	D	D	D	O	O
Function	carrier	ISG/IFN ↑	ISG/IFN ↑	block translation	block transcription	inflammation

Figure 1.1 Summary of dsRNA sensors involved in inducing antiviral responses. The dsRNA-binding region is indicated as a light region within the structure. *The structure presented is that of SR-AI, as a representative of the SR-A group that includes SR-AI, -AII, MARCO, Scara3, 4 and 5. ** The structure of OAS1 is included as a representative of the OAS group that includes OAS1, 2, 3 and L. Nalp3 has yet to be shown to bind dsRNA directly, therefore no binding information is included. [1]Murphy *et al.* (2005), DeWitte-Orr *et al.* (2010). [2]Fukuda *et al.* (2008), Liu *et al.* (2008). [3]Barral *et al.* (2009), Li *et al.* (2009). [4]Saunders and Barber (2003), Sadler and Williams (2007), Lemaire *et al.* (2008). [5]Minks *et al.* (1979), Hartmann *et al.* (2003), Kodym *et al.* (2009). [6]Kanneganti *et al.* (2006a,b), Owen and Gale (2009). CTD, C terminal domain; D, dimerization; DRBD, dsRNA binding domain; ECD, ectodomain; N/E, not yet elucidated; O, oligomerization.

C-terminal domain (CTD; Takahasi *et al.*, 2009). The ligand specificity observed between RIG-I and MDA5 may be a result of a phenylalanine change within the CTD (Takahasi *et al.*, 2008). Host cells use these dsRNA-binding proteins as sentinels to detect viral infections. Indeed, it is the presence and quantity of dsRNA binding proteins that highlight the importance of dsRNA molecules in the innate immune response.

Sensing at the cell surface

For extracellular dsRNA to have an effect it must be taken up at the cell surface. Class A scavenger receptors (SR-As) are integral membrane proteins, which bind a number of ligands including nucleic acids (Murphy *et al.*, 2005). Recently, SR-As were found to be the surface receptor for extracellular dsRNA, mediating dsRNA-induced type I IFN (DeWitte-Orr *et al.*, 2010) and pro-inflammatory (Limmon *et al.*, 2008) responses. SR-As were found to be essential mediators of extracellular dsRNA entry via clathrin-mediated endocytosis, delivering dsRNA to the intracellular dsRNA sensors, namely the TLRs and RLRs (DeWitte-Orr *et al.*, 2010). Functionally, less IFN-stimulated gene (ISG) expression (DeWitte-Orr *et al.*, 2010) and lower pro-inflammatory cytokine levels (Limmon *et al.*, 2008) were found in the lungs of poly-IC stimulated SR-AI/II knockout mice compared with wild-type mice. Also, SR-AI/II knockout mice show increased susceptibility to HSV-1 infection (Suzuki *et al.*, 1997). Although SR-As are mainly thought to function as a 'carrier', binding extracellular dsRNA at the cell surface and delivering it to intracellular sensors, it is possible that SR-As themselves could be contributing to intracellular signalling pathways. In spite of a short cytoplasmic tail, SR-A ligand binding triggers a series of intracellular signalling pathways, including tyrosine phosphorylation of phospholipase

C-γ1 (PLC-γ1) and phosphatidylinositol 3-kinase (PI 3-kinase) activation, as well as activation of pathways involving protein kinase C (PKC) and mitogen-activated protein kinases (MAPKs; Murphy *et al.*, 2005). The mechanisms behind these signalling pathways and the link between endosomal dsRNA and cytosolic sensors are currently under investigation.

Although SR-As bind polyanionic ligands such as dsRNA via their collagenous domain (Murphy *et al.*, 2005), their ligand specificity is poorly understood, as negative charge is essential but not sufficient to define an SR-A ligand (Brown and Goldstein, 1983). Thus SR-A RNA ligands are not limited to duplexed RNA; SR-As can bind ssRNA molecules as well (Murphy *et al.*, 2005). This means that under non-homeostatic conditions (such as uncontrolled cell lysis), endogenous RNA molecules containing single-stranded or double-stranded motifs, could be acting as extracellular signalling molecules, delivered to endosomal TLRs by SR-As. Although not as stable as viral dsRNA, host ssRNA could be acting as a 'danger signal' in a more limited or local capacity. The possibility of endogenous RNA molecules acting as DAMPs will be discussed further in the autoimmunity section of this book chapter, but relevant to this section, it can be concluded that SR-As could be involved in mediating a network of extracellular signals carried by RNA, either single-stranded or double-stranded in nature.

Intracellular sensing and signalling

Once dsRNA enters the cell, intracellular PRRs are activated. The intracellular dsRNA sensor group contains the 'trinity' of innate immune sensors, including the TLRs (TLR3), RLRs (RIG-I and MDA5) and the NLRs (NLRP3), in addition to other important sensors such as PKR and the OAS (oligoadenylate synthetase)/RNaseL detection system. These sensors will be discussed in more depth in later chapters, thus only aspects of these pathways relevant to extracellular dsRNA will be discussed here.

TLR3 is a member of the TLR family, an evolutionarily conserved group of membrane bound receptors that recognize PAMPs and initiate innate immune responses. The RLR family of cytoplasmic dsRNA sensors consists of three members: RIG-I (retinoic acid-inducible gene I), MDA5 (melanoma differentiation-associated gene 5) and LGP2 (laboratory of genetics and physiology-2; Saito and Gale, 2008). Both RIG-I and MDA5 bind dsRNA and trigger downstream antiviral responses, while LGP2 can bind dsRNA but not signal. The NLRs are a family of 23 members in humans and at least 34 in mice; however, only a few NLRs have characterized PRR functions. NLRP3 is an NLR that is activated by dsRNA and was found to be important in limiting influenza A viral infection (Allen *et al.*, 2009; Thomas *et al.*, 2009). Importantly, although studies show dsRNA induces NLRP3 activation, none have shown that NLRP3 functions as a PRR for viral RNA species (Owen and Gale, 2009). PKR is a member of a small family of kinases that contains two dsRNA binding domains (Sadler and Williams, 2007), while OAS detects dsRNA through electrostatic interactions between the negatively charged nucleic acid and a positively charged groove in OAS (Hartmann *et al.*, 2003).

The mechanism of extracellular dsRNA activating these intracellular PRRs is not well understood. In resting cells TLR3 is found on the endoplasmic reticulum (Kim *et al.*, 2008) and on the cell surface of some cell types (Matsumoto *et al.*, 2002). During a viral infection it is recruited to the endosome, where upon endosomal acidification it can bind dsRNA (de Bouteiller *et al.*, 2005; Leonard *et al.*, 2008). Therefore it can be postulated that SR-A mediated uptake of dsRNA would deliver dsRNA to the endosomal compartment, and when acidified, TLR3 could bind the dsRNA and be activated. Less clear is how dsRNA escapes the endosome and activates the cytoplasmic sensors, the RLRs, NLRs, PKR and OAS/RNaseL. What is clear is that extracellular dsRNA can activate cytoplasmic sensors, including RIG-I and MDA5 (DeWitte-Orr *et al.*, 2010) and NLRP3 (Nalp3; Kanneganti *et al.*, 2006b).

Once dsRNA enters the cell, the intracellular sensors are activated and stimulate a number of signalling pathways. TLR3 and RLR signalling culminates in the activation of the transcription factor IFN regulatory factor (IRF)-3. When phosphorylated, IRF-3 dimerizes and translocates to the nucleus where it induces type I IFN production (Kawai and Akira, 2009). Type III IFN (λ) is also induced by dsRNA (Ank *et al.*, 2008). Type I IFNs induce the transcription of IFN-stimulated genes (ISGs), including IRF-7, which collectively with IRF-3 is important for up-regulation of IFNα/β by a positive feedback loop. These actions culminate with the establishment of an antiviral state, and active repression of viral replication (Taniguchi and Takaoka, 2002; Randall and Goodbourn, 2008). IFN λ signals through its cognate receptor, made of IL10Rβ and IL28Rα, to transduce signals similar to type I IFNs (Kotenko *et al.*, 2003). In contrast to type I IFNs, which are able to stimulate a response in nearly all nucleated cells, the cell types able to respond to type III IFNs are more restricted (Ank *et al.*, 2008). Instead of inducing IFNs, NLRP3 activated by dsRNA is involved in the assembly of the 'inflammasome', which is responsible for caspase 1 activation and the subsequent release of pro-inflammatory cytokines (IL-1β and IL18; Martinon and Tschopp, 2007). When bound to dsRNA, PKR dimerizes and autophosphorylates (Sadler and Williams, 2007). Activated PKR limits viral replication by inhibiting protein synthesis (via eIF2α phosphorylation), as well as controlling gene expression through NF-κB and IRF1 (Kumar *et al.*, 1997), and contributing to dsRNA-induced apoptosis (Gil and Esteban, 2000). Finally, when cytoplasmic OAS binds viral dsRNA, it oligomerizes, activates and produces 2′ to 5′-linked oligoadenylates from ATP, which are bound by latent RNaseL, causing its activation. Active RNaseL degrades ssRNA molecules, both host and virus derived, thereby controlling viral replication at the level of transcription (Silverman, 2007).

Effects of dsRNA in non-mammalian systems

Double-stranded RNA-induced type I IFN responses

The dsRNA-induced responses outlined thus far have focused on what is known in mammals; however, it is important to note that non-mammalian vertebrates appear to express conserved dsRNA-mediated antiviral pathways, similar to mammals. Despite the fact that IFN activity was discovered using an avian model (embryonated chicken eggs), antiviral responses in non-mammalian animal species are poorly understood. In fact, molecular characterization of non-mammalian IFNs began only 15 years ago (Sekellick, 1994). Birds, teleost (jawed) fish and reptiles all produce IFNs or IFN-like molecules with antiviral activity (Schultz *et al.*, 2004). In many cases, dsRNA-induced IFN-like activity was reported many years prior to molecular characterization of the IFNs themselves. For example, IFN activity was identified in rainbow trout (*Oncorhynchus mykiss*) infected with Egtved virus in 1973 (de Kinkelin and Dorson, 1973) and treated with poly-IC in 1990 (Eaton, 1990). After 1990, many studies have shown dsRNA-induced ISG production and establishment of an antiviral state in rainbow trout, similar to that in mammals, but it wasn't until 2007 that a rainbow trout type I IFN sequence was first identified (Zou *et al.*, 2007). The difficulty in identifying IFN homologues has been its low sequence similarity between species; for example, chicken and fish type I IFNs show 20–25% and 15% similarity, respectively, to their mammalian homologues (Schultz *et al.*, 2004).

DsRNA sensors are also not well understood in non-mammalian vertebrates. Very little is known of SR-A expression in species other than placental vertebrates (Bowdish and Gordon, 2009); however, SR-like behaviour has been demonstrated in chickens (He *et al.*, 2009), teleost fish (Frøystad *et al.*, 2002; Kleveland *et al.*, 2006) and a turtle species, *Chrysemys picta* (Duggan *et al.*, 2002). Extended further, protein coding sequences containing a SR cysteine-rich domain (SRCR) have been identified in sea urchin, the sponge *Geodia cydonium* and the unicellular green alga *Chlamydomonas reinhardtii* (Bowdish and Gordon, 2009). It is unclear whether the SRCR containing proteins expressed in these lower species have any innate immune function. TLR3 is expressed and functions similarly to its mammalian counterparts in both chickens (Karpala *et al.*, 2008) and teleost fish (Oshiumi *et al.*, 2008). Interestingly, teleosts appear to express another TLR whose ligand is dsRNA, TLR22, which does not have a mammalian orthologue. TLR22 mediates dsRNA-induced type I IFN production similar to TLR3 via TRIF; however, it differs from TLR3 in its ligand selection, with a binding preference for long dsRNA molecules, and cellular localization, being expressed only on the cell surface (Matsuo *et al.*, 2008). Using *in silico* analysis, RLRs were identified in a number of non-mammalian vertebrates. RIG-I sequences were found in amphibians and teleost fish, but not chickens, while MDA5 and LGP2 homologues could be identified in birds, fish, and amphibians (Zou *et al.*, 2009). Interestingly, RIG-I-like sequences were also identified in sea urchins and the sea anemone *Nematostella vectensis*. Further studies are required to elucidate the function of these RLR homologues. Clearly further research is necessary to understand how dsRNA-induced, type I IFN-mediated antiviral responses compare between non-mammalian vertebrates.

Antiviral RNAi responses

The dsRNA-induced responses described above, namely type I IFN and ISG induction, are important antiviral mechanisms in vertebrates. In contrast, invertebrates and plants use RNA interference (RNAi) mechanisms as a dsRNA-mediated antiviral response. RNAi refers to the mechanism of short RNA molecules recognizing and manipulating complementary RNA sequences. There are a number of excellent review papers discussing RNAi and the antiviral response in detail (Alvarado and Scholthof, 2009; Obbard *et al.*, 2009; Umbach and Cullen, 2009), and thus RNAi will be discussed only briefly in this section. Simply put, RNAi pathways are used by plants and invertebrates in two ways. First, the host RNAi pathways can detect long, viral dsRNA molecules, and use these molecules as templates to target and degrade viral RNA thus limiting viral replication. Second, both host and viruses can use RNAi to manipulate each other's gene expression (Obbard *et al.*, 2009). In the first instance, long, viral dsRNA is cleaved by the cytoplasmic ribonuclease, Dicer, to generate siRNA molecules (also referred to as viRNAs). One strand of the viRNA is incorporated into the RISC (RNA-induced silencing complex). The RISC finds complementary RNA strands, either viral RNA transcripts or genomes, and mediates their degradation. In plants and nematodes, RNA-dependent RNA polymerases (RdRPs) can amplify this response by producing more copies of the viRNAs. Also, in plants, nematodes (Obbard *et al.*, 2009) and *Drosophila* (Saleh *et al.*, 2009) the viRNAs can spread throughout the organism to mediate a systemic response to infection. There has yet to be a convincing example of the viRNA pathway in a mammalian system (Matranga and Zamore, 2007). Of course viruses encode many proteins that block this RNAi pathway, which are collectively referred to as viral suppressors of RNAi (VSRs). Many of these VSRs are dsRNA-binding proteins (Li and Ding, 2006).

The second RNAi pathway utilized as an antiviral mechanism is the miRNA pathway. In this pathway the source of the dsRNA is not long, viral dsRNA but host or viral encoded sequences, which when transcribed produce hairpin structures that are cleaved by the nuclear ribonuclease, Drosha, to produce short, duplexed RNAs that are processed by Dicer and subsequently loaded into RISC to mediate degradation of complementary sequences. The miRNA pathway is utilized by both plants and animals to regulate endogenous gene expression (Bushati and Cohen, 2007; Pillai *et al.*, 2007). Viruses encode miRNAs that modify host gene expression, and similarly, the host produces miRNAs that target viral protein expression (Obbard *et al.*, 2009). As opposed to the RNAi pathway, which is only well established in plants and invertebrates, there are some examples of miRNA pathways manipulating antiviral responses in mammals as well (Lecellier *et al.*, 2005; Stern-Ginossar *et al.*, 2007). Therefore, while dsRNA-mediated antiviral responses are predominantly type I IFN mediated in vertebrates, dsRNA mediates an alternative antiviral response predominantly in plants and invertebrates using the RNAi pathway.

As an interesting aside, there have been a few studies suggesting an antiviral role for Dicer independent of its role in RNAi. In addition to its RNase III domain, Dicer contains a dsRNA binding domain and a DExD/H-box RNA helicase domain, similar to that found in the RLRs, RIG-I and MDA-5 (Aliyari and Ding, 2009). While Dicer has been almost exclusively studied for its role in the RNAi pathway, a recent report shows that, in addition to RNAi, drosophila Dicer-2 functions to induce genes involved in the antiviral response (Deddouche *et al.*, 2008). In support of the hypothesis that Dicer is a dsRNA sensor involved in triggering the antiviral response, Dicer knockdown in a monkey kidney cell line led to an increase in influenza virus replication (Matskevich and Moelling, 2007). Further research is required to elucidate a possible role for Dicer as a mediator of dsRNA-induced antiviral responses.

Double-stranded RNA and adaptive immunity

The effect of dsRNA on adaptive immune cells

Not only does dsRNA have potent effects on the innate immune response, it also acts as a bridge to induce adaptive immune responses (Fig. 1.2). dsRNA appears to have both direct and indirect effects on adaptive immune cells. Its indirect effects are better studied, where dsRNA is sensed by macrophages, dendritic cells (DCs) or non-immune cells such as fibroblasts and stromal cells, resulting in cellular activation and production of cytokines, which are in turn sensed by the adaptive immune cells. Much of the indirect effects of dsRNA are via type I IFN expression (Le Bon and Tough, 2002; Hoebe and Beutler, 2004;). Recent studies have nicely shown that poly-IC-stimulated production of type I IFNs by monocytes, stromal cells and/or DCs results in DC maturation and CD4+ immunity (Trumpfheller *et al.*, 2008; Longhi *et al.*, 2009), CD8+ T cell differentiation (Ngoi *et al.*, 2008) and memory CD8+ T cell generation (Wang *et al.*, 2010). The intermediary cells appear to be sensing dsRNA via classic dsRNA sensors, such as TLR3 and MDA5 (Schulz *et al.*, 2005; Wang *et al.*, 2010). In addition to type I IFNs, other dsRNA-induced cytokines can also influence adaptive immunity. For example,

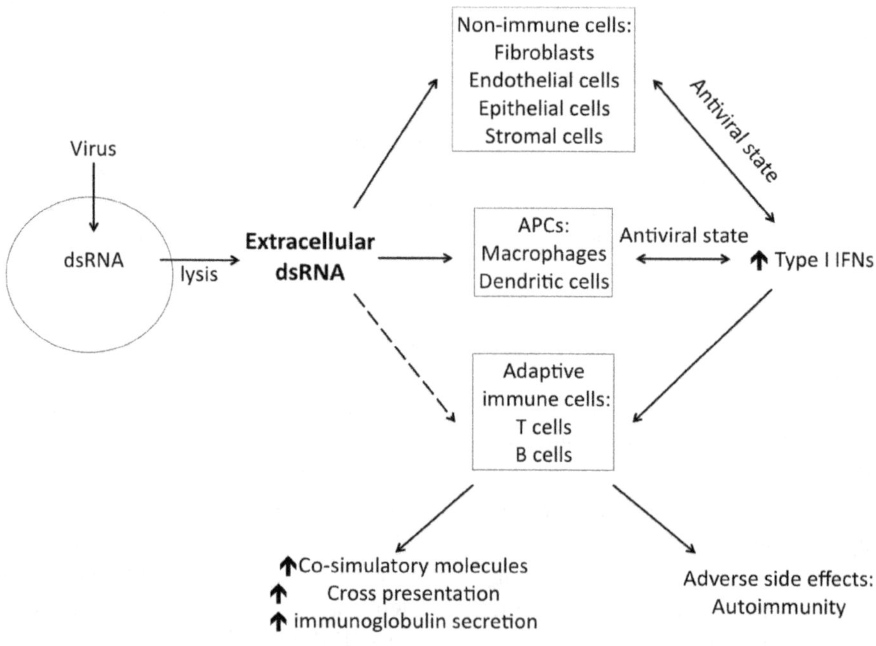

Figure 1.2 The pleiotropic effects of extracellular dsRNA in vertebrates. DsRNA is generated intracellularly during a viral infection and is released into the extracellular space during cell lysis. Its effects are cell-type dependent. Non-immune cells and APCs take up extracellular dsRNA and respond by up-regulating expression of ISGs and type I IFNs, culminating in the establishment of an antiviral state. The type I IFNs produced by these cells activates adaptive immune cells, T cells and B cells, causing up-regulation of co-stimulatory molecules, cross-presentation and immunoglobulin secretion. The dashed line indicates that a subset of adaptive immune cells can sense and respond to extracellular dsRNA itself. Sometimes extracellular dsRNA can induce adverse side-effects within the host, including autoimmunity.

poly-IC can cause the expansion of regulatory T (Treg) cells indirectly through activation of DCs and their production of IL-6. In this study, IFN-α was shown to have a suppressive effect (Kubota *et al.*, 2010). Although type I IFN and IL-6 production by plasmacytoid DCs promotes plasma cell differentiation and immunoglobulin secretion, this system is stimulated by TLR7/8 and TLR9 ligands but not TLR3 ligands (Douagi *et al.*, 2009).

The direct effects of dsRNA on adaptive immune cells are difficult to elucidate and are not as well reported. The best way to determine a role for dsRNA itself in adaptive immune responses is to investigate the expression of dsRNA sensors in adaptive immune cells. To the authors' knowledge the role of SR-As in dsRNA mediated responses in either T cells or B cells have not been elucidated. T cells express TLR3 (Kabelitz, 2007) and RLRs (Anz *et al.*, 2010) and are able to respond to extracellular dsRNA. Poly-IC has been shown to act directly on naïve CD4+ T cells causing their differentiation into an IL-21 producing phenotype (Holm *et al.*, 2009), and to induce IFN-γ in γδ T cells but not αβ T cells when a TCR stimulus is also present (Wesch *et al.*, 2006). In this second study, TLR3 was identified as a co-stimulatory receptor in γδ T cells. It was also found that the loss of Treg cell function during EMCV infection was dependent on expression of MDA5 (Anz *et al.*, 2010). With regards to B cells, naïve and memory B cells do not express TLR3 but terminally differentiated plasma cells do (Dorner *et al.*, 2009). Although B cells have been shown to respond to extracellular dsRNA directly, to the authors' knowledge, RLR expression in B cells has yet to be elucidated. A subset of mucosal B cells express TLR3 and in response to dsRNA undergoes class-switch DNA recombination in a T cell-independent manner (Xu *et al.*, 2008). Also, the production and secretion of immunoglobulin is elevated in differentiated plasma cells following poly-IC treatment (Dorner *et al.*, 2009), as well as nitric oxide production in B-1 cells (Tumurkhuu *et al.*, 2010). What is interesting is that even though adaptive immune cells express dsRNA sensors, their responses are similar to TCR or BCR stimulation as opposed to the type I IFN expression found in innate immune cells. Therefore, the cumulative effect of dsRNA on adaptive immunity is induction of co-stimulatory molecules, enhanced antigen presentation, polarization of T cell responses and reversal of suppressive Treg activity, and in some cases breaking of central tolerance and autoimmunity (von Herrath and Bot, 2003; Hoebe and Beutler, 2004; Kabelitz, 2007).

Double-stranded RNA and autoimmunity

As mentioned above, in some cases dsRNA can induce breaking of central tolerance and establishment of an autoimmune response. Therefore dsRNA is being investigated as a causative agent of autoimmune diseases. Under experimental conditions, poly-IC but not CpG DNA can induce T cell expression of IL17A and IL21, two cytokines associated with regulating autoimmunity (Holm *et al.*, 2009). Also, injections of poly-IC can cause arthritis in murine knee joints (Zare *et al.*, 2004), while injections of poly-IC and a self-insulin peptide can cause the production of insulin autoantibodies and the development of diabetes in mice (Moriyama *et al.*, 2002). Although RIG-I null mice develop colitis, an autoimmune disease (Wang *et al.*, 2007), dsRNA-mediated autoimmunity via SR-As or RLRs has yet to be elucidated. TLRs appear to play a role in regulating autoimmunity and TLR3 is particularly important; its activation is associated with chronic inflammation in cases of rheumatoid arthritis (RA; Brentano *et al.*, 2005), and type I IFNs and activated TLR3 were found in RA synovium (Roelofs *et al.*, 2009). Also, TLR3, in combination with TLR7 and TLR9, have been found to be important for the generation of anti-nucleic acid autoantibodies in systemic lupus erythematosus (SLE; Kono *et al.*, 2009).

There are two hypotheses for the source of the TLR3 ligand in autoimmune conditions. The first could be endogenous ligands released from necrotic cells. Recently, Cavassani *et al.* (2008) showed that an endogenous ligand stimulates TLR3 mediating inflammatory responses during necrosis, independent of a viral infection. Furthermore, it has been shown that *in vitro* transcribed mRNA and RNA from necrotic cells acts as a TLR3 ligand and causes DCs to produce type I IFNs (Karikó *et al.*, 2004). Following this,

Brentano *et al.* (2005) demonstrated that RA synovial tissue expressed higher levels of TLR3 than osteoarthritis synovial tissue, and that TLR3 was mostly expressed in RA synovial fibroblasts. These fibroblasts produced type I IFNs in response to poly-IC and necrotic synovial fluid cells from patients with RA. In addition to long endogenous RNA molecules, small nuclear ribonucleoproteins, which act as autoantigens in SLE patients, contain stretches of ss and dsRNA and may also function as a TLR3 ligand (Rahman and Eisenberg, 2006). An RNP, U1, which contains a substantial double-stranded secondary structure, was found to induce cytokine production via TLR3 (Hoffman *et al.*, 2004) and TLR7 (Kelly *et al.*, 2006). These data suggest that endogenous RNA molecules can act as a TLR3 ligand, perhaps acting as a stimulus for an autoimmune response. It is also possible that these endogenous TLR3 ligands are functioning to induce pro-inflammatory responses in any non-homeostatic condition, where uncontrolled cell death occurs.

The second source of TLR3 ligand could be dsRNA generated during virus infections, which may initiate or facilitate development of autoimmunity. There are a number of viruses whose infections have been associated with arthritis, including parvovirus B19, retroviruses and the herpesviruses cytomegalovirus and Epstein–Barr virus (EBV; Hyrich and Inman, 2001). More specifically, EBV infections have been associated with the onset of multiple sclerosis (Pender and Greer, 2007) and SLE and RA (Niller *et al.*, 2008). The mechanism of EBV-mediated TLR3 activation is probably via dsRNA-like, EBV-encoded small RNAs that are generated at high amounts during an EBV infection and are able to stimulate type I IFN production via TLR3 (Iwakiri *et al.*, 2009). It is likely that both endogenous and exogenous TLR3 ligands contribute to dsRNA-mediated autoimmunity and it is a matter of time before other dsRNA sensors are found to be involved in triggering these pathways.

Therapeutic applications

With its ability not only to induce innate but also adaptive immune responses, dsRNA appears to be a likely candidate for therapeutics. Because of its potency, poly-IC was hailed as a possible 'wonder drug' in the 1960s; however, as mentioned above, poly-IC's potency causes adverse physiological effects and cannot be used as a therapeutic agent 'as is' (Robinson *et al.*, 1976). Two poly-IC-based drugs, Ampligen and Hiltonol, have been tested in clinical trials. Ampligen is a mismatched dsRNA, poly(I)·poly(C_{12}U), which because of its mismatch is less stable *in vivo* resulting in favourable dsRNA mediated effects with little toxicity (Carter *et al.*, 1985). The second drug is Hiltonol (poly-IC–LC), a poly-IC molecule stabilized with poly-L-lysine and carboxymethyl cellulose (LC). It is able to induce type I IFNs and, in contrast with Ampligen, is more resistant to nuclease degradation than poly-IC. Perhaps because of its resistance, Hiltonol has demonstrated what are considered mild to moderate severe side-effects during phase I and II clinical trials (Champney *et al.*, 1979; Durie *et al.*, 1985).

Ampligen has recently been shown to have promise as both an adjuvant and an antiviral. With respect to its adjuvanticity, mice were protected against subtype H5N1 infection when vaccinated intranasally using Ampligen as an adjuvant (Ichinohe *et al.*, 2009). Also, Ampligen is able to function as an adjuvant for anti-tumour vaccines, as it is able to promote DC maturation and a Th1 response against tumour antigens (Jasani *et al.*, 2009). With respect to its antiviral activity, Ampligen, but not ribavirin, was able to protect mice against a mouse-adapted strain of SARS-CoV (Day *et al.*, 2009). Its antiviral activity appears to be TLR3 mediated; however, in contrast to poly-IC, Ampligen cannot enter the cytoplasm and trigger the cytoplasmic sensors (Gowen *et al.*, 2007). There is evidence that Ampligen is effective in controlling chronic viral infections thought to contribute to chronic fatigue syndrome (Suhadolnik *et al.*, 1994; Behan *et al.*, 2002). However, these results are controversial, and in December 2009, the FDA refused to approve Ampligen as a treatment for this syndrome.

Hiltonol, like poly-IC and Ampligen, is a TLR3 ligand able to stimulate type I IFN production and activate NK cells and macrophages. It is able to function as an antiviral drug and mediates complete protection against both seasonal and avian influenza virus infections in mice (Wong *et al.*, 2007), as well as subtype H1N1 (Wong *et al.*, 1995), respiratory syncytial virus and SARS (Wong *et al.*, 2009), western equine encephalitis virus (Wong *et al.*, 2005), yellow fever (Stephen *et al.*, 1977) and Rift Valley fever virus (Kende, 1985). In the influenza studies, liposome encapsulated Hiltonol demonstrated a more robust and prolonged antiviral response compared with naked Hiltonol treatments. A recent study investigated the immune response against protein-based vaccines in rhesus macaques, using Hiltonol and Ampligen as adjuvants (Stahl-Hennig *et al.*, 2009). Both drugs were found to be effective adjuvants, inducing cellular immunity against a model protein. Furthermore, Hiltonol but not CpG DNA (a TLR9 ligand) was able to induce a human papillomavirus 16-specific Th1 immune response. These data suggest that both dsRNA-based drugs have similar adjuvant capabilities *in vivo*.

When considering future directions for dsRNA research, novel dsRNA-based therapeutics should be the goal. Although the therapeutic potential of Ampligen and Hiltonol appears promising, both drugs have been through clinical trials, and neither has been approved by the FDA for general use. One problem with these drugs is that they are based on poly-IC, an unnatural nucleic acid combination, whose host-induced responses do not always align with those of viral dsRNA. The next generation dsRNA-based therapeutic should utilize a sequence and subsequent secondary structure that is more representative of an actual native viral dsRNA. Better dsRNA-based therapeutics would be valuable, as they have the capacity to provide broad-spectrum antiviral immunity. This is of particular interest for RNA viruses, which cause catastrophic pandemics, and as they demonstrate considerable divergence and antigenic shifts, it has proven difficult to produce effective vaccine and antiviral strategies.

Conclusions

Over 55 years of evidence has shown that viral dsRNA is an important modulator of host antiviral responses and viral pathogenesis. Whether it is present in intracellular or extracellular compartments, dsRNA can trigger both innate and adaptive immune responses, which are vital for controlling viral infections. Its effects are well conserved, affecting antiviral pathways in vertebrates, invertebrates and plants. Its stability and potency are profound, and the influence of this unique nucleic acid should be considered when studying viral pathogenesis, and should be manipulated to develop novel antiviral therapeutics.

References

Anon. (1967). Hope to end virus disease. Sci. News *92*, 173–174.

Aliyari, R., and Ding, S.-W. (2009). RNA-based viral immunity initiated by the Dicer family of host immune receptors. Immunol. Rev. *227*, 176–188.

Allen, I., Scull, M., Moore, C., Holl, E., McElvania-TeKippe, E., Taxman, D., Guthrie, E., Pickles, R., and Ting, J. (2009). The NLRP3 inflammasome mediates *in vivo* innate immunity to influenza A virus through recognition of viral RNA. Immunity *30*, 556–565.

Alvarado, V., and Scholthof, H.B. (2009). Plant responses against invasive nucleic acids: RNA silencing and its suppression by plant viral pathogens. Semin. Cell Dev. Biol. *20*, 1032–1040.

Ank, N., Iversen, M.B., Bartholdy, C., Staeheli, P., Hartmann, R., Jensen, U.B., Dagnaes-Hansen, F., Thomsen, A.R., Chen, Z., Haugen, H., *et al.* (2008). An important role for type III interferon (IFN-lambda/IL-28) in TLR-induced antiviral activity. J. Immunol. *180*, 2474–2485.

Anz, D., Koelzer, V.H., Moder, S., Thaler, R., Schwerd, T., Lahl, K., Sparwasser, T., Besch, R., Poeck, H., Hornung, V., *et al.* (2010). Immunostimulatory RNA blocks suppression by regulatory T cells. J. Immunol. *184*, 939–946.

Barral, P.M., Sarkar, D., Su, Z.-z., Barber, G.N., Desalle, R., Racaniello, V.R., and Fisher, P.B. (2009). Functions of the cytoplasmic RNA sensors RIG-I and MDA-5: key regulators of innate immunity. Pharmacol. Ther. *124*, 219–234.

Behan, M., Chaudhuri, A., Gow, J., Behan, P., and McKay, I. (2002). Antiviral pathway activation in chronic fatigue syndrome and acute infection. Clin. Infect. Dis. *34*, 1420–1422.

Bobst, A., Torrence, P., Kouidou, S., and Witkop, B. (1976). Dependence of interferon induction on nucleic acid conformation. Proc. Natl. Acad. Sci. U.S.A. *73*, 3788–3792.

Borst, P., and Weissmann, C. (1965). Replication of viral RNA. 8. Studies on the enzymatic mechanism of replication of MS2 RNA. Proc. Natl. Acad. Sci. U.S.A. *54*, 982–987.

de Bouteiller, O., Merck, E., Hasan, U., Hubac, S., Benguigui, B., Trinchieri, G., Bates, E., and Caux, C. (2005). Recognition of double-stranded RNA by human toll-like receptor 3 and downstream receptor signaling requires multimerization and an acidic pH. J. Biol. Chem. *280*, 38133–38145.

Bowdish, D.M.E., and Gordon, S. (2009). Conserved domains of the class A scavenger receptors: evolution and function. Immunol. Rev. *227*, 19–31.

Brentano, F., Schorr, O., Gay, R.E., Gay, S., and Kyburz, D. (2005). RNA released from necrotic synovial fluid cells activates rheumatoid arthritis synovial fibroblasts via Toll-like receptor 3. Arthr. Rheum. *52*, 2656–2665.

Brown, M., and Goldstein, J. (1983). Lipoprotein metabolism in the macrophage: implications for cholesterol deposition in atherosclerosis. Annu. Rev. Biochem. *52*, 223–261.

Bushati, N., and Cohen, S.M. (2007). MicroRNA functions. Annu. Rev. Cell Dev. Biol. *23*, 175–205.

Carter, W., and De Clercq, E. (1974). Viral infection and host defense. Science *186*, 1172–1178.

Carter, W.A., Strayer, D.R., Hubbell, H.R., and Brodsky, I. (1985). Preclinical studies with ampligen (mismatched double-stranded RNA). J. Biol. Response Mod. *4*, 495–502.

Cavassani, K., Ishii, M., Wen, H., Schaller, M., Lincoln, P., Lukacs, N., Hogaboam, C., and Kunkel, S. (2008). TLR3 is an endogenous sensor of tissue necrosis during acute inflammatory events. J. Exp. Med. *205*, 2609–2621.

Champney, K.J., Levine, D.P., Levy, H.B., and Lerner A.M. (1979). Modified polyriboinosinic-polyribocytidylic acid complex: sustained interferonemia and its physiological associates in humans. Infect. Immun. *25*, 831–837.

Cox, H. (1953). Viral and rickettsial toxins. Annu. Rev. Microbiol. *7*, 197–218.

Day, C.W., Baric, R., Cai, S.X., Frieman, M., Kumaki, Y., Morrey, J.D., Smee, D.F., and Barnard, D.L. (2009). A new mouse-adapted strain of SARS-CoV as a lethal model for evaluating antiviral agents *in vitro* and *in vivo*. Virology *395*, 210–222.

Deddouche, S., Matt, N., Budd, A., Mueller, S., Kemp, C., Galiana-Arnoux, D., Dostert, C., Antoniewski, C., Hoffmann, J.A., and Imler, J.-L. (2008). The DExD/H-box helicase Dicer-2 mediates the induction of antiviral activity in Drosophila. Nat. Immunol. *9*, 1425–1432.

DeWitte-Orr, S.J., Collins, S.E., Bauer, C.M.T., Bowdish, D.M., and Mossman, K.L. (2010). An accessory to the 'Trinity': SR-As are essential pathogen sensors of extracellular dsRNA, mediating entry and leading to subsequent type I IFN responses. PLoS Pathog. *6*, e1000829.

DeWitte-Orr, S.J., Mehta, D.R., Collins, S.E., Suthar, M.S., Gale, M., and Mossman, K.L. (2009). Long double-stranded RNA induces an antiviral response independent of IFN regulatory factor 3, IFN-beta promoter stimulator 1, and IFN. J. Immunol. *183*, 6545–6553.

Dorner, M., Brandt, S., Tinguely, M., Zucol, F., Bourquin, J.-P., Zauner, L., Berger, C., Bernasconi, M., Speck, R.F., and Nadal, D. (2009). Plasma cell toll-like receptor (TLR) expression differs from that of B cells, and plasma cell TLR triggering enhances immunoglobulin production. Immunology *128*, 573–579.

Douagi, I., Gujer, C., Sundling, C., Adams, W.C., Smed-Sörensen, A., Seder, R.A., Karlsson Hedestam, G.B., and Loré, K. (2009). Human B cell responses to TLR ligands are differentially modulated by myeloid and plasmacytoid dendritic cells. J. Immunol. *182*, 1991–2001.

Duggan, A.E., Marie, R.S., and Callard, I.P. (2002). Expression of SR-BI (Scavenger Receptor Class B Type I) in turtle (Chrysemys picta) tissues and other nonmammalian vertebrates. J. Exp. Zool. *292*, 430–434.

Durie, B.G., Levy, H.B., Voakes, J., Jett, J.R., and Levine, A.S. (1985). Poly(I,C)-LC as an interferon inducer in refractory multiple myeloma. J. Biol. Response Mod. *4*, 518–524.

Eaton, W.D. (1990). Anti-viral activity in four species of salmonids following exposure to poly-inosinic:cytidylic acid. Dis. Aquat. Org. *9*, 193–198.

Fang, J., Bredow, S., Taishi, P., Majde, J., and Krueger, J. (1999). Synthetic influenza viral double-stranded RNA induces an acute-phase response in rabbits. J. Med. Virol. *57*, 198–203.

Frøystad, M.K., Volden, V., Berg, T., and Gjøen, T. (2002). Metabolism of oxidized and chemically modified low density lipoproteins in rainbow trout--clearance via scavenger receptors. Dev. Comp. Immunol. *26*, 723–733.

Fukuda, K., Watanabe, T., Tokisue, T., Tsujita, T., Nishikawa, S., Hasegawa, T., Seya, T., and Matsumoto, M. (2008). Modulation of double-stranded RNA recognition by the N-terminal histidine-rich region of the human toll-like receptor 3. J. Biol. Chem., *283*, 22787–22794.

Gast, F.U., and Sänger, H.L. (1994). Gel dependence of electrophoretic mobilities of double-stranded and viroid RNA and estimation of the contour length of a viroid by gel electrophoresis. Electrophoresis *15*, 1493–1498.

Gil, J., and Esteban, M. (2000). Induction of apoptosis by the dsRNA-dependent protein kinase (PKR): mechanism of action. Apoptosis *5*, 107–114.

Gowen, G.G., Wong, M.H., Jung, K.H., Sanders, A.B., Mitchell, W.M., Alexopoulou,L., Flavell, R.A., and Sidwell, R.W. (2007). TLR3 is essential for the induction of protective immunity against toro virus infection by the double-stranded RNA (dsRNA), poly(I:C12U), but not poly(I:C): differential recognition of synthetic dsRNA molecules. J. Immunol. *178*, 5200–5208.

Hartmann, R., Justesen, J., Sarkar, S.N., Sen, G.C., and Yee, V.C. (2003). Crystal structure of the 2′-specific and double-stranded RNA-activated interferon-induced antiviral protein 2′–5′-oligoadenylate synthetase. Mol. Cell *12*, 1173–1185.

He, H., MacKinnon, K.M., Genovese, K.J., Nerren, J.R., Swaggerty, C.L., Nisbet, D.J., and Kogut, M.H. (2009). Chicken scavenger receptors and their ligand-induced cellular immune responses. Mol. Immunol. *46*, 2218–2225.

von Herrath, M., and Bot, A. (2003). Immune responsiveness, tolerance and dsRNA: implications for traditional paradigms. Trends Immunol. *24*, 289–292.

Hoebe, K., and Beutler, B. (2004). LPS, dsRNA and the interferon bridge to adaptive immune responses: Trif,

Tram, and other TIR adaptor proteins. J. Endotoxin Res. *10*, 130–136.

Hoffman, R.W., Gazitt, T., Foecking, M.F., Ortmann, R.A., Misfeldt, M., Jorgenson, R., Young, S.L., and Greidinger, E.L. (2004). U1 RNA induces innate immunity signaling. Arthritis Rheum. *50*, 2891–2896.

Holm, C.K., Petersen, C.C., Hvid, M., Petersen, L., Paludan, S.R., Deleuran, B., and Hokland, M. (2009). TLR3 ligand polyinosinic:polycytidylic acid induces IL-17A and IL-21 synthesis in human Th cells. J. Immunol. *183*, 4422–4431.

Hyrich, K.L., and Inman, R.D. (2001). Infectious agents in chronic rheumatic diseases. Curr. Opin. Rheumatol. *13*, 300–304.

Ichinohe, T., Ainai, A., Tashiro, M., Sata, T., Hasegawa, H. (2009). PolyI:PolyC12U adjuvant-combined intranasal vaccine protects mice against highly pathogenic H5N1 influenza virus variants. Vaccine *27*, 6276–6279.

Isaacs, A., and Lindemann, J. (1957). Virus interference. I. The interferon. Proc. Roy. Soc. B Biol. Sci. *147*, 258–267.

Isaacs, A. Cox, R.S., and Rolem, Z. (1963). Foreign nucleic acids as the stimulus to make interferon. Lancet *2*, 113–116.

Iwakiri, D., Zhou, L., Samanta, M., Matsumoto, M., Ebihara, T., Seya, T., Imai, S., Fujieda, M., Kawa, K., and Takada, K. (2009). Epstein–Barr virus (EBV)-encoded small RNA is released from EBV-infected cells and activates signaling from Toll-like receptor 3. J. Exp. Med. *206*, 2091–2099.

Jacobs, B., and Langland, J. (1996). When two strands are better than one: the mediators and modulators of the cellular responses to double-stranded RNA. Virology *219*, 339–349.

Jasani, B., Navabi, H., and Adams, M. (2009). A potential toll-like 3 receptor adjuvant for immunotherapy of cancer. Vaccine *27*, 3401–3404.

Jelinek, W., and Darnell, J.E. (1972). Double-stranded regions in heterogeneous nuclear RNA from Hela cells. Proc. Natl. Acad. Sci. U.S.A. *69*, 2537–2541.

Kabelitz, D. (2007). Expression and function of Toll-like receptors in T lymphocytes. Curr. Opin. Immunol. *19*, 39–45.

Kanneganti, T., Body-Malapel, M., Amer, A., Park, J., Whitfield, J., Franchi, L., Taraporewala, Z., Miller, D., Patton, J., Inohara, N., *et al.* (2006a). Critical role for cryopyrin/Nalp3 in activation of caspase-1 in response to viral infection and double-stranded RNA. J. Biol. Chem. *281* 36560–36568.

Kanneganti, T., Ozoren, N., Body-Malapel, M., Amer, A., Park, J., Franchi, L., Whitfield, J., Barchet, W., Colonna, M., Vandenabeele, P., *et al.* (2006b). Bacterial RNA and small antiviral compounds activate caspase-1 through cryopyrin/Nalp3. Nature *440*, 233–236.

Karikó, K., Ni, H., Capodici, J., Lamphier, M., and Weissman, D. (2004). mRNA is an endogenous ligand for Toll-like receptor 3. J. Biol. Chem. *279*, 12542–12550.

Karpala, A.J., Lowenthal, J.W., and Bean, A.G. (2008). Activation of the TLR3 pathway regulates IFNbeta production in chickens. Dev. Comp. Immunol. *32*, 435–444.

Kato, H., Takeuchi, O., Mikamo-Satoh, E., Hirai, R., Kawai, T., Matsushita, K., Hiiragi, A., Dermody, T., Fujita, T., and Akira, S. (2008). Length-dependent recognition of double-stranded ribonucleic acids by retinoic acid-inducible gene-I and melanoma differentiation-associated gene 5. J. Exp. Med. *205*, 1601–1610.

Kawai, T., and Akira, S. (2009). The roles of TLRs, RLRs and NLRs in pathogen recognition. Int. Immunol. *21*, 317–337.

Kelly, K.M., Zhuang, H., Nacionales, D.C., Scumpia, P.O., Lyons, R., Akaogi, J., Lee, P., Williams, B., Yamamoto, M., Akira, S., *et al.* (2006). 'Endogenous adjuvant' activity of the RNA components of lupus autoantigens Sm/RNP and Ro 60. Arthritis Rheum. *54*, 1557–1567.

Kende, M. (1985). Prophylactic and therapeutic efficacy of poly-(I,C)-LC against Rift Valley fever virus infection in mice. J. Biol. Response Mod. *4*, 503–511.

Kim, Y., Brinkmann, M., Paquet, M., and Ploegh, H. (2008). UNC93B1 delivers nucleotide-sensing toll-like receptors to endolysosomes. Nature *452*, 234–238.

de Kinkelin, P., and Dorson, M. (1973). Interferon production in rainbow trout (*Salmo gairdneri*) experimentally infected with Egtved virus. J. Gen. Virol. *19*, 125–127.

Kleveland, E.J., Syvertsen, B.L., Ruyter, B., Vegusdal, A., Jørgensen, S.M., and Gjøen, T. (2006). Characterization of scavenger receptor class B, type I in Atlantic salmon (*Salmo salar* L.). Lipids *41*, 1017–1027.

Kodym, R., Kodym, E., and Story, M.D. (2009). 2′-5′-Oligoadenylate synthetase is activated by a specific RNA sequence motif. Biochem. Biophys. Res. Commun. *388*, 317–322.

Kono, D.H., Haraldsson, M.K., Lawson, B.R., Pollard, K.M., Koh, Y.T., Du, X., Arnold, C.N., Baccala, R., Silverman, G.J., Beutler, B.A., *et al.* (2009). Endosomal TLR signaling is required for anti-nucleic acid and rheumatoid factor autoantibodies in lupus. Proc. Natl. Acad. Sci. U.S.A. *106*, 12061–12066.

Kotenko, S.V., Gallagher, G., Baurin, V.V., Lewis-Antes, A., Shen, M., Shah, N.K., Langer, J.A., Sheikh, F., Dickensheets, H., and Donnelly, R.P. (2003). IFN-lambdas mediate antiviral protection through a distinct class II cytokine receptor complex. Nat. Immunol. *4*, 69–77.

Kubota, N., Ebihara, T., Matsumoto, M., Gando, S., and Seya, T. (2010). IL-6 and IFN-alpha from dsRNA-stimulated dendritic cells control expansion of regulatory T cells. Biochem. Biophys. Res. Commun. *391*, 1421–1426.

Kumar, A., Yang, Y.L., Flati, V., Der, S., Kadereit, S., Deb, A., Haque, J., Reis, L., Weissmann, C., and Williams, B.R. (1997). Deficient cytokine signaling in mouse embryo fibroblasts with a targeted deletion in the PKR gene: role of IRF-1 and NF-kappaB. EMBO J. *16*, 406–416.

Kumar, H., Koyama, S., Ishii, K., Kawai, T., and Akira, S. (2008). Cutting edge: cooperation of IPS-1- and TRIF-dependent pathways in poly-IC-enhanced antibody production and cytotoxic T cell responses. J. Immunol. *180*, 683–687.

Kumar, M., and Carmichael, G.G. (1998). Antisense RNA: function and fate of duplex RNA in cells of higher eukaryotes. Microbiol. Mol. Biol. Rev. *62*, 1415–1434.

Lampson, G.P., Tytell, A.A., Field, A.K., Nemes, M.M., and Hilleman, M.R. (1967). Inducers of interferon and host resistance. I. Double-stranded RNA from extracts of *Penicillium funiculosum*. Proc. Natl. Acad. Sci. U.S.A. *58*, 782–789.

Le Bon, A., and Tough, D.F. (2002). Links between innate and adaptive immunity via type I interferon. Curr. Opin. Immunol. *14*, 432–436.

Lecellier, C.-H., Dunoyer, P., Arar, K., Lehmann-Che, J., Eyquem, S., Himber, C., Saïb, A., and Voinnet, O. (2005). A cellular microRNA mediates antiviral defense in human cells. Science *308*, 557–560.

Lehner, B., Williams, G., Campbell, R.D., and Sanderson, C.M. (2002). Antisense transcripts in the human genome. Trends Genet. *18*, 63–65.

Lemaire, P.A., Anderson, E., Lary, J., and Cole, J.L. (2008). Mechanism of PKR activation by dsRNA. J. Mol. Biol. *381*, 351–360.

Leonard, J., Ghirlando, R., Askins, J., Bell, J., Margulies, D., Davies, D., and Segal, D. (2008). The TLR3 signaling complex forms by cooperative receptor dimerization. Proc. Natl. Acad. Sci. U.S.A. *105*, 258–263.

Li, F., and Ding, S.-W. (2006). Virus counterdefense: diverse strategies for evading the RNA-silencing immunity. Annu. Rev. Microbiol. *60*, 503–531.

Li, X., Lu, C., Stewart, M., Xu, H., Strong, R.K., Igumenova, T., and Li, P. (2009). Structural basis of double-stranded RNA recognition by the RIG-I like receptor MDA5. Arch. Biochem. Biophys. *488*, 23–33.

Libonati, M., Carsana, A., and Furia, A. (1980). Double-stranded RNA. Mol. Cell. Biochem. *31*, 147–164.

Limmon, G., Arredouani, M., McCann, K., Minor, R., Kobzik, L., and Imani, F. (2008). Scavenger receptor class-A is a novel cell surface receptor for double-stranded RNA. FASEB J. *22*, 159–167.

Liu, L., Botos, I., Wang, Y., Leonard, J., Shiloach, J., Segal, D., and Davies, D. (2008). Structural basis of Toll-like receptor 3 signaling with double-stranded RNA. Science *320*, 379–381.

Longhi, M., Trumpfheller, C., Idoyaga, J., Caskey, M., Matos, I., Kluger, C., Salazar, A., Colonna, M., and Steinman, R. (2009). Dendritic cells require a systemic type I interferon response to mature and induce CD4+ Th1 immunity with poly-IC as adjuvant. J. Exp. Med. *206*, 1589–1602.

Majde, J. (2000). Viral double-stranded RNA, cytokines and the flu. J. Interferon Cytokine Res. *20* 259–272.

Majde, J., Brown, R., Jones, M., Dieffenbach, C., Maitra, N., Krueger, J., Cady, A., Smitka, C., and Maassab, H. (1991). Detection of toxic viral-associated double-stranded RNA (dsRNA) in influenza-infected lung. Microb. Pathog. *10*, 105–115.

Majde, J., Guha-Thakurta, N., Chen, Z., Bredow, S., and Krueger, J. (1998). Spontaneous release of stable viral double-stranded RNA into the extracellular medium by influenza virus-infected MDCK epithelial cells: implications for the viral acute phase response. Arch. Virol. *143*, 2371–2380.

Marcus, P.I., Sekellick, M.J. (1977). Defective interfering particles with covalently linked [+/–]RNA induce interferon. Nature *266*, 815–819.

Marques, J.T., Devosse, T., Wang, D., Zamanian-Daryoush, M., Serbinowski, P., Hartmann, R., Fujita, T., Behlke, M.A., and Williams, B.R.G. (2006). A structural basis for discriminating between self and nonself double-stranded RNAs in mammalian cells. Nat. Biotechnol. *24*, 559–565.

Martinon, F., and Tschopp, J. (2007). Inflammatory caspases and inflammasomes: master switches of inflammation. Cell Death Differ. *14*, 10–22.

Matranga, C., and Zamore, P.D. (2007). Small silencing RNAs. Curr. Biol. *17*, R789–793.

Matskevich, A.A., and Moelling, K. (2007). Dicer is involved in protection against influenza A virus infection. J. Gen. Virol. *88*, 2627–2635.

Matsumoto, M., Kikkawa, S., Kohase, M., Miyake, K., and Seya, T. (2002). Establishment of a monoclonal antibody against human Toll-like receptor 3 that blocks double-stranded RNA-mediated signaling. Biochem. Biophys. Res. Commun. *293*, 1364–1369.

Matsuo, A., Oshiumi, H., Tsujita, T., Mitani, H., Kasai, H., Yoshimizu, M., Matsumoto, M., and Seya, T. (2008). Teleost TLR22 recognizes RNA duplex to induce IFN and protect cells from birnaviruses. J. Immunol. *181*, 3474–3485.

McCartney, S., Vermi, W., Gilfillan, S., Cella, M., Murphy, T.L., Schreiber, R.D., Murphy, K.M., and Colonna, M. (2009). Distinct and complementary functions of MDA5 and TLR3 in poly(I:C)-mediated activation of mouse NK cells. J. Exp. Med. *206*, 2967–2976.

Minks, M.A., West, D.K., Benvin, S., and Baglioni, C. (1979). Structural requirements of double-stranded RNA for the activation of 2′, 5′-oligo(A) polymerase and protein kinase of interferon-treated HeLa cells. J. Biol. Chem. *254*, 10180–10183.

Montagnier, L., and Sanders, F.K. (1963). Replicative form of encephalomyocarditis virus ribonucleic acid. Nature *199*, 644–667.

Moriyama, H., Wen, L., Abiru, N., Liu, E., Yu, L., Miao, D., Gianani, R., Wong, F.S., and Eisenbarth, G.S. (2002). Induction and acceleration of insulitis/diabetes in mice with a viral mimic (polyinosinic-polycytidylic acid) and an insulin self-peptide. Proc. Natl. Acad. Sci. U.S.A. *99*, 5539–5544.

Murphy, J., Tedbury, P., Homer-Vanniasinkam, S., Walker, J., and Ponnambalam, S. (2005). Biochemistry and cell biology of mammalian scavenger receptors. Atherosclerosis *182*, 1–15.

Nallagatla, S.R., Toroney, R., and Bevilacqua, P.C. (2008). A brilliant disguise for self RNA: 5′-end and internal modifications of primary transcripts suppress elements of innate immunity. RNA Biol. *5*, 140–144.

Ngoi, S.M., Tovey, M.G., and Vella, A.T. (2008). Targeting poly(I:C) to the TLR3-independent pathway boosts effector CD8 T cell differentiation through IFN-alpha/beta. J. Immunol. *181*, 7670–7680.

Nicholson, A.W. (2006). Structure, reactivity, and biology of double-stranded RNA. Prog. Nucleic Acid Res. Mol. Biol. *52*, 1–65.

Niller, H.H., Wolf, H., and Minarovits, J. (2008). Regulation and dysregulation of Epstein–Barr virus latency: implications for the development of autoimmune diseases. Autoimmunity *41*, 298–328.

Obbard, D., Gordon, K., Buck, A., and Jiggins, F. (2009). The evolution of RNAi as a defence against viruses and transposable elements. Philos. Trans. R. Soc. Biol. Sci. *364* 99–115.

Oberg, B., and Philipson, L. (1971). Replicative structures of poliovirus RNA *in vivo*. J. Mol. Biol. *58*, 725–737.

Oshiumi, H., Matsuo, A., Matsumoto, M., and Seya, T. (2008). Pan-vertebrate toll-like receptors during evolution. Curr. Genomics *9*, 488–493.

Owen, D., and Gale, M. (2009). Fighting the flu with inflammasome signaling. Immunity *30* 476–478.

Pender, M.P., and Greer, J.M. (2007). Immunology of multiple sclerosis. Curr. Allergy Asthma Rep. *7*, 285–292.

Pillai, R.S., Bhattacharyya, S.N., and Filipowicz, W. (2007). Repression of protein synthesis by miRNAs: how many mechanisms? Trends Cell. Biol. *17*, 118–126.

Pirher, N., Ivicak, K., Pohar, J., Bencina, M., and Jerala, R. (2008). A second binding site for double-stranded RNA in TLR3 and consequences for interferon activation. Nat. Struct. Mol. Biol. *15*, 761–763.

Rahman, A.H., and Eisenberg, R.A. (2006). The role of toll-like receptors in systemic lupus erythematosus. Springer Semin. Immunopathol. *28*, 131–143.

Randall, R., and Goodbourn, S. (2008). Interferons and viruses: an interplay between induction, signalling, antiviral responses and virus countermeasures. J. Gen. Virol. *89*, 1–47.

Robinson, R.A., DeVita, V.T., Levy, H.B., Baron, S., Hubbard, S.P., and Levine, A.S. (1976). A phase I-II trial of multiple dose polyriboinosinic-polyribocytidylic acid in patients with leukemia or solid tumors. J. Natl. Cancer Inst. *57*, 599–602.

Roelofs, M.F., Wenink, M.H., Brentano, F., Abdollahi-Roodsaz, S., Oppers-Walgreen, B., Barrera, P., Van Riel, P.L.C.M., Joosten, L.A.B., Kyburz, D., Van Den Berg, W.B., *et al.* (2009). Type I interferons might form the link between Toll-like receptor (TLR) 3/7 and TLR4-mediated synovial inflammation in rheumatoid arthritis (RA). Ann. Rheum. Dis. *68*, 1486–1493.

Sadler, A.J., and Williams, B.R.G. (2007). Structure and function of the protein kinase R. Curr. Top. Microbiol. Immunol. *316*, 253–292.

Saito, T., and Gale, M. (2008). Differential recognition of double-stranded RNA by RIG-I-like receptors in antiviral immunity. J. Exp. Med. *205*, 1523–1527.

Saleh, M.-C., Tassetto, M., van Rij, R.P., Goic, B., Gausson, V., Berry, B., Jacquier, C., Antoniewski, C., and Andino, R. (2009). Antiviral immunity in Drosophila requires systemic RNA interference spread. Nature *458*, 346–350.

Saunders, L.R., and Barber, G.N. (2003). The dsRNA binding protein family: critical roles, diverse cellular functions. FASEB J. *17*, 961–983.

Schonborn, J., Oberstrass, J., Breyel, E., Tittgen, J., Schumacher, J., and Lukacs, N. (1991). Monoclonal antibodies to double-stranded RNA as probes of RNA structure in crude nucleic acid extracts. Nucleic Acids Res. *19*, 2993–3000.

Schultz, U., Kaspers, B., and Staeheli, P. (2004). The interferon system of non-mammalian vertebrates. Dev. Comp. Immunol. *28*, 499–508.

Schulz, O., Diebold, S.S., Chen, M., Näslund, T.I., Nolte, M.A., Alexopoulou, L., Azuma, Y.-T., Flavell, R.A., Liljeström, P., and Reis e Sousa, C. (2005). Toll-like receptor 3 promotes cross-priming to virus-infected cells. Nature *433*, 887–892.

Sekellick, M.J., Ferrandino, A.F., Hopkins, D.A., Marcus, P.I. (1994). Chicken interferon genes: cloning, expression, and analysis. J. Interferon Res. *14*, 71–79.

Silverman, R.H. (2007). Viral encounters with 2′,5′-oligoadenylate synthetase and RNase L during the interferon antiviral response. J. Virol. *81*, 12720–12729.

Stahl-Hennig, C., Eisenblätter, M., Jasny, E., Rzehak, T., Tenner-Racz, K., Trumpfheller, C., Salazar, A.M., Uberla, K., Nieto, K., Kleinschmidt, J., *et al.* (2009). Synthetic double-stranded RNAs are adjuvants for the induction of T helper 1 and humoral immune responses to human papillomavirus in rhesus macaques. PLoS Pathog. *5*, e1000373.

Stephen, E.L., Sammons, M.L., Pannier, W.L., Baron, S., Spertzel, R.O. (1977). Effect of a nuclease-resistant derivative of a polyriboinosinicpolyribocytidylic acid complex on yellow fever in rhesus monkeys (*Macaca mulatta*). J. Infect. Dis. *136*, 122–126.

Stern-Ginossar, N., Elefant, N., Zimmermann, A., Wolf, D.G., Saleh, N., Biton, M., Horwitz, E., Prokocimer, Z., Prichard, M., Hahn, G., *et al.* (2007). Host immune system gene targeting by a viral miRNA. Science *317*, 376–381.

Suhadolnik, R.J., Rechenbach, N.L., Hitzges, P., Sobol, R.W., Peterson, D.L., Henry, B., Ablashi, D.V., Müller, W.E., Schröder, H.C., and Carter, W.A. (1994). Upregulation of the 2–5A synthetase/RNaseL antiviral pathway associated with chronic fatigue syndrome. Clin. Infect. Dis. *18*, S96–104.

Suzuki, H., Kurihara, Y., Takeya, M., Kamada, N., Kataoka, M., Jishage, K., Ueda, O., Sakaguchi, H., Higashi, T., Suzuki, T., *et al.* (1997). A role for macrophage scavenger receptors in atherosclerosis and susceptibility to infection. Nature *386*, 292–296.

Takahasi, K., Kumeta, H., Tsuduki, N., Narita, R., Shigemoto, T., Hirai, R., Yoneyama, M., Horiuchi, M., Ogura, K., Fujita, T., *et al.* (2009). Solution structures of cytosolic RNA sensor MDA5 and LGP2 C-terminal domains: identification of the RNA recognition loop in RIG-I-like receptors. J. Biol. Chem. *284*, 17465–17474.

Takahasi, K., Yoneyama, M., Nishihori, T., Hirai, R., Kumeta, H., Narita, R., Gale, M., Inagaki, F., and Fujita, T. (2008). Nonself RNA-sensing mechanism of RIG-I helicase and activation of antiviral immune responses. Mol. Cell *29*, 428–440.

Tang, R.S., and Draper, D.E. (1994). On the use of phasing experiments to measure helical repeat and bulge loop-associated twist in RNA. Nucleic Acids Res. *22*, 835–841.

Taniguchi, T., and Takaoka, A. (2002). The interferon-alpha/beta system in antiviral responses: a multimodal machinery of gene regulation by the IRF family of transcription factors. Curr. Opin. Immunol. *14*, 111–116.

Thomas, P., Dash, P., Aldridge, Jr, Ellebedy, A., Reynolds, C., Funk, A., Martin, W., Lamkanfi, M., Webby, R., Boyd, K., *et al.* (2009). The intracellular sensor NLRP3 mediates key innate and healing responses to influenza A virus via the regulation of caspase-1. Immunity *30*, 566–575.

Trumpfheller, C., Caskey, M., Nchinda, G., Longhi, M.P., Mizenina, O., Huang, Y., Schlesinger, S.J., Colonna, M., and Steinman, R.M. (2008). The microbial mimic poly-IC induces durable and protective CD4+ T cell immunity together with a dendritic cell targeted vaccine. Proc. Natl. Acad. Sci. U.S.A. *105*, 2574–2579.

Tumurkhuu, G., Koide, N., Dagvadorj, J., Noman, A.S.M., Khuda, I.I.-E., Naiki, Y., Komatsu, T., Yoshida, T., and Yokochi, T. (2010). B1 cells produce nitric oxide in response to a series of toll-like receptor ligands. Cellular Immunology *261*, 122–127.

Umbach, J.L., and Cullen, B.R. (2009). The role of RNAi and microRNAs in animal virus replication and antiviral immunity. Genes Dev. *23*, 1151–1164.

Urbano, P., and Urbano, F.G. (1994). The Reoviridae family. Comp. Immunol. Microbiol. Infect. Dis. *17*, 151–161.

Van Den Hout, M., Vilfan, I.D., Hage, S., and Dekker, N.H. (2010). Direct force measurements on double-stranded RNA in solid-state nanopores. Nano Lett. *10*, 701–707.

Wang, Q., and Carmichael, G. (2004). Effects of length and location on the cellular response to double-stranded RNA. Microbiol. Mol. Biol. Rev. *68*, 432–452.

Wang, Y., Cella, M., Gilfillan, S., and Colonna, M. (2010). Cutting edge: polyinosinic:polycytidylic acid boosts the generation of memory CD8 T cells through melanoma differentiation-associated protein 5 expressed in stromal cells. J. Immunol. *184*, 2751–2755.

Wang, Y., Zhang, H.-X., Sun, Y.-P., Liu, Z.-X., Liu, X.-S., Wang, L., Lu, S.-Y., Kong, H., Liu, Q.-L., Li, X.-H., *et al.* (2007). Rig-I−/− mice develop colitis associated with downregulation of G alpha i2. Cell Res. *17*, 858–868.

Weber, F., Wagner, V., Rasmussen, S., Hartmann, R., and Paludan, S. (2006). Double-stranded RNA is produced by positive-strand RNA viruses and DNA viruses but not in detectable amounts by negative-strand RNA viruses. J. Virol. *80*, 5059–5064.

Werner, A. (2005). Natural antisense transcripts. RNA Biol. *2*, 53–62.

Werner, A., Carlile, M., and Swan, D. (2009). What do natural antisense transcripts regulate? RNA Biol. *6*, 43–48.

Wesch, D., Beetz, S., Oberg, H.-H., Marget, M., Krengel, K., and Kabelitz, D. (2006). Direct costimulatory effect of TLR3 ligand poly(I:C) on human gamma delta T lymphocytes. J. Immunol. *176*, 1348–1354.

Wong, J.P., Christopher, M.E., Salazar, A.M., Dale, R.M., Sun, L.Q., and Wang, M. (2007). Nucleic acid-based antiviral drugs against seasonal and avian influenza viruses. Vaccine *25*, 3175–3178.

Wong, J.P., Christopher, M.E., Viswanathan, S., Dai, X., Salazar, A.M., Sun, L.Q., and Wang, M. (2009). Antiviral role of toll-like receptor-3 agonists against seasonal and avian influenza viruses. Curr. Pharm. Des. *15*, 1269–1274.

Wong, J.P., Nagata, L.P., Christopher, M.E., Salazar, A.M., and Dale, R.M.K. (2005). Prophylaxis of acute respiratory virus infection using nucleic acid-based drugs. Vaccine *23*, 2266–2268.

Wong, J.P., Saravolac, E.G., Sabuda, D., Levy, H.B., and Kende, M. (1995). Prophylactic and therapeutic efficacies of poly(IC.LC) against respiratory influenza A virus infection in mice. Antimicrob. Agents Chemother. *39*, 2574–2576.

Xu, W., Santini, P.A., Matthews, A.J., Chiu, A., Plebani, A., He, B., Chen, K., and Cerutti, A. (2008). Viral double-stranded RNA triggers Ig class switching by activating upper respiratory mucosa B cells through an innate TLR3 pathway involving BAFF. J. Immunol. *181*, 276–287.

Zare, F., Bokarewa, M., Nenonen, N., Bergström, T., Alexopoulou, L., Flavell, R.A., and Tarkowski, A. (2004). Arthritogenic properties of double-stranded (viral) RNA. J. Immunol. *172*, 5656–5663.

Zou, J., Chang, M., Nie, P., and Secombes, C.J. (2009). Origin and evolution of the RIG-I like RNA helicase gene family. BMC Evol. Biol. *9*, 85.

Zou, J., Tafalla, C., Truckle, J., and Secombes, C.J. (2007). Identification of a second group of type I IFNs in fish sheds light on IFN evolution in vertebrates. J. Immunol. *179*, 3859–3871.

Type I Interferon Production by Viruses

2

Kazuhide Onoguchi, Kiyohiro Takahasi, Mitsutoshi Yoneyama and Takashi Fujita

Abstract

Type I interferon (IFN) is produced in variety of tissues in the body. It has been known that viral infection efficiently induces type I IFN. Bacterial endotoxin and double-stranded (ds) RNA are representative non-viral inducers. Recent works revealed that cellular receptors for pathogen-associated molecular patterns (PAMPs) are responsible for triggering IFN production. In the case of virus infection, RNA molecules encoded by viruses are sensed by the PAMPs receptors. Different viruses preferentially activate different sensor molecules. Current knowledge on virus- or RNA pathogen-specificity as well as structure–function relationship of RNA sensing is summarized. Furthermore, numerous signalling adaptors are reported to participate in the regulation of IFN gene activation.

Type I interferon production by viruses

Within hours of viral infection, cells in our body activate a group of genes including type I and type III IFNs. These IFNs secondarily activate so-called interferon-stimulated genes (ISGs). The products of these genes contribute to the establishment antiviral state in which various steps of viral replication are inhibited. The antiviral innate immune responses are triggered by the activation of sensors for viral (or non-self) RNA patterns.

Cell type-specific mechanism of IFN production: recognition of viral RNA in endosomes

Toll-like receptors (TLRs) recognize PAMPs of bacterial and viral origin at the cell surface or within endosomes (Akira *et al.*, 2006). Single-stranded RNA such as poly-(rU) is recognized by TLR7 and 8 (Diebold *et al.*, 2004; Heil *et al.*, 2004). Unmethylated DNA (CpG DNA) is recognized by TLR9 (Hemmi *et al.*, 2000). It is worth noting that low molecular weight synthetic compounds (imiquimod and R848) act as ligands for TLR7 (Hemmi *et al.*, 2002). TLR7, 8, and 9 are expressed in special cell types including plasmacytoid dendritic cells (pDCs), which are responsible for the systemic production of type I IFN (Kato *et al.*, 2005). Double-stranded (ds) RNA including synthetic poly-rI: poly-rC (poly-I:C) is recognized by TLR3 (Alexopoulou *et al.*, 2001). TLR3 is also expressed in a cell type-specific manner. External dsRNA is incorporated by endocytosis and activates TLR3 in the late endosome (Matsumoto *et al.*, 2003). The mechanism of dsRNA's recognition by TLR3 is described in Figure 2.1. The optimal pH for TLR3's recognition of dsRNA is 5.5, which is consistent with the process taking place in the late endosome. In fact, prevention of the acidification of endosomes by drugs such as chloroquine inhibits TLR3-induced production of IFN (de Bouteiller *et al.*, 2005; Kumar *et al.*, 2006) but enhances the incorporation of dsRNA

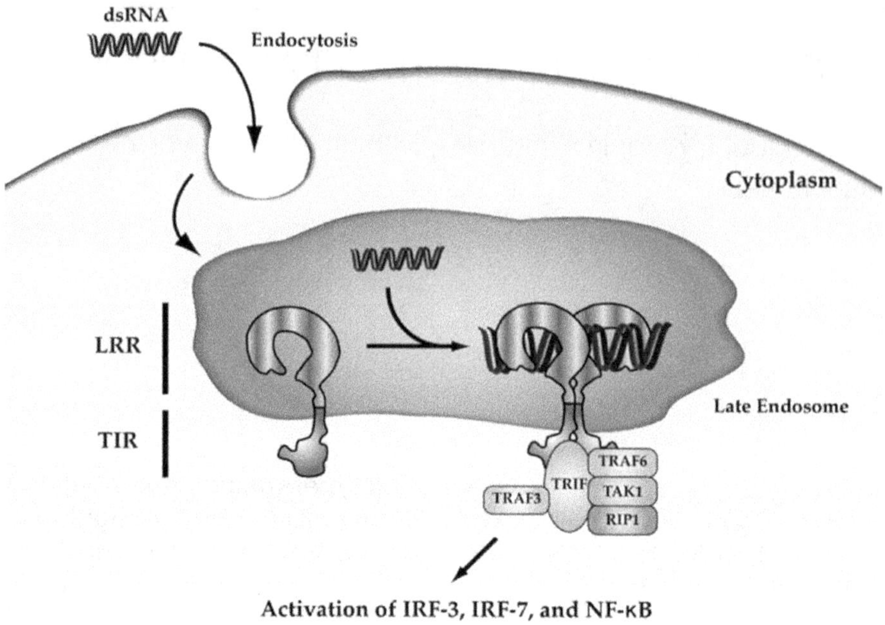

Figure 2.1 Activation of TLR3 by dsRNA. dsRNA present outside the cell is taken up by endocytosis likely mediated by scavenger receptors. At low pH (~pH 5.5) in the late endosome, dsRNA binds with two molecules of TLR3 through the leucine-rich region (LRR). A closely associated cytoplasmic TIR domain recruits adaptors, such as TRIF, TRAF3, TRAF6 and RIP1, activating the regulatory protein kinases IRF-3, IRF-7 and NF-κB.

into the cytoplasm and subsequent production of IFN induced by cytoplasmic sensors (see below). The minimal length of dsRNA recognizable by TLR3 appears to be 40–50 bp. The crystal structure of the TLR3/dsRNA complex revealed that a single molecule of dsRNA is recognized by two molecules of TLR3, thereby facilitating the formation of a TLR3 dimer (Liu *et al.*, 2008). This dimerization facilitates association with the cytoplasmic portion of TLR3, termed the Toll Interleukin 1 receptor (TIR) domain, triggering the signal. TIR recruits signalling adaptors TRIF/TICAM1, TRAF3, and TRAF6 (Yamamoto *et al.*, 2002, 2003; Oshiumi *et al.*, 2003; Sasai *et al.*, 2005; Hacker *et al.*, 2006; Oganesyan *et al.*, 2006), resulting in the activation of protein kinases TBK1, IKK-i (also known as IKK-ε), and IKK complex (Fitzgerald *et al.*, 2003; Sharma *et al.*, 2003; Hemmi *et al.*, 2004; McWhirter *et al.*, 2004; Perry *et al.*, 2004). These protein kinases activate transcription factors including interferon regulatory factors (IRFs) and NF-κB.

Common mechanism of IFN production: cytoplasmic sensing of viral RNA

Almost all cell lines and cultured primary cells except undifferentiated embryonic carcinoma cells are capable of producing IFN upon viral infection (Burke, 1978). Viral replication is required for efficient IFN production. Cells with a defective TLR3 are capable of producing IFN after transfection of dsRNA (Edelmann *et al.*, 2004), suggesting the presence of a cytoplasmic viral RNA sensor. RIG-I, MDA5, and LGP2 (collectively termed the RIG-I-Like Receptors, RLRs) are a family of DExD/H box helicases that function as cytoplasmic sensors for viral RNA (Yoneyama *et al.*, 2004; Fig. 2.2). RLRs share a DExD/H helicase domain and a C-terminal domain (CTD; Yoneyama *et al.*, 2005; Cui *et al.*, 2008; Takahasi *et al.*, 2008, 2009). RIG-I and MDA5, but not LGP2, contain a tandem caspase activation and recruitment domain (CARD) in the N-terminal region.

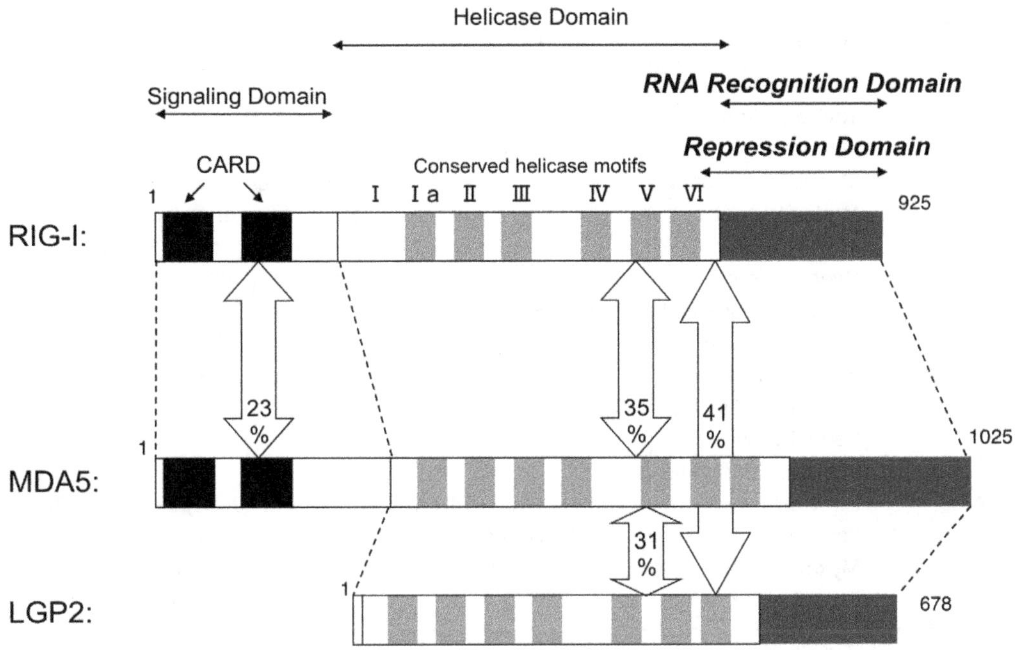

Figure 2.2 Structure of RLRs. Schematic representation of RLR domains. The caspase activation and recruitment domain (CARD), Helicase domain and RNA recognition domain are indicated. Conserved helicase motifs (I to VI) are shown. The repression domain (hRIG-I: aa. 935–925; hLGP2: aa. 476–678) and RNA recognition domain (hRIG-I: aa. 799–925; hMDA5: aa. 896–1025; hLGP2: aa. 546–678) are shown. Conservation of the amino acid sequence in the indicated region among human RLRs is indicated in percentage terms.

Specificity of RLRs

The specificity with which RLRs recognize viral RNA was revealed using knockout or knockdown technology (results summarized in Table 2.1). For example, cells devoid of RIG-I are highly susceptible to infection by a variety of viruses (Kato *et al.*, 2006). A RIG-I deficiency does not influence the replication of *Picornaviridae*, but MDA5 plays a critical role in protection against these viruses (Kato *et al.*, 2006). Some viruses including West Nile virus are recognized by both RIG-I and MDA5 (Fredericksen *et al.*, 2008). Overexpression of LGP2 in cultured cells dominantly inhibits IFN gene expression induced by viruses (Rothenfusser *et al.*, 2005; Yoneyama *et al.*, 2005; Komuro and Horvath, 2006). Similarly, artificial deletion of the CARD of RIG-I rendered it a dominant inhibitor (Yoneyama *et al.*, 2004). Mapping of the inhibitory domain revealed a repression domain in the C-terminal region of both RIG-I and LGP2 (Saito *et al.*, 2007). However, the corresponding region of MDA5 does not exhibit repression (Komuro and Horvath, 2006; Saito *et al.*, 2007). Interestingly, mice devoid of LGP2 and dendritic cells derived from these mice exhibit impaired IFN production upon infection by EMCV, VSV and NDV but not by influenza virus (Venkataraman *et al.*, 2007; Satoh *et al.*, 2010). These results indicate that LGP2 plays a positive role *in vitro* by detecting some viral infections to promote IFN production.

Recognition of viral RNA by RLRs

The differential viral recognition by RIG-I and MDA5 is due at least in part to the differential recognition of distinct types of RNA patterns by these receptors (Fig. 2.3). Commercial poly-I:C activates MDA5 but not RIG-I (Kato *et al.*, 2006). Double-stranded RNA produced by annealing complementary T7 transcripts preferentially activates RIG-I (Kato *et al.*, 2006). The length of dsRNA is one of the determinants for this

Table 2.1 Virus-specific recognition by RLRs

Genome	Type species	RLR	Reference
(+)ssRNA	Encephalomyocarditis virus	MDA5	Kato *et al.* (2006), Gitlin *et al.* (2006)
	Theiler's virus		
	Mengo virus		
	Murine norovirus-1		McCartney *et al.* (2008)
	Murine hepatitis virus		Roth-Cross *et al.* (2008)
	West Nile virus	MDA5/RIG-1	Fredericksen *et al.* (2008)
	Dengue virus		Loo *et al.* (2008)
	Japanese encephalitis virus	RIG-I	Kato *et al.* (2006)
	Hepatitis C virus		Saito *et al.* (2008)
(–)ssRNA	Influenza A virus		Kato *et al.* (2006)
	Newcastle disease virus		
	Sendai virus		
	Vesicular stomatitis virus		
dsRNA	Reovirus	MDA5/RIG-1	Kato *et al.* (2008), Loo *et al.* (2008)
dsDNA	Vaccinia virus	MDA5	Pichlmair *et al.* (2009)
	Myxoma virus	RIG-I	Wang *et al.* (2008)
	Epstein–Barr virus		Samanta *et al.* (2006)

Summary of the specificity with which RLRs recognize a variety of viruses. Family and type of genomic nucleic acids are indicated.

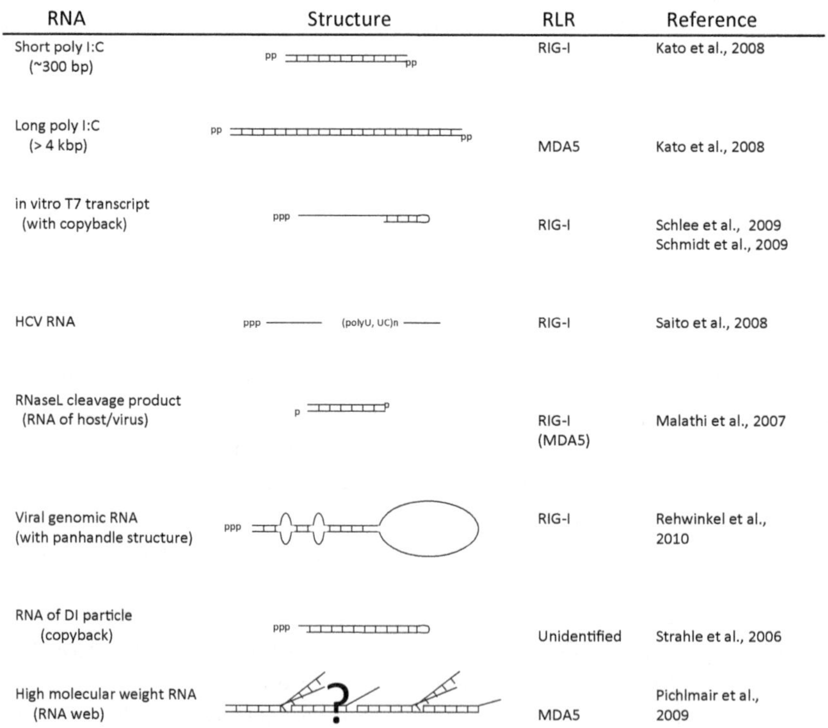

Figure 2.3 Viral RNA patterns and RLR recognition. Reported synthetic and viral RNA patterns and recognizing RLR are summarized. Schematic structure of each RNA pattern is indicated. The precise structure of the RNA web is unknown (indicated by '?').

specificity: short (~300 bp) and long (>4 kbp) poly-I:C activate RIG-I and MDA5, respectively (Kato *et al.*, 2008). Although it was reported that 5′-triphosphate (5′-ppp) containing single-stranded (ss) RNA is the ligand for RIG-I (Hornung *et al.*, 2006; Pichlmair *et al.*, 2006), later it was discovered that RNA produced by *in vitro* transcription contains a partial double-stranded structure a result of 'copy-back' activity of the phage polymerase (Schlee *et al.*, 2009; Schmidt *et al.*, 2009). Importantly, chemically synthesized ssRNA with 5′-ppp is inactive towards RIG-I. It is proposed that RIG-I recognizes most efficiently 5′-ppp-containing RNA with a partial double-stranded structure (Schlee *et al.*, 2009). Natural RNA of this category includes viral genomic panhandle RNA such as that of influenza A virus (Rehwinkel *et al.*, 2010) and copy-back RNA contained in defective interfering particles of Sendai virus (Strahle *et al.*, 2006). Mapping of HCV RNA revealed that a poly-(rU) or (rA) tract is responsible for activating RIG-I (Saito *et al.*, 2008). In this case, 5′-ppp is indispensable. It is implied that host RNaseL activated by the antiviral response participates in producing the RLR ligand (dsRNA with 3′ monophosphate) by digesting host or viral RNA (Malathi *et al.*, 2007). EMCV and vaccinia virus produce high molecular weight RNA (RNA web) with a single and double-stranded structure and this RNA complex is responsible for the activation of MDA5 (Pichlmair *et al.*, 2009). However, the molecular mechanism by which MDA5 but not RIG-I recognizes this complex is unknown.

Structure of the RNA recognition domain of RLRs

The RNA recognition domain of RIG-I as well as LGP2 exhibits high affinity for dsRNA, however that of MDA5 binds with significantly less efficiency (Yoneyama *et al.*, 2005). Although the overall structure of the RLR CTD is similar, the 3-dimensional structure of the CTD exhibits significant differences (Cui *et al.*, 2008; Takahasi *et al.*, 2008, 2009; Pippig *et al.*, 2009; Fig. 2.4). The CTDs of RIG-I and LGP2 share a basic cleft, which interacts with dsRNA as demonstrated by NMR's titration by dsRNA. The corresponding cleft of MDA5 is more extended. In silico docking of the RIG-I RNA-binding domain and dsRNA revealed that there is a conserved loop containing phenylalanine and lysine, which interacts with dsRNA (Takahasi *et al.*, 2009). This phenylalanine is conserved in LGP2, but the corresponding position in MDA5 has a cysteine. In summary, the CTD of RIG-I and LGP2 is responsible for the recognition of dsRNA, however how MDA5 recognizes long dsRNA is not clear.

Model for triggering antiviral signalling by RLRs

All RLRs retain a RNA helicase domain and recombinant RIG-I catalyses the unwinding of dsRNA *in vitro* in a 3′ overhang-dependent manner (Takahasi *et al.*, 2008). However, dsRNA with a 3′ overhang is a poor inducer of IFN production compared with dsRNA with a 5′ overhang or blunt end, suggesting that the unwinding reaction per se is not involved in the signalling (Takahasi *et al.*, 2008). Artificial overexpression of the CARD of RIG-I constitutively activates signalling without a viral infection, however full-length RIG-I exhibits little basal activity (Yoneyama *et al.*, 2004). Full-length RIG-I is hypothesized to form a closed structure through intramolecular interaction between the repression domain, CARD and the helicase linker region. This autorepression is reversed by a conformational change induced by the binding of the ligand RNA, the subsequent binding to the helicase domain and hydrolysis of ATP (Fig. 2.5). This model is consistent with the mutation of the ATP-binding domain of RIG-I rendering it a dominant inhibitor. It is reported that after binding to dsRNA, RIG-I can slide along the RNA chain without ATP hydrolysis (Myong *et al.*, 2009). However, the physiological significance of this remains unclear.

Virus-induced signalling cascade and IFN gene activation

After RIG-I or MDA5 detects viral RNA, the tandem CARDs are unmasked and interact with IPS-1 (also called MAVS, VISA, or Cardif), which contains one CARD, by CARD–CARD interaction (Kawai *et al.*, 2005; Meylan *et al.*, 2005; Seth *et al.*, 2005; Xu *et al.*, 2005). IPS-1 is located on the outer membrane of mitochondria and this specific localization is critical for signalling, since mislocalization to other membrane compartments

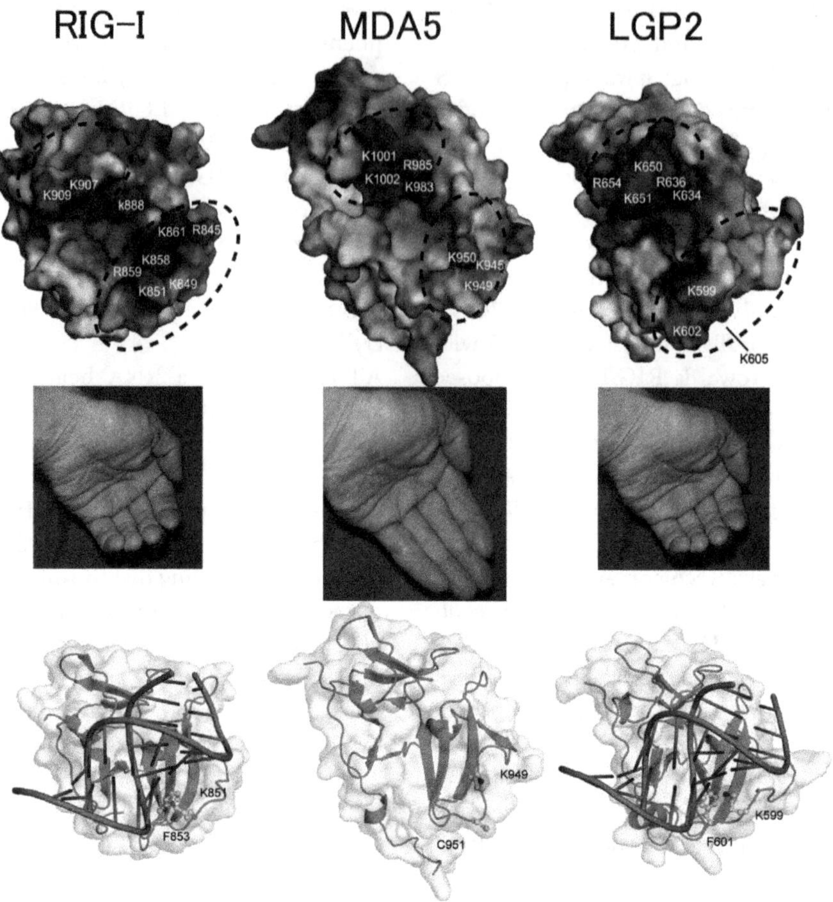

Figure 2.4 Structure of the RNA recognition domain of RLRs. Model of the RNA recognition domain present in the C-terminal region of RLRs (Top), their representation by a palm model (middle), and in silico docking with the dsRNA A helix (bottom). The RNA recognition domain is backed by basic amino acids. F853/K851 and F601/K599 of RIG-I and LGP2, respectively, exhibit potential interaction with dsRNA.

or release into the cytosol impairs its signalling potential (Seth *et al.*, 2005). The formation of the RLR–IPS-1 complex results in the recruitment of numerous signalling adaptors as summarized in Fig. 2.6.

It has been revealed that tumour necrosis factor (TNF) receptor-associated factor (TRAF) family members are involved in this signalling cascade. TRAF3, known to be an E3 ligase for Lys63-linked polyubiquitination, interacts directly with the TRAF-interacting motif (TIM) in the proline-rich region of IPS-1 (Oganesyan *et al.*, 2006; Saha *et al.*, 2006). TRAF3 KO cells show substantial impairment of virus-induced signalling, indicating an essential role for TRAF3 in RLR signalling. Similarly, interactions between IPS-1 and TRAF2 or TRAF6 were demonstrated (Xu *et al.*, 2005). Recruited TRAF family members transmit signals to downstream protein kinases, inhibitor of NF-κB (IκB) kinase (IKK) family members, which are essential for the activation of the transcription factors IRF-3, IRF-7 and NF-κB. The canonical IKK complex, consisting of IKK-α, IKK-β, and the regulatory subunit NF-κB essential modulator (NEMO; also termed IKK-γ), phosphorylates IκB, and subsequent proteasome-dependent degradation of IκB allows functional NF-κB to translocate to the nucleus (Karin and Ben-Neriah, 2000). By contrast, non-canonical IKKs, TRAF family associated NF-κB activator

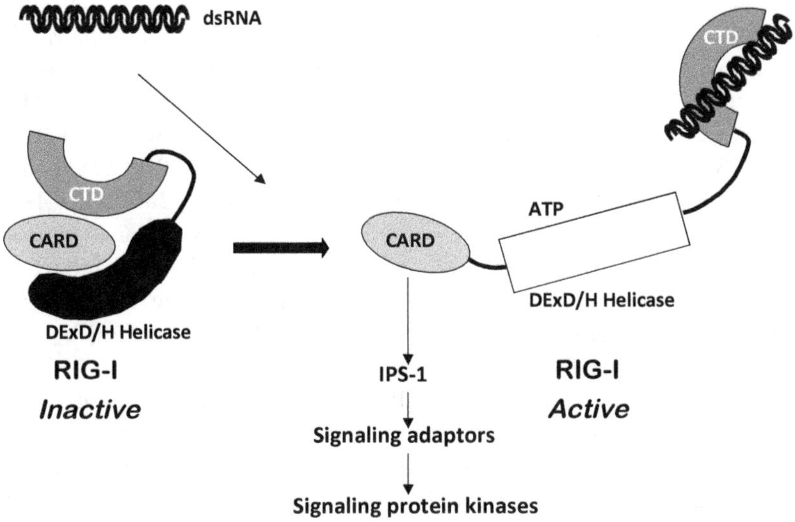

Figure 2.5 Model for the activation of RIG-I. In unstimulated cells, RIG-I forms a closed structure in which the repression domain in the C-terminal domain (CTD) interacts with the CARD and helicase linker region (Saito *et al.*, 2007; Takahasi *et al.*, 2008). Upon binding of dsRNA with the RNA-binding domain (also within the CTD), the conformation is changed by the action of the helicase domain. The binding and/or hydrolysis of ATP is required for the conformational change. The activated RIG-I releases CARD which interacts with the adaptor, IPS-1, to transmit signal.

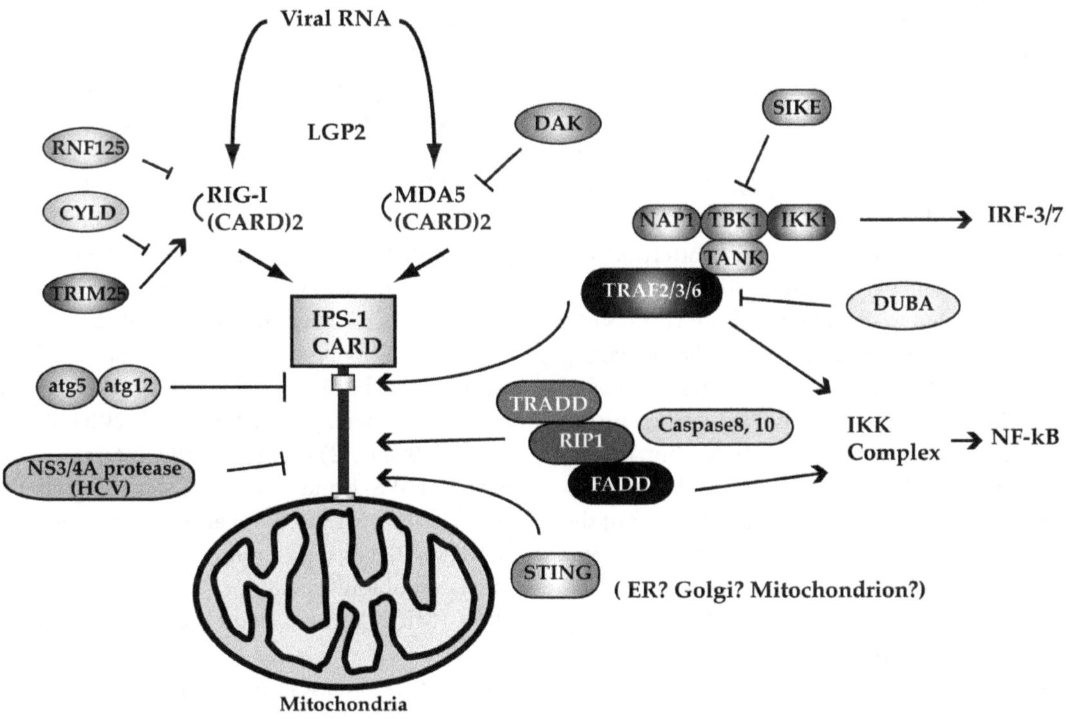

Figure 2.6 RLR-mediated signalling and its regulation. Activated RIG-I and MDA5 interact with interferon beta promoter stimulator 1 (IPS-1) and transmit signal through the indicated signalling molecules (see text for details).

(TANK)-binding kinase 1 (TBK1) and IKK-i (also termed IKKε), activate the signal-dependent phosphorylation of IRF-3 and IRF-7 (Fitzgerald *et al.*, 2003; Sharma *et al.*, 2003; Hemmi *et al.*, 2004; McWhirter *et al.*, 2004; Perry *et al.*, 2004). Interaction between IPS-1 and TRAF3 is essential for the recruitment of both IKK complexes, whereas TRAF2/6 is likely to be responsible for NF-κB's activation (Xu *et al.*, 2005; Saha *et al.*, 2006). As a regulatory component, TANK and NAK-associated protein 1 (NAP1) are commonly involved in the TBK-1/IKK-i-mediated activation of IRFs (Sasai *et al.*, 2006; Guo and Cheng, 2007). NEMO, which is known to be a regulatory subunit for the canonical IKK complex, plays a role in the TBK-1 / IKK-i-mediated activation of IRFs (Zhao *et al.*, 2007), suggesting crosstalk between these two IKK complexes in RLR-mediated signalling. The involvement of Fas-associated death domain (FADD) and receptor interacting protein 1 (RIP1), which are both death domain (DD)-containing molecules, in the IPS-1 signalling complex was demonstrated (Balachandran *et al.*, 2004; Kawai *et al.*, 2005). These molecules interact with the C-terminal portion of IPS-1 and activate the NF-κB-activating axis via interaction with and the activation of caspase-8 and caspase-10 (Kawai *et al.*, 2005; Takahashi *et al.*, 2006). Furthermore, recent reports demonstrated that TNFR-associated DD (TRADD) forms a complex with IPS-1, TRAF3, TANK, FADD, and RIP1, and leads to the activation of both IRF3 and NF-κB (Michallet *et al.*, 2008). This observation is closely related to signalling mediated by TNFR, which is known to form part of a similar signalling complex (Micheau and Tschopp, 2003); however, it remains unknown why IRF-3 is activated by the IPS-1-containing signalling complex but not by TNFR-mediated signalling. A plausible explanation is that the mitochondrial localization of the former may recruit a differential signalling complex. Another signalling adaptor termed STING (also called MITA) is involved in RLR signalling (Ishikawa and Barber, 2008; Zhong *et al.*, 2008), however its localization and molecular mechanism of action are controversial.

Tripartite motif protein 25 (TRIM25), an E3 ubiquitin ligase, specifically interacts with the RIG-I CARD and conjugates Lys-63-linked ubiquitin at the Lys172 residue of the CARD (Gack *et al.*, 2007). Interestingly, TRIM25 does not induce the ubiquitination of MDA5, suggesting the differential regulation of CARDs in RLRs. By contrast, ubiquitination of RIG-I by the E3 ligase ring finger 125 (RNF125) had the opposite effect (Arimoto *et al.*, 2007). Lys-48-linked ubiquitination in the N-terminal portion of RIG-I, and also presumably MDA5, induces proteasomal degradation, suggesting the negative regulation of RIG-I-mediated signalling by RNF125. Furthermore, the conjugation of ubiquitin-like ISG15 proteins to RIG-I has also been described for the negative regulation of RIG-I-mediated signalling (Zhao *et al.*, 2005; Kim *et al.*, 2008). De-ubiquitinating enzyme A (DUBA), an ovarian tumour domain (OTU)-containing enzyme, interacts directly with TRAF3 and catalyses the cleavage of Lys-63-linked ubiquitin chains of TRAF3. This de-ubiquitination by DUBA induces the dissociation of TBK1 from TRAF3, resulting in the sequestration of RLR-mediated signalling (Kayagaki *et al.*, 2007). Furthermore, another OTU-containing de-ubiquitinating enzyme, A20, which is known to be involved in the activation of NF-κB, shows a negative regulatory effect on RLR signalling (Wang *et al.*, 2004; Saitoh *et al.*, 2005; Lin *et al.*, 2006). In this case, the C-terminal ubiquitin ligase domain of A20, rather than the N-terminal de-ubiquitination domain, is critical to the inhibitory activity (Lin *et al.*, 2006), suggesting that the degradation of uncharacterized targets is involved in the signalling cascade. Another OTU-containing de-ubiquitinating enzyme, CYLD (cylindromatosis), has also been reported to be a negative regulator of RIG-I-mediated signalling (Friedman *et al.*, 2008; Zhang *et al.*, 2008). CYLD physically interacts with RIG-I or IPS-1 and inactivates signalling via the dissociation of Lys-63-linked ubiquitin moieties of RIG-I, suggesting a counteraction to TRAM25-mediated ubiquitination of RIG-I. Suppressor of IKKε (SIKE) forms a complex with TBK1 and/or IKK-i in the absence of viral stimulation and inhibits these kinases from interacting with RIG-I or IRF3 (Huang *et al.*, 2005). Upon viral infection, SIKE dissociates from the TBK1/IKK-i complex via an unknown mechanism and the released kinases can be incorporated into the signalling complex. The Atg5–Atg12 conjugate, an essential regulator of the autophagic machinery, directly associates with the CARD of RIG-I and IPS-1 and inhibits the interaction between them (Jounai *et al.*, 2007). A member of the NLR family, NLRX1, has also been identified as a negative regulator of RLR-mediated antiviral signalling (Moore *et al.*, 2008). As an MDA5-specific negative regulator, dihydroxyacetone kinase (DAK) has been identified as a protein-interacting with MDA5 (Diao *et al.*, 2007). Data indicate that DAK specifically associates with MDA5, but not with RIG-I, and inhibits polyI:C-induced IFN production.

Collectively, although recent reports strongly suggest positive and negative regulation of RLRs and related signalling molecules via post-translational modifications, the precise regulatory mechanisms remain to be elucidated.

References

Akira, S., Uematsu, S., and Takeuchi, O. (2006). Pathogen recognition and innate immunity. Cell *124*, 783–801.

Alexopoulou, L., Holt, A.C., Medzhitov, R., and Flavell, R.A. (2001). Recognition of double-stranded RNA and activation of NF-kappaB by Toll-like receptor 3. Nature *413*, 732–738.

Arimoto, K., Takahashi, H., Hishiki, T., Konishi, H., Fujita, T., and Shimotohno, K. (2007). Negative regulation of the RIG-I signaling by the ubiquitin ligase RNF125. Proc. Natl. Acad. Sci. U.S.A. *104*, 7500–7505.

Balachandran, S., Thomas, E., and Barber, G.N. (2004). A FADD-dependent innate immune mechanism in mammalian cells. Nature *432*, 401–405.

de Bouteiller, O., Merck, E., Hasan, U.A., Hubac, S., Benguigui, B., Trinchieri, G., Bates, E.E., and Caux, C. (2005). Recognition of double-stranded RNA by human toll-like receptor 3 and downstream receptor signaling requires multimerization and an acidic pH. J. Biol. Chem. *280*, 38133–38145.

Burke, D.C. (1978). Appearance of interferon inducibility and sensitivity during differentiation of murine teratocarcinoma cells *in vitro*. Cell *13*, 243–248.

Cui, S., Eisenacher, K., Kirchhofer, A., Brzozka, K., Lammens, A., Lammens, K., Fujita, T., Conzelmann, K.K., Krug, A., and Hopfner, K.P. (2008). The C-terminal regulatory domain is the RNA 5′-triphosphate sensor of RIG-I. Mol Cell *29*, 169–179.

Diao, F., Li, S., Tian, Y., Zhang, M., Xu, L.G., Zhang, Y., Wang, R.P., Chen, D., Zhai, Z., Zhong, B., *et al.* (2007). Negative regulation of MDA5- but not RIG-I-mediated innate antiviral signaling by the dihydroxyacetone kinase. Proc. Natl. Acad. Sci. U.S.A. *104*, 11706–11711.

Diebold, S.S., Kaisho, T., Hemmi, H., Akira, S., and Reis e Sousa, C. (2004). Innate antiviral responses by means of TLR7-mediated recognition of single-stranded RNA. Science *303*, 1529–1531.

Edelmann, K.H., Richardson-Burns, S., Alexopoulou, L., Tyler, K.L., Flavell, R.A., and Oldstone, M.B. (2004). Does Toll-like receptor 3 play a biological role in virus infections? Virology *322*, 231–238.

Fitzgerald, K.A., McWhirter, S.M., Faia, K.L., Rowe, D.C., Latz, E., Golenbock, D.T., Coyle, A.J., Liao, S.M., and Maniatis, T. (2003). IKKepsilon and TBK1 are essential components of the IRF3 signaling pathway. Nature Immunol. *4*, 491–496.

Fredericksen, B.L., Keller, B.C., Fornek, J., Katze, M.G., and Gale, M., Jr. (2008). Establishment and maintenance of the innate antiviral response to West Nile Virus involves both RIG-I and MDA5 signaling through IPS-1. J. Virol. *82*, 609–616.

Friedman, C.S., O'Donnell, M.A., Legarda-Addison, D., Ng, A., Cardenas, W.B., Yount, J.S., Moran, T.M., Basler, C.F., Komuro, A., Horvath, C.M., *et al.* (2008). The tumour suppressor CYLD is a negative regulator of RIG-I-mediated antiviral response. EMBO Rep *9*, 930–936.

Gack, M.U., Shin, Y.C., Joo, C.H., Urano, T., Liang, C., Sun, L., Takeuchi, O., Akira, S., Chen, Z., Inoue, S., *et al.* (2007). TRIM25 RING-finger E3 ubiquitin ligase is essential for RIG-I-mediated antiviral activity. Nature *446*, 916–920.

Guo, B., and Cheng, G. (2007). Modulation of the interferon antiviral response by the TBK1/IKKi adaptor protein TANK. J. Biol. Chem. *282*, 11817–11826.

Hacker, H., Redecke, V., Blagoev, B., Kratchmarova, I., Hsu, L.C., Wang, G.G., Kamps, M.P., Raz, E., Wagner, H., Hacker, G., *et al.* (2006). Specificity in Toll-like receptor signalling through distinct effector functions of TRAF3 and TRAF6. Nature *439*, 204–207.

Heil, F., Hemmi, H., Hochrein, H., Ampenberger, F., Kirschning, C., Akira, S., Lipford, G., Wagner, H., and Bauer, S. (2004). Species-specific recognition of single-stranded RNA via toll-like receptor 7 and 8. Science *303*, 1526–1529.

Hemmi, H., Kaisho, T., Takeuchi, O., Sato, S., Sanjo, H., Hoshino, K., Horiuchi, T., Tomizawa, H., Takeda, K., and Akira, S. (2002). Small anti-viral compounds activate immune cells via the TLR7 MyD88-dependent signaling pathway. Nature Immunol. *3*, 196–200.

Hemmi, H., Takeuchi, O., Kawai, T., Kaisho, T., Sato, S., Sanjo, H., Matsumoto, M., Hoshino, K., Wagner, H., Takeda, K., *et al.* (2000). A Toll-like receptor recognizes bacterial DNA. Nature *408*, 740–745.

Hemmi, H., Takeuchi, O., Sato, S., Yamamoto, M., Kaisho, T., Sanjo, H., Kawai, T., Hoshino, K., Takeda, K., and Akira, S. (2004). The roles of two IkappaB kinase-related kinases in lipopolysaccharide and double-stranded RNA signaling and viral infection. J. Exp. Med. *199*, 1641–1650.

Hornung, V., Ellegast, J., Kim, S., Brzozka, K., Jung, A., Kato, H., Poeck, H., Akira, S., Conzelmann, K.K., Schlee, M., *et al.* (2006). 5′-Triphosphate RNA is the ligand for RIG-I. Science *314*, 994–997.

Huang, J., Liu, T., Xu, L.G., Chen, D., Zhai, Z., and Shu, H.B. (2005). SIKE is an IKK epsilon/TBK1-associated suppressor of TLR3- and virus-triggered IRF-3 activation pathways. EMBO J. *24*, 4018–4028.

Ishikawa, H., and Barber, G.N. (2008). STING is an endoplasmic reticulum adaptor that facilitates innate immune signalling. Nature *455*, 674–678.

Jounai, N., Takeshita, F., Kobiyama, K., Sawano, A., Miyawaki, A., Xin, K.Q., Ishii, K.J., Kawai, T., Akira, S., Suzuki, K., *et al.* (2007). The Atg5 Atg12 conjugate associates with innate antiviral immune responses. Proc. Natl. Acad. Sci. U.S.A. *104*, 14050–14055.

Karin, M., and Ben-Neriah, Y. (2000). Phosphorylation meets ubiquitination: the control of NF-κB activity. Ann. Rev. Immunol. *18*, 621–663.

Kato, H., Sato, S., Yoneyama, M., Yamamoto, M., Uematsu, S., Matsui, K., Tsujimura, T., Takeda, K., Fujita, T., Takeuchi, O., *et al.* (2005). Cell type-specific involvement of RIG-I in antiviral response. Immunity *23*, 19–28.

Kato, H., Takeuchi, O., Mikamo-Satoh, E., Hirai, R., Kawai, T., Matsushita, K., Hiiragi, A., Dermody, T.S., Fujita, T., and Akira, S. (2008). Length-dependent recognition of double-stranded ribonucleic acids by retinoic acid-inducible gene-I and melanoma differentiation-associated gene 5. J. Exp. Med. *205*, 1601–1610.

Kato, H., Takeuchi, O., Sato, S., Yoneyama, M., Yamamoto, M., Matsui, K., Uematsu, S., Jung, A., Kawai, T., Ishii, K.J., *et al.* (2006). Differential roles of MDA5 and RIG-I helicases in the recognition of RNA viruses. Nature *441*, 101–105.

Kawai, T., Takahashi, K., Sato, S., Coban, C., Kumar, H., Kato, H., Ishii, K.J., Takeuchi, O., and Akira, S. (2005). IPS-1, an adaptor triggering RIG-I- and Mda5-mediated type I interferon induction. Nature Immunol. *6*, 981–988.

Kayagaki, N., Phung, Q., Chan, S., Chaudhari, R., Quan, C., O'Rourke, K.M., Eby, M., Pietras, E., Cheng, G., Bazan, J.F., *et al.* (2007). DUBA: a deubiquitinase that regulates type I interferon production. Science *318*, 1628–1632.

Kim, M.J., Hwang, S.Y., Imaizumi, T., and Yoo, J.Y. (2008). Negative feedback regulation of RIG-I-mediated antiviral signaling by interferon-induced ISG15 conjugation. J. Virol. *82*, 1474–1483.

Komuro, A., and Horvath, C.M. (2006). RNA- and virus-independent inhibition of antiviral signaling by RNA helicase LGP2. J. Virol. *80*, 12332–12342.

Kumar, A., Zhang, J., and Yu, F.S. (2006). Toll-like receptor 3 agonist poly(I:C)-induced antiviral response in human corneal epithelial cells. Immunology *117*, 11–21.

Lin, R., Yang, L., Nakhaei, P., Sun, Q., Sharif-Askari, E., Julkunen, I., and Hiscott, J. (2006). Negative regulation of the retinoic acid-inducible gene I-induced antiviral state by the ubiquitin-editing protein A20. J. Biol. Chem. *281*, 2095–2103.

Liu, L., Botos, I., Wang, Y., Leonard, J.N., Shiloach, J., Segal, D.M., and Davies, D.R. (2008). Structural basis of toll-like receptor 3 signaling with double-stranded RNA. Science *320*, 379–381.

Malathi, K., Dong, B., Gale, M., Jr., and Silverman, R.H. (2007). Small self-RNA generated by RNase L amplifies antiviral innate immunity. Nature *448*, 816–819.

Matsumoto, M., Funami, K., Tanabe, M., Oshiumi, H., Shingai, M., Seto, Y., Yamamoto, A., and Seya, T. (2003). Subcellular localization of Toll-like receptor 3 in human dendritic cells. J. Immunol. *171*, 3154–3162.

McWhirter, S.M., Fitzgerald, K.A., Rosains, J., Rowe, D.C., Golenbock, D.T., and Maniatis, T. (2004). IFN-regulatory factor 3-dependent gene expression is defective in Tbk1-deficient mouse embryonic fibroblasts. Proc. Natl. Acad. Sci. U.S.A. *101*, 233–238.

Meylan, E., Curran, J., Hofmann, K., Moradpour, D., Binder, M., Bartenschlager, R., and Tschopp, J. (2005). Cardif is an adaptor protein in the RIG-I antiviral pathway and is targeted by hepatitis C virus. Nature *437*, 1167–1172.

Michallet, M.C., Meylan, E., Ermolaeva, M.A., Vazquez, J., Rebsamen, M., Curran, J., Poeck, H., Bscheider, M., Hartmann, G., Konig, M., *et al.* (2008). TRADD protein is an essential component of the RIG-like helicase antiviral pathway. Immunity *28*, 651–661.

Micheau, O., and Tschopp, J. (2003). Induction of TNF receptor I-mediated apoptosis via two sequential signaling complexes. Cell *114*, 181–190.

Moore, C.B., Bergstralh, D.T., Duncan, J.A., Lei, Y., Morrison, T.E., Zimmermann, A.G., Accavitti-Loper, M.A., Madden, V.J., Sun, L., Ye, Z., *et al.* (2008). NLRX1 is a regulator of mitochondrial antiviral immunity. Nature *451*, 573–577.

Myong, S., Cui, S., Cornish, P.V., Kirchhofer, A., Gack, M.U., Jung, J.U., Hopfner, K.P., and Ha, T. (2009). Cytosolic viral sensor RIG-I is a 5′-triphosphate-dependent translocase on double-stranded RNA. Science *323*, 1070–1074.

Oganesyan, G., Saha, S.K., Guo, B., He, J.Q., Shahangian, A., Zarnegar, B., Perry, A., and Cheng, G. (2006). Critical role of TRAF3 in the Toll-like receptor-dependent and -independent antiviral response. Nature *439*, 208–211.

Oshiumi, H., Matsumoto, M., Funami, K., Akazawa, T., and Seya, T. (2003). TICAM-1, an adaptor molecule that participates in Toll-like receptor 3-mediated interferon-beta induction. Nature Immunol. *4*, 161–167.

Perry, A.K., Chow, E.K., Goodnough, J.B., Yeh, W.C., and Cheng, G. (2004). Differential requirement for TANK-binding kinase-1 in type I interferon responses to toll-like receptor activation and viral infection. J. Exp. Med. *199*, 1651–1658.

Pichlmair, A., Schulz, O., Tan, C.P., Naslund, T.I., Liljestrom, P., Weber, F., and Reis e Sousa, C. (2006). RIG-I-mediated antiviral responses to single-stranded RNA bearing 5′-phosphates. Science *314*, 997–1001.

Pichlmair, A., Schulz, O., Tan, C.P., Rehwinkel, J., Kato, H., Takeuchi, O., Akira, S., Way, M., Schiavo, G., and Reis e Sousa, C. (2009). Activation of MDA5 requires higher-order RNA structures generated during virus infection. J. Virol. *83*, 10761–10769.

Pippig, D.A., Hellmuth, J.C., Cui, S., Kirchhofer, A., Lammens, K., Lammens, A., Schmidt, A., Rothenfusser, S., and Hopfner, K.P. (2009). The regulatory domain of the RIG-I family ATPase LGP2 senses double-stranded RNA. NAR *37*, 2014–2025.

Rehwinkel, J., Tan, C.P., Goubau, D., Schulz, O., Pichlmair, A., Bier, K., Robb, N., Vreede, F., Barclay, W., Fodor, E., *et al.* (2010). RIG-I detects viral genomic RNA during negative-strand RNA virus infection. Cell *140*, 397–408.

Rothenfusser, S., Goutagny, N., DiPerna, G., Gong, M., Monks, B.G., Schoenemeyer, A., Yamamoto, M., Akira, S., and Fitzgerald, K.A. (2005). The RNA helicase Lgp2 inhibits TLR-independent sensing of viral replication by retinoic acid-inducible gene-I. J. Immunol. *175*, 5260–5268.

Saha, S.K., Pietras, E.M., He, J.Q., Kang, J.R., Liu, S.Y., Oganesyan, G., Shahangian, A., Zarnegar, B., Shiba, T.L., Wang, Y., *et al.* (2006). Regulation of antiviral responses by a direct and specific interaction between TRAF3 and Cardif. EMBO J. *25*, 3257–3263.

Saito, T., Hirai, R., Loo, Y.M., Owen, D., Johnson, C.L., Sinha, S.C., Akira, S., Fujita, T., and Gale, M., Jr. (2007). Regulation of innate antiviral defenses through a shared repressor domain in RIG-I and LGP2. Proc. Natl. Acad. Sci. U.S.A. *104*, 582–587.

Saito, T., Owen, D.M., Jiang, F., Marcotrigiano, J., and Gale, M., Jr. (2008). Innate immunity induced by composition-dependent RIG-I recognition of hepatitis C virus RNA. Nature *454*, 523–527.

Saitoh, T., Yamamoto, M., Miyagishi, M., Taira, K., Nakanishi, M., Fujita, T., Akira, S., Yamamoto, N., and Yamaoka, S. (2005). A20 is a negative regulator of IFN regulatory factor 3 signaling. J. Immunol. *174*, 1507–1512.

Sasai, M., Oshiumi, H., Matsumoto, M., Inoue, N., Fujita, F., Nakanishi, M., and Seya, T. (2005). Cutting edge: NF-kappaB-activating kinase-associated protein 1 participates in TLR3/Toll-IL-1 homology domain-containing adapter molecule-1-mediated IFN regulatory factor 3 activation. J. Immunol. *174*, 27–30.

Sasai, M., Shingai, M., Funami, K., Yoneyama, M., Fujita, T., Matsumoto, M., and Seya, T. (2006). NAK-associated protein 1 participates in both the TLR3 and the cytoplasmic pathways in type I IFN induction. J. Immunol. *177*, 8676–8683.

Satoh, T., Kato, H., Kumagai, Y., Yoneyama, M., Sato, S., Matsushita, K., Tsujimura, T., Fujita, T., Akira, S., and Takeuchi, O. (2010). LGP2 is a positive regulator of RIG-I- and MDA5-mediated antiviral responses. Proc. Natl. Acad. Sci. U.S.A. *107*, 1512–1517.

Schlee, M., Roth, A., Hornung, V., Hagmann, C.A., Wimmenauer, V., Barchet, W., Coch, C., Janke, M., Mihailovic, A., Wardle, G., *et al.* (2009). Recognition of 5′ triphosphate by RIG-I helicase requires short blunt double-stranded RNA as contained in panhandle of negative-strand virus. Immunity *31*, 25–34.

Schmidt, A., Schwerd, T., Hamm, W., Hellmuth, J.C., Cui, S., Wenzel, M., Hoffmann, F.S., Michallet, M.C., Besch, R., Hopfner, K.P., *et al.* (2009). 5′-triphosphate RNA requires basepaired structures to activate antiviral signaling via RIG-I. Proc. Natl. Acad. Sci. U.S.A. *106*, 12067–12072.

Seth, R.B., Sun, L., Ea, C.K., and Chen, Z.J. (2005). Identification and characterization of MAVS, a mitochondrial antiviral signaling protein that activates NF-kappaB and IRF 3. Cell *122*, 669–682.

Sharma, S., tenOever, B.R., Grandvaux, N., Zhou, G.P., Lin, R., and Hiscott, J. (2003). Triggering the interferon antiviral response through an IKK-related pathway. Science *300*, 1148–1151.

Strahle, L., Garcin, D., and Kolakofsky, D. (2006). Sendai virus defective-interfering genomes and the activation of interferon-beta. Virology *351*, 101–111.

Takahashi, K., Kawai, T., Kumar, H., Sato, S., Yonehara, S., and Akira, S. (2006). Roles of caspase-8 and caspase-10 in innate immune responses to double-stranded RNA. J. Immunol. *176*, 4520–4524.

Takahasi, K., Kumeta, H., Tsuduki, N., Narita, R., Shigemoto, T., Hirai, R., Yoneyama, M., Horiuchi, M., Ogura, K., Fujita, T., *et al.* (2009). Solution structures of cytosolic RNA sensor MDA5 and LGP2 C-terminal domains: identification of the RNA recognition loop in RIG-I-like receptors. J. Biol. Chem. *284*, 17465–17474.

Takahasi, K., Yoneyama, M., Nishihori, T., Hirai, R., Kumeta, H., Narita, R., Gale, M., Jr., Inagaki, F., and Fujita, T. (2008). Nonself RNA-sensing mechanism of RIG-I helicase and activation of antiviral immune responses. Mol Cell *29*, 428–440.

Venkataraman, T., Valdes, M., Elsby, R., Kakuta, S., Caceres, G., Saijo, S., Iwakura, Y., and Barber, G.N. (2007). Loss of DExD/H box RNA helicase LGP2 manifests disparate antiviral responses. J. Immunol. *178*, 6444–6455.

Wang, Y.Y., Li, L., Han, K.J., Zhai, Z., and Shu, H.B. (2004). A20 is a potent inhibitor of TLR3- and Sendai virus-induced activation of NF-kappaB and ISRE and IFN-beta promoter. FEBS Lett. *576*, 86–90.

Xu, L.G., Wang, Y.Y., Han, K.J., Li, L.Y., Zhai, Z., and Shu, H.B. (2005). VISA is an adapter protein required for virus-triggered IFN-beta signaling. Mol Cell *19*, 727–740.

Yamamoto, M., Sato, S., Hemmi, H., Hoshino, K., Kaisho, T., Sanjo, H., Takeuchi, O., Sugiyama, M., Okabe, M., Takeda, K., *et al.* (2003). Role of adaptor TRIF in the MyD88-independent toll-like receptor signaling pathway. Science *301*, 640–643.

Yamamoto, M., Sato, S., Mori, K., Hoshino, K., Takeuchi, O., Takeda, K., and Akira, S. (2002). Cutting edge: a novel Toll/IL-1 receptor domain-containing adapter that preferentially activates the IFN-beta promoter in the Toll-like receptor signaling. J. Immunol. *169*, 6668–6672.

Yoneyama, M., Kikuchi, M., Matsumoto, K., Imaizumi, T., Miyagishi, M., Taira, K., Foy, E., Loo, Y.M., Gale, M., Jr., Akira, S., *et al.* (2005). Shared and Unique Functions of the DExD/H-Box Helicases RIG-I, MDA5, and LGP2 in Antiviral Innate Immunity. J. Immunol. *175*, 2851–2858.

Yoneyama, M., Kikuchi, M., Natsukawa, T., Shinobu, N., Imaizumi, T., Miyagishi, M., Taira, K., Akira, S., and Fujita, T. (2004). The RNA helicase RIG-I has an essential function in double-stranded RNA-induced innate antiviral responses. Nature Immunol. *5*, 730–737.

Zhang, M., Wu, X., Lee, A.J., Jin, W., Chang, M., Wright, A., Imaizumi, T., and Sun, S.C. (2008). Regulation of IkappaB kinase-related kinases and antiviral responses by tumor suppressor CYLD. J. Biol. Chem. *283*, 18621–18626.

Zhao, C., Denison, C., Huibregtse, J.M., Gygi, S., and Krug, R.M. (2005). Human ISG15 conjugation targets both IFN-induced and constitutively expressed proteins functioning in diverse cellular pathways. Proc. Natl. Acad. Sci. U.S.A. *102*, 10200–10205.

Zhao, T., Yang, L., Sun, Q., Arguello, M., Ballard, D.W., Hiscott, J., and Lin, R. (2007). The NEMO adaptor bridges the nuclear factor-kappaB and interferon regulatory factor signaling pathways. Nature Immunol. *8*, 592–600.

Zhong, B., Yang, Y., Li, S., Wang, Y.Y., Li, Y., Diao, F., Lei, C., He, X., Zhang, L., Tien, P., *et al.* (2008). The adaptor protein MITA links virus-sensing receptors to IRF3 transcription factor activation. Immunity *29*, 538–550.

Type III Interferons in Antiviral Immunity

3

Srikanth Chiliveru and Søren R. Paludan

Abstract

Virus infections stimulate host immune responses characterized by production of interferons (IFNs). The identification of IFN-λ (alternatively termed interleukin 28A/B, 29 or type III IFNs) has revealed that the immune response to viruses has more components than the type I IFNs known for more than 50 years. IFN-λs are known to have type-I-IFN-like biological activities, but our understanding of these novel players in the antiviral response is still under development. In this chapter, we describe the current knowledge on expression and function of type III IFNs in innate antiviral immune defences and discuss recent findings proposing IFN-λ to shape the adaptive immune response. We suggest that type III IFNs are key antiviral cytokines exerting direct antiviral functions at epithelial surfaces at the early stages of infection, and that this class of cytokines also promotes antigen-specific cytotoxic activity of CD8+ T lymphocytes at later stages of the antiviral immune response.

Introduction

The first line of defence against infections relies on the innate immune system, which uses germ-line-encoded pattern recognition receptors (PRRs) to recognize molecular structures conserved among classes of pathogens (e.g. viral double-stranded RNA) or abnormal location of molecules shared between microbes and host cells (e.g. cytoplasmic localization of DNA; Mogensen, 2009). In the case of virus infections, the production of type I interferons (IFNs), or IFN-α/β, has long been appreciated as a key innate antiviral defence response (Garcia-Sastre and Biron, 2006). The type I IFNs exert direct antiviral activity by inhibiting viral replication, and also support cellular immune functions of both the innate and adaptive immune system, hence contributing to both early restriction of virus and long-term immunity (Garcia-Sastre and Biron, 2006). The type I IFNs were discovered in 1957 as a substance interfering with replication of influenza virus (Isaacs and Lindemann, 1957). Since then, our knowledge on the biology of type I IFNs has grown immensely, and we now have much information on the mechanisms of induction and action of type I IFNs as well as the strategies used by viruses to evade the type I IFN response (Garcia-Sastre and Biron, 2006; Bowie and Unterholzner, 2008). In addition, we have taken advantage of the biological activities of type I IFNs to use this natural antiviral cytokine as a drug to treat chronic viral infections. Type III IFNs were discovered as interleukin (IL)-28A/B, 29 or IFN-λ1–3 and demonstrated to share many functions with type I IFN (Kotenko *et al.*, 2003; Sheppard *et al.*, 2003). In addition, innate antiviral activities directly mediated by PRRs, and independent of the whole IFN system have been reported (Paladino *et al.*, 2006). Together, these findings paint a picture illustrating that the early events in host-defence against viral infections are more complex than previously thought, and highlight that we still have a lot to learn on the early events in virus–host interaction. In this chapter we describe how the type III IFNs were discovered, and discuss what has been learned on their role and mechanism of action in antiviral defence. Finally, we suggest future directions for the research on type III IFN biology.

Discovery

Type III IFNs, which were first described in 2003, were discovered by two independent groups by the use of bioinformatics (Kotenko *et al.*, 2003; Sheppard *et al.*, 2003). Kotenko *et al.* were conducting a study aiming at identifying novel members of the class II cytokine receptor family, and used TBLASTN search of the GenBank and ENSEMBL databases to identify a gene called CRF2–12, which is now known as IL-28Rα or IFN-λR1. The subsequent identification of the cytokine stimulating this receptor led to the description of IFN-λs, which were found to be virus-inducible and able to induce signalling and gene expression similar to type I IFNs (Kotenko *et al.*, 2003). The second report came from a group at ZymoGenetics, which was interested in finding novel cytokines. They used a combination of bioinformatics tools to identify two sequences with homology to the IL-10 and type I IFN helical cytokine families. These cytokine were virus inducible and exhibited type I IFN-like antiviral activity (Sheppard *et al.*, 2003). The IFN-λ receptor was identified through scanning of translated human genomic sequences for sequences related to class II cytokine receptors. The two papers were published back-to-back in Nature Immunology and together demonstrated the existence of a new antiviral type-I-IFN-like cytokine system, which is induced by viruses and stimulates similar activities, but acts through a distinct receptor system. Type I IFNs had been known since 1957 (Isaacs and Lindemann, 1957) and were generally believed to be the sole cytokine-mediator of innate antiviral defence. Therefore, the description of an additional type of innate antiviral cytokines raised a number of questions and urged for understanding of the role of IFN-λs in antiviral defence.

IFN-λ signal transduction and activation of gene expression

The biological activities of cytokines are mediated through engagement with specific receptors, which induce intracellular activities via signal transduction pathways. Type I IFNs signal through the IFNAR complex consisting of IFNAR1 and 2 (Fig. 3.1). IFN–α/β–IFNAR interactions lead to activation of the receptor interacting tyrosine kinases Jak1 and Tyk2, which in turn phosphorylate members of the STAT (signal transducer and activator of transcription) family, leading to STAT dimerization and activation of their transcription factor activities (Stark *et al.*, 1998). STAT1 and 2 are the main IFN-activated transcription factors, which together with IRF9 form a trimeric complex, ISGF3, which drives transcription of IFN-stimulated genes (ISGs; Stark *et al.*, 1998). Other STATs activated by type I IFN are STAT3, 4, and 5 (Ank *et al.*, 2006a). IFN-λs utilize an entirely different receptor complex to trigger their biological activities. This involves the IL-10R2 chain, which is shared with IL-10 and several other members of the IL-10 superfamily (Donnelly *et al.*, 2004), and the IFN-λ-specific IL-28RA chain (Kotenko *et al.*, 2003; Sheppard *et al.*, 2003). Despite the lack of similarities between the receptor systems used by type I and type III IFNs, the intracellular signalling programme activated by IFN-λ is very similar, if not identical, to what is activated upon IFN-α/β stimulation (Kotenko *et al.*, 2003; Dumoutier *et al.*, 2004; Fig. 3.1). This observation is supported by cellular gene expression profiles induced after stimulation with IFN-α versus IFN-λ. Three studies have compared the gene expression profiles of cells stimulated with IFN-λ versus IFN-α (Doyle *et al.*, 2006; Marcello *et al.*, 2006; Zhou *et al.*, 2007). Marcello *et al.* examined the IFN response in hepatitis C virus (HCV) replicon-harbouring Huh.7.5 cells and found that IFN-α and IFN-λ induced a highly overlapping set of genes, with IFN-λ generally inducing gene expression with a slower kinetics than IFN-α, thus potentially stimulating a more prolonged antiviral response (Marcello *et al.*, 2006). Doyle *et al.* used HepG2 cells to address the same question (Doyle *et al.*, 2006). Using a very stringent data analysis strategy, these authors found that 35 genes were highly induced after treatment with both IFN-α and -λ, and that the cytokines induced gene expression in this cell type with identical kinetics (Doyle *et al.*, 2006). Finally, Hartmann and colleagues compared the IFN-λ versus IFN-α response in the B cell line Raji and found that the response to IFN-α

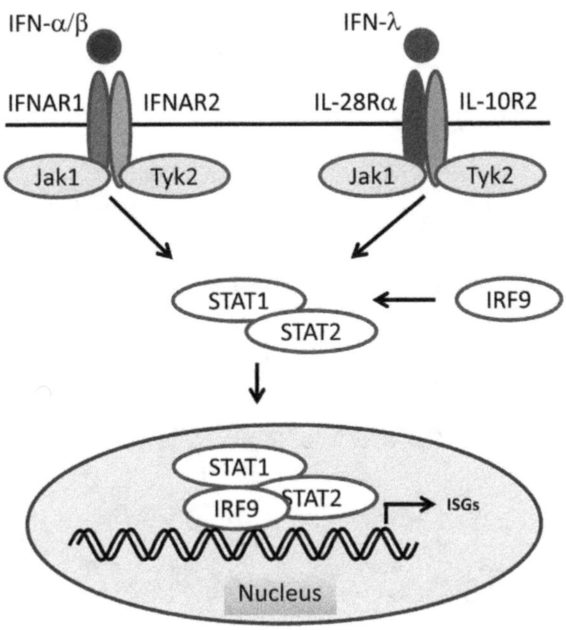

Figure 3.1 Signal transduction from the type I and III IFN receptors. Type I and III IFNs act through the IFNAR1/IFNAR2 and IL-28Rα/IL-10R2 receptor complexes, respectively. The intracellular signalling pathways activated by the two different receptor complexes are very similar, with particularly the IFN-activated trimeric transcription factor ISGF3 (STAT1:STAT2:IRF9) being responsible for expression of the majority of IFN-stimulated genes.

was much more potent than the response to IFN-λ and importantly, that all IFN-λ-inducible genes were also induced by IFN-α (Zhou *et al.*, 2007). Thus, IFN-λ employs a type I IFN-like signalling and gene induction programme to induce antiviral activity. It remains unresolved whether IFN-λ has functions not shared with type I IFN and also whether these two types of IFNs exert antiviral activity with different kinetics.

Expression of type I and III IFNs during viral infection

Type III IFNs are induced by most, if not all, classes of viruses as well as by some bacterial products (Kotenko *et al.*, 2003; Sheppard *et al.*, 2003; Ank *et al.*, 2006b; Zhang *et al.*, 2007). Importantly, type I and III IFNs display very similar patterns of expression, indicating similar mechanisms of induction. Both classes of IFNs are induced through transcriptional mechanisms involving the transcription factors IFN regulatory factors (IRFs) and nuclear factor (NF) κB (Fig. 3.2). However, as will be discussed below, despite these apparent similarities there are important differences in the mechanisms governing expression of type I and III IFNs.

A key feature of IFN expression is the requirement for detection of microbes by intracellular receptors. This explains why viruses are the most potent inducers of IFN responses but also why several bacteria and protozoa with an intracellular habitat induce these responses. Nucleotides represent the most potent inducers of IFN responses (Fig. 3.3). In the cytoplasm double-stranded (ds) and 5′-triphosphorylated RNAs are detected by the DEXD/H-box helicases RIG-I and MDA5 (Yoneyama *et al.*, 2004; Kato *et al.*, 2006), while the localization of dsRNA and ssRNA in endosomes is detected by Toll-like receptor (TLR) 3 and 7/8 (Alexopoulou *et al.*, 2001; Diebold *et al.*, 2004; Heil *et al.*, 2004). DNA is also a potent inducer of IFN production, but the receptor systems mediating these responses are not well established. It has been reported that DAI (DNA-dependent activator of IFN-regulatory factors) detects cytosolic B-DNA and induces IFN expression (Takaoka *et al.*, 2007). Subsequent generation of DAI$^{-/-}$ mice

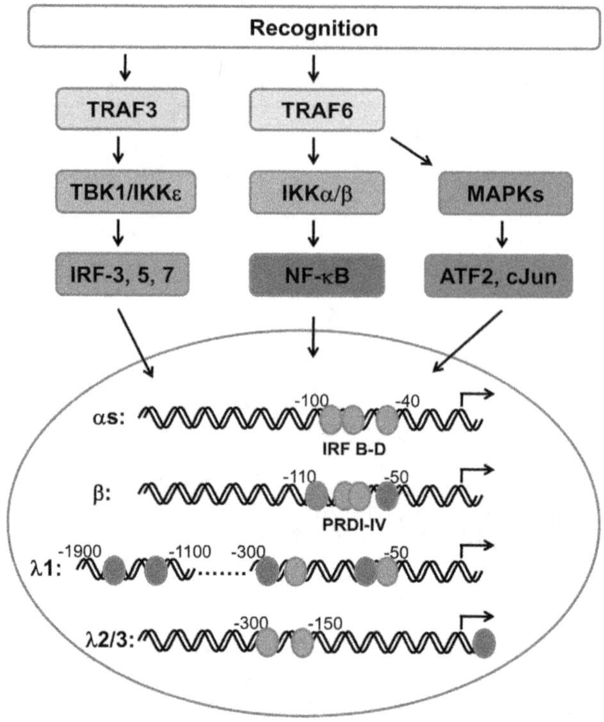

Figure 3.2 Signalling pathways and promoter regions controlling IFN expression. Recognition of viruses and other intracellular infections activates signalling to TRAF3, which in turn stimulates activation of the kinases TBK1/IKKε and the transcription factors IRF-3/7. Activation of NF-κB and ATF2/cJun proceeds through signalling pathways generally relying on TRAF6 and the specific activating kinases, IKKα/β and MAPKs, respectively. The type I and III IFN promoter regions are activated by similar sets of transcription factors but through different mechanisms. IFN-α promoters contain at least 3 active binding sites for IRF-3 and 7, while the IFN-β promoter is activated through coordinated action of IRFs, NF-κB, and ATF2/c-Jun. The IFN-λ1 promoter contains proximal binding sites for NF-κB and IRF3 and a distal cluster of NF-κB binding sites. These two regions independently activate IFN-λ1 transcription. The IFN-λ2/3 promoters contain binding sites for IRFs and NF-κB, both of which are required for transcription.

revealed no apparent phenotype with respect to the response to DNA (Ishii *et al.*, 2008), but a recent report suggests a role for DAI in IRF-3 activation in response to human cytomegalovirus DNA (Defilippis *et al.*, 2009). Interestingly, there is cross-talk between the RNA- and DNA-sensing systems, since some DNAs can serve as templates for RNA polymerase III and lead to *de novo* production of 5′-triphosphorylated RNAs with double-stranded structures, which potently activate RIG-I (Ablasser *et al.*, 2009; Chiu *et al.*, 2009). Finally, it has recently become apparent that non-RNA/DNA PAMPs (pathogen-associated molecular patterns) reaching the intracellular milieu are also capable of inducing IFN expression, but the receptors and signalling mechanisms driving these responses are poorly characterized (Barbalat *et al.*, 2009; Coulombe *et al.*, 2009; McWhirter *et al.*, 2009). Although it is unlikely we have full understanding of the requirement for intracellular localization of PAMPs for induction of IFN expression, it is clear that part of the explanation is owing to the essential role for the signalling molecule TRAF3 (Hacker *et al.*, 2005; Oganesyan *et al.*, 2005), which does not localize to the plasma membrane but to specific pleomorphic structures scattered throughout the cytosol (Kagan *et al.*, 2008). All together, intracellular sensing of microbes through different receptor systems of the innate immune system triggers signalling to NF-κB and IRFs, which leads to expression of type I and III IFNs.

Since the discovery of IFN-λs and the initial characterization of their pattern of expression

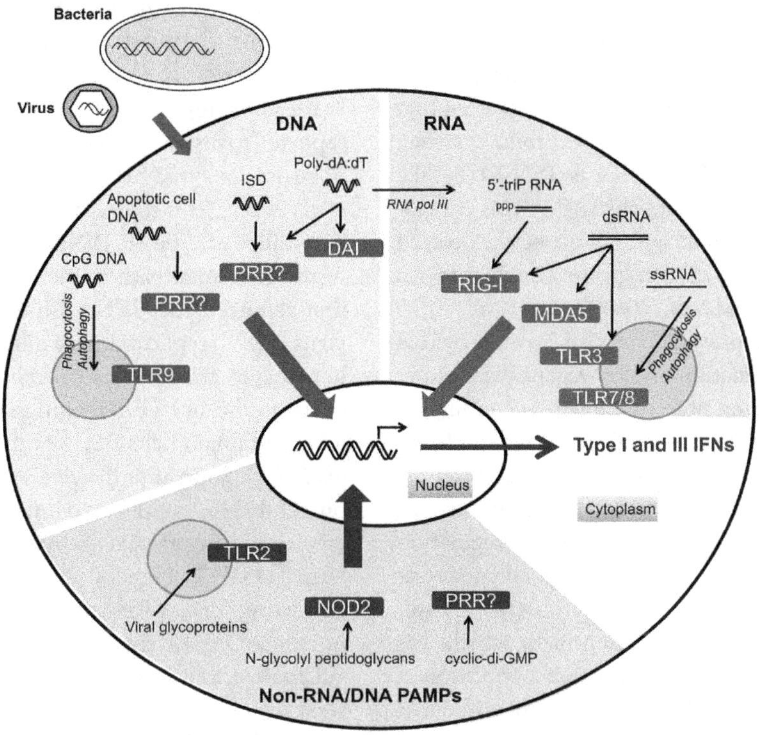

Figure 3.3 Pattern recognition receptors inducing IFN expression. Intracellular pattern recognition receptors detect nucleic acid structures and various microbe-specific molecular structures in the intracellular environment and initiate signalling to the IFN promoters.

(Kotenko *et al.*, 2003; Sheppard *et al.*, 2003; Coccia *et al.*, 2004), there has been an interest for the mechanisms that govern this expression. This includes identification of potential differences between type I and III IFNs, which would contribute to the understanding of the requirement for two seemingly similar biological systems. IFN-α promoters all contain a cluster of IRF-binding sites (Fig. 3.2), which bind IRF-3 and -7 with different affinities, and are responsible for inducible gene expression (Genin *et al.*, 2009). Since the majority of IFN-α promoters favour IRF-7 binding, this transcription factor is absolutely essential for IFN-α expression (Honda *et al.*, 2005). The IFN-β promoter is complex and very well characterized (Maniatis *et al.*, 1998). It is activated by the concerted action of an enhanceosome containing IRFs, NF-κB, ATF2/c-Jun, as well as basal transcription factors (Maniatis *et al.*, 1998). This requirement or high order organization of the enhanceosome makes IFN-β expression a stochastic event taking place in only a sub-fraction of virus-infected cells (Apostolou and Thanos, 2008). The IRF-binding site in the IFN-β promoter is activated well by both IRF-3 and 7, which has impact on the kinetics of expression. IRF-3 is expressed constitutively, whereas IRF-7 is an IFN-inducible gene. Therefore, IFN-β is an early response gene, while the IFN-αs are generally expressed in higher amounts but with a delayed kinetics (Marie *et al.*, 1998; Honda *et al.*, 2005).

The promoters of the IFN-λ genes are not well characterized compared with the type I IFN promoters. Both the IFN-λ1 and IFN-λ2/3 promoters contain binding sites for IRFs and NF-κB, and their initial characterization suggested identical mechanisms of activation (Onoguchi *et al.*, 2007; Osterlund *et al.*, 2007). Dr Fujita first reported that both type I and III IFNs are activated through the IRF and NF-κB pathways, a finding in agreement with recent reports (see discussion below; Onoguchi *et al.*, 2007). The group of Dr. Julkunen took an approach with over-expression of transcription factors and pattern recognition receptors and demonstrated that both NF-κB and IRF3/7 activate transcription of IFN-λ1 and IFN-λ2/3 (Osterlund *et al.*, 2007). Importantly, based on their findings the authors suggested that the IFN-λ1 gene, like IFN-β, exhibits a strong dependence on IRF-3, whereas IFN-λ2/3, like IFN-αs, are induced mainly by IRF-7. This conclusion was in agreement with an observed delayed kinetics of the IFN-λ2/3 response compared with IFN-λ1 (Osterlund *et al.*, 2007).

The reports presented above have provided important information on IFN-λ gene expression, but have not been able to explain accumulating evidence of differential expression of type I and III IFNs under certain conditions (Mihm *et al.*, 2004; Wang *et al.*, 2009). Such findings suggest that important differences exist in the regulation of these gene promoters. In a recent characterization of the human IFN-λ1 promoter, a cluster of distal NF-κB sites was identified, which was essential for full IFN-λ1 induction (Thomson *et al.*, 2009). These NF-κB sites, which originated from transposable elements, activated the IFN-λ1 gene promoter independently of IRF-3/7. Thus, this study demonstrated that IRFs and NF-κB activate gene expression independently in the IFN-γ1 promoter via spatially separated promoter elements. This is in contrast to the concerted action of the transcription factors in the IFN-β enhanceosome. Based on the presented data, Udalova and associates suggested independent evolution of the IFN-λ1 and IFN-β promoters. Data from this laboratory support an essential role for NF-κB in IFN-λ expression. We have reported that inhibition of the NF-κB pathway in dendritic cells (DCs) and in whole animals strongly inhibits induction of IFN-λ expression but only has a minor effect on type I IFN expression (Iversen *et al.*, 2010). Together, these findings suggest that although the same set of transcription factors are involved in activation of expression of type I and III IFNs, the IRFs play a dominant role in IFN-α/β expression whereas NF-κB is a key regulator of IFN-λ expression. They also suggest that IFN-λ promoters are more flexible than the IFN-β promoter because of the independent action of NF-κB and IRF-3/7, and hence are able to receive distinct signals, which independently can activate type III IFN expression.

Functions of type III IFN in innate antiviral immune responses

In the initial reports on IFN-λs, the cytokines were reported to display antiviral activity against a panel of viruses *in vitro* (Kotenko *et al.*, 2003; Sheppard *et al.*, 2003). This rapidly raised the question as to the ability of type III IFNs to restrict replication of major human pathogenic viruses. Robek *et al.* first reported that IFN-γ2 inhibited hepatitis B virus (HBV) replication in a differentiated murine hepatocyte cell line and blocked replication of a subgenomic and a full-length genomic HCV replicon in human hepatocyte Huh7 cells (Robek *et al.*, 2005). Soon after that, we showed that recombinant IFN-λ1 and -λ2 reduced the replication efficiency and cytotoxic activity of herpes simplex virus (HSV) in HepG2 cells (Ank *et al.*, 2006b). Following this, Mordstein *et al.* demonstrated potent antiviral activity of exogenous IFN-λ2 and -λ3 against influenza A virus in the lung (Mordstein *et al.*, 2008). Most recently, it has been shown that IFN-λ1 and -λ2 inhibit replication of both laboratory-adapted and clinical strains of HIV-1 in human macrophages (Hou *et al.*, 2009). Thus, replication of structurally and functionally diverse human pathogenic viruses is inhibited by IFN-λ in a range of different systems.

The above-described antiviral activity of IFN-λs against human pathogenic viruses strongly resembles that of type I IFNs, which could indicate that IFN-α/β and -λ have identical functions in the immune system. However, the two types of IFNs differ substantially with respect to which cells they target. Whereas IFNARs are expressed ubiquitously, IL-28Rα is expressed only in a few cell types, notably epithelial cells and some classes of leucocytes including macrophages, plasmacytoid DCs and conventional DCs, and hence the cellular response to type III IFN is restricted to a narrow range of cell types and tissues (Kotenko *et al.*, 2003; Sheppard *et al.*, 2003; Lasfar *et al.*, 2006; Mennechet and Uze, 2006; Zhou *et al.*, 2007; Ank *et al.*, 2008; Sommereyns *et al.*, 2008). As will be discussed later in this chapter, the differential expression of type I versus type III IFN receptors has profound impact in on the biological activities of these functionally related cytokines in the antiviral response in living organisms.

Following the early characterization of the antiviral activity of IFN-λ *in vitro*, several studies with recombinant IFN-λ in mice were conducted (Bartlett *et al.*, 2005; Ank *et al.*, 2006b; Mordstein *et al.*, 2008). These reports showed that recombinant IFN-λ added exogenously or expressed from a recombinant virus potently inhibited replication of HSV-2, vaccinia virus and influenza A virus at epithelial surfaces (Bartlett *et al.*, 2005; Ank *et al.*, 2006b; Mordstein *et al.*, 2008). These studies established that type III IFN does indeed posses innate antiviral activity *in vivo* and that epithelial surfaces are key sites of action of IFN-λ in the innate response. To evaluate the role of endogenous type III IFN in antiviral defence, we generated IL-28RA$^{-/-}$ mice and challenged them with a wide range of viruses (Ank *et al.*, 2008). Surprisingly, in none of the infection models did the lack of responsiveness to IFN-λ affect the antiviral response. This outcome was in sharp contrast to the significant antiviral activity of recombinant IFN-λ *in vivo* (Bartlett *et al.*, 2005; Ank *et al.*, 2006b; Mordstein *et al.*, 2008) and also the impairment of antiviral activity in IFNAR$^{-/-}$ mice after infection with the same viruses (Ank *et al.*, 2008; Table 3.1). The only essential role for IFN-λ in antiviral defence was observed when agonists for TLR3 and 9 were used as topical microbicides. In this case, IL-28RA$^{-/-}$ mice failed to protect against a subsequent vaginal HSV infection (Ank *et al.*, 2008). Together, these findings suggest IFN-λ to be redundant for the innate antiviral defence against a broad range of viruses, and that no essential role for type III IFNs can be revealed in the presence of a fully operative type I IFN system. This hypothesis was subsequently tested by Staeheli and associates, who generated mice lacking both the type I and III receptor systems on an Mx$^{+/+}$ genetic background and challenged the mice with influenza A virus (Mordstein *et al.*, 2008). Interestingly, the authors found that whereas the IFNAR$^{-/-}$ and IL-28RA$^{-/-}$ mice both

Table 3.1 Role of type I and III IFNs in antiviral defence in mice

		Mouse strain (phenotype)				
Virus	Route of infection	IFNAR$^{-/-}$	IL-28RA$^{-/-}$	IFNAR$^{-/-}$, IL-28RA$^{-/-}$	STAT1$^{-/-}$	References
HSV-2	i.vag.	Virus ↑↑	Virus ↔	NR	Virus ↑	Ank *et al.* (2008), Halford *et al.* (2006)
EMCV	i.p.	Virus ↑↑↑	Virus ↔	NR	NR	Ank *et al.* (2008)
LCMV	i.v.	Virus ↑↑	Virus ↔	NR	Virus ↑↑	Ank *et al.* (2008), Miyagi *et al.* (2007)
VSV	i.v.	Virus ↑↑↑	Virus↔	NR	Virus ↑↑↑	Ank *et al.* (2008), Durbin *et al.* (1996)
IAV	i.n.	Virus ↑	Virus ↑	Virus ↑	Virus ↑↑↑	Ank *et al.* (2008), Mordstein *et al.* (2008), Umbach *et al.* (2008)
IBV	i.n.	Virus ↑↑	Virus ↔	Virus ↑↑↑	NR	Mordstein *et al.* (2010)
HMPV	i.n.	Virus ↔	NR	Virus ↑↑	NR	Mordstein *et al.* (2010)
THOV	i.p.	Virus ↑↑	NR	Virus ↑↑	NR	Mordstein *et al.* (2008)
RVFV	i.p.	Virus ↑↑	NR	Virus ↑↑	NR	Mordstein *et al.* (2008)
RSV	i.n.	Virus ↔	Virus ↔	Virus ↑↑↑	Virus ↑↑	Johnson *et al.* (2005), Mordstein *et al.* (2010)
SARS	i.n.	Virus ↑	Virus ↑	Virus ↑↑↑	Virus ↑	Mordstein *et al.* (2010), Hogan *et al.* (2004)
MNV-1	p.o.	Virus ↔	NR	NR	Virus ↑↑	Karst *et al.* (2003)

EMCV, encephalomyocarditis virus; HMPV, human metapneumovirus; IBV, influenza B virus; i.n., intranasal; i.p., intraperitoneal; i.vag., intravaginal; LCMV, lymphocytic choriomeningitis virus; NR, not reported; p.o., per oral; RSV, respiratory syncytial virus; RVFV, Rift Valley fever virus; SARS, severe acute respiratory syndrome coronavirus; THOV, thogotovirus; VSV, vesicular stomatitis virus.

displayed only marginally elevated viral titres in the lungs, the IFNAR/IL-28RA-double-deficient mice were highly compromised in their antiviral response, were unable to control the virus and hence succumbed to infection, even with attenuated influenza A virus strains (Mordstein *et al.*, 2008). These data suggest that IFN-λs do indeed contribute to innate defence against some viral infections and act in parallel with type I IFN to mount the first line of antiviral defence. In an extension of the initial work on influenza A virus, Staheli and co workers recently reported that IFNAR/IL-28RA-double-deficient mice are highly sensitive to a range of human pathogenic viruses after intranasal infection (Mordstein *et al.*, 2010; Table 3.1). Based on these findings it was concluded that type III IFNs render epithelial cells of the respiratory and gastrointestinal tracts resistant to viral infections (Mordstein *et al.*, 2010). It is noteworthy that there are several reports in the literature of $STAT1^{-/-}$ mice displaying a more profound phenotype than $IFNAR^{-/-}$ mice after viral challenge, a finding that was previously left unexplained, or alternatively suggested to be explained by a role for IFN-γ. With the identification of IFN-λ acting through STAT1, and the recent discovery of redundancy between type I and III IFN in some disease models, it seems likely that important antiviral activities of IFN-λ account for the additional impairment of the antiviral response in $STAT1^{-/-}$ mice compared with $IFNAR^{-/-}$ mice. Recent data with respiratory syncytial virus support this idea (Johnson *et al.*, 2005; Mordstein *et al.*, 2010), and may also explain the discrepancy between the phenotypes of the $STAT1^{-/-}$ and $IFNAR^{-/-}$ mice in control of murine norovirus 1, which causes gastroenteritis after oral infection (Karst *et al.*, 2003).

Perhaps the best evidence in favour of a role of a host molecule in antiviral defence is the identification of viral strategies to inhibit its action. It has long been known that largely all known pathogenic viruses inhibit the type I IFN response in some way or another, and we now know that many of these evasion strategies also impair the type III IFN response (Katze *et al.*, 2002). Specifically, viruses targeting the molecules essential for expression (e.g. IRF-3) and function (e.g. STAT1/2) of type I and III IFNs, will evade the antiviral activities of both classes of IFNs, as seen for a range of both RNA and DNA viruses (Katze *et al.*, 2002). The role of IFN-λ in protective antiviral immunity has recently been supported by the discovery that the poxvirus Yaba-like disease virus, which causes vesicular skin lesions in primates that can be transmitted to humans, encodes a secreted protein (Y136) which binds both type I and III IFNs, thus inhibiting their biological activities (Huang *et al.*, 2007). Altogether, type III IFNs play an important role in innate antiviral immunity, notable at epithelial surfaces, and are targeted by viruses to allow establishment of infection.

Functions of type III IFN in adaptive immune responses

Early after the initial discovery of type III IFNs, reports suggested that these cytokines may possess activities other than classical antiviral type I IFN-like functions. Some of these functions seem to impact on the nature of the adaptive immune response, hence demonstrating that type III IFNs are yet another aspect of the innate immune system contributing to shaping of the adaptive immune response. The first indication of this was provided by Mennechet and Uze who reported that DCs acquire the IFN-λ receptor complex during differentiation and that stimulation through this receptor specifically induces proliferation of tolerogenic $CD4^{+}CD25^{+}Foxp3^{+}$ regulatory T (Treg) cells (Mennechet and Uze, 2006). Based on these findings the authors speculated that type III IFNs, in contrast to the Th1-promoting type I IFNs, may direct DCs into a maturation stage that promotes regulatory T-cell expansion. Subsequent studies in more physiological systems, however, suggest that IFN-λ reduces Treg activity during development of an adaptive immune response (Morrow *et al.*, 2009), and may thus play a constructive role in the adaptive immune response. This latter idea is supported by a series of findings.

The group of Grant Gallagher first reported that IFN-λ1 induced expression of proinflammatory cytokines in human peripheral blood mononuclear cells, primarily owing to targeting of monocytes (Jordan *et al.*, 2007a). The same group subsequently showed that IFN-λ also targets helper T cells and down-regulates IL-4 and IL-13 expression without affecting the IFN-γ response (Jordan *et al.*, 2007b; Dai *et al.*, 2009), hence skewing the Th1/Th2 profile towards a Th1 phenotype. The authors found that IL-28Rα was expressed on both naïve and memory human CD4+ T cells and responded to IFN-λ stimulation by down-regulating GATA3 and IL-4Rα (Dai *et al.*, 2009). Finally, IFN-λ inhibited the conversion of central memory T cells into effector memory T cells. A parallel publication demonstrated that use of IFN-λ3 (IL28B) as an adjuvant in a DNA vaccine led to development of an adaptive immune response with increased numbers of splenic CD8+ T cells compared with when IL-12 was used as adjuvant, and that these cells were more granular and exhibited higher antigen-specific degranulation (Morrow *et al.*, 2009). Together with a recent report of IFNs being able to re-programme Th2 cells to a stable GATA3^{+}Tbet^{+} cell subset (Hegazy *et al.*, 2010), these results suggest that type III IFNs may play a key role in achieving the most beneficial T cell environment to combat viral infections by prevention of Th2 differentiation, 'redirection' of the existing Th cell phenotypic profile and stimulation of potent antigen-specific cytotoxic activity. All together, the recent description of IFN-λ playing a part in activation, shaping and maintenance of the optimal adaptive antiviral immune response adds new insight into IFN-λ biology. Moreover, this could also indicate that dysregulation of type III IFN may screw the CD4 T cell response towards a Th2 phenotype with implications on development of asthma. Fig. 3.4 summarizes the current understanding of IFN-λ in innate and adaptive immunology.

IFN-λ and human diseases

A central question relates to the role of type III IFNs in host defence in humans. Based on the findings in mice and cell culture, the prediction would be that IFN-λ protects epithelial surfaces against viral infections and supports development and maintenance of Th1 T cell responses. Indeed, it has been reported that human bronchial epithelial cells respond to IFN-λ treatment and restrict rhinovirus replication (Contoli *et al.*, 2006) and that deficient IFN-λ production in response to rhinovirus correlates with asthma exacerbation and elevated virus load (Contoli *et al.*, 2006). In support of this, human alveolar type II epithelial cells respond to influenza A virus infection with a potent type III IFN response, and recombinant IFN-γ1 displayed potent antiviral activity *in vitro* (Wang *et al.*, 2009). This suggests that IFN-γ contributes to innate host defence in humans against viral infections at mucosal surfaces.

Most of the current knowledge on the role of IFN-λ in antiviral defence is based on studies in

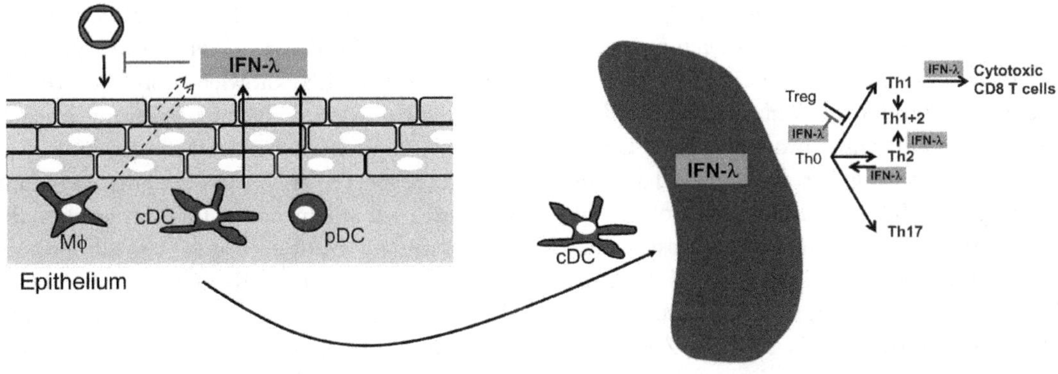

Figure 3.4 Roles of type III IFN in innate and adaptive immune responses. Virus infection induces expression of type III IFN, with DC cell populations being important sources of the cytokines. The IFN in turn induces expression of ISGs in epithelial cells leading to the classical antiviral state in these cells. Type III IFNs also modulate the adaptive immune response at various levels. Notably, these IFNs suppress Th2 responses and promote Th1-driven antigen-specific cytotoxicity of CD8+ T lymphocytes.

mice (Ank *et al.*, 2006b, 2008; Mordstein *et al.*, 2008; Sommereyns *et al.*, 2008). While the murine system allows detailed analysis of the mechanisms of action of IFN-λ under physiological conditions, it is important to bear in mind that significant differences may exist between the human and murine systems. At least one such difference seems to exist in that the IL-28Rα chain is expression in human hepatocytes (Doyle *et al.*, 2006), which are responsive to IFN-λ (Doyle *et al.*, 2006; Lazaro *et al.*, 2007), whereas the murine liver seems not to respond to IFN-λ (Mordstein *et al.*, 2008). The importance of the IFN-λ system in defence against viral infections in the liver in humans was recently emphasized by reports demonstrating that single nucleotide polymorphisms (SNPs) in the IFN-λ3 (*IL28B*) gene region are associated with increased and decreased frequency of spontaneous clearance of HCV infection (Thomas *et al.*, 2009; Rauch *et al.*, 2010). The frequencies of these alleles differ between ethnic groups. For instance the rs12979860 SNP associated with the increased spontaneous clearance allele is fixed throughout East Asia, has an intermediate frequency in Europe and is the minor allele in Africa (Thomas *et al.*, 2009).

IFN-α is currently used as a drug to treat chronic HBV and HCV infections in combination with other therapeutics. However, IFN-α has a number of severe side-effects and is not effective in all patients. For this reason, and the above-described findings that IFN-γ blocks HCV infection in human hepatocytic cell lines (Robek *et al.*, 2005; Marcello *et al.*, 2006), IFN-λ has been proposed, and already tested, as a treatment of hepatitis C (Dodds *et al.*, 2009; Muir *et al.*, 2009). In a phase 1B trial of patients with chronic hepatitis C who are responders or relapsers, the drug had a robust activity against HCV and limited toxicity (Dodds *et al.*, 2009; Shiffman *et al.*, 2009), most likely a result of the restricted expression of IL-28Rα (Kotenko *et al.*, 2003; Sheppard *et al.*, 2003; Lasfar *et al.*, 2006; Mennechet and Uze, 2006; Zhou *et al.*, 2007; Ank *et al.*, 2008; Sommereyns *et al.*, 2008). Thus, IFN-λ is a promising molecule for the future treatment of hepatitis C, and a dose-ranging phase 2 trial is currently planned.

Collectively, there is now good evidence for a protective role of IFN-λ in the antiviral response at epithelial surfaces in humans. In addition, IFN-λ plays a key role in the spontaneous clearance of HCV infections although the underlying mechanisms remain unknown. Finally, we still remain to learn how the reported role of IFN-λ in development and maintenance of the Th1 response impacts on the adaptive immune response to viral infections

Future trends

Based on the literature on IFN-λ, it is now well-established that this cytokine functions in the innate immune response at mucosal surfaces. Moreover, recent data suggest an emerging role for the type III IFNs in the adaptive immune response. Finally, IFN-λs have shown promising results as a future drug against chronic viral infections. This field is thus at a stage where there is an urgent need for further understanding of both basic biology and therapeutic antiviral activity of type III IFNs.

First, we still lack important pieces of information on basic IFN-λ function. What is the molecular nature of the interaction between the cytokine and the receptor? Better insight into the early interactions between IFN-λ and the receptor complex may provide information on what turns on signalling and could also allow for development of cytokines with altered function. At the level of signal transduction, the current dogma is that type I and III IFNs induce identical or very similar signalling pathways. Although this is certainly true for the Jak1/Tyk2-ISGF3 pathway, we have very limited knowledge on other IFN-λ-activated pathways, which may have large impacts on the biological activities of these cytokines.

With respect to IFN-λ and the innate immune response, we still need to further develop the understanding of the relative contribution of the type I versus type III IFN system in host defence. For instance, the recent reports on IFN-λ transcription being strongly regulated by NF-κB (Thomson *et al.*, 2009; Iversen *et al.*, 2010), suggests that type III IFN could play a key role in innate defence at mucosal surfaces against viruses targeting the IRF system. The role and mechanism of action of IFN-λ in the adaptive immune response needs to be described in more details.

Which cells of the adaptive immune system respond to IFN-λ? Is endogenous IFN-λ essential for development and maintenance of optimal adaptive immune responses to viral infections? Such information is of utmost importance, and may have implications for our understanding of the mechanism of action of therapeutic IFN-λ and the potential use of IFN-λs as vaccine adjuvants. Of relevance for all aspects of IFN-λ biology is the question as to which cells that produce the cytokine at different stages of infection and whether this always coincides with expression of type I IFNs. Finally, if further clinical tests of IFN-λ1 as a biological antiviral drug live up to the promising results of the early trials, we need to deepen our understanding of the antiviral mechanisms of IFN-λ in living organisms, and also which mechanisms viruses have evolved to circumvent the type III IFN system. Answers to some of the questions raised above will undoubtedly allow for a better understanding of IFN-λ as a virus-induced cytokine with important function in innate and adaptive antiviral immune responses.

Concluding remarks

Since the discovery of IFN-λ in 2003 as a novel type-I-IFN-like cytokine, our understanding has grown to a stage where we now know that the type III IFNs play important roles both as an innate cytokine at mucosal surfaces and as a stimulator of a Th1 response with potent antigen-specific cytotoxic activity of CD8+ T cells. Ten years ago, virologists and immunologists thought that type I IFNs were the only mediator of the classical 'antiviral state' in innate antiviral defence. Since then, type III IFNs have emerged as a more specialized virus-induced IFN with functions both overlapping with and distinct from type I IFN. Knowledge on the functions of and interplay between the different antiviral systems in the immune system is essential for understanding of how the organism eliminates invading viral pathogens, and establishes long-lasting immunity without excessive activation of inflammation.

References

Ablasser, A., Bauernfeind, F., Hartmann, G., Latz, E., Fitzgerald, K.A., and Hornung, V. (2009). RIG-I-dependent sensing of poly(dA:dT) through the induction of an RNA polymerase III-transcribed RNA intermediate. Nat. Immunol. *10*, 1065–1072.

Alexopoulou, L., Holt, A.C., Medzhitov, R., and Flavell, R.A. (2001). Recognition of double-stranded RNA and activation of NF-kappaB by Toll-like receptor 3. Nature *413*, 732–738.

Ank, N., Iversen, M.B., Bartholdy, C., Staeheli, P., Hartmann, R., Jensen, U.B., Dagnaes-Hansen, F., Thomsen, A.R., Chen, Z., Haugen, H., Klucher, K., and Paludan, S.R. (2008). An important role for Type III interferon (IFN-lambda/IL-28) in Toll-like receptor-induced antiviral activity. J. Immunol. *180*, 2474–2485.

Ank, N., West, H., and Paludan, S.R. (2006a). IFN-lambda: novel antiviral cytokines. J. Interferon Cytokine Res. *26*, 373–379.

Ank, N., West, H., Bartholdy, C., Eriksson, K., Thomsen, A.R., and Paludan, S.R. (2006b). Lambda interferon (IFN-lambda), a type III IFN, is induced by viruses and IFNs and displays potent antiviral activity against select virus infections *in vivo*. J. Virol. *80*, 4501–4509.

Apostolou, E., and Thanos, D. (2008). Virus infection induces NF-kappaB-dependent interchromosomal associations mediating monoallelic IFN-beta gene expression. Cell *134*, 85–96.

Barbalat, R., Lau, L., Locksley, R.M., and Barton, G.M. (2009). Toll-like receptor 2 on inflammatory monocytes induces type I interferon in response to viral but not bacterial ligands. Nat. Immunol. *10*, 1200–1207.

Bartlett, N.W., Buttigieg, K., Kotenko, S.V., and Smith, G.L. (2005). Murine interferon lambdas (type III interferons) exhibit potent antiviral activity *in vivo* in a poxvirus infection model. J. Gen. Virol. *86*, 1589–1596.

Bowie, A.G., and Unterholzner, L. (2008). Viral evasion and subversion of pattern-recognition receptor signalling. Nat. Rev. Immunol. *8*, 911–922.

Chiu, Y.H., Macmillan, J.B., and Chen, Z.J. (2009). RNA Polymerase III detects cytosolic DNA and induces type I interferons through the RIG-I pathway. Cell *138*, 576–591.

Coccia, E.M., Severa, M., Giacomini, E., Monneron, D., Remoli, M.E., Julkunen, I., Cella, M., Lande, R., and Uze, G. (2004). Viral infection and Toll-like receptor agonists induce a differential expression of type I and lambda interferons in human plasmacytoid and monocyte-derived dendritic cells. Eur. J. Immunol. *34*, 796–805.

Contoli, M., Message, S.D., Laza-Stanca, V., Edwards, M.R., Wark, P.A., Bartlett, N.W., Kebadze, T., Mallia, P., Stanciu, L.A., Parker, H.L., *et al.* (2006). Role of deficient type III interferon-lambda production in asthma exacerbations. Nat. Med. *12*, 1023–1026.

Coulombe, F., Divangahi, M., Veyrier, F., de Leseleuc, L., Gleason, J.L., Yang, Y., Kelliher, M.A., Pandey, A.K., Sassetti, C.M., Reed, M.B., and Behr, M.A. (2009). Increased NOD2-mediated recognition of N-glycolyl muramyl dipeptide. J. Exp. Med. *206*, 1709–1716.

Dai, J., Megjugorac, N.J., Gallagher, G.E., Yu, R.Y., and Gallagher, G. (2009). IFN-lambda1 (IL-29) inhibits GATA3 expression and suppresses Th2 responses in human naive and memory T cells. Blood *113*, 5829–5838.

Defilippis, V.R., Alvarado, D., Sali, T., Rothenburg, S., and Fruh, K. (2009). Human cytomegalovirus induces the interferon response via the DNA sensor ZBP1. J. Virol. *84*, 585–598.

Diebold, S.S., Kaisho, T., Hemmi, H., Akira, S., and Reis e Sousa, C. (2004). Innate antiviral responses by means of TLR7-mediated recognition of single-stranded RNA. Science *303*, 1529–1531.

Dodds, M. G., Hausman, D. F., and Miller, D. M. (2009). Viral kinetic modeling during treatment with interferon lambda-1a in genotype 1 chronic Hepatitis C patients. EASL, 2009, Copenhagen. J. Hepatol. *50*, S342.

Donnelly, R.P., Sheikh, F., Kotenko, S.V., and Dickensheets, H. (2004). The expanded family of class II cytokines that share the IL-10 receptor-2 (IL-10R2) chain. J. Leukoc. Biol. *76*, 314–321.

Doyle, S.E., Schreckhise, H., Khuu-Duong, K., Henderson, K., Rosler, R., Storey, H., Yao, L., Liu, H., Barahmand-pour, F., Sivakumar, P., *et al.* (2006). Interleukin-29 uses a type 1 interferon-like program to promote antiviral responses in human hepatocytes. Hepatology *44*, 896–906.

Dumoutier, L., Tounsi, A., Michiels, T., Sommereyns, C., Kotenko, S.V., and Renauld, J.C. (2004). Role of the interleukin (IL)-28 receptor tyrosine residues for antiviral and antiproliferative activity of IL-29/interferon-lambda 1: similarities with type I interferon signaling. J. Biol. Chem. *279*, 32269–32274.

Durbin, J.E., Hackenmiller, R., Simon, M.C., and Levy, D.E. (1996). Targeted disruption of the mouse Stat1 gene results in compromised innate immunity to viral disease. Cell *84*, 443–450.

Garcia-Sastre, A., and Biron, C.A. (2006). Type 1 interferons and the virus–host relationship: a lesson in detente. Science *312*, 879–882.

Genin, P., Vaccaro, A., and Civas, A. (2009). The role of differential expression of human interferon-a genes in antiviral immunity. Cytokine Growth Factor Rev. *20*, 283–295.

Hacker, H., Redecke, V., Blagoev, B., Kratchmarova, I., Hsu, L.C., Wang, G.G., Kamps, M.P., Raz, E., Wagner, H., Hacker, G., Mann, M., and Karin, M. (2005). Specificity in Toll-like receptor signalling through distinct effector functions of TRAF3 and TRAF6. Nature 439, 204–207.

Halford, W.P., Weisend, C., Grace, J., Soboleski, M., Carr, D.J., Balliet, J.W., Imai, Y., Margolis, T.P., and Gebhardt, B.M. (2006). ICP0 antagonizes Stat 1-dependent repression of herpes simplex virus: implications for the regulation of viral latency. Virol. J. 3, 44.

Hegazy, A.N., Peine, M., Helmstetter, C., Panse, I., Frohlich, A., Bergthaler, A., Flatz, L., Pinschewer, D.D., Radbruch, A., and Lohning, M. (2010). Interferons direct Th2 cell reprogramming to generate a stable GATA-3(+)T-bet(+) cell subset with combined Th2 and Th1 cell functions. Immunity 32, 116–128.

Heil, F., Hemmi, H., Hochrein, H., Ampenberger, F., Kirschning, C., Akira, S., Lipford, G., Wagner, H., and Bauer, S. (2004). Species-specific recognition of single-stranded RNA via toll-like receptor 7 and 8. Science *303*, 1526–1529.

Hogan, R.J., Gao, G., Rowe, T., Bell, P., Flieder, D., Paragas, J., Kobinger, G.P., Wivel, N.A., Crystal, R.G., Boyer, J., Feldmann, H., Voss, T.G., and Wilson, J.M. (2004). Resolution of primary severe acute respiratory syndrome-associated coronavirus infection requires Stat1. J. Virol. *78*, 11416–11421.

Honda, K., Yanai, H., Negishi, H., Asagiri, M., Sato, M., Mizutani, T., Shimada, N., Ohba, Y., Takaoka, A., Yoshida, N., and Taniguchi, T. (2005). IRF-7 is the master regulator of type-I interferon-dependent immune responses. Nature *434*, 772–777.

Hou, W., Wang, X., Ye, L., Zhou, L., Yang, Z.Q., Riedel, E., and Ho, W.Z. (2009). Lambda interferon inhibits human immunodeficiency virus type 1 infection of macrophages. J. Virol. *83*, 3834–3842.

Huang, J., Smirnov, S.V., Lewis-Antes, A., Balan, M., Li, W., Tang, S., Silke, G.V., Pütz, M.M., Smith, G.L., and Kotenko, S.V. (2007). Inhibition of type I and type III interferons by a secreted glycoprotein from Yaba-like disease virus. Proc. Natl. Acad. Sci. U.S.A. *104*, 9822–9827.

Isaacs, A., and Lindemann, J.J. (1957). Virus interference: the interferon. Proc. R. Soc. London *147*, 258–267.

Ishii, K.J., Kawagoe, T., Koyama, S., Matsui, K., Kumar, H., Kawai, T., Uematsu, S., Takeuchi, O., Takeshita, F., Coban, C., and Akira, S. (2008). TANK-binding kinase-1 delineates innate and adaptive immune responses to DNA vaccines. Nature *451*, 725–729.

Iversen, M.B., Ank, N., Melchjorsen, J., and Paludan, S.R. (2010). Expression of type III IFN in the vaginal mucosa is mediated primarily by DCs and displays stronger dependence on NF-kB than type I IFNs. J. Virol. *84*, 4579–4586.

Johnson, T.R., Mertz, S.E., Gitiban, N., Hammond, S., Legallo, R., Durbin, R.K., and Durbin, J.E. (2005). Role for innate IFNs in determining respiratory syncytial virus immunopathology. J. Immunol. *174*, 7234–7241.

Jordan, W.J., Eskdale, J., Boniotto, M., Rodia, M., Kellner, D., and Gallagher, G. (2007a). Modulation of the human cytokine response by interferon lambda-1 (IFN-lambda1/IL-29). Genes Immun. *8*, 13–20.

Jordan, W.J., Eskdale, J., Srinivas, S., Pekarek, V., Kelner, D., Rodia, M., and Gallagher, G. (2007b). Human interferon lambda-1 (IFN-lambda1/IL-29) modulates the Th1/Th2 response. Genes Immun. *8*, 254–261.

Kagan, J.C., Su, T., Horng, T., Chow, A., Akira, S., and Medzhitov, R. (2008). TRAM couples endocytosis of Toll-like receptor 4 to the induction of interferon-beta. Nat. Immunol. *9*, 361–368.

Karst, S.M., Wobus, C.E., Lay, M., Davidson, J., and Virgin, H.W. (2003). STAT1-dependent innate immunity to a Norwalk-like virus. Science *299*, 1575–1578.

Kato, H., Takeuchi, O., Sato, S., Yoneyama, M., Yamamoto, M., Matsui, K., Uematsu, S., Jung, A., Kawai, T., Ishii, K.J., Yamaguchi, O., Otsu, K., Tsujimura, T., Koh, C.S., Reis e Sousa, C., Matsuura, Y., Fujita, T., and Akira, S. (2006). Differential roles of MDA5 and RIG-I helicases in the recognition of RNA viruses. Nature *441*, 101–105.

Katze, M.G., He, Y., and Gale, M., Jr. (2002). Viruses and interferon: a fight for supremacy. Nat. Rev. Immunol. *2*, 675–687.

Kotenko, S.V., Gallagher, G., Baurin, V.V., Lewis-Antes, A., Shen, M., Shah, N.K., Langer, J.A., Sheikh, F., Dickensheets, H., and Donnelly, R.P. (2003). IFN-lambdas mediate antiviral protection through a distinct class II cytokine receptor complex. Nat. Immunol. *4*, 69–77.

Lasfar, A., Lewis-Antes, A., Smirnov, S.V., Anantha, S., Abushahba, W., Tian, B., Reuhl, K., Dickensheets, H., Sheikh, F., Donnelly, R.P., Raveche, E., and Kotenko, S.V. (2006). Characterization of the mouse IFN-lambda ligand-receptor system: IFN-lambdas exhibit antitumor activity against B16 melanoma. Cancer Res. *66*, 4468–4477.

Lazaro, C.A., Chang, M., Tang, W., Campbell, J., Sullivan, D.G., Gretch, D.R., Corey, L., Coombs, R.W., and Fausto, N. (2007). Hepatitis C virus replication in transfected and serum-infected cultured human fetal hepatocytes. Am. J. Pathol. *170*, 478–489.

Maniatis, T., Falvo, J.V., Kim, T.H., Kim, T.K., Lin, C.H., Parekh, B.S., and Wathelet, M.G. (1998). Structure and function of the interferon-beta enhanceosome. Cold Spring Harb. Symp. Quant. Biol. *63*, 609–620.

Marcello, T., Grakoui, A., Barba-Spaeth, G., Machlin, E.S., Kotenko, S.V., MacDonald, M.R., and Rice, C.M. (2006). Interferons alpha and lambda inhibit hepatitis C virus replication with distinct signal transduction and gene regulation kinetics. Gastroenterology *131*, 1887–1898.

Marie, I., Durbin, J.E., and Levy, D.E. (1998). Differential viral induction of distinct interferon-alpha genes by positive feedback through interferon regulatory factor-7. EMBO J. *17*, 6660–6669.

McWhirter, S.M., Barbalat, R., Monroe, K.M., Fontana, M.F., Hyodo, M., Joncker, N.T., Ishii, K.J., Akira, S., Colonna, M., Chen, Z.J., *et al.* (2009). A host type I interferon response is induced by cytosolic sensing of the bacterial second messenger cyclic-di-GMP. J. Exp. Med. *206*, 1899–1911.

Mennechet, F.J., and Uze, G. (2006). Interferon-lambda-treated dendritic cells specifically induce proliferation of FOXP3-expressing suppressor T cells. Blood *107*, 4417–4423.

Mihm, S., Frese, M., Meier, V., Wietzke-Braun, P., Scharf, J.G., Bartenschlager, R., and Ramadori, G. (2004). Interferon type I gene expression in chronic hepatitis C. Lab Invest *84*, 1148–1159.

Miyagi, T., Gil, M.P., Wang, X., Louten, J., Chu, W.M., and Biron, C.A. (2007). High basal STAT4 balanced by STAT1 induction to control type 1 interferon effects in natural killer cells. J. Exp. Med. *204*, 2383–2396.

Mogensen, T.H. (2009). Pathogen recognition and inflammatory signaling in innate immune defenses. Clin. Microbiol. Rev. *22*, 240–273.

Mordstein, M., Kochs, G., Dumoutier, L., Renauld, J.C., Paludan, S.R., Klucher, K., and Staeheli, P. (2008). Interferon-lambda contributes to innate immunity of mice against influenza A virus but not against hepatotropic viruses. PLoS. Pathog. *4*, e1000151.

Mordstein, M., Neugebauer, E., Ditt, V., Jessen, B., Rieger, T., Falcone, V., Sorgeloos, F., Ehl, S., Mayer, D., Kochs, G., *et al.* (2010). Interferon-lambda renders epithelial cells of respiratory and gastrointestinal tract resistant to viral infections. J. Virol. *84*, 5670–5677.

Morrow, M.P., Pankhong, P., Laddy, D.J., Schoenly, K.A., Yan, J., Cisper, N., and Weiner, D.B. (2009). Comparative ability of IL-12 and IL-28B to regulate Treg populations and enhance adaptive cellular immunity. Blood *113*, 5868–5877.

Muir, A.J., Shiffman, M.L., Zaman, A., Yoffe, B., Lopez-Talavera, J.A., Souza, S., Hausman, D.F., Hunder, N.N., Fontana, D., and Lawitz, E. (2009). A phase 1b dose-ranging study of 4 weeks of PEG-interferon (IFN) lambda (PEG-rIL-29) in combination with ribavirin (RBV) in patients with chronic genotype 1 Hepatitis C virus (HCV) infection. 60th Annual Meeting of the American Association for the Study of Liver Diseases, Boston, MA, Hynes Convention Center, 30 October to 3 November 2009.

Oganesyan, G., Saha, S.K., Guo, B., He, J.Q., Shahangian, A., Zarnegar, B., Perry, A., and Cheng, G. (2005). Critical role of TRAF3 in the Toll-like receptor-dependent and -independent antiviral response. Nature *439*, 211.

Onoguchi, K., Yoneyama, M., Takemura, A., Akira, S., Taniguchi, T., Namiki, H., and Fujita, T. (2007). Viral infections activate types I and III interferon genes through a common mechanism. J. Biol. Chem. *282*, 7576–7581.

Osterlund, P.I., Pietila, T.E., Veckman, V., Kotenko, S.V., and Julkunen, I. (2007). IFN regulatory factor family members differentially regulate the expression of type III IFN (IFN-lambda) genes. J. Immunol. *179*, 3434–3442.

Paladino, P., Cummings, D.T., Noyce, R.S., and Mossman, K.L. (2006). The IFN-independent response to virus particle entry provides a first line of antiviral defense that is independent of TLRs and retinoic acid-inducible gene I. J. Immunol. *177*, 8008–8016.

Rauch, A., Kutalik, Z., Descombes, P., Cai, T., Di Iulio, J., Mueller, T., Bochud, M., Battegay, M., Bernasconi, E., Borovicka, J., *et al.* (2010). Genetic variation in IL28B is associated with chronic Hepatitis C and treatment failure: a genome-wide association study. Gastroenterology *138*, 1338–1345.

Robek, M.D., Boyd, B.S., and Chisari, F.V. (2005). Lambda interferon inhibits hepatitis B and C virus replication. J. Virol. *79*, 3851–3854.

Sheppard, P., Kindsvogel, W., Xu, W., Henderson, K., Schlutsmeyer, S., Whitmore, T.E., Kuestner, R., Garrigues, U., Birks, C., Roraback, J., *et al.* (2003). IL-28, IL-29 and their class II cytokine receptor IL-28R. Nat. Immunol. *4*, 63–68.

Shiffman, M.L., Lawitz, E., Zaman, A., Vierling, J., Yoffe, B., Souza, S., Freeman, J., Gray, T., Hausman, D.F., and Muir, A.J. (2009). PEG-IFN-lambda: antiviral activity and safety profile in a 4-week phase 1b study in relapsed genotype 1 hepatitis C infection. 60th Annual Meeting of the American Association for the Study of Liver Diseases. Boston, MA, Hynes Convention Center, 30 October to 3 November 2009.

Sommereyns, C., Paul, S., Staeheli, P., and Michiels, T. (2008). IFN-lambda (IFN-lambda) is expressed in a tissue-dependent fashion and primarily acts on epithelial cells *in vivo*. PLoS. Pathog. *4*, e1000017.

Stark, G.R., Kerr, I.M., Williams, B.R., Silverman, R.H., and Schreiber, R.D. (1998). How cells respond to interferons. Annu. Rev. Biochem. *67*, 227–264.

Takaoka, A., Wang, Z., Choi, M.K., Yanai, H., Negishi, H., Ban, T., Lu, Y., Miyagishi, M., Kodama, T., Honda, K., Ohba, Y., and Taniguchi, T. (2007). DAI (DLM-1/ZBP1) is a cytosolic DNA sensor and an activator of innate immune response. Nature *448*, 501–505.

Thomas, D.L., Thio, C.L., Martin, M.P., Qi, Y., Ge, D., O'Huigin, C., Kidd, J., Kidd, K., Khakoo, S.I., Alexander, G., *et al.* (2009). Genetic variation in IL28B and spontaneous clearance of hepatitis C virus. Nature *461*, 798–801.

Thomson, S.J., Goh, F.G., Banks, H., Krausgruber, T., Kotenko, S.V., Foxwell, B.M., and Udalova, I.A. (2009). The role of transposable elements in the regulation of IFN-lambda1 gene expression. Proc. Natl. Acad. Sci. U.S.A. *106*, 11564–11569.

Umbach, J.L., Kramer, M.F., Jurak, I., Karnowski, H.W., Coen, D.M., and Cullen, B.R. (2008). MicroRNAs expressed by herpes simplex virus 1 during latent infection regulate viral mRNAs. Nature *454*, 780–783.

Wang, J., Oberley-Deegan, R., Wang, S., Nikrad, M., Funk, C.J., Hartshorn, K.L., and Mason, R.J. (2009). Differentiated human alveolar type II cells secrete antiviral IL-29 (IFN-lambda 1) in response to influenza A infection. J. Immunol. *182*, 1296–1304.

Yoneyama, M., Kikuchi, M., Natsukawa, T., Shinobu, N., Imaizumi, T., Miyagishi, M., Taira, K., Akira, S., and Fujita, T. (2004). The RNA helicase RIG-I has an essential function in double-stranded RNA-induced innate antiviral responses. Nat. Immunol. *5*, 730–737.

Zhang, S.Y., Jouanguy, E., Ugolini, S., Smahi, A., Elain, G., Romero, P., Segal, D., Sancho-Shimizu, V., Lorenzo, L., Puel, A., *et al.* (2007). TLR3 deficiency in patients with herpes simplex encephalitis. Science *317*, 1522–1527.

Zhou, Z., Hamming, O.J., Ank, N., Paludan, S.R., Nielsen, A.L., and Hartmann, R. (2007). Type III interferon (IFN) induces a type I IFN-like response in a restricted subset of cells though signaling pathways involving both Jak-STAT pathway and the mitogen-activated protein kinases. J. Virol. *81*, 7749–7758.

Antiviral Function of Interferons

4

Marisela Rodriguez, Jessica A. Campbell and Deborah J. Lenschow

Abstract

The type I interferon (IFN) system plays a critical role in limiting the spread of viral infection. Viruses induce the production of IFN-α and -β, which bind to the IFN-α/β receptor (IFNAR) and trigger the JAK/STAT signalling cascade. The ensuing induction of IFN-stimulated genes (ISGs) inhibits viral replication by targeting multiple points in the viral life cycle. ISGs exert their antiviral function through diverse mechanisms, including activities directly targeting the virus such as the degradation of viral RNA, the inhibition of translation, the blockade of virion release, and actions that modify the host response including regulation of the IFN response, NF-κB signalling, and apoptosis, among others. This chapter reviews several ISGs that have been shown to mediate antiviral activity either *in vitro* or *in vivo*, and in some cases, both. The mechanisms by which individual ISGs confer cellular protection are summarized, although the effector pathways of certain ISGs are still being delineated. The study of ISGs continues to provide important contributions to our understanding of the host–virus interface and the cellular antiviral response.

Interferon-stimulated genes impacting viral nucleic acid synthesis and stability

A successful viral infection requires the proper utilization of both host and viral nucleic acids. Viral transcripts must be produced and allowed to interact with the translational machinery; host RNAs that are required for the viral life cycle, such as rRNAs, must be intact; and for RNA viruses, viral RNA must be produced and packaged into new virions. Some ISGs induce the degradation of RNA or mutate RNA in a manner that reduces its stability or introduces frameshift mutations. The ISGs that mediate these activities include oligoadenylate synthetase (OAS), ribonuclease L (RNase L), adenosine deaminase that acts on dsRNA (ADAR1), the Mx GTPases, the apolipoprotein B messenger RNA editing factor with cytidine deamination activity (APOBEC), and the zinc finger antiviral protein (ZAP), and are covered in this section.

Oligoadenylate synthetase (OAS)/ ribonuclease L (RNaseL)

The 2′,5′-OAS/RNase L pathway is a dsRNA-activated RNA degradation system. Upon binding dsRNA, OAS enzymes catalyse the synthesis of unique 2′–5′-linked phosphodiester bonds that polymerize ATP into oligoadenylates collectively known as 2–5A, with the general formula pppA(2′5′A)$_n$ (Kerr and Brown, 1978). In cells, pppA(2′5′A)$_3$ is the principal oligomer present (Knight *et al.*, 1980). 2–5A then activates the constitutively expressed RNase L to degrade RNA (Baglioni *et al.*, 1978; Clemens and Williams, 1978; Hovanessian *et al.*, 1979; Williams *et al.*, 1979; Hovanessian and Wood, 1980).

OAS is expressed constitutively at low levels and is induced by IFN-α, -β, and -γ (Flenniken *et al.*, 1988; Marie *et al.*, 1990; Khanna *et al.*, 1991). Differential induction, localization, and expression suggest that the multiple OAS molecules may be specialized responders to different stimuli in different cellular and subcellular compartments (Marie *et al.*, 1990; Kakuta *et al.*, 2002). Humans encode three *OAS* genes (*OAS1–3*) that produce splice variants, along with an OAS-related sequence known as *OASL* (Hovnanian *et al.*, 1998, 1999). In addition to *OAS2*, *OAS3*, and *OASL*, mice encode eight *OAS1* genes for a total of 11 orthologues (Kakuta *et al.*, 2002). OAS is activated by binding dsRNA in a groove characterized by multiple solvent-exposed basic residues (Sarkar *et al.*, 1999; Hartmann *et al.*, 2003). Analysis of multiple OAS variants has revealed a domain known as an OAS unit that contains highly conserved stretches of 7–14 amino acids (Rebouillat and Hovanessian, 1999). All OAS proteins contain one or more OAS units, but OASL is unable to synthesize 2–5A (Hartmann *et al.*, 1998; Rebouillat *et al.*, 1998; Eskildsen *et al.*, 2003), suggesting that this protein may have an RNase L-independent function. OASL is unique in containing a C-terminal diubiquitin-like sequence, which is required for its antiviral activity against encephalomyocarditis virus (EMCV; Marques *et al.*, 2008).

Although constitutively expressed, RNase L is an inactive monomer until it binds 2–5A, which leads to its activation and homodimerization (Zhou *et al.*, 1993; Dong and Silverman, 1995). Dimerization relieves repression of the nuclease domains, allowing RNase L to cleave ssRNA on the 3′ side of UpAp and UpUp dinucleotides (Floyd-Smith *et al.*, 1981; Wreschner *et al.*, 1981b; Dong *et al.*, 2001). Viral and self RNAs are cleaved by RNase L, including intact rRNAs (Wreschner *et al.*, 1981a; Silverman *et al.*, 1983; Li *et al.*, 1998; Malathi *et al.*, 2007).

In addition to cleaving viral RNA, RNase L is also involved in the IFN-β amplification loop (Malathi *et al.*, 2007). The cleavage products of RNase L contain 3′-monophosphoryl groups that can bind to and activate the cytoplasmic dsRNA sensors MDA5 and RIG-I, resulting in IFN-β induction. This function of RNase L is apparent during EMCV and Sendai virus (SeV) infection of RNase L-deficient mice, which have significantly lower serum IFN-β levels than wild-type mice (Malathi *et al.*, 2007).

RNase L deficient mice have been used to assess the antiviral role of the OAS/RNase L system. The protective effects of OAS/RNase L are generally more pronounced for RNA viruses compared with DNA viruses. This probably reflects the variety of antiviral mediators that detect different viral species. Compared with wild-type mice, RNase L$^{-/-}$ mice have greater disease severity and defective responses towards the RNA viruses EMCV, coxsackievirus, West Nile virus (WNV), and mouse hepatitis virus (MHV; Zhou *et al.*, 1997; Flodstrom-Tullberg *et al.*, 2005; Samuel *et al.*, 2006; Ireland *et al.*, 2009). *In vitro*, RNase L deficiency enhances vesicular stomatitis virus (VSV) replication (Maitra and Silverman, 1998; Khabar *et al.*, 2003). Interestingly, RNase L acts in a cell type specific manner to control a neurotropic strain of MHV (Ireland *et al.*, 2009). RNase L-deficient mice are significantly more susceptible to MHV despite displaying similar CNS viral titres as control mice throughout the time course of infection. RNase L deficiency did not significantly alter inflammation in the CNS or inhibit type I IFN production, but immunohistochemical analysis revealed that brain stems in mice lacking RNase L were severely demyelinated compared with the minimal demyelination seen in wild-type animals. These experiments illustrate that RNase L can provide cell type-specific protection from viral infection, even when viral replication is not overtly inhibited.

The role of RNase L during DNA viral infection is less clear. RNase L deficient mice are not more susceptible to the DNA virus vaccinia, but contrasting results have been reported for herpes simplex virus (HSV) infection (Zheng *et al.*, 2001; Xiang *et al.*, 2002; Al-Khatib *et al.*, 2005; Duerst and Morrison, 2007). Thus, RNase L provides protection against some DNA viruses under certain conditions, but is ineffective against others.

Although OAS deficient mice have not been reported, genetic analysis has uncovered a critical role for this molecule in flavivirus infection. The murine locus *flv* determines susceptibility to Murray Valley encephalitis virus and yellow fever virus (YFV) in an autosomal dominant manner (Sangster *et al.*, 1993; Sangster *et al.*, 1994). *Flv* has been mapped to a missense mutation in *OAS1b*, which yields a truncated protein (Mashimo *et al.*, 2002; Perelygin *et al.*, 2002). Transgenic expression of the resistant *OAS1b* allele in susceptible mice confers resistance to YFV infection (Scherbik *et al.*, 2007). For these reasons, *OAS* polymorphisms in humans were examined for their contribution to the response against WNV, a related flavivirus. Analysis of human WNV seroconverters revealed that the frequency of a hypofunctional *OAS1* allele was increased in seropositive versus seronegative patients (Lim *et al.*, 2009). Thus, OAS plays an important role in controlling flavivirus infection in mice and potentially humans. Additional analysis will be needed to determine the function of OAS during other viral infections.

Adenosine deaminase that acts on dsRNA (ADAR1)

Posttranscriptional modifications can alter the function of cellular and viral RNAs, significantly impacting biological and viral activities. One such enzyme, ADAR1, can induce RNA mutations by deamination of adenosines to produce inosines (A-to-I mutations; Bass and Weintraub, 1988; Wagner *et al.*, 1989). ADAR1 was originally identified as a factor capable of unwinding dsRNA in *Xenopus* oocytes (Rebagliati and Melton, 1987; Bass and Weintraub, 1988; Wagner *et al.*, 1989). Subsequent cloning of cDNAs from a natural killer cell library identified ADAR1 as the protein responsible for the adenosine deamination activity observed in bovine liver extracts (Kim *et al.*, 1994a,b). Human ADAR1 is a 1226 aa protein induced by IFN-α and IFN-γ (Kim *et al.*, 1994b; Patterson and Samuel, 1995; Patterson *et al.*, 1995). However, later studies revealed that ADAR1 exists in two forms inside the cell: a constitutively active 100 kDa form (p110) localized exclusively to the nucleus, and an IFN-inducible 150 kDa form (p150) that is found in both the nucleus and cytoplasm (Patterson and Samuel, 1995; George and Samuel, 1999a,b). While both forms have adenosine deaminase activity, owing to its induction by IFN and localization to the cytoplasm, the p150 form is thought to be responsible for the A-to-I changes seen in viral RNA.

ADAR1 consists of a C-terminal catalytic domain responsible for deaminase activity and two nucleic acid binding domains for interaction with dsRNA and Z-DNA (Lai *et al.*, 1995; Liu and Samuel, 1996; Maas *et al.*, 1996). Three highly conserved copies of dsRNA binding motifs ($dsRBM_{I-III}$) are found in ADAR1. Mutation of the $dsRBM_{III}$ completely abolished ADAR1 deaminase activity while mutation of $dsRBM_{I}$ only partially diminished its deaminase activity and $dsRBM_{II}$ was dispensable (Lai *et al.*, 1995; Liu and Samuel, 1996; Liu *et al.*, 1997). These results indicate that these dsRBM are not functionally homologous, and that $dsRBM_{III}$ is required for both dsRNA binding and deaminase activity.

ADAR1 has been shown to not only hypermutate host RNA but also viral RNAs. Multiple viruses accumulate A-to-I hypermutations during infection including measles virus (MV), hepatitis C virus (HCV), human parainfluenza virus, respiratory syncytial virus (RSV), lymphocytic choriomeningitis virus (LCMV), and human immunodeficiency virus (HIV; O'Hara *et al.*, 1984; Cattaneo *et al.*, 1988; Murphy *et al.*, 1991; Cattaneo, 1994; Hajjar and Linial, 1995; Kumar and Carmichael, 1997; Jayan and Casey, 2002a,b; Gandy *et al.*, 2007; Zahn *et al.*, 2007; Doria *et al.*, 2009; George *et al.*, 2009). For example, full length gene transcripts of MV cloned from brain biopsies of subacute sclerosing panencephalitis patients contained numerous A-to-I hypermutations with seven A-to-G amino acid residue mutations identified in the MV M gene alone (Cattaneo *et al.*, 1988). Thus, ADAR1-mediated hypermutations can accumulate in viral RNA over the course of an infection, and may lead to increased RNA instability or protein coding changes during translation, both of which could impact overall viral fitness. Indeed, the LCMV genomic sequence corresponding to its glycoprotein (GP) acquired numerous ADAR1-dependent A-to-G mutations that modified one or more amino acids in the translated GP protein (Zahn *et al.*, 2007). Many of these mutant LCMV GP clones had diminished GP expression on the surface of infected cells. While not absolute, the lack of LCMV GP expression on the cell surface correlated with decreased LCMV infectivity (Zahn *et al.*, 2007) indicating that ADAR1-dependent hypermutations could affect viral propagation by modifying sequences of a translated viral protein. Further evidence for ADAR1 functioning as an antiviral molecule includes the observation that multiple viruses, including vaccinia virus and adenovirus, contain ADAR1 immune evasion factors capable of inhibiting ADAR1-mediated hypermutations (Lei *et al.*, 1998; Liu *et al.*, 2001; Taylor *et al.*, 2005).

A surprising result has been a recent series of studies revealing that ADAR1 may function as a proviral molecule by negatively regulating the function of another known ISG, dsRNA-dependent protein kinase (PKR). During viral infection PKR is activated and induces the phosphorylation of the alpha subunit of eukaryotic initiation factor 2 (eIF2α), leading to inhibition of protein translation (Mellor *et al.*, 1994; Lu *et al.*, 1999; Garaigorta and Chisari, 2009). Gene overexpression studies revealed that ADAR1 variants p110 and p150 interact with PKR (Nie *et al.*, 2007). While this interaction was also observed with endogenous proteins in Jurkat cells, PKR only interacted with the IFN inducible p150 form during HIV infection (Clerzius *et al.*, 2009). Increased HIV-1 virion release was observed in cells overexpressing ADAR1 (Clerzius *et al.*, 2009; Doria *et al.*, 2009), and this correlated with an inhibition of the phosphorylation of PKR and its downstream target, eIF-2α. The overexpression of ADAR1 in PKR sufficient cells also resulted in increased susceptibility of these cells to VSV infection (Nie *et al.*, 2007). This increased susceptibility was PKR-dependent since the overexpression of ADAR1 in PKR-/- cells had no significant effect on VSV replication (Nie *et al.*, 2007). Furthermore, in cells in which ADAR1 had been stably knocked down, increased phosphorylation of PKR and eIF2α was observed (Li *et al.*, 2010). In these cells the growth of MV lacking accessory proteins V or C was diminished, and the ability of IFN treatment to induce an antiviral state during VSV infection was enhanced (Toth *et al.*, 2009; Li *et al.*, 2010). Taken together, these results demonstrate that ADAR1 also mediates proviral effects through inhibition of the PKR antiviral pathway. Whether the proviral function of ADAR1 requires its editing activity is still debated. Overexpression of an ADAR1 mutant lacking editing function was unable to increase HIV virion release or p24 expression in the supernatant of infected cells (Phuphuakrat *et al.*, 2008; Doria *et al.*, 2009). However, overexpression of a naturally existing ADAR1 variant lacking editing activity was still able to inhibit activation of eIF2α and increase VSV virus titres (Nie *et al.*, 2007; Doria *et al.*, 2009). Further studies will be needed to define the exact mechanism by which ADAR1 functions as a proviral molecule in these systems.

Mx family GTPases

The Mx family proteins are some of the best characterized antiviral ISGs. They belong to the class of dynamin-like large guanosine triphosphatases (GTPases). The two human Mx proteins, MxA (662 aa) and MxB (715 aa), are localized to chromosome 21 while the mouse Mx1 (631 aa) and Mx2 (665 aa) are located on chromosome 16 (Horisberger *et al.*, 1988; Aebi *et al.*, 1989). Both human and mouse Mx proteins are induced by type I IFNs, while MxA is also induced by IFN-γ (Krug *et al.*, 1985; Aebi *et al.*, 1989; Jin *et al.*, 1999). Antiviral activity has been described for murine Mx1, which resides in the nucleus, and Mx2, which resides in the cytoplasm, with the specificity of their antiviral activity reflecting their subcellular localization (Dreiding *et al.*, 1985; Staeheli *et al.*, 1986). In humans, both Mx GTPases reside in the cytoplasm (Aebi *et al.*, 1989; Jin *et al.*, 1999). MxA has wide antiviral activity, while MxB has no detectable antiviral activity. Mx2 is non-functional in laboratory mouse strains owing to a translation frameshift mutation caused by an insertion; however, feral mice NJL and SPR do express a functional Mx2 gene that is induced following type I IFN stimulation (Jin *et al.*, 1999; Sadler and Williams, 2008). Overexpression of the SPR-derived Mx2 gene in 3T3 cells significantly diminished VSV viral titres indicating that Mx2 is antiviral and can confer some protection against viruses (Jin *et al.*, 1999; Sadler and Williams, 2008).

Mx proteins consist of a large N-terminal GTPase domain, a central interacting domain (CID) and a C-terminal leucine zipper (LZ) domain. The N-terminal GTP motif is highly conserved while the C-terminus is divergent. Self-assembly into high molecular weight homo-oligomers protects Mx proteins from rapid degradation and enhances their GTPase activity. The GTPase domain is important for Mx antiviral function since mutagenesis studies revealed that mutants lacking the GTPase domain cannot effectively control virus infections (Pitossi *et al.*, 1993;

Ponten *et al.*, 1997). The LZ domain of MxA is also important for viral target recognition, since point mutations disrupt this activity and mutants lacking the C-terminal LZ domain were unable to inhibit Thogoto virus antigen accumulation in infected cells (Ponten *et al.*, 1997).

The literature regarding the antiviral role of Mx proteins is plentiful and dates back to the 1960s. Studies initially characterizing murine orthomyxovirus infection reported that A2G inbred mice were selectively resistant to various influenza strains (Lindenmann *et al.*, 1963). This phenotype was attributed to a dominant gene termed Mx, which was absent from various inbred strains susceptible to influenza infection such as A/J and LMH mice (Lindenmann *et al.*, 1963; Haller *et al.*, 1979). Subsequent cloning and characterization of the Mx proteins enabled scientists to investigate the spectrum of their antiviral activities (Horisberger *et al.*, 1988; Aebi *et al.*, 1989). Mx1 confers resistance to viruses that replicate in the nucleus, such as influenza and Thogoto virus, while Mx2 inhibits cytoplasmic viruses including members of the Rhabdoviridae and Bunyaviridae family (Haller *et al.*, 1995). In contrast, MxA has broad antiviral activity against viruses replicating in both the cytoplasm and nucleus including Rhabdoviridae, Togoviridae, Bunyaviridae, Orthomyxoviridae, Paramyxoviridae, Picornaviridae, and Hepadnaviridae (Pavlovic *et al.*, 1990; Pavlovic and Staeheli, 1991; Schnorr *et al.*, 1993; Schneider-Schaulies *et al.*, 1994; Pavlovic *et al.*, 1995; Hefti *et al.*, 1999; Chieux *et al.*, 2001; Samuel, 2001; Andersson *et al.*, 2004; Sadler and Williams, 2008).

The mechanism(s) by which Mx proteins exert their antiviral function is still incompletely understood. Mx GTPases act early in the viral life cycle, before amplification of the genome. The general mechanism by which they inhibit viruses is to bind viral nucleocapsid structures and to sequester them in locations that prevent them from being accessible for new virion production. For example, MxA was shown to specifically bind Thogoto viral nucleocapsids and this interaction was dependent on its GTPase activity (Kochs and Haller, 1999a). MxA inhibited the nuclear import of Thogoto virus nucleocapsids and blocked viral genome replication (Kochs and Haller, 1999b). Both MxA and Mx1 can interact with the influenza PB2 and NP proteins; however, it is thought that MxA inhibits viral replication by blocking the nuclear import of influenza nucleocapsids, while Mx1 inhibits viral RNA synthesis, potentially by sequestering viral mRNA or impacting the stability of viral mRNA transcripts (Krug *et al.*, 1985; Turan *et al.*, 2004). MxA also binds to the nucleocapsid protein of bunyaviruses forming copolymers that then accumulate near the nuclear membrane (Kochs *et al.*, 2002). This accumulation, and subsequent mislocalization of viral components, inhibits viral replication (Kochs *et al.*, 2002). MxA is hypothesized to regulate the replication of cytoplasmic viruses by redirecting their nucleocapsid structures to alternative sites within the cytoplasm.

Studies in mice have demonstrated the importance of the Mx proteins to host antiviral defence. Transgenic mice expressing MxA or Mx1 rescue susceptible mice from viral infection. More importantly, MxA was able to confer protection against Thogoto, La Crosse virus (LACV), and Semliki Forest virus (SFV) infection in the absence of IFN signalling in IFNAR1$^{-/-}$ mice (Hefti *et al.*, 1999). In this case a single IFN effector molecule was sufficient to protect mice from viral infection.

The APOBEC family of cytidine deaminases

The first APOBEC protein was cloned from a rat small intestine cDNA library and was described as an apolipoprotein B messenger RNA editing factor with cytidine deamination activity (Navaratnam *et al.*, 1993; Teng *et al.*, 1993). APOBEC proteins are members of the cytidine deaminase superfamily and include APOBEC1, APOBEC2, APOBEC3A-H, APOBEC4 and activation-induced deaminase (AID; Jarmuz *et al.*, 2002). These proteins contain one or more Zn^{2+} binding motifs that enable cytidine deamination (i.e. the removal of the amine-NH_2 group) from cytosines, converting them to uracils (Navaratnam *et al.*, 1993; Jarmuz *et al.*, 2002). While APOBEC1, APOBEC2 and AID exhibit no antiviral activity (Zheng *et al.*, 2004; Doehle *et al.*, 2005a; Dang *et al.*, 2006), APOBEC3 proteins can confer immunity to a number of viruses, including retroviruses and hepatitis B virus (HBV). The APOBEC3 proteins contain one or two consensus cytidine deaminase active sites, and each has been shown to function as cytidine deaminases. At least one APOBEC3 protein (A3G) is induced by type I IFNs, although in a cell type specific manner, including hepatic cells and macrophages, but not in lymphoid cells (Sarkis *et al.*, 2006).

The antiviral role of APOBEC proteins has been extensively characterized for HIV. Replication of HIV in monocytes and CD4+ lymphocytes requires expression of the viral protein virion activity factor (Vif; Fisher *et al.*, 1987; Strebel *et al.*, 1987). A subtractive cDNA library comprised of HIV-permissive CEM-SS cells and CEM non-permissive cells revealed that expression of APOBEC3G transcripts inversely correlated with the infectivity of cell with HIV deficient in Vif (HIVΔVif; Sheehy *et al.*, 2002). Human A3G cDNA was sufficient to confer the non-permissive phenotype when expressed in the permissive CEM-SS cells (Sheehy *et al.*, 2002). Additional members of the APOBEC3 family can also inhibit HIV-1 replication (Cullen, 2006). A3F, similar to A3G, inhibits HIV-1 replication resulting in a 10–20-fold decrease in replication (Zheng *et al.*, 2004). A3B and A3C were also incorporated into the virion and inhibited replication, although in the case of A3B only 5–10-fold, and with A3C only 2-fold inhibition was observed (Cullen, 2006). The antiviral activity of all of the APOBEC3 family members can be antagonized by Vif, with the exception of A3B, which is not sensitive to Vif (Liddament *et al.*, 2004; Zheng *et al.*, 2004; Doehle *et al.*, 2005a; Dang *et al.*, 2006). Vif antagonizes the activity of A3G by binding to it and recruiting the E3 ubiquitin ligase complex consisting of Cullin 5, Elongin B/C and Ring Box 1, reviewed elsewhere (Mariani *et al.*, 2003; Takeuchi and Matano, 2008; Romani *et al.*, 2009). Vif functions as an 'F-box' protein acting as a bridge between the E3 ligase complex and the target protein. The interaction of Vif with the APOBEC3 family members results in their ubiquitination and subsequent degradation by the proteosome (Sheehy *et al.*, 2003).

The mechanism by which APOBEC3 proteins restrict HIV replication has been extensively studied for A3G. In the absence of Vif, A3G is incorporated into the virion. It does this by binding to the nucleocapsid domain of the Gag polyprotein during assembly (Schafer *et al.*, 2004; Zennou *et al.*, 2004). The Gag–hA3G interaction is dependent upon RNA being present (Svarovskaia *et al.*, 2004; Khan *et al.*, 2005). It then interferes with replication by several potential mechanisms. First, A3G induces deamination of dC residues, causing guanosine to adenosine hypermutations in the proviral DNA, resulting in a failure to produce infectious particles (Sheehy *et al.*, 2002; Harris *et al.*, 2003; Lecossier *et al.*, 2003; Mangeat *et al.*, 2003; Zhang *et al.*, 2003). While G-to-A hypermutations in HIV-1 proviruses have been detected in HIV-1 infected patients, there are examples of A3G mutants lacking cytidine deaminase activity that were still able to significantly inhibit HIVΔVif viral infectivity, suggesting that A3G enzymatic activity is not required for all of its antiretroviral function (Newman *et al.*, 2005). A second hypothesis is that A3G may inhibit tRNA annealing or processing during reverse transcription. HIV-1 utilizes the $tRNA_3^{lys}$ as a primer for synthesizing the strong stop minus strand cDNA, which is required for reverse transcription of the viral genome. Studies showed that $tRNA_3^{lys}$ reverse transcription was significantly diminished in Vif-deficient HIV virions derived from H9 non-permissive cells (Dettenhofer *et al.*, 2000). Subsequent experiments revealed that the reduction in $tRNA_3^{lys}$–mediated reverse transcription was a result of A3G, independent of its cytidine deaminase activity (Guo *et al.*, 2006). Finally, it has been suggested that A3G may inhibit viral plus-strand cDNA transfer.

In addition to HIV-1, the APOBEC3 family members inhibit additional lentiviruses, retroviruses, and HBV. The overexpression of all primate A3Gs, as well as human APOBEC3B, APOBEC3C, APOBEC3F and APOBEC3G inhibit SIVΔVif replication (Dang *et al.*, 2006). The impact of the APOBEC family on retroviruses is less well characterized. Human T-lymphotropic virus (HTLV-I) can infect human lymphoid cells, which are known to express A3G and A3F at levels capable of inhibiting HIV-1 (Navarro *et al.*, 2005; Cullen, 2006). It is currently unknown how HTLV-I escapes from APOBEC3 restriction. Foamy virus can also replicate in cells expressing the APOBEC3 family, although in this case the viral protein Bet has been shown to antagonize

A3G antiviral activity by unclear mechanisms (Russell *et al.*, 2005). Murine leukaemia virus (MLV) incorporates human hA3G into its virion and hA3G can inhibit its replication. In contrast, MLV does not incorporate murine A3 (the only APOBEC3 gene) into its virion and is not inhibited by mA3 (Doehle *et al.*, 2005b). This is a result of a failure of mA3 binding to MLV Gag proteins.

Finally, hA3G as well as other A3 family members have been shown to inhibit HBV replication (Turelli *et al.*, 2004). The mechanism of this inhibition is debated with one group reporting that inhibition of HBV infectivity was not a result of DNA editing. They saw little evidence of DNA mutations, since hA3G mutants lacking the cytidine deaminase activity could still inhibit HBV replication (Turelli *et al.*, 2004). In a separate report APOBEC3 family members were reported to be incorporated into the HBV capsid and G-to-A mutations were detected, indicating DNA editing had occurred (Suspene *et al.*, 2005). Therefore, while APOBEC3 family members can inhibit the replication of lentiviruses, retroviruses and HBV, further studies are needed to define its mechanisms of action.

Zinc finger antiviral protein (ZAP)

ZAP is an IFN-stimulated CCCH zinc finger protein that was identified during screenings of mammalian cDNA libraries for novel proteins that inhibit retroviral replication (Gao *et al.*, 2002). Overexpression of the rat ZAP N-terminus region (NZAP) in rat fibroblast cells inhibited replication of Moloney murine leukaemia virus (MMLV; Gao *et al.*, 2002). This antiviral effect also extended beyond retroviruses as NZAP overexpression inhibited the replication of multiple members of the genus *Alphavirus*, SFV, Ross river virus (RRV), Venezuelan equine encephalitis virus (VEE), Sindbis virus (SINV) and filoviruses, and resulted in significantly decreased Ebola and Marburg virus titres 5–7 days post infection (Bick *et al.*, 2003; Muller *et al.*, 2007). The heterologous expression of ZAP in a Sindbis double subgenomic promoter virus also rendered neonatal mice resistance to infection, suggesting that ZAP may have antiviral activity *in vivo* (Muller *et al.*, 2007; Zhang *et al.*, 2007). While active against several viruses, its effects are not global since its overexpression had no effect on the replication of YFV, HSV, and VSV (Bick *et al.*, 2003).

Studies examining the mechanism by which ZAP functions as an antiviral molecule suggested that ZAP inhibits viral gene expression by inhibiting virus-specific RNA expression. ZAP did not inhibit MMLV reverse transcription or nuclear entry, but instead prevented the accumulation of viral mRNA transcripts in the cytoplasm of infected cells (Gao *et al.*, 2002). ZAP also inhibited Sindbis viral RNA expression in infected cells as early as 6 hours post infection and Ebola virus NP mRNA transcripts were undetectable in infected cells overexpressing NZAP (Bick *et al.*, 2003; Muller *et al.*, 2007). Structure–function analysis of ZAP revealed that the four CCCH-type zinc finger motifs located in the first 254 amino acids of ZAP are critical for its antiviral activity (Guo *et al.*, 2004; Muller *et al.*, 2007). Mutation of the second and fourth zinc finger motifs significantly diminished its antiviral activity rendering cells susceptible to Sindbis, Ebola and Marburg virus infection (Guo *et al.*, 2004; Muller *et al.*, 2007).

Given the ability of ZAP to reduce viral RNA levels, studies were undertaken to assess whether ZAP recognized virus-specific RNA. Fragments of genomic sequences corresponding to each of the Sindbis virus genes were cloned into a luciferase expression reporter plasmid. Transfection of these reporter plasmids into cells overexpressing NZAP identified a Sindbis genome fragment (nucleotides 6062–6715) susceptible to ZAP-mediated inhibition of RNA expression suggesting that viral genomic sequence motifs exist that are somehow specifically recognized by ZAP (Guo *et al.*, 2004). Similar studies performed using Ebola Zaire and Marburg Musoke sequences showed that regions coding for the L protein also contain features that allow ZAP to inhibit RNA expression. These interactions were dependent on the ZAP zinc finger motifs because mutation of the second zinc finger motif abolished binding to ZAP-sensitive sequences from the Sindbis genome (Guo *et al.*, 2004). Binding of ZAP to the specific viral RNA sequences prevented its accumulation in the cytoplasm and led to its eventual degradation. This activity was dependent upon the recruitment and binding of the RNA exosome and the p72 DEAD box RNA helicase (Guo *et al.*, 2007; Chen *et al.*, 2008). Depletion of p72 greatly diminished the antiviral activity of ZAP (Chen *et al.*, 2008). Therefore, the current model is that ZAP, through its zinc finger motifs, binds to specific sequences within the viral genome (termed ZAP responsive elements) and recruits the RNA exosome and p72 RNA helicase. This complex results in the degradation of specific viral RNA to inhibit viral replication.

Interferon-stimulated genes affecting mRNA translation

To impede viral replication, some antiviral proteins target mRNA translation to prevent the synthesis of viral proteins and subsequent assembly of productive virions. The cytoplasmic antiviral dsRNA-dependent protein kinase (PKR) accomplishes this by globally shutting down mRNA translation during certain RNA viral infections by activating the eukaryotic initiation factor 2 (eIF2α). The ISG56 family of genes can also globally disrupt mRNA translation, however, this is accomplished by either disrupting eIF3-mediated stabilization of the ternary translation complex or by blocking eIF3 from forming the 48S pre-initiation complex. In this section, we shall discuss the antiviral role of PKR and the ISG56 family and detail the mechanisms by which they restrict viral spread.

Protein kinase R

Protein kinase R (PKR) is a cytoplasmic serine-threonine kinase whose best-characterized response to viral infection is phosphorylation of eIF2α, which results in global translational arrest. Although PKR is constitutively expressed, its promoter contains an IFN-stimulated response element (ISRE), allowing for inducible expression by type I IFN (Tanaka and Samuel, 1994; Kuhen and Samuel, 1997; Ward and Samuel, 2002; Ward *et al.*, 2002). Its promoter also contains a p53RE element that allows for p53-mediated upregulation (Yoon *et al.*, 2009). PKR is composed of two functionally distinct domains. The N-terminal regulatory dsRNA-binding domain (RBD) binds activating ligands while the C-terminal catalytic domain exhibits kinase activity (Meurs *et al.*, 1990; Feng *et al.*, 1992). The RBD utilizes two dsRNA-binding motifs (RBMs) to recognize any dsRNA of a sufficient length with A-form helical structure (Patel and Sen, 1994; Bevilacqua *et al.*, 1998; Nanduri *et al.*, 1998). It also binds to ssRNA with 5′-triphosphates and short stem–loop structures (Nallagatla *et al.*, 2007). In addition to nucleic acids, PKR binds and is activated by the polyanionic molecules heparin and ceramide, and by the cellular protein activator of the IFN-inducible protein kinase (PACT; Patel and Sen, 1998; Bergeron *et al.*, 2000; Patel *et al.*, 2000). PACT itself is activated by indicators of cellular stress, including serum starvation, sodium arsenite, and peroxide (Patel *et al.*, 2000).

In healthy cells, unbound PKR is an inactive monomer owing to an autoinhibitory interaction between the second RBM and the catalytic domain (Nanduri *et al.*, 2000). Binding to an activating ligand induces a conformational change in PKR that allows ATP to bind the kinase domain and permits both homodimerization and autophosphorylation (Galabru and Hovanessian, 1987; Patel *et al.*, 1994, 1995; Romano *et al.*, 1998). This activated form of PKR then phosphorylates eIF2α, which is the rate-limiting step in translation when phosphorylated (Mellor *et al.*, 1994; Lu *et al.*, 1999). Phospho-eIF2α binds and sequesters eIF2B, the guanine-nucleotide exchange factor required for eIF2 recycling. By preventing exchange of GDP for GTP on eIF2, PKR ultimately inhibits viral replication by blocking global cellular translation, although some IRES-driven viral transcripts are still translated (Garaigorta and Chisari, 2009).

PKR phosphorylates a number of substrates other than eIF2α, including the transcription factor NP90, the tumour suppressor p53 and the PKR regulator PERK, the phosphorylation of which is required for full activation of PKR (Cuddihy *et al.*, 1999; Langland *et al.*, 1999; Bennett *et al.*, 2004). PKR is also involved in the activation of NF-κB via indirect phosphorylation of IKKβ (Maran *et al.*, 1994; Yang *et al.*, 1995; Kumar *et al.*, 1997; Taddeo *et al.*, 2003). Through the adaptors TRAF2 and TRAF5, PKR interacts with the IKK complex in a NEMO-dependent fashion to activate NF-κB (Gil *et al.*, 1999; Bonnet *et al.*, 2006).

The involvement of PKR in antiviral responses is underscored by the increased susceptibility of

PKR$^{-/-}$ mice to various viral infections. Mice with a targeted deletion of the PKR catalytic domain have significantly decreased resistance to certain viral infections, including VSV and influenza A/WSN (Balachandran *et al.*, 2000; Stojdl *et al.*, 2000). The alphavirus SFV replicates to higher titres in PKR$^{-/-}$ mice compared with wild type, although PKR deficiency does not affect mouse survival (Barry *et al.*, 2009). However, PKR deficiency does not increase susceptibility of mice to all viruses, including HSV, EMCV, and influenza strain W29 (Friedman *et al.*, 1984; Abraham *et al.*, 1999; Leib *et al.*, 2000). This is most likely a result of either virally encoded PKR antagonists that incapacitate PKR or of the involvement of other antiviral molecules (Leib *et al.*, 2000). HSV-1, for example, encodes a negative regulator of PKR called ICP34.5, which binds to a protein phosphatase and causes it to dephosphorylate eIF2α (He *et al.*, 1997). Since the downstream effect of PKR activation and eIF2α phosphorylation is already antagonized by the virus, there is no effect of PKR deficiency during infection with wild-type HSV-1 (Leib *et al.*, 2000). However, infection of PKR$^{-/-}$ mice with ΔICP34.5 HSV-1 restored virulence of the ΔICP34.5 virus (which is attenuated in wild-type mice), showing the specificity of ICP34.5 for the PKR pathway (Leib *et al.*, 2000). Based on infections of PKR$^{-/-}$ fibroblasts, PKR is also implicated in the response to HCV, Sindbis, foot-and-mouth disease virus, and WNV (Chinsangaram *et al.*, 2001; Gorchakov *et al.*, 2004; Chang *et al.*, 2006; Samuel *et al.*, 2006). Finally, PKR may be implicated in the control of HCV in humans. Genetic analysis of human PKR polymorphisms revealed that certain allelic variants are more likely to be expressed by HCV-infected patients compared with uninfected controls (Samuel *et al.*, 2006).

Apoptosis is an important process in limiting viral replication and spread. Because MEFs from PKR$^{-/-}$ mice are resistant to apoptosis induced by TNF, dsRNA, and LPS, investigations into the role of PKR-dependent pro-apoptotic signalling were undertaken (Der *et al.*, 1997). Through an incompletely defined mechanism, PKR triggers cell death via FADD-dependent activation of caspase 8 (Lee and Esteban, 1994; Balachandran *et al.*, 1998; Donze *et al.*, 1999; Gil *et al.*, 1999; Gil and Esteban, 2000a,b). PKR-induced apoptosis is also partially controlled by NF-κB activation described above (Donze *et al.*, 1999; Gil *et al.*, 1999; Gil and Esteban, 2000a). Taken together, PKR appears to participate in antiviral defence through multiple mechanisms. The relative importance and utilization of each pathway probably varies with respect to cell type and viral species.

The ISG56 family

In humans the ISG56 family of genes is comprised of four members (ISG54/IFIT2, ISG56/IFIT1, ISG58/IFIT5, and ISG60/IFIT4), while in the mouse there are three members (ISG54/IFIT2, ISG56/IFIT1, and ISG49/IFIT3). These proteins were initially found to be strongly induced by type I IFN, but additional stimuli including viral infection and LPS can also upregulate their expression (Levy *et al.*, 1986; Wathelet *et al.*, 1986; Lund *et al.*, 2006; Fensterl *et al.*, 2008). ISG56 family members contain multiple tetratricopeptide motifs, which mediate protein–protein interactions through their helix–turn–helix structures (Cortajarena and Regan, 2006).

One of the first functions described for ISG56 family members was the inhibition of translation. The addition of either human or mouse ISG54 or ISG56 inhibits *in vitro* translation systems, as well as translation of reporter constructs in whole cells (Guo *et al.*, 2000; Hui *et al.*, 2003; Hui *et al.*, 2005; Terenzi *et al.*, 2006; Fensterl *et al.*, 2008). ISG56 overexpression also suppresses translation from the HCV IRES sequence (Wacher *et al.*, 2007). It appears that the mechanism by which translation is inhibited differs between family members. Human ISG54 and ISG56 associate with the e subunit of the multiprotein complex eukaryotic initiation factor 3 (eIF3). This interaction prevents eIF3 from stabilizing the ternary complex of eIF2, GTP, and Met-tRNA, resulting in translation inhibition (Guo *et al.*, 2000; Hui *et al.*, 2003; Terenzi *et al.*, 2006). In contrast, murine ISG54, murine ISG56, and human ISG54 interact with the c subunit of eIF3, blocking eIF3-mediated formation of the 48S pre-initiation complex, which is composed of the 40S and 20S ribosomal subunits, the ternary complex, eIF4F and mRNA (Hui *et al.*, 2005; Terenzi *et al.*, 2006). Thus, human ISG54 inhibits two steps of eIF3-mediated translation.

Murine ISG49 also binds to eIF3c; however, it does not appear to inhibit protein synthesis (Fensterl *et al.*, 2008). Additional work is needed to determine the function of murine ISG49 and human ISG58 and ISG60.

There is evidence that ISG56 can function as an antiviral molecule, although the precise mechanism by which it is functioning is still debated. In addition to its role in regulating protein translation, ISG56 has also been identified as a negative regulator of the IFN signalling pathway by binding the mitochondrial membrane adaptor protein MITA (Li *et al.*, 2009). Upon binding to their ligands, the intracellular RNA sensors, RIG-I and MDA5, induce type I IFN. RIG-I and MDA5 bind to the adapter protein MAVS/IPS-1/VISA/Cardif, which then signals though MITA to recruit both TBK1 and IRF3 (Ishikawa and Barber, 2008; Zhong *et al.*, 2008). Binding of ISG56 to MITA prevented MITA from interacting with both MAVS and TBK1, disrupting signalling (Li *et al.*, 2009). Accordingly, the overexpression of human ISG56 reduced virally induced IFN-β expression (Li *et al.*, 2009). ISG56 also appears to play a role in regulating human papillomavirus (HVP) infection. Human ISG56 was found to directly interact with the E1 phosphoprotein of HPV which is required for its replication (Terenzi *et al.*, 2008). Recombinant human ISG56 inhibited both E1-dependent *in vitro* viral DNA replication and E1 DNA helicase activity (Terenzi *et al.*, 2008). The overexpression of ISG56 inhibited HPV-replication, while ISG56-knockdown ablated IFN-β-mediated inhibition of viral replication (Terenzi *et al.*, 2008). Further studies are needed to determine if this antiviral activity is a result of ISG56 binding to the HVP E1 protein and disrupting viral replication, ISG56 binding to eIF3 and disrupting translation, or ISG56 regulation of the IFN signalling pathway.

Thus, ISG56 and potentially additional ISG56 family members appear to function in diverse pathways that limit translation, viral replication, and host innate immune signalling pathways. Elucidation of which pathway(s) contribute to the antiviral activity of ISG56 family members will require loss-of-function mutants and knockout mice, but it is plausible that these proteins perform multiple regulatory functions in cell type-, virus-, and temporal-specific manners.

Interferon-stimulated genes that disrupt virion assembly and release

To produce disease, viruses must successfully assemble progeny virions that eventually exit the host cell to infect new cells. Without the ability to spread to uninfected cells, the initial inoculum is less likely to establish an infection. Some ISGs counteract these vital steps of the viral life cycle by perturbing virion assembly or preventing release of assembled virions from the plasma membrane. This section discusses the ISGs tetherin and viperin.

Tetherin/BST-2

Tetherin (also known as HM1.24, CD317 and bone marrow stromal antigen, BST-2) was identified a few years ago by two independent groups as the IFN induced host restriction factor that tethered HIV virions to the cell surface, and whose activity was antagonized by the HIV viral protein Vpu (Neil *et al.*, 2008; Van Damme *et al.*, 2008). In addition to inhibiting the release of HIV-1, tetherin also restricts the release of other retroviruses, lentiviruses, arenaviruses (Lassa), filoviruses (Marburg), and herpesviruses (Kaposi's sarcoma-associated herpesvirus – KSHV; Sauter *et al.*, 2010). Tetherin is only expressed in plasmacytoid dendritic cells, differentiated B cells, and bone marrow stromal cells at baseline. Its expression is strongly induced in many cell types by type I IFN. Tetherin is a 30–36 kDa dimeric type II membrane protein with a unique topology (Fig. 4.1). It contains an N-terminal cytoplasmic domain followed by a hydrophobic transmembrane domain, a coiled-coil extracellular ectodomain, and a C-terminal glycosylphosphatidylinositol (GPI) anchor suggesting that tetherin is a membrane spanning protein (Kupzig *et al.*, 2003; Tokarev *et al.*, 2009). Tetherin's antiviral activity is lost with either the removal of the cytoplasmic tail or the GPI anchor (Neil *et al.*, 2008). The ectodomain is predicted to have two α-helices separated by a short loop and these α-helices are thought to form a dimeric coiled-coiled structure (Kupzig *et al.*, 2003; Tokarev *et al.*, 2009). The tetherin

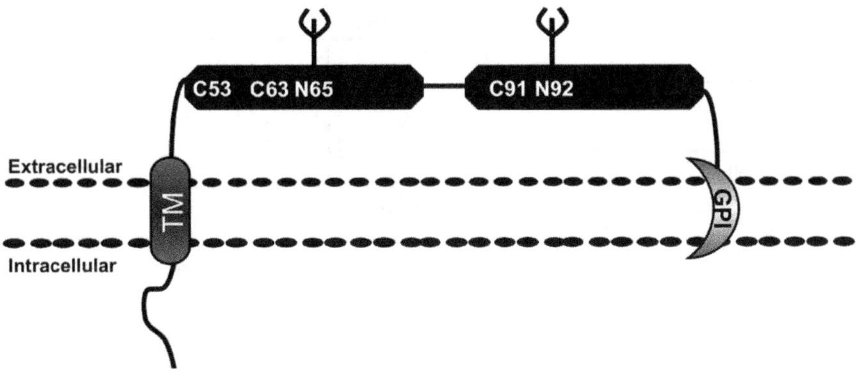

Figure 4.1 Diagram depicting the topology of tetherin. Tetherin consists of an N-terminal transmembrane domain, an extracellular ectodomain comprised of two α-helices and a C-terminal GPI anchor. The cysteine residues thought to engage in disulfide bonding and N-glycosylated asparagine residues are shown in bold.

ectodomain includes two conserved N-glycosylated asparagine residues (N65 and N92) and three conserved cysteine residues (C53, C63 and C91), the latter of which are important for dimerization of tetherin molecules (Ohtomo *et al.*, 1999; Kupzig *et al.*, 2003). Mutagenesis studies showed that N-glycosylation of the asparagine residues is dispensable for HIV virion release (Andrew *et al.*, 2009). In contrast, tetherin dimerization appears to be important for inhibiting HIV virion release (Andrew *et al.*, 2009), but dispensable for filovirus (Lassa and Marburg virus-like particle) release (Sakuma *et al.*, 2009). Finally, it appears that the unusual structure of tetherin, and not specific sequences, mediate its antiviral activity. Engineered proteins with different amino acids sequences, but conserved domain structure, maintained anti-viral activity (Perez-Caballero *et al.*, 2009). These results, along with confocal microscopy studies that localized tetherin to the plasma membrane and in association with lipid rafts, and co-localized it with HIV gag protein, Ebola virus VP40, and Marburg virus VP40 at the cell surface, suggest that tetherin may directly tether viruses to the plasma membrane or each other (Kupzig *et al.*, 2003; Rollason *et al.*, 2007; Dube *et al.*, 2009; Jouvenet *et al.*, 2009; Masuyama *et al.*, 2009; Mitchell *et al.*, 2009; Tokarev *et al.*, 2009). Electron microscopy studies support this model, finding tetherin associated with HIV virions at the cell surface and in some cases finding it localized in between the plasma membrane and the virion (Perez-Caballero *et al.*, 2009; Fitzpatrick *et al.*, 2010).

Various models have been proposed to explain how tetherin retains virions on the cell surface (Fig. 4.2). In the 'ectodomain self-interaction' model, tetherin molecules are embedded in both the viral and plasma membrane and inhibit virion release by intermolecular disulfide binding (Perez-Caballero *et al.*, 2009; Fitzpatrick *et al.*, 2010). This model is supported by evidence demonstrating that tetherin dimerization is important in retention of HIV virions at the cell surface (Andrew *et al.*, 2009). Contrary to this model is the finding that treatment with the reducing agent DTT was unable to inhibit tetherin-mediated retention of HIV and Ebola VLP budding (Kaletsky *et al.*, 2009). An alternative model is the 'membrane spanning' model whereby dimerized tetherin molecules embed one end into the plasma membrane and the other end into the virion in either a parallel or antiparallel orientation (Perez-Caballero *et al.*, 2009; Fitzpatrick *et al.*, 2010). Treatment of HIVΔ Vpu-infected HeLa cells with phosphatidylinositol-specific phospholipase C (PI-PLC), which cleaves GPI anchors, did not restore virion release (Fitzpatrick *et al.*, 2010). Moreover, treatment of infected HeLa cells with both PI-PLC and DTT also did not impact tetherin-mediated retention of HIV virions (Fitzpatrick *et al.*, 2010). These studies suggest that tetherin restricts virion release by an unknown mechanism that is not accounted

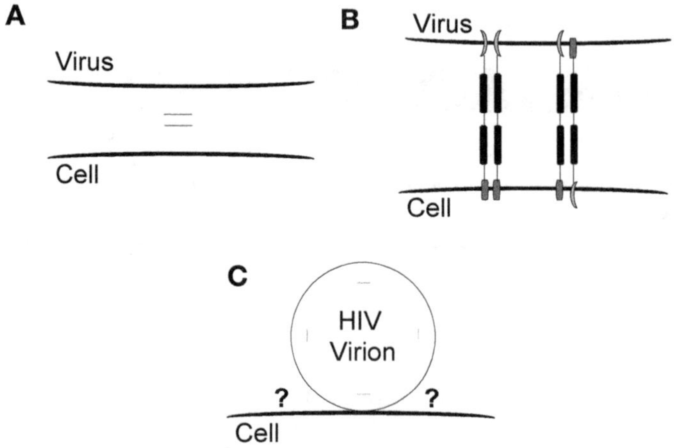

Figure 4.2 Potential models proposed as to how tetherin retains virions on the cell surface. (A) The ectodomain interaction model: Tetherin molecules embedded in the virion and in the cell retain virions at the plasma membrane through intermolecular disulfide bonds or interactions between α-helices on two opposing molecules. (B) The membrane spanning model: Tetherin dimers retain virions at the cell surface by embedding one end into the virion and the other end into the cell membrane in a parallel or antiparallel orientation. (C) Recent studies suggest that tetherin molecules located inside a budding virion may somehow retain viruses at the cell surface by an unidentified mechanism.

for by either the 'ectodomain self-interaction' or 'membrane spanning' model. Electron microscopy revealed that tetherin is also found within HIV virions; thus, it has been proposed that tetherin somehow retains virions at the cell surface from inside the virus (Fitzpatrick *et al.*, 2010). Further studies will be needed to decipher the mechanism by which tetherin functions against the diverse number of viruses it appears to restrict.

To date there has been no description of a tetherin knockout mouse to ascertain the importance of this molecule *in vivo*. However, many viruses have evolved distinct strategies to antagonize its function. Studies elucidating the antiviral function of tetherin were driven by investigations into the function of the HIV protein, Vpu. HIV virion release was enhanced in the presence of Vpu (Cohen *et al.*, 1988; Strebel *et al.*, 1988), but in its absence, HIV virions were not released, but were found bound to the cell surface and/or concentrated in endosomal vesicles (Klimkait *et al.*, 1990). Vpu antagonizes tetherin's activity through a direct interaction with tetherin's transmembrane domain, with a single point mutant in this domain making tetherin resistant to the actions of Vpu (Gupta *et al.*, 2009; McNatt *et al.*, 2009) Vpu targets tetherin to the trans-Golgi for either proteosomal or lysosomal degradation (Gupta *et al.*, 2009). Most primate lentiviruses do not contain a Vpu gene. Instead they antagonize the actions of tetherin with their Nef protein. Nef targets the cytoplasmic tail of tetherin to reduce its expression on the cell surface (Jia *et al.*, 2009; Zhang *et al.*, 2009). Additional immune evasion strategies have been developed by other viruses to antagonize tetherin's activity. HIV-2 utilizes its Env protein to sequester tetherin from sites of assembly (Le Tortorec and Neil, 2009; Neil and Bieniasz, 2009). The Ebola virus glycoprotein binds to tetherin and sequesters it from sites of budding (Kaletsky, 2009). In addition, the herpes virus KSHV utilizes its K5 protein, a RING containing E3 ubiquitin ligase, to ubiquitinate tetherin and induce its degradation (Bartee *et al.*, 2006). To date, most of the research on tetherin has been performed in transfected cells and has not used primary cells for which the viruses are tropic. Future studies using intact viruses, primary cells and animal models will be critical to fully understanding the importance of this protein in the host antiviral response.

Viperin

Virus inhibitory protein, endoplasmic reticulum-associated IFN-inducible (viperin), which is also known as RSAD2 and cig5 in humans, and vig1 in mice, is strongly induced in response to types I and II IFN, LPS, and infection by several viruses (Boudinot *et al.*, 2000; Chin and Cresswell, 2001; Olofsson *et al.*, 2005). Interestingly, viperin is also induced in an IRF1-dependent but IFNAR-independent pathway, which may ensure viperin expression even during infection with viruses that counteract IFN induction or function (Boudinot *et al.*, 2000; Stirnweiss *et al.*, 2010). Viperin is evolutionarily conserved across mammals and lower vertebrates, and its cellular overexpression or silencing either inhibits or enhances, respectively, the replication of many human DNA and RNA viruses, including human cytomegalovirus (HCMV), HCV, HIV, and alphaviruses (Chin and Cresswell, 2001; Helbig *et al.*, 2005; Rivieccio *et al.*, 2006; Fink *et al.*, 2007; Zhang *et al.*, 2007). The mechanisms by which viperin exerts this activity has been explored for a limited number of viruses.

Viperin disrupts the formation of lipid rafts on the plasma membrane (Wang *et al.*, 2007). This impairs the cellular release of influenza virus, which buds from lipid rafts (Wang *et al.*, 2007). Lipid raft disruption occurs through an association between viperin and a key enzyme in the isoprenoid biosynthetic pathway, farnesyl diphosphate synthase (FPPS; Wang *et al.*, 2007). By inhibiting the enzymatic activity of FPPS, viperin presumably blocks the formation of lipid raft component precursors (Fig. 4.3). The specific molecule(s) downstream of FPPS that are affected by the FPPS–viperin association remain unknown, as does the mechanism of inhibition. However, the recent finding that viperin is a radical *S*-adenosylmethionine enzyme may provide insight into discovering the downstream metabolite(s) of FPPS that mediate viperin's antiviral activity (Jiang *et al.*, 2008; Duschene and Broderick, 2010; Shaveta *et al.*, 2010).

Viperin regulation of HCV has also been explored. By virtue of an amphipathic α-helix, viperin localizes to the cytosolic face of the ER

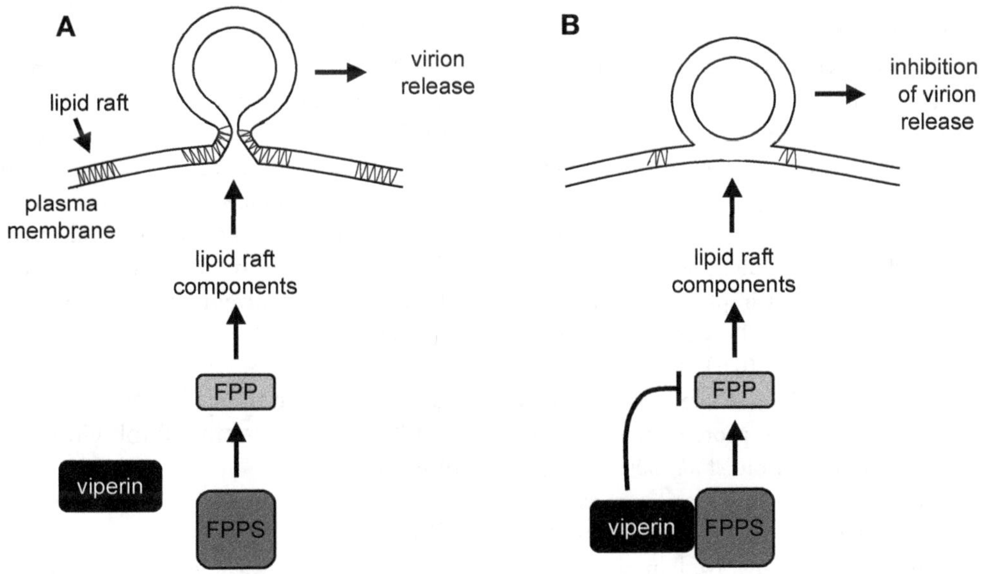

Figure 4.3 Proposed mechanism of viperin-mediated inhibition of influenza virus release from the plasma membrane. (A) When not associated with viperin, FPPS synthesizes FPP, a precursor of lipid raft components. Influenza virions subsequently bud from the plasma membrane in a lipid raft-dependent manner. (B) Association with viperin inhibits the enzymatic activity of FPPS, leading to disruption of lipid rafts. The reduction of lipid rafts diminishes influenza release.

and to intracellular structures that are thought to harbour HCV replication, called lipid droplets (Hinson and Cresswell, 2009a,b). Viperin overexpression inhibits RNA expression by HCV replicons, but this inhibition is much less efficient during expression of a viperin deletion mutant that lacks the domain necessary for lipid droplet localization (Jiang *et al.*, 2008). The molecular mechanism of this lipid droplet-dependent HCV inhibition is unclear.

Viperin-deficient mice have been generated but their sensitivity to viral infection, in particular influenza, has not been reported (Qiu *et al.*, 2009). Examination of these mice revealed that stimulation of viperin-deficient CD4 T cells results in defective NF-κB and AP-1 activation, impaired T_H2 differentiation, and reduced levels of the T_H2 cytokines IL-4, IL-5, and IL-13 (Qiu *et al.*, 2009). The basis of these defects is unknown, but it has been suggested that differences in lipid raft-dependent signals (such as PKC-θ) may be responsible. Assessing the responses of these knockout mice to viral infection will determine viperin's importance in host and viral responses.

The tripartite motif (TRIM) family of proteins

The TRIM family consists of a large number of proteins (72 human TRIM genes) that regulate a diverse array of biological processes including cellular proliferation, differentiation, oncogenesis, apoptosis and immune signalling (Nisole *et al.*, 2005; Ozato *et al.*, 2008; Nakayama and Shioda, 2010). In addition, an increasing number of these proteins have been implicated in antiviral responses (Nisole *et al.*, 2005; Ozato *et al.*, 2008; Nakayama and Shioda, 2010). In some cases, these proteins directly inhibit viral replication, as described for many of the proteins that regulate retroviral replication (Nisole *et al.*, 2005; Nakayama and Shioda, 2010). There is also increasing evidence that members of this family regulate the signalling pathways involved in the innate immune response, including pathways that sense the presence of pathogens (RLR and TLR pathways) and innate cytokine pathways (type I IFN, NF-κB signalling) that regulate host responses (Ozato *et al.*, 2008). In addition, some of these family members, through genetic linkage analysis or other approaches, are associated with a variety of autoinflammatory and autoimmune disorders (Ozato *et al.*, 2008; Kastner *et al.*, 2010). Further functional characterization of these family members will probably reveal important insights into the host antiviral response.

TRIM proteins have a conserved architecture that contains a TRIM or RBCC motif which is composed of a RING finger domain, one or two B-box domains, a coiled-coiled domain and a variable C-terminus. The TRIM proteins are classified into 11 subclasses (C-I to C-XI) based upon the nature of their C terminal domains as defined by Short and Cox, or are unclassified based on the lack of a RING finger domain (Short and Cox, 2006). The RING domain is a zinc-binding motif that has been reported for many family members to function as an E3 ligase for ubiquitin (TRIM5a, TRIM8, TRIM11, TRIM21, TRIM22, TRIM25) or other ubiquitin-like proteins (TRIM25 [ISG15], TRIM19 [PML] and TRIM63 [SUMO]). The other two domains, the B-box and coiled-coiled domains, are involved in protein interactions and homo/heterodimerization.

Many TRIM proteins are upregulated by type I and II IFNs (Rajsbaum *et al.*, 2008; Carthagena *et al.*, 2009), although it's important to note that the upregulation of only a limited number of these TRIM genes has been confirmed at the protein level. In this section we will discuss in greater detail the IFN-induced TRIM genes that have been identified as potential antiviral molecules, either through their direct activity on viral replication or their involvement in regulating innate immune responses.

Interferon induced TRIM family members that directly inhibit viral replication

TRIM5α

In 2004, TRIM5α was identified as a major host factor responsible for the host range restriction of HIV. It is a member of the C-III subfamily of TRIMs proteins, containing a C-terminal SPRY domain. It is robustly induced by type I IFNs (Carthagena *et al.*, 2009). Owing to multiple splice sites, TRIM5 exists in multiple forms, with the α isoform being the longest (Nisole *et al.*, 2005). TRIM5 is located on chromosome 11 in a locus that contains additional TRIM family members (TRIM6, TRIM34, and TRIM22).

TRIM5α restricts retroviral replication at a post-entry, pre-reverse transcriptase stage in the viral life cycle, but in a species-specific manner (Shibata *et al.*, 1995; Himathongkham and Luciw, 1996). Both rhesus and cynomolgus monkey TRIM5α restrict HIV-1, but not macaque SIV (SIVmac; Stremlau *et al.*, 2004). In contrast, human TRIM5α restricts N-tropic MLV, but not HIV (Sakuma *et al.*, 2007; Carthagena *et al.*, 2008). Through mutational analysis it is now known that both the B-box2 and coiled-coiled domains are required for TRIM5α antiviral activity through the formation of higher ordered structures (B-Box 2 domain) and homo-oligomerization (coil-coil domain), respectively. Blockade in the formation of these higher ordered structures results in diminished antiviral activity (Javanbakht *et al.*, 2005; Javanbakht *et al.*, 2006; Li and Sodroski, 2008). The SPRY domain is responsible for the species specificity of TRIM5α activity. A single mutation at position 332 in the V1 region of human TRIM5α (Pro-to-Arg) leads to restriction of HIV-1 (Stremlau *et al.*, 2005; Yap *et al.*, 2005). The SPRY domain binds to the viral core, probably through recognition of the viral capsid protein (Sebastian and Luban, 2005; Stremlau *et al.*, 2006). This binding is thought to require proper processing of the capsid into its assembled form. Finally the RING domain of TRIM5α functions as an E3 ubiquitin ligase. During viral infection, TRIM5α is auto-ubiquitinated and then targeted for proteosomal degradation (Diaz-Griffero *et al.*, 2006; Rold and Aiken, 2008). RING mutants abrogate this auto-ubiquitination, but only partially attenuate its ability to restrict viral infection (Javanbakht *et al.*, 2005). One model for a potential mechanism by which TRIM5α restricts retroviral replication is through binding the capsid/viral core, resulting in its insertion into the capsid lattice. Following ubiquitination of TRIM5α, it is targeted to the proteosome for degradation, resulting in disruption of the lattice, interfering with uncoating and subsequent replication (Nakayama and Shioda, 2010). However, given the finding that mutations in the RING domain only partially attenuate its antiviral activity, TRIM5α may also function via a proteosome-independent mechanism. Additional studies will provide further details into the mechanism by which TRIM5α inhibits retroviral replication and may provide important new drug targets.

TRIM19/PML

TRIM19/PML was originally identified in patients with promyelocytic leukaemia (PML) as a part of the fusion protein with retinoic acid receptor alpha (RARα) that results from a chromosomal translocation. It is also known to be a critical component of PML nuclear bodies (NBs), which are also known as nuclear dot 10 (ND10) or PML oncogenic domains (PODs; Ishov *et al.*, 1999; Zhong *et al.*, 2000). PML NBs appear to regulate a wide variety of cellular processes including oncogenesis, DNA damage responses, apoptosis, regulation of both transcription and translation, senescence, signal transduction, viral infection and the antiviral response. In this section we will briefly highlight the complex role of PML during viral infection. A comprehensive review of this topic can also be found in other references (Everett and Chelbi-Alix, 2007).

TRIM19/PML belongs to the C-V subfamily of TRIM proteins. The N-terminal region of PML contains the RBCC motif. Many different isoforms of PML are expressed and classified into seven distinct groups (PML I-VII) based upon expression of distinct C-terminal domains (Jensen *et al.*, 2001). These different isoforms have different localization patterns within cells and probably have different functions, making studies into the mechanism of PML action quite complex (Beech *et al.*, 2005; Condemine *et al.*, 2006). PML is expressed at a low level in cells independent of type I IFNs. However, upon IFN stimulation, PML expression and PML NBs number and size are greatly enhanced (Chelbi-Alix *et al.*, 1995). The protein constituents of PML NBs are quite complex and appear to change depending upon the cell type being examined and the activation status of the cell (Beech *et al.*, 2005; Condemine *et al.*, 2006). It is known that PML is required for PML NBs to form, since in the absence of PML, such as in PML-/- cells, PML NBs are absent

(Ishov *et al.*, 1999; Zhong *et al.*, 2000). The localization of PML to the NBs requires SUMO modification, with unmodified PML remaining in the nucleoplasm (Seeler and Dejean, 2001).

Given the robust induction of PML by IFNs, the role of PML and PML NBs as antiviral effectors has been speculated. Despite an extensive amount of investigation, this topic remains hotly debated in the field. The contradictions probably arise from the large number of PML isoforms that are expressed in distinct cell types and the changing protein makeup of the PML NBs depending upon the stimulating conditions utilized. This topic has been extensively reviewed (Everett and Chelbi-Alix, 2007).

Based largely upon *in vitro* studies, PML is thought to function as an antiviral molecule against RNA viruses by both p53-dependent and -independent mechanisms. In p53 defective cells, overexpression of PML decreased human foamy virus (HFV) replication by inhibiting viral transcription (Regad *et al.*, 2001). In this case PML interacted with the viral transactivator, Tas, interfering with its ability to bind viral DNA and activate gene expression (Regad *et al.*, 2001). This required the RING domain and the localization of PML to the nucleus. In addition, the overexpression of PML III decreased infection by VSV and influenza A viruses at both the mRNA and protein expression level (Chelbi-Alix *et al.*, 1998). Furthermore, depletion of all PML isoforms by siRNA resulted in increased viral replication (Iki *et al.*, 2005). The mechanism by which PML functions in these systems remains unclear. In some cases, PML's antiviral activity is mediated through its association with p53. Poliovirus infection activates p53 in a PML dependent manner leading to apoptosis and inhibition of viral replication (Pampin *et al.*, 2006). These effects are lost when either p53 or PML is down-regulated by siRNA (Pampin *et al.*, 2006).

The situation with DNA viruses is more complex since the viral genome/replication complex is often tightly associated with PML NB. However, as recently reviewed in Everett *et al.* there are several reasons to believe PML NBs contribute to the host antiviral response (Everett and Chelbi-Alix, 2007):

1. viral DNA replication sites are closely associated with PML NBs;
2. viruses encode proteins that target and often disrupt the NBs; and
3. disruption of PML NBs correlates with increased viral gene expression.

The overexpression of PML inhibits adenovirus and HCMV infection *in vitro* (Ahn and Hayward, 2000). Knockdown of PML expression by siRNA also enhanced both wild-type and IE72 mutant HCMV and ICPO null HSV-1, further supporting its role in host defence (Tavalai *et al.*, 2006).

As a testament to the importance of PML NBs in the host response to viral infection, both RNA and DNA viruses encode proteins to target and often disrupt them. The most studied of these proteins is the ICP0 protein of HSV-1. It contains a RING finger domain that functions as an E3 ubiquitin ligase. ICP0 induces the degradation of PML and SUMO modification of sp100, resulting in the destruction of PML NBs (Boutell *et al.*, 2002). In addition, rabies virus encodes phosphoprotein P and 4 amino-terminally truncated products (P2, P3, P4, and P5) which cause reorganization of PML NBs (Blondel *et al.*, 2002). Similar effects are also seen with LCMV Z protein (Borden *et al.*, 1998).

Finally, PML$^{-/-}$ mice have been generated and characterized, and in many cases the results have been contradictory. In certain cell types, PML appears to be required for the IFN-mediated antiviral effects against HSV. PML$^{-/-}$ cells infected with an HSV-1 ICP0 null virus were found to be less sensitive to IFN antiviral effects (Chee *et al.*, 2003). However, despite these *in vitro* findings, there was no difference in pathogenicity of either wild-type or ICP0 null HSV-1 viruses in PML$^{-/-}$ mice (Halford *et al.*, 2006). While the overexpression of PML had no effect on rabies virus, LCMV, and Lassa virus, they were all found to replicate to higher levels in PML$^{-/-}$ cells (Djavani *et al.*, 2001; Asper *et al.*, 2004), and PML$^{-/-}$ mice were found to be more susceptible to LCMV and VSV infection (Bonilla *et al.*, 2002). So while there is increasing evidence that PML functions as part of the host antiviral defence mechanism, many questions still remain.

TRIM22

TRIM22 is classified in the C-III subfamily of TRIM proteins. Previous studies have shown that TRIM22 is induced by both type I and type II IFNs in both human and murine cell lines (Rajsbaum *et al.*, 2008; Carthagena *et al.*, 2009). It has recently been found to exhibit antiviral activity against several viruses, but the most extensively studied has been its role during HIV-1 infection. TRIM22 was initially identified as an HIV restriction factor when it was found to repress transcription mediated by the HIV-1 long terminal repeat (LTR; Nisole *et al.*, 2005). It has also been found to inhibit HIV replication through its disruption of HIV budding (Barr *et al.*, 2008). TRIM22 interacts with the Gag protein of HIV-1, but not MLV or equine infectious anaemia virus (EIAV), disrupting proper trafficking of the Gag proteins within cells (Barr *et al.*, 2008). Cells overexpressing TRIM22 failed to accumulate Gag protein at the plasma membrane. This was dependent upon the E3 ligase activity of the RING domain since mutation of the catalytic cysteine resulted in a loss of HIV inhibition (Barr *et al.*, 2008). While HIV-Gag proteins can be co-immunoprecipitated with TRIM22, to date there is no evidence that Gag is ubiquitinated. Therefore, the target of TRIM22 E3 ligase activity is unknown. TRIM22 can also exert antiviral activity against ECMV and HBV. Overexpression of TRIM22 resulted in a modest decrease in ECMV replication, which was dependent upon the RING domain of TRIM22 (Eldin *et al.*, 2009). The target of this activity is believed to be the ECMV protease 3C, which is ubiquitinated by TRIM22 (Eldin *et al.*, 2009). TRIM22 was also noted to inhibit HBV gene expression and replication. In this case TRIM22 was found to inhibit the activity of the HBV core promoter, resulting in decreased transcription (Gao *et al.*, 2009). The C-terminal SPRY domain was required to localize TRIM22 to the nucleus to mediate its activity (Gao *et al.*, 2009). Mutations in the RING domain, including mutation of the catalytic cysteine, did not alter its localization but did abolish its antiviral activity (Gao *et al.*, 2009). Therefore, TRIM22 appears to function as an antiviral molecule against several viruses through the utilization of distinct mechanisms.

Interferon induced TRIM proteins that regulate host innate immune responses

TRIM25

TRIM25 is classified in the C-IV subfamily of TRIM proteins. In both human and murine cells TRIM25 is upregulated by stimuli that generate a type I IFN response (Rajsbaum *et al.*, 2008; Carthagena *et al.*, 2009). TRIM25 functions as an E3 ligase for both ubiquitin and ISG15; in both cases, TRIM25 targets the cell cycle regulator, 14–3–3ς (Zou and Zhang, 2006; Gack *et al.*, 2007). TRIM25 was also shown to play an important role in regulating the host antiviral response. The SPRY domain of TRIM25 interacts with the first CARD domain of the intracellular viral sensor RIG-I and mediates K63-linked ubiquitination of the second N-terminal CARD domain of RIG-I (Gack *et al.*, 2007). This ubiquitin modification is required for RIG-I to interact with MAVS and induce IFN-β production in response to RIG-I ligand binding (Gack *et al.*, 2007). The importance of TRIM25 in RIG-I signalling was confirmed by evaluating TRIM25$^{-/-}$ MEFs. IFN-β induced by SeV or VSV infection was nearly undetectable in cells lacking TRIM25, resulting in more robust viral growth (Gack *et al.*, 2007). Further confirmation of its importance in the host's innate immune response is found in recent studies showing that viruses have targeted TRIM25 as an immune evasion strategy. The NS1/A protein of influenza A virus has been shown to bind the coil-coiled domain of TRIM25 and inhibit its ability to ubiquitinate RIG-I (Gack *et al.*, 2009). A recent siRNA screen also identified TRIM25 as an antiviral molecule for its ability to inhibit both HIV and MLV late stage replication (Uchil *et al.*, 2008). Additional studies will be needed to determine its mechanism of action during retroviral replication.

TRIM20/MEFV/pyrin

Positional cloning identified *TRIM20* (MEFV or pyrin) as the gene that is mutated in the systemic autoinflammatory syndrome Familial Mediterranean Fever (FMF; Kastner *et al.*, 2010). This disorder is characterized by self-limited recurrent fevers and serosal synovial inflammation. TRIM20 remains an unclassified member of the TRIM family owing to its lack of a RING finger domain. In its place it has a pyrin domain, which is important for its interaction with the inflammasome. Studies in both murine and human macrophages have shown TRIM20 is induced by IFN stimulation (Rajsbaum *et al.*, 2008; Carthagena *et al.*, 2009).

TRIM20 regulates both IL-1 and pro-inflammatory cytokine production through its interactions with various components of the inflammasome and NF-κB signalling pathway (Kastner *et al.*, 2010). The inflammasome is a macromolecular complex that results in the activation of IL-1β. It is composed of NLRP3, which is linked to caspase 1 via the adaptor protein, ASC. Expression of full length TRIM20 has been shown to decrease IL-1 production, while overexpression of the various mutants identified in FMF patients results in less inhibition of IL-1 production (Chae *et al.*, 2003, 2006). Several mechanisms by which TRIM20 regulates these pathways have been proposed. First, the N-terminal pyrin domain of TRIM20 binds to ASC and may function as a negative regulator of the IL-1 pathway through competition for ASC binding to NLRP3 and caspase 1 (Richards *et al.*, 2001; Chae *et al.*, 2003). Second, the B30.2 domain, which contains many of the mutations identified in FMF, interacts directly with caspase 1 to regulate its activity (Chae *et al.*, 2006). Finally, TRIM20 was recently identified as a target of caspase 1, being cleaved at Asp 330 (Chae *et al.*, 2008). The N-terminal pyrin fragment was found to localize in the nucleus of cells and bind to the p65 subunit of NF-κB leading to enhanced degradation of IκBα (Chae *et al.*, 2008). The N-terminal fragment of TRIM20 enhanced NF-κB activity, while full length or the C-terminal fragment did not mediate this effect (Chae *et al.*, 2008). Therefore, in addition to regulating the IL-1 pathway, TRIM20 also regulates the NF-κB signalling pathway. With the recent finding that the inflammasome can regulate host antiviral responses, it will also be important to determine if TRIM20 contributes to the regulation of these responses.

TRIM21

TRIM21, or Ro52, has been known as a target of autoantibodies in systemic lupus erythematosus (SLE) and Sjögren's syndrome for many years. Polymorphisms in TRIM21 have also been linked to both SLE and Sjögren's syndrome (Frank *et al.*, 1993; Nakken *et al.*, 2001). TRIM21 is classified in the C-III subfamily, containing a C-terminal domain with both a PRY and SPRY domain. It was shown in both human and murine macrophages to be induced by type I and type II IFNs (Rajsbaum *et al.*, 2008; Carthagena *et al.*, 2009). TRIM21 resides in the cytoplasm in resting cells, but is relocalized to the nucleus upon stimulation with IFN. The RING domain of TRIM21 contains E3 ubiquitin ligase activity and has been shown to target IRF3, IRF5 and IRF8 for polyubiquitination. Among these targets, IRF3 has been most well characterized. TRIM21 interacts with IRF3 following the stimulation of cells with IFN. In one report, TRIM21 was found to negatively regulate IRF3 resulting in diminished IFN-β activity when overexpressed (Higgs *et al.*, 2008). This was a result of TRIM21-mediated polyubiquitination of IRF3 via its RING domain, leading to its proteosomal degradation. In contrast, another study showed that TRIM21 functioned as a positive regulator of IRF3, stabilizing its activity (Yang *et al.*, 2009). TRIM21, via its B30.2 domain and not its RING domain, bound IRF3 and interfered with the ability of PinI to phosphorylate IRF3 (Yang *et al.*, 2009). This lack of phosphorylation resulted in decreased ubiquitination and diminished degradation. In this study, overexpressing TRIM21 resulted in increase IFN-β promoter activity and siRNA knockdown resulted in diminished IFN-β, ISG56, and Rantes production (Yang *et al.*, 2009). Both of these studies showed no impact of TRIM21 on NF-κB activity.

Two different groups have subsequently generated TRIM21 knockout mice. In both cases the development of immune cells was unaffected by a lack of TRIM21 (Yoshimi *et al.*, 2009). The levels of ubiquitination of both IRF3 and IRF8 were diminished in TRIM21 knockout mice, supporting them as targets of TRIM21 E3 ligase activity *in vivo* (Espinosa *et al.*, 2009; Yoshimi *et al.*, 2009). In unmanipulated mice, a lack of TRIM21 resulted in no differences in IFN-α or IFN-β transcripts or ISG transcripts following poly(I:C) stimulation. Further, TRIM21$^{-/-}$ and TRIM21$^{+/+}$ BMDCs had similar responses to CpG and viral infection

(Espinosa *et al.*, 2009; Yoshimi *et al.*, 2009). The lack of a robust phenotype in the upmaniupulated mice may be secondary to compensation by other TRIM proteins, with support for this hypothesis coming from the observation that in the TRIM21$^{-/-}$ cells, other TRIM genes were upregulated (Yoshimi *et al.*, 2009). The only phenotype observed in TRIM21$^{-/-}$ unmanipulated mice was that embryonic fibroblasts produced increased levels of IL-1, TNF, IL-6, and CSCL10 following LPS or poly(I:C) stimulation, suggesting that the NF-κB pathway was negatively regulated by TRIM21 (Espinosa *et al.*, 2009; Yoshimi *et al.*, 2009). Also noted was increased NF-κB promoter activity in TRIM21 knockout cells (Yoshimi *et al.*, 2009). These results indicated a redundancy in TRIM21 activity in regards to regulation of IRF3 activity, but also implicated TRIM21 in the regulation of the NF-κB pathway for the first time. However, it was noted that following trauma from ear tagging, 100% of the TRIM21$^{-/-}$ mice developed a severe and expanding dermatitis (Espinosa *et al.*, 2009). Furthermore the TRIM21$^{-/-}$ mice that developed dermatitis went on to develop a systemic autoimmune responses including the development of ANA, hypergammaglobulinaemia, Ig deposition in the glomeruli of the kidney leading to proteinuria, as well as an increase in the production of several proinflammatory cytokines including IL-6, IL-12, IL-17, and IL-23 (Espinosa *et al.*, 2009). The development of both dermatitis and systemic autoimmunity was dependent upon IL-23 since breeding the TRIM21$^{-/-}$ mice onto the IL-23$^{-/-}$ background abrogated disease (Espinosa *et al.*, 2009). These studies support an *in vivo* role for TRIM21 as a negative regulator of pro-inflammatory cytokine production, particularly those controlled by the NF-κB pathway. Further investigation will increase our understanding of the mechanism by which TRIM21 regulates the NF-κB pathway.

TRIM30α

TRIM30α is induced by TLR agonists including LPS, poly(I:C), and CpG (Rajsbaum *et al.*, 2008; Shi *et al.*, 2008). No human orthologue exists for TRIM30. In the murine system TRIM30 has been demonstrated to play an important role in negatively regulating TLR signalling. TRIM30α binds directly to TAK1 and TAB2, therefore allowing it to interact with the TAK1/TAB2/TAB3 complex (Shi *et al.*, 2008). Binding to TAK1 was found to be dependent upon the TRIM30 RING domain since either deletion of this motif or mutation of the active site cysteine resulted in decreased binding (Shi *et al.*, 2008). In contrast, binding to TAB2 did not require the RING domain. Despite the requirement of the RING domain for TAK1 binding, to date there is no evidence that the RING domain of TRIM30α has E3 ubiquitin ligase activity. The functional outcome of TRIM30α binding to this complex was the degradation of TAB2 and TAB3 via lysosomal mediated degradation, which prevented TRAF6 autoubiquitination, resulting in decreased IκBα phosphorylation and negative regulation of NF-κB activation (Shi *et al.*, 2008). This was confirmed *in vivo* with TRIM30α transgenic mice, which were found to be resistant to endotoxin shock (Shi *et al.*, 2008).

TRIM8

TRIM8 belongs to the C-V subfamily of TRIM proteins. Its C-terminal domain has no identified motifs or domains. In murine macrophages TRIM8 is upregulated by both LPS, poly(I:C), and both type I and type II IFNs (Toniato *et al.*, 2002; Rajsbaum *et al.*, 2008). A yeast two-hybrid screen identified TRIM8 as an interacting partner with the suppressor of cytokine signalling (SOCS) proteins (Toniato *et al.*, 2002). Signalling through the IFN receptors are regulated by several mechanisms including a negative feedback regulation by SOCS1. TRIM8 was found to interact with the SH2 and SOCs box domain of SOCS1. Co-expression of SOCS1 and TRIM8 decreased the stability of SOCS1 and resulted in a diminished ability of SOCS1 to inhibit IFN-γ signalling (Toniato *et al.*, 2002). It has yet to be determined if this activity is dependent upon the E3 ubiquitin ligase activity and/or the RING domain of TRIM8.

Additional TRIMs

As we learn more about the TRIM family of proteins we are finding that they play a significant role in regulating the host innate immune response along with responses to various pathogens. In addition to the IFN-induced TRIM proteins described above, there is a subset of TRIM proteins that have previously been found to play a role in regulating the innate immune response (TRIM1/MID2, TRIM11, TRIM27, TRIM28, TRIM62) or are implicated in inhibition of various stages of the retroviral life cycle (TRIM11, TRIM32, TRIM15), but are not regulated by IFNs (Nisole *et al.*, 2005; Zhang *et al.*, 2006; Nakayama and Shioda, 2010). It also remains to be determined if the TRIM genes that are induced by IFN (TRIM2, TRIM3, TRIM10, TRIM13, TRIM18, TRIM23, TRIM26, TRIM31, TRIM34, TRIM38, TRIM45, TRIM46, TRIM48, TRIM56, TRIM62, and TRIM69; Rajsbaum *et al.*, 2008; Carthagena *et al.*, 2009) will exert antiviral activity and/or play a role in regulating the innate immune response. As observed with TRIM21/Ro52 and TRIM20/MEFV/pyrin, the dysregulation of these proteins may have a profound effect on maintaining an appropriate host response to pathogenic challenge or other danger signals.

Interferon-stimulated genes with undefined mechanisms of action

Several ISGs have been identified as antiviral molecules; however, the mechanism by which they function remains undefined. The analysis of mice lacking the ubiquitin like molecule, ISG15, has identified it as a key player in the host response to several viruses. The mechanism by which it exerts this activity is still active under investigation. The 3′,5′ exonuclease, ISG20, has been shown to confer protection against certain RNA viral infections both *in vitro* and *in vivo*. Recent evidence also demonstrated that the newly identified IFITM proteins can inhibit the cellular replication of multiple RNA viruses. While the mechanisms by which these ISGs combat viruses are not clearly defined, their involvement in antiviral responses is readily apparent. This section is devoted to discussions on the recent literature implicating these ISGs as antiviral effectors and their potential mechanisms of action.

Interferon-stimulated gene 15

Interferon-stimulated gene 15 (ISG15) is one of the most abundantly induced transcripts upon IFN treatment (Der *et al.*, 1997). The appearance of a 15 kDa protein in IFN-treated cell lysates was first noted in 1984 and was soon after identified as having sequence homology to ubiquitin (Korant *et al.*, 1984; Haas *et al.*, 1987). Indeed, the crystal structure of ISG15 revealed that it is composed of two domains, each of which assumes a β-grasp fold similar to the ubiquitin structure (Vijay-Kumar *et al.*, 1987; Narasimhan *et al.*, 2005). ISG15 is synthesized as a 17 kDa precursor that is proteolytically processed into a 15-kDa form (Knight *et al.*, 1988). Free ISG15 is detectable in cell lysates within two hours after IFN treatment and precedes conjugate formation, which occurs 12–16 hours later (Haas *et al.*, 1987; Loeb and Haas, 1992). Like ubiquitin, the mature form of ISG15 contains a C-terminal LRLRGG motif that mediates its covalent conjugation to target proteins (Haas *et al.*, 1987; Loeb and Haas, 1992; Narasimhan *et al.*, 1996). ISG15 conjugation is achieved through a parallel but distinct mechanism to that of the ubiquitin cascade (Fig. 4.4). The activating E1 enzyme, Ube1L, forms an ATP-dependent thioester bond with ISG15 (Yuan and Krug, 2001). Following activation, ISG15 is transferred to the active-site cysteine of the conjugating E2 enzyme UbcM8/UbcH8 (Kim *et al.*, 2004; Zhao *et al.*, 2004). The E3 ligases HERC5/Ceb1 (human), HERC6 (mouse), HHARI, and Efp transfer ISG15 to lysine residues on specific substrates (Dastur *et al.*, 2006; Nakasato *et al.*, 2006; Takeuchi *et al.*, 2006; Wong *et al.*, 2006; Zou and Zhang, 2006; Versteeg *et al.*, 2010). HERC5 appears to function as a dominant E3 ligase for the ISG15 system. Knockdown of HERC5 expression with siRNA inhibited greater than 90% of the global levels of ISG15 conjugates (Dastur *et al.*, 2006). This process, known as ISGylation, is reversible because of specific removal of ISG15 from conjugated proteins by the deconjugating enzyme Ubp43 (Malakhov *et al.*, 2002). A screen for additional deubiquitinating enzymes that could target ISG15 conjugates identified USP2, USP5, USP13, and USP14 as potential candidates, whose significance remain to be determined *in vivo* (Catic *et al.*, 2007). Numerous screens have identified nearly 200 candidate ISG15 targets which span a diverse array of biological processes (Malakhov *et al.*, 2003; Giannakopoulos *et al.*,

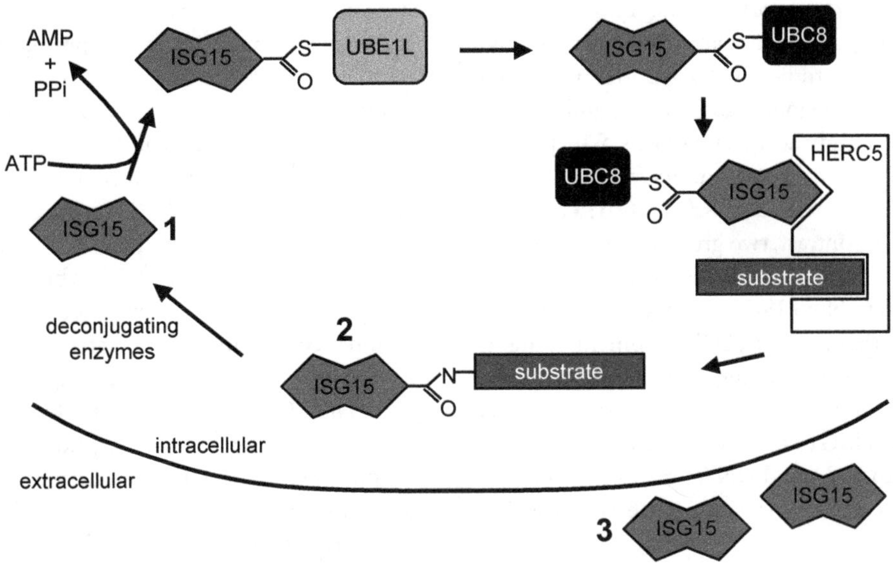

Figure 4.4 Schematic of the ISG15 conjugation cascade. The activating E1 enzyme UBE1L forms a thioester bond with ISG15 in an ATP-dependent manner. ISG15 is then transferred to the E2 conjugating enzyme UBC8. The E3 ligase HERC5 catalyzes the conjugation of ISG15 to lysine residues of target substrate proteins. Deconjugating enzymes reverse the conjugation by removing ISG15 from target proteins. ISG15 exists in three forms: free intracellular (1), conjugated intracellular (2) and free extracellular (3).

2005; Zhao *et al.*, 2005; Pattyn *et al.*, 2008). It is still largely unknown what impact ISG15 modification has on a target protein. Unlike ubiquitination, ISGylation does not appear to directly target proteins for proteasome-mediated degradation (Liu *et al.*, 2003). Insights from the limited number of targets that have been analysed will be discussed below. Interestingly, like ISG15 itself, all members of the ISG15 conjugation cascade are upregulated by type I IFNs. Many of the target proteins identified by proteomic screen include proteins that play an important roles in immunity, such as STAT1, JAK1, RIG-1, ISG56, PKR, and MxA (Malakhov *et al.*, 2003; Giannakopoulos *et al.*, 2005; Zhao *et al.*, 2005; Pattyn *et al.*, 2008). These findings suggest that the ISG15 pathway could play a significant role in the host antiviral response. Studies utilizing mice deficient in ISG15 confirmed this hypothesis.

The analysis of ISG15$^{-/-}$ mice revealed a critical antiviral role for this protein. The initial characterization of these mice found no defect in their development, intact IFN signalling and normal resistance to VSV and LCMV infection (Osiak *et al.*, 2005). However, subsequent experiments using a different set of viruses found that the ISG15$^{-/-}$ mice, compared with wild type mice, were more susceptible to infection with Sindbis virus, influenza A and B virus, HSV-1, and murine gammaherpesvirus 68 (Lenschow *et al.*, 2005; Lenschow *et al.*, 2007). While both free ISG15 and ISG15 conjugates were detected during infection, ISGylation was shown to be required for ISG15-dependent resistance to Sindbis and influenza virus *in vivo* (Lenschow *et al.*, 2007; Giannakopoulos *et al.*, 2009; Lai *et al.*, 2009). The increased lethality seen in mice infected with Sindbis virus could be rescued by a double subgenomic Sindbis virus expressing wild type ISG15, but was not rescued by viruses containing mutations in the terminal LRLRGG motif, which is required for ISG15 conjugation to occur (Lenschow *et al.*, 2007; Giannakopoulos *et al.*, 2009). In addition, mice lacking the ISG15 E1 enzyme, UbE1L, which still express free ISG15, but fail to form ISG15 conjugates, were also susceptible to both Sindbis virus and influenza virus infection (Giannakopoulos *et al.*, 2009; Lai *et al.*,

2009). Further support for the importance of ISG15 conjugation in mediating its antiviral activity come from studies that have identified viral proteins that antagonize ISG15. The influenza B NS1 protein and the vaccinia virus E3L protein bind to ISG15 and prevent its conjugation to target proteins (Yuan and Krug, 2001; Guerra *et al.*, 2008). In contrast, two groups of RNA viruses, nairoviruses and arteriviruses, utilize a different mechanism to blunt ISG15 activity. These viruses encode ovarian tumour (OTU) domain-containing proteases that deconjugate both ubiquitin and ISG15 from target proteins (Frias-Staheli *et al.*, 2007). Thus, ISG15 has exerted selective pressure against both RNA and DNA viruses in a manner that has selected for viruses capable of subverting conjugation.

In addition to these *in vivo* results, several viruses are restricted by ISG15 in *in vitro* infection systems. These viruses include HIV-1, Japanese encephalitis virus (JEV), vaccinia, Newcastle disease virus, SeV, VSV and Ebola (Okumura *et al.*, 2006; Guerra *et al.*, 2008; Malakhova and Zhang, 2008; Okumura *et al.*, 2008; Hsiao *et al.*, 2010; Shi *et al.*, 2010). In the case of HIV-1, ISG15 contributes to its restriction by inhibiting the essential ubiquitination of two proteins involved in virion release: the viral protein Gag and the endosomal sorting complex component Tsg101 (Okumura *et al.*, 2006). How ISG15 prevents Gag and Tsg101 ubiquitination is unclear, though it may involve ISG15-dependent inhibition of ubiquitin ligases (Takeuchi *et al.*, 2005; Zou *et al.*, 2005). During Ebola infection, ISG15, independent of its conjugation system, can bind to and inhibits the Nedd4 ubiquitin ligase that is required for viral-like particle budding (Malakhova and Zhang, 2008; Okumura *et al.*, 2008). Thus, both free and conjugated ISG15 can exert loss-of-function activities that impact viral replication and budding *in vitro*. The significance of this *in vivo* remains to be determined.

With few exceptions, the mechanisms by which ISG15 exerts its antiviral activity are unclear. This is due, in part, to the unknown fate of the vast majority of ISGylated target proteins. Recently, the mechanism by which ISG15 alters the function of several substrates has been identified, but whether these targets contribute to the antiviral activity of ISG15 remains to be tested. For example, ISG15 negatively regulates the scaffold protein filamin B (Jeon *et al.*, 2009). Unmodified filamin B facilitates type I IFN-induced apoptosis by recruiting RAC1, MEKK1, MKK4, and JNK1 (Jeon *et al.*, 2008). ISGylation of filamin B, however, releases RAC1, MEKK1, and MKK4 from the scaffold, preventing IFN-α-dependent JNK activation and apoptosis (Jeon *et al.*, 2009). It has been suggested that this mechanism may protect uninfected bystander cells from IFN-mediated apoptosis, but this hypothesis awaits further experimentation (Jeon *et al.*, 2009). In contrast, ISG15 conjugation positively regulates the transcription factor IRF3 that promotes type I IFN induction (Shi *et al.*, 2010). ISGylation of IRF3 abolishes its binding to Pin1, a protein that promotes IRF3 ubiquitination and subsequent degradation. ISG15-dependent inhibition of the IRF3–Pin1 interaction correlates with decreased IRF3 degradation and increased IFN-β induction. Another example of a positively regulated ISGylated protein is the mRNA 5′ cap structure-binding protein 4EHP (Okumura *et al.*, 2007). 4EHP binds mRNA in a manner that prevents translation. ISGylated 4EHP exhibits even greater cap structure binding than unmodified 4EHP, leading to an increase in translational inhibition. The contributions of these ISG15 modifications to the antiviral activity of ISG15 are still unclear. Nevertheless, progress is being made towards understanding these important aspects of ISG15 biology and their involvement in host defence.

In addition to host proteins being the target of ISGylation, it has recently been demonstrated that viral proteins can also be modified by ISG15. In the case of influenza A, the IFN antagonist NS1 is the target of ISGylation during infection (Tang *et al.*, 2010; Zhao *et al.*, 2010). Two groups have reported this finding, although the lysine residues that were identified are distinct. In one case, ISGylation of NS1 disrupted its association with importin-α, a protein required for nuclear import of NS1 (Tang *et al.*, 2010; Zhao *et al.*, 2010). The generation of recombinant viruses in which the lysine residue(s) that were targeted for ISGylation were mutated, such that NS1 was no longer modified, demonstrated that the ISGylation of NS1 inhibited replication of influenza in IFN-β-treated cells (Zhao *et al.*, 2010). Future experiments are needed to determine whether NS1 is the only viral protein modified during influenza virus infection. It also remains unknown if this modification accounts for the entire ISG15$^{-/-}$ phenotype seen in mice, or if the modification of host proteins is also contributing to the phenotype. Regulation of the host antiviral response by ISG15 is complex, and will probably vary for different viruses depending upon the proteins that are targeted for modification. Future studies will be important in determining how broadly this pathway impacts the host response to viral and bacterial pathogens, and whether the modification of host proteins will have important consequences for other biological pathways.

ISG20

ISG20 (also termed HEM45) was identified by two independent groups as a gene induced following IFN or oestrogen stimulation (Gongora *et al.*, 1997; Pentecost, 1998). As a member of the DEDD exonuclease superfamily, whose members include RNases D and T, ISG20 contains the characteristic 150 aa exonuclease domain in which four acidic amino acids (1 glutamate, 3 aspartates) are distributed throughout three exonuclease motifs (Exo I-III; Gongora *et al.*, 1997; Moser *et al.*, 1997). Biochemical analyses demonstrated that ISG20 has 3′,5′exonuclease activity with a strong preference for ssRNA over ssDNA (Nguyen *et al.*, 2001). ISG20 is composed of a unique exonuclease domain with no regulatory domains, suggesting that it may interact with other proteins to regulate its function and prevent non-specific damage.

Its role as an antiviral molecule has been largely supported by *in vitro* studies. Numerous pathogens [CMV, HIV, Epstein–Barr virus (EBV), HBV, *Listeria, Toxoplasma gondii*) and several PAMPs (CpG, LPS, poly(I:C)] induce ISG20 in a variety of cell types (Degols *et al.*, 2007). Further support has come from overexpression studies where it was shown that the overexpression of ISG20 protected against VSV infection and to a lesser extent influenza and EMCV infection (Espert *et al.*, 2003). This antiviral response was dependent on the 3′,5′exonuclease activity of ISG20 since an ISG20 mutant lacking enzymatic activity did not protect cells from viral infection (Espert *et al.*, 2003). ISG20 overexpression also conferred resistance to HCV infection in a tetracycline-inducible cell line (Jiang *et al.*, 2008) and significantly decreased Sindbis virus titres 12 and 24 hours post infection in tetracycline-inducible MEFs (Zhang *et al.*, 2007). To confirm the antiviral effects of ISG20 *in vivo*, investigators generated a double subgenomic Sindbis virus expressing ISG20. Infection of neonatal mice with this virus significantly increased their per cent survival and average survival time (Zhang *et al.*, 2007). Confirmation of the antiviral activity of ISG20 awaits the production and characterization of ISG20 knockout mice, and further studies are needed to determine the mechanism by which ISG20 targets viral replication. Its 3′,5′ RNAse activity could directly target viral RNA to inhibit replication or ISG20 could target cellular RNAs that are required for viral replication. In either case, the mechanism at play and the factors that mediate this specificity need to be elucidated.

The IFITM protein family

The IFN-induced transmembrane (IFITM) proteins were identified in a screen for mRNAs upregulated in IFN-treated neuroblastoma cells (Friedman *et al.*, 1984). Humans and mice encode IFITM1 (9–27/Leu-13/fragilis 2/CD225), IFITM2 (1–8D/fragilis 3), IFITM3 (1–8U/fragilis) and IFITM5 (fragilis 4), while mice also encode IFITM6 (fragilis 5). IFITM proteins contain two transmembrane domains that both localize to the interior of the cell (Brem *et al.*, 2003; Smith *et al.*, 2006). Accordingly, IFITM proteins were found in membrane but not cytoplasmic fractions, and immunofluorescence microscopy localized IFITM proteins to both the cell surface and interior vesicles, potentially exosomes (Brem *et al.*, 2003).

IFITM proteins were identified in a large-scale siRNA screen as molecules that confer resistance to infection with influenza A (H1N1) in cell culture (Brass *et al.*, 2009). Silencing of IFITM1, -2 or -3 enhanced cellular susceptibility to influenza, while overexpression of these molecules increased resistance to infection. WNV and dengue virus were also controlled by IFITM1, 2, and 3 in this report. Mice with a targeted deletion of the entire IFITM gene cluster have been generated, but studies reporting their susceptibility to viral infection have not been described (Lange *et al.*, 2008).

The mechanism by which IFITM proteins mediate cellular resistance to these viruses is unknown, although the block appears to be at an early step between virus–host cell binding and virus entry into the cytosol (Brass *et al.*, 2009). The only reported ligand for an IFITM member is osteopontin (OPN), a secreted glycoprotein that promotes anchorage-independent growth and metastasis (El-Tanani *et al.*, 2010). IFITM3–OPN interactions reduced OPN expression and cellular proliferation, as well as inhibited anchorage-independent growth. This correlates with previous studies showing that IFITM proteins transmit anti-proliferative signals (Evans *et al.*, 1990; Deblandre *et al.*, 1995; Brem *et al.*, 2001). IFITM1, 2, and 3 have been demonstrated to associate with the B cell signalling complex of CD81, CD21 and CD19 (Takahashi *et al.*, 1990; Bradbury *et al.*, 1992; Matsumoto *et al.*, 1993; Smith *et al.*, 2006). Whether these aspects of IFITM biology are involved in viral restriction remains to be tested, but it is likely that viral infection will be a useful probe into the cell biology of IFITM proteins.

ISG12

ISG12 (p27) belongs to the family of small ISGs and is part of the ISG12 group that includes 6–16, ISG12, and ISG12-S based upon nucleic acid homology. Human ISG12 was initially cloned from a cDNA library derived from oestradiol-treated MCF7 breast cancer cells while screening for oestrogen-induced genes (Rasmussen *et al.*, 1993). However, the presence of ISG12 mRNA transcripts in multiple human breast cancer tumour lines did not correlate with oestrogen receptor expression. Instead, ISG12 mRNA expression was induced in human cells following treatment with IFN-α (Rasmussen *et al.*, 1993; Gjermandsen *et al.*, 2000; Martensen *et al.*, 2001). ISG12 is also induced by CMV and poly(I:C), and is slightly induced by IFNγ. Subsequent studies revealed that ISG12 was localized to the nuclear envelope; however, the relevance of its association with the nuclear membrane is unknown (Martensen *et al.*, 2001).

Despite its robust induction by IFN, very little is known about the antiviral function of ISG12. The only report suggesting it may function as an antiviral molecule comes from a search for genes that mediate age-dependent resistance to alphavirus encephalitis in mice. This study revealed that ISG12 mRNA expression was significantly higher in the brains of Sindbis virus resistant 4-week old mice in comparison to susceptible 1-day-old neonates (Labrada *et al.*, 2002). ISG12 was subsequently expressed heterologously in the subgenomic dsTE12Q Sindbis virus to determine whether ISG12 could protect neonatal mice from virus-induced mortality. Infection of 1 day old neonates with dsTE12Q-ISG12 increased their mean survival time by 2 days in comparison with dsTE12Q-control infected mice (Labrada *et al.*, 2002). It is unclear mechanistically how ISG12 delayed Sindbis virus-induced mortality since no differences in viral titres, viral antigen staining or virus-induced apoptosis were detected in the brains of dsTE12Q-ISG12 and dsTE12Q-control infected mice (Labrada *et al.*, 2002). Further experiments are needed to fully evaluate its role as an antiviral molecule, including the production and evaluation of ISG12 knockout mice.

Future trends

IFN plays a critical role regulating the host response to viral infection through the establishment of a cellular antiviral state and by modulating the host's innate and adaptive immune response. These activities are mediated by hundreds of ISGs that are induced by IFNs. This chapter outlined our current understanding of the IFN-stimulated genes that have been characterized as antiviral proteins and contribute to the host's ability to control viral infection. It was found several years ago that mice lacking all three of the best characterized ISGs (PKR, RNaseL, Mx) still have IFN induced antiviral activity, supporting the idea that additional antiviral ISGs are yet to be identified. The recent characterization of several new antiviral ISGs including ISG15, viperin, tetherin, and ISG20 confirm this idea. Future studies of protein families, such as the TRIM family, will probably uncover additional IFN-induced antiviral molecules.

Despite the prospect of novel antiviral ISGs, there are still many unanswered questions concerning the current antiviral molecules. For several ISGs, the mechanism of action is poorly understood. It is still uncertain as to how the ISG15 pathway regulates the host antiviral response. This will require studies evaluating which proteins (host and viral) are modified during viral infection and validating these endogenous targets proteins in cells. Studies determining the fate of an ISGylated protein are also needed. Finally, given the recent observation that viral proteins are ISGylated, we will need to evaluate the overall impact of viral vs. host modifications during viral infection. Despite the importance of the Mx proteins, we still do not have a clear understanding of how they exert their antiviral function. Additional studies are also needed to uncover the mechanism by which several of the newly identified antiviral ISGs including ISG20, ISG12, IFITM, and several of the TRIM proteins perturb viral propagation and spread.

It is also likely that as proteins are further characterized novel mechanisms of action will be uncovered. For example, ADAR1 functioning as an adenosine deaminase was expected to be detrimental to the virus. However, studies in ADAR1-depleted cells demonstrated it was proviral and led to the identification of a second function for this protein as a negative regulator of PKR (Nie *et al.*, 2007; Clerzius *et al.*, 2009; Doria *et al.*, 2009). ISG56 functions to inhibit translation, but additional studies revealed it also negatively regulates IFN signalling through its interaction with MITA (Li *et al.*, 2009). Finally, PKR phosphorylates eIF2α, but the impact of PKR's other substrates on the antiviral and host immune response is unknown. Additional insight will probably be gained as the already defined ISGs are further characterized.

Understanding the function of these ISGs in the context of viral infection in either primary cells or within an intact host will also be an important future direction for ISG research. Most studies have utilized either overexpression systems or siRNA studies in fibroblasts. There is increasing evidence that ISGs are induced by different stimuli and expressed in a cell type-specific manner. The expression of ISG56 and ISG54 *in vivo* revealed that depending upon the stimuli being used, there was differential expression of these mRNAs and proteins in different cell types (Terenzi *et al.*, 2007). TRIM proteins are also differentially expressed in macrophages, dendritic cells and lymphocytes at rest and following IFN stimulation (Rajsbaum *et al.*, 2008). It suggests that our understanding of how these various ISGs regulate antiviral responses *in vivo* is just in its infancy.

IFNs, in addition to their role to induce an antiviral state, also play a critical role in regulating the innate and adaptive immune response through their impact upon NK cell activation, dendritic cell maturation and antigen presentation, stimulating antibody production from B cells and enhancing CD8 T cell memory and effector function. Future studies will need to take into account the impact of these ISGs on the regulation of the immune response. Increasingly, IFNs are being appreciated for their role in the pathogenesis of autoimmune disease and potential impact upon malignancy. It is likely that many of the ISGs discussed in this chapter will impact upon the development and/or propagation of these diseases because of their impact on the host response to viral infection and modulation of the immune response.

References

Abraham, N., Stojdl, D.F., Duncan, P.I., Methot, N., Ishii, T., Dube, M., Vanderhyden, B.C., Atkins, H.L., Gray, D.A., McBurney, M.W., *et al.* (1999). Characterization of transgenic mice with targeted disruption of the catalytic domain of the double-stranded RNA-dependent protein kinase, PKR. J. Biol. Chem. *274*, 5953–5962.

Aebi, M., Fah, J., Hurt, N., Samuel, C.E., Thomis, D., Bazzigher, L., Pavlovic, J., Haller, O., and Staeheli, P. (1989). cDNA structures and regulation of two interferon-induced human Mx proteins. Mol. Cell. Biol. *9*, 5062–5072.

Ahn, J.H., and Hayward, G.S. (2000). Disruption of PML-associated nuclear bodies by IE1 correlates with efficient early stages of viral gene expression and DNA replication in human cytomegalovirus infection. Virology *274*, 39–55.

Al-Khatib, K., Williams, B.R., Silverman, R.H., Halford, W., and Carr, D.J. (2005). Dichotomy between survival and lytic gene expression in RNase L- and PKR-deficient mice transduced with an adenoviral vector expressing murine IFN-beta following ocular HSV-1 infection. Exp. Eye Res. *80*, 167–173.

Andersson, I., Bladh, L., Mousavi-Jazi, M., Magnusson, K.E., Lundkvist, A., Haller, O., and Mirazimi, A.

(2004). Human MxA protein inhibits the replication of Crimean-Congo hemorrhagic fever virus. J. Virol. *78*, 4323–4329.

Andrew, A.J., Miyagi, E., Kao, S., and Strebel, K. (2009). The formation of cysteine-linked dimers of BST-2/tetherin is important for inhibition of HIV-1 virus release but not for sensitivity to Vpu. Retrovirology *6*, 80.

Asper, M., Sternsdorf, T., Hass, M., Drosten, C., Rhode, A., Schmitz, H., and Gunther, S. (2004). Inhibition of different Lassa virus strains by alpha and gamma interferons and comparison with a less pathogenic arenavirus. J. Virol. *78*, 3162–3169.

Baglioni, C., Minks, M.A., and Maroney, P.A. (1978). Interferon action may be mediated by activation of a nuclease by pppA2′p5′A2′p5′A. Nature *273*, 684–687.

Balachandran, S., Kim, C.N., Yeh, W.C., Mak, T.W., Bhalla, K., and Barber, G.N. (1998). Activation of the dsRNA-dependent protein kinase, PKR, induces apoptosis through FADD-mediated death signaling. EMBO J. *17*, 6888–6902.

Balachandran, S., Roberts, P.C., Brown, L.E., Truong, H., Pattnaik, A.K., Archer, D.R., and Barber, G.N. (2000). Essential role for the dsRNA-dependent protein kinase PKR in innate immunity to viral infection. Immunity *13*, 129–141.

Barr, S.D., Smiley, J.R., and Bushman, F.D. (2008). The interferon response inhibits HIV particle production by induction of TRIM22. PLoS Pathog. *4*, e1000007.

Barry, G., Breakwell, L., Fragkoudis, R., Attarzadeh-Yazdi, G., Rodriguez-Andres, J., Kohl, A., and Fazakerley, J.K. (2009). PKR acts early in infection to suppress Semliki Forest virus production and strongly enhances the type I interferon response. J. Gen. Virol. *90*, 1382–1391.

Bartee, E., McCormack, A., and Fruh, K. (2006). Quantitative membrane proteomics reveals new cellular targets of viral immune modulators. PLoS Pathog. *2*, e107.

Bass, B.L., and Weintraub, H. (1988). An unwinding activity that covalently modifies its double-stranded RNA substrate. Cell *55*, 1089–1098.

Beech, S.J., Lethbridge, K.J., Killick, N., McGlincy, N., and Leppard, K.N. (2005). Isoforms of the promyelocytic leukemia protein differ in their effects on ND10 organization. Exp Cell Res *307*, 109–117.

Bennett, R.L., Blalock, W.L., and May, W.S. (2004). Serine 18 phosphorylation of RAX, the PKR activator, is required for PKR activation and consequent translation inhibition. J. Biol. Chem. *279*, 42687–42693.

Bergeron, J., Benlimame, N., Zeng-Rong, N., Xiao, D., Scrivens, P.J., Koromilas, A.E., and Alaoui-Jamali, M.A. (2000). Identification of the interferon-inducible double-stranded RNA-dependent protein kinase as a regulator of cellular response to bulky adducts. Cancer Res. *60*, 6800–6804.

Bevilacqua, P.C., George, C.X., Samuel, C.E., and Cech, T.R. (1998). Binding of the protein kinase PKR to RNAs with secondary structure defects: role of the tandem A-G mismatch and noncontiguous helixes. Biochemistry *37*, 6303–6316.

Bick, M.J., Carroll, J.W., Gao, G., Goff, S.P., Rice, C.M., and MacDonald, M.R. (2003). Expression of the zinc-finger antiviral protein inhibits alphavirus replication. J. Virol. *77*, 11555–11562.

Blondel, D., Regad, T., Poisson, N., Pavie, B., Harper, F., Pandolfi, P.P., De The, H., and Chelbi-Alix, M.K. (2002). Rabies virus P and small P products interact directly with PML and reorganize PML nuclear bodies. Oncogene *21*, 7957–7970.

Bonilla, W.V., Pinschewer, D.D., Klenerman, P., Rousson, V., Gaboli, M., Pandolfi, P.P., Zinkernagel, R.M., Salvato, M.S., and Hengartner, H. (2002). Effects of promyelocytic leukemia protein on virus-host balance. J. Virol. *76*, 3810–3818.

Bonnet, M.C., Daurat, C., Ottone, C., and Meurs, E.F. (2006). The N-terminus of PKR is responsible for the activation of the NF-kappaB signaling pathway by interacting with the IKK complex. Cell Signal *18*, 1865–1875.

Borden, K.L., Campbelldwyer, E.J., Carlile, G.W., Djavani, M., and Salvato, M.S. (1998). Two RING finger proteins, the oncoprotein PML and the arenavirus Z protein, colocalize with the nuclear fraction of the ribosomal P proteins. J. Virol. 72, 3819–3826.

Boudinot, P., Riffault, S., Salhi, S., Carrat, C., Sedlik, C., Mahmoudi, N., Charley, B., and Benmansour, A. (2000). Vesicular stomatitis virus and pseudorabies virus induce a vig1/cig5 homologue in mouse dendritic cells via different pathways. J. Gen. Virol. *81*, 2675–2682.

Boutell, C., Sadis, S., and Everett, R.D. (2002). Herpes simplex virus type 1 immediate-early protein ICP0 and is isolated RING finger domain act as ubiquitin E3 ligases *in vitro*. J. Virol. *76*, 841–850.

Bradbury, L.E., Kansas, G.S., Levy, S., Evans, R.L., and Tedder, T.F. (1992). The CD19/CD21 signal transducing complex of human B lymphocytes includes the target of antiproliferative antibody-1 and Leu-13 molecules. J. Immunol. *149*, 2841–2850.

Brass, A.L., Huang, I.C., Benita, Y., John, S.P., Krishnan, M.N., Feeley, E.M., Ryan, B.J., Weyer, J.L., van der Weyden, L., Fikrig, E., *et al.* (2009). The IFITM proteins mediate cellular resistance to influenza A H1N1 virus, West Nile virus, and dengue virus. Cell *139*, 1243–1254.

Brem, R., Oraszlan-Szovik, K., Foser, S., Bohrmann, B., and Certa, U. (2003). Inhibition of proliferation by 1–8U in interferon-alpha-responsive and non-responsive cell lines. Cell Mol. Life Sci. *60*, 1235–1248.

Carthagena, L., Bergamaschi, A., Luna, J.M., David, A., Uchil, P.D., Margottin-Goguet, F., Mothes, W., Hazan, U., Transy, C., Pancino, G., *et al.* (2009). Human TRIM gene expression in response to interferons. PLoS One *4*, e4894.

Carthagena, L., Parise, M.C., Ringeard, M., Chelbi-Alix, M.K., Hazan, U., and Nisole, S. (2008). Implication of TRIM alpha and TRIMCyp in interferon-induced anti-retroviral restriction activities. Retrovirology *5*, 59.

Catic, A., Fiebiger, E., Korbel, G.A., Blom, D., Galardy, P.J., and Ploegh, H.L. (2007). Screen for ISG15-crossreactive deubiquitinases. PLoS One *2*, e679.

Cattaneo, R. (1994). Biased (A-->I) hypermutation of animal RNA virus genomes. Curr. Opin. Genet. Dev. *4*, 895–900.

Cattaneo, R., Schmid, A., Eschle, D., Baczko, K., ter Meulen, V., and Billeter, M.A. (1988). Biased hypermutation and other genetic changes in defective measles viruses in human brain infections. Cell *55*, 255–265.

Chae, J.J., Komarow, H.D., Cheng, J., Wood, G., Raben, N., Liu, P.P., and Kastner, D.L. (2003). Targeted disruption of pyrin, the FMF protein, causes heightened sensitivity to endotoxin and a defect in macrophage apoptosis. Mol. Cell *11*, 591–604.

Chae, J.J., Wood, G., Masters, S.L., Richard, K., Park, G., Smith, B.J., and Kastner, D.L. (2006). The B30.2 domain of pyrin, the familial Mediterranean fever protein, interacts directly with caspase-1 to modulate IL-1beta production. Proc. Natl. Acad. Sci. U.S.A. *103*, 9982–9987.

Chae, J.J., Wood, G., Richard, K., Jaffe, H., Colburn, N.T., Masters, S.L., Gumucio, D.L., Shoham, N.G., and Kastner, D.L. (2008). The familial Mediterranean fever protein, pyrin, is cleaved by caspase-1 and activates NF-kappaB through its N-terminal fragment. Blood *112*, 1794–1803.

Chang, K.S., Cai, Z., Zhang, C., Sen, G.C., Williams, B.R., and Luo, G. (2006). Replication of hepatitis C virus (HCV) RNA in mouse embryonic fibroblasts: protein kinase R (PKR)-dependent and PKR-independent mechanisms for controlling HCV RNA replication and mediating interferon activities. J. Virol. *80*, 7364–7374.

Chee, A.V., Lopez, P., Pandolfi, P.P., and Roizman, B. (2003). Promyelocytic leukemia protein mediates interferon-based anti-herpes simplex virus 1 effects. J. Virol. *77*, 7101–7105.

Chelbi-Alix, M.K., Pelicano, L., Quignon, F., Koken, M.H., Venturini, L., Stadler, M., Pavlovic, J., Degos, L., and de The, H. (1995). Induction of the PML protein by interferons in normal and APL cells. Leukemia *9*, 2027–2033.

Chelbi-Alix, M.K., Quignon, F., Pelicano, L., Koken, M.H., and de The, H. (1998). Resistance to virus infection conferred by the interferon-induced promyelocytic leukemia protein. J. Virol. *72*, 1043–1051.

Chen, G., Guo, X., Lv, F., Xu, Y., and Gao, G. (2008). p72 DEAD box RNA helicase is required for optimal function of the zinc-finger antiviral protein. Proc. Natl. Acad. Sci. U.S.A. *105*, 4352–4357.

Chieux, V., Chehadeh, W., Harvey, J., Haller, O., Wattre, P., and Hober, D. (2001). Inhibition of coxsackievirus B4 replication in stably transfected cells expressing human MxA protein. Virology *283*, 84–92.

Chin, K.C., and Cresswell, P. (2001). Viperin (cig5), an IFN-inducible antiviral protein directly induced by human cytomegalovirus. Proc. Natl. Acad. Sci. U.S.A. *98*, 15125–15130.

Chinsangaram, J., Koster, M., and Grubman, M.J. (2001). Inhibition of L-deleted foot-and-mouth disease virus replication by alpha/beta interferon involves double-stranded RNA-dependent protein kinase. J. Virol. *75*, 5498–5503.

Clemens, M.J., and Williams, B.R. (1978). Inhibition of cell-free protein synthesis by pppA2′p5′A2′p5′A: a novel oligonucleotide synthesized by interferon-treated L cell extracts. Cell *13*, 565–572.

Clerzius, G., Gelinas, J.F., Daher, A., Bonnet, M., Meurs, E.F., and Gatignol, A. (2009). ADAR1 interacts with PKR during human immunodeficiency virus infection of lymphocytes and contributes to viral replication. J. Virol. *83*, 10119–10128.

Cohen, E.A., Terwilliger, E.F., Sodroski, J.G., and Haseltine, W.A. (1988). Identification of a protein encoded by the vpu gene of HIV-1. Nature *334*, 532–534.

Condemine, W., Takahashi, Y., Zhu, J., Puvion-Dutilleul, F., Guegan, S., Janin, A., and de The, H. (2006). Characterization of endogenous human promyelocytic leukemia isoforms. Cancer Res *66*, 6192–6198.

Cortajarena, A.L., and Regan, L. (2006). Ligand binding by TPR domains. Protein Sci. *15*, 1193–1198.

Cuddihy, A.R., Li, S., Tam, N.W., Wong, A.H., Taya, Y., Abraham, N., Bell, J.C., and Koromilas, A.E. (1999). Double-stranded-RNA-activated protein kinase PKR enhances transcriptional activation by tumor suppressor p53. Mol. Cell Biol. *19*, 2475–2484.

Cullen, B.R. (2006). Role and mechanism of action of the APOBEC3 family of antiretroviral resistance factors. J. Virol. *80*, 1067–1076.

Dang, Y., Wang, X., Esselman, W.J., and Zheng, Y.H. (2006). Identification of APOBEC3DE as another antiretroviral factor from the human APOBEC family. J. Virol. *80*, 10522–10533.

Dastur, A., Beaudenon, S., Kelley, M., Krug, R.M., and Huibregtse, J.M. (2006). Herc5, an interferon-induced HECT E3 enzyme, is required for conjugation of ISG15 in human cells. J. Biol. Chem. *281*, 4334–4338.

Deblandre, G.A., Marinx, O.P., Evans, S.S., Majjaj, S., Leo, O., Caput, D., Huez, G.A., and Wathelet, M.G. (1995). Expression cloning of an interferon-inducible 17-kDa membrane protein implicated in the control of cell growth. J. Biol. Chem. *270*, 23860–23866.

Degols, G., Eldin, P., and Mechti, N. (2007). ISG20, an actor of the innate immune response. Biochimie *89*, 831–835.

Der, S.D., Yang, Y.L., Weissmann, C., and Williams, B.R. (1997). A double-stranded RNA-activated protein kinase-dependent pathway mediating stress-induced apoptosis. Proc. Natl. Acad. Sci. U.S.A. *94*, 3279–3283.

Dettenhofer, M., Cen, S., Carlson, B.A., Kleiman, L., and Yu, X.F. (2000). Association of human immunodeficiency virus type 1 Vif with RNA and its role in reverse transcription. J. Virol. *74*, 8938–8945.

Diaz-Griffero, F., Li, X., Javanbakht, H., Song, B., Welikala, S., Stremlau, M., and Sodroski, J. (2006). Rapid turnover and polyubiquitylation of the retroviral restriction factor TRIM5. Virology *349*, 300–315.

Djavani, M., Rodas, J., Lukashevich, I.S., Horejsh, D., Pandolfi, P.P., Borden, K.L., and Salvato, M.S. (2001). Role of the promyelocytic leukemia protein PML in the interferon sensitivity of lymphocytic choriomeningitis virus. J. Virol. *75*, 6204–6208.

Doehle, B.P., Schafer, A., and Cullen, B.R. (2005a). Human APOBEC3B is a potent inhibitor of HIV-1 infectivity and is resistant to HIV-1 Vif. Virology *339*, 281–288.

Doehle, B.P., Schafer, A., Wiegand, H.L., Bogerd, H.P., and Cullen, B.R. (2005b). Differential sensitivity of murine leukemia virus to APOBEC3-mediated inhibition is governed by virion exclusion. J. Virol. *79*, 8201–8207.

Dong, B., Niwa, M., Walter, P., and Silverman, R.H. (2001). Basis for regulated RNA cleavage by functional analysis of RNase L and Ire1p. Rna *7*, 361–373.

Dong, B., and Silverman, R.H. (1995). 2-5A-dependent RNase molecules dimerize during activation by 2-5A. J. Biol. Chem. *270*, 4133–4137.

Donze, O., Dostie, J., and Sonenberg, N. (1999). Regulatable expression of the interferon-induced double-stranded RNA dependent protein kinase PKR induces apoptosis and fas receptor expression. Virology *256*, 322–329.

Doria, M., Neri, F., Gallo, A., Farace, M.G., and Michienzi, A. (2009). Editing of HIV-1 RNA by the double-stranded RNA deaminase ADAR1 stimulates viral infection. Nucleic Acids Res *37*, 5848–5858.

Dreiding, P., Staeheli, P., and Haller, O. (1985). Interferon-induced protein Mx accumulates in nuclei of mouse cells expressing resistance to influenza viruses. Virology *140*, 192–196.

Dube, M., Roy, B.B., Guiot-Guillain, P., Mercier, J., Binette, J., Leung, G., and Cohen, E.A. (2009). Suppression of Tetherin-restricting activity upon human immunodeficiency virus type 1 particle release correlates with localization of Vpu in the trans-Golgi network. J. Virol. *83*, 4574–4590.

Duerst, R.J., and Morrison, L.A. (2007). Herpes simplex virus type 2-mediated disease is reduced in mice lacking RNase L. Virology *360*, 322–328.

Duschene, K.S., and Broderick, J.B. (2010). The antiviral protein viperin is a radical SAM enzyme. FEBS Lett. *584*, 1263–1267.

El-Tanani, M.K., Jin, D., Campbell, F.C., and Johnston, P.G. (2010). Interferon-induced transmembrane 3 binds osteopontin *in vitro*: expressed *in vivo* IFITM3 reduced OPN expression. Oncogene *29*, 752–762.

Eldin, P., Papon, L., Oteiza, A., Brocchi, E., Lawson, T.G., and Mechti, N. (2009). TRIM22 E3 ubiquitin ligase activity is required to mediate antiviral activity against encephalomyocarditis virus.J. Gen. Virol. *90*, 536–545.

Eskildsen, S., Justesen, J., Schierup, M.H., and Hartmann, R. (2003). Characterization of the 2′-5′-oligoadenylate synthetase ubiquitin-like family. Nucleic Acids Res *31*, 3166–3173.

Espert, L., Degols, G., Gongora, C., Blondel, D., Williams, B.R., Silverman, R.H., and Mechti, N. (2003). ISG20, a new interferon-induced RNase specific for single-stranded RNA, defines an alternative antiviral pathway against RNA genomic viruses. J. Biol. Chem. *278*, 16151–16158.

Espinosa, A., Dardalhon, V., Brauner, S., Ambrosi, A., Higgs, R., Quintana, F.J., Sjostrand, M., Eloranta, M.L., Ni Gabhann, J., Winqvist, O., *et al.* (2009). Loss of the lupus autoantigen Ro52/Trim21 induces tissue inflammation and systemic autoimmunity by disregulating the IL-23-Th17 pathway. J. Exp. Med. *206*, 1661–1671.

Evans, S.S., Lee, D.B., Han, T., Tomasi, T.B., and Evans, R.L. (1990). Monoclonal antibody to the interferon-inducible protein Leu-13 triggers aggregation and inhibits proliferation of leukemic B cells. Blood *76*, 2583–2593.

Everett, R.D., and Chelbi-Alix, M.K. (2007). PML and PML nuclear bodies: implications in antiviral defence. Biochimie *89*, 819–830.

Feng, G.S., Chong, K., Kumar, A., and Williams, B.R. (1992). Identification of double-stranded RNA-binding domains in the interferon-induced double-stranded RNA-activated p68 kinase. Proc. Natl. Acad. Sci. U.S.A. *89*, 5447–5451.

Fensterl, V., White, C.L., Yamashita, M., and Sen, G.C. (2008). Novel characteristics of the function and induction of murine p56 family proteins. J. Virol. *82*, 11045–11053.

Fink, J., Gu, F., Ling, L., Tolfvenstam, T., Olfat, F., Chin, K.C., Aw, P., George, J., Kuznetsov, V.A., Schreiber, M., *et al.* (2007). Host gene expression profiling of dengue virus infection in cell lines and patients. PLoS Negl Trop Dis *1*, e86.

Fisher, A.G., Ensoli, B., Ivanoff, L., Chamberlain, M., Petteway, S., Ratner, L., Gallo, R.C., and Wong-Staal, F. (1987). The sor gene of HIV-1 is required for efficient virus transmission *in vitro*. Science *237*, 888–893.

Fitzpatrick, K., Skasko, M., Deerinck, T.J., Crum, J., Ellisman, M.H., and Guatelli, J. (2010). Direct restriction of virus release and incorporation of the interferon-induced protein BST-2 into HIV-1 particles. PLoS Pathog. *6*, e1000701.

Flenniken, A.M., Galabru, J., Rutherford, M.N., Hovanessian, A.G., and Williams, B.R. (1988). Expression of interferon-induced genes in different tissues of mice. J. Virol. *62*, 3077–3083.

Flodstrom-Tullberg, M., Hultcrantz, M., Stotland, A., Maday, A., Tsai, D., Fine, C., Williams, B., Silverman, R., and Sarvetnick, N. (2005). RNase L and double-stranded RNA-dependent protein kinase exert complementary roles in islet cell defense during coxsackievirus infection. J. Immunol. *174*, 1171–1177.

Floyd-Smith, G., Slattery, E., and Lengyel, P. (1981). Interferon action: RNA cleavage pattern of a (2′-5′) oligoadenylate--dependent endonuclease. Science *212*, 1030–1032.

Frank, M.B., Itoh, K., Fujisaku, A., Pontarotti, P., Mattei, M.G., and Neas, B.R. (1993). The mapping of the human 52-kD Ro/SSA autoantigen gene to human chromosome 11, and its polymorphisms. Am. J. Hum. Genet. *52*, 183–191.

Frias-Staheli, N., Giannakopoulos, N.V., Kikkert, M., Taylor, S.L., Bridgen, A., Paragas, J., Richt, J.A., Rowland, R.R., Schmaljohn, C.S., Lenschow, D.J., *et al.* (2007). Ovarian tumor domain-containing viral proteases evade ubiquitin- and ISG15-dependent innate immune responses. Cell Host Microbe *2*, 404–416.

Friedman, R.L., Manly, S.P., McMahon, M., Kerr, I.M., and Stark, G.R. (1984). Transcriptional and posttranscriptional regulation of interferon-induced gene expression in human cells. Cell *38*, 745–755.

Gack, M.U., Albrecht, R.A., Urano, T., Inn, K.S., Huang, I.C., Carnero, E., Farzan, M., Inoue, S., Jung, J.U., and Garcia-Sastre, A. (2009). Influenza A virus NS1 targets the ubiquitin ligase TRIM25 to evade recognition by the host viral RNA sensor RIG-I. Cell Host Microbe *5*, 439–449.

Gack, M.U., Shin, Y.C., Joo, C.H., Urano, T., Liang, C., Sun, L., Takeuchi, O., Akira, S., Chen, Z., Inoue, S., *et al.* (2007). TRIM25 RING-finger E3 ubiquitin ligase is essential for RIG-I-mediated antiviral activity. Nature *446*, 916–920.

Galabru, J., and Hovanessian, A. (1987). Autophosphorylation of the protein kinase dependent on double-stranded RNA. J. Biol. Chem. *262*, 15538–15544.

Gandy, S.Z., Linnstaedt, S.D., Muralidhar, S., Cashman, K.A., Rosenthal, L.J., and Casey, J.L. (2007). RNA editing of the human herpesvirus 8 kaposin transcript eliminates its transforming activity and is induced during lytic replication. J. Virol. *81*, 13544–13551.

Gao, B., Duan, Z., Xu, W., and Xiong, S. (2009). Tripartite motif-containing 22 inhibits the activity of hepatitis B virus core promoter, which is dependent on nuclear-located RING domain. Hepatology *50*, 424–433.

Gao, G., Guo, X., and Goff, S.P. (2002). Inhibition of retroviral RNA production by ZAP, a CCCH-type zinc finger protein. Science *297*, 1703–1706.

Garaigorta, U., and Chisari, F.V. (2009). Hepatitis C virus blocks interferon effector function by inducing protein kinase R phosphorylation. Cell Host Microbe *6*, 513–522.

George, C.X., Li, Z., Okonski, K.M., Toth, A.M., Wang, Y., and Samuel, C.E. (2009). Tipping the balance: antagonism of PKR kinase and ADAR1 deaminase functions by virus gene products. J. Interferon Cytokine Res. *29*, 477–487.

George, C.X., and Samuel, C.E. (1999a). Characterization of the 5′-flanking region of the human RNA-specific adenosine deaminase ADAR1 gene and identification of an interferon-inducible ADAR1 promoter. Gene *229*, 203–213.

George, C.X., and Samuel, C.E. (1999b). Human RNA-specific adenosine deaminase ADAR1 transcripts possess alternative exon 1 structures that initiate from different promoters, one constitutively active and the other interferon inducible. Proc. Natl. Acad. Sci. U.S.A. *96*, 4621–4626.

Giannakopoulos, N.V., Arutyunova, E., Lai, C., Lenschow, D.J., Haas, A.L., and Virgin, H.W. (2009). ISG15 Arg151 and the ISG15-conjugating enzyme UbE1L are important for innate immune control of Sindbis virus. J. Virol. *83*, 1602–1610.

Giannakopoulos, N.V., Luo, J.K., Papov, V., Zou, W., Lenschow, D.J., Jacobs, B.S., Borden, E.C., Li, J., Virgin, H.W., and Zhang, D.E. (2005). Proteomic identification of proteins conjugated to ISG15 in mouse and human cells. Biochem. Biophys. Res. Commun. *336*, 496–506.

Gil, J., Alcami, J., and Esteban, M. (1999). Induction of apoptosis by double-stranded-RNA-dependent protein kinase (PKR) involves the alpha subunit of eukaryotic translation initiation factor 2 and NF-kappaB. Mol. Cell. Biol. *19*, 4653–4663.

Gil, J., and Esteban, M. (2000a). Induction of apoptosis by the dsRNA-dependent protein kinase (PKR): mechanism of action. Apoptosis *5*, 107–114.

Gil, J., and Esteban, M. (2000b). The interferon-induced protein kinase (PKR), triggers apoptosis through FADD-mediated activation of caspase 8 in a manner independent of Fas and TNF-alpha receptors. Oncogene *19*, 3665–3674.

Gjermandsen, I.M., Justesen, J., and Martensen, P.M. (2000). The interferon-induced gene ISG12 is regulated by various cytokines as the gene 6–16 in human cell lines. Cytokine *12*, 233–238.

Gongora, C., David, G., Pintard, L., Tissot, C., Hua, T.D., Dejean, A., and Mechti, N. (1997). Molecular cloning of a new interferon-induced PML nuclear body-associated protein. J. Biol. Chem. *272*, 19457–19463.

Gorchakov, R., Frolova, E., Williams, B.R., Rice, C.M., and Frolov, I. (2004). PKR-dependent and -independent mechanisms are involved in translational shutoff during Sindbis virus infection. J. Virol. *78*, 8455–8467.

Guerra, S., Caceres, A., Knobeloch, K.P., Horak, I., and Esteban, M. (2008). Vaccinia virus E3 protein prevents the antiviral action of ISG15. PLoS Pathog. *4*, e1000096.

Guo, F., Cen, S., Niu, M., Saadatmand, J., and Kleiman, L. (2006). Inhibition of formula-primed reverse transcription by human APOBEC3G during human immunodeficiency virus type 1 replication. J. Virol. *80*, 11710–11722.

Guo, J., Hui, D.J., Merrick, W.C., and Sen, G.C. (2000). A new pathway of translational regulation mediated by eukaryotic initiation factor 3. EMBO J. *19*, 6891–6899.

Guo, X., Carroll, J.W., Macdonald, M.R., Goff, S.P., and Gao, G. (2004). The zinc finger antiviral protein directly binds to specific viral mRNAs through the CCCH zinc finger motifs. J. Virol. *78*, 12781–12787.

Guo, X., Ma, J., Sun, J., and Gao, G. (2007). The zinc-finger antiviral protein recruits the RNA processing exosome to degrade the target mRNA. Proc. Natl. Acad. Sci. U.S.A. *104*, 151–156.

Gupta, R.K., Hue, S., Schaller, T., Verschoor, E., Pillay, D., and Towers, G.J. (2009). Mutation of a single residue renders human tetherin resistant to HIV-1 Vpu-mediated depletion. PLoS Pathog. *5*, e1000443.

Haas, A.L., Ahrens, P., Bright, P.M., and Ankel, H. (1987). Interferon induces a 15-kilodalton protein exhibiting marked homology to ubiquitin. J. Biol. Chem. *262*, 11315–11323.

Hajjar, A.M., and Linial, M.L. (1995). Modification of retroviral RNA by double-stranded RNA adenosine deaminase. J. Virol. *69*, 5878–5882.

Halford, W.P., Weisend, C., Grace, J., Soboleski, M., Carr, D.J., Balliet, J.W., Imai, Y., Margolis, T.P., and Gebhardt, B.M. (2006). ICP0 antagonizes Stat 1-dependent repression of herpes simplex virus: implications for the regulation of viral latency. Virol. J. 3, 44.

Haller, O., Arnheiter, H., Gresser, I., and Lindenmann, J. (1979). Genetically determined, interferon-dependent resistance to influenza virus in mice. J. Exp. Med. 149, 601–612.

Haller, O., Frese, M., Rost, D., Nuttall, P.A., and Kochs, G. (1995). Tick-borne thogoto virus infection in mice is inhibited by the orthomyxovirus resistance gene product Mx1. J. Virol. 69, 2596–2601.

Harris, R.S., Bishop, K.N., Sheehy, A.M., Craig, H.M., Petersen-Mahrt, S.K., Watt, I.N., Neuberger, M.S., and Malim, M.H. (2003). DNA deamination mediates innate immunity to retroviral infection. Cell 113, 803–809.

Hartmann, R., Justesen, J., Sarkar, S.N., Sen, G.C., and Yee, V.C. (2003). Crystal structure of the 2′-specific and double-stranded RNA-activated interferon-induced antiviral protein 2′-5′-oligoadenylate synthetase. Mol Cell 12, 1173–1185.

Hartmann, R., Olsen, H.S., Widder, S., Jorgensen, R., and Justesen, J. (1998). p59OASL, a 2′-5′ oligoadenylate synthetase like protein: a novel human gene related to the 2′-5′ oligoadenylate synthetase family. Nucleic Acids Res. 26, 4121–4128.

He, B., Gross, M., and Roizman, B. (1997). The gamma(1)34.5 protein of herpes simplex virus 1 complexes with protein phosphatase 1alpha to dephosphorylate the alpha subunit of the eukaryotic translation initiation factor 2 and preclude the shutoff of protein synthesis by double-stranded RNA-activated protein kinase. Proc. Natl. Acad. Sci. U.S.A. 94, 843–848.

Hefti, H.P., Frese, M., Landis, H., Di Paolo, C., Aguzzi, A., Haller, O., and Pavlovic, J. (1999). Human MxA protein protects mice lacking a functional alpha/beta interferon system against La crosse virus and other lethal viral infections. J. Virol. 73, 6984–6991.

Helbig, K.J., Lau, D.T., Semendric, L., Harley, H.A., and Beard, M.R. (2005). Analysis of ISG expression in chronic hepatitis C identifies viperin as a potential antiviral effector. Hepatology 42, 702–710.

Higgs, R., Ni Gabhann, J., Ben Larbi, N., Breen, E.P., Fitzgerald, K.A., and Jefferies, C.A. (2008). The E3 ubiquitin ligase Ro52 negatively regulates IFN-beta production post-pathogen recognition by polyubiquitin-mediated degradation of IRF3. J. Immunol. 181, 1780–1786.

Himathongkham, S., and Luciw, P.A. (1996). Restriction of HIV-1 (subtype B) replication at the entry step in rhesus macaque cells. Virology 219, 485–488.

Hinson, E.R., and Cresswell, P. (2009a). The antiviral protein, viperin, localizes to lipid droplets via its N-terminal amphipathic alpha-helix. Proc. Natl. Acad. Sci. U.S.A. 106, 20452–20457.

Hinson, E.R., and Cresswell, P. (2009b). The N-terminal amphipathic alpha-helix of viperin mediates localization to the cytosolic face of the endoplasmic reticulum and inhibits protein secretion. J. Biol. Chem. 284, 4705–4712.

Horisberger, M.A., Wathelet, M., Szpirer, J., Szpirer, C., Islam, Q., Levan, G., Huez, G., and Content, J. (1988). cDNA cloning and assignment to chromosome 21 of IFI-78K gene, the human equivalent of murine Mx gene. Somat Cell Mol Genet 14, 123–131.

Hovanessian, A.G., Wood, J., Meurs, E., and Montagnier, L. (1979). Increased nuclease activity in cells treated with pppA2′p5′A2′p5′ A. Proc. Natl. Acad. Sci. U.S.A. 76, 3261–3265.

Hovanessian, A.G., and Wood, J.N. (1980). Anticellular and antiviral effects of pppA(2′p5′A)n. Virology 101, 81–90.

Hovnanian, A., Rebouillat, D., Levy, E.R., Mattei, M.G., and Hovanessian, A.G. (1999). The human 2′,5′-oligoadenylate synthetase-like gene (OASL) encoding the interferon-induced 56-kDa protein maps to chromosome 12q24.2 in the proximity of the 2′,5′-OAS locus. Genomics 56, 362–363.

Hovnanian, A., Rebouillat, D., Mattei, M.G., Levy, E.R., Marie, I., Monaco, A.P., and Hovanessian, A.G. (1998). The human 2′,5′-oligoadenylate synthetase locus is composed of three distinct genes clustered on chromosome 12q24.2 encoding the 100-, 69-, and 40-kDa forms. Genomics 52, 267–277.

Hsiao, N.W., Chen, J.W., Yang, T.C., Orloff, G.M., Wu, Y.Y., Lai, C.H., Lan, Y.C., and Lin, C.W. (2010). ISG15 over-expression inhibits replication of the Japanese encephalitis virus in human medulloblastoma cells. Antiviral Res 85, 504–511.

Hui, D.J., Bhasker, C.R., Merrick, W.C., and Sen, G.C. (2003). Viral stress-inducible protein p56 inhibits translation by blocking the interaction of eIF3 with the ternary complex eIF2.GTP.Met-tRNAi. J. Biol. Chem. 278, 39477–39482.

Hui, D.J., Terenzi, F., Merrick, W.C., and Sen, G.C. (2005). Mouse p56 blocks a distinct function of eukaryotic initiation factor 3 in translation initiation. J. Biol. Chem. 280, 3433–3440.

Iki, S., Yokota, S., Okabayashi, T., Yokosawa, N., Nagata, K., and Fujii, N. (2005). Serum-dependent expression of promyelocytic leukemia protein suppresses propagation of influenza virus. Virology 343, 106–115.

Ireland, D.D., Stohlman, S.A., Hinton, D.R., Kapil, P., Silverman, R.H., Atkinson, R.A., and Bergmann, C.C. (2009). RNase L mediated protection from virus induced demyelination. PLoS Pathog. 5, e1000602.

Ishikawa, H., and Barber, G.N. (2008). STING is an endoplasmic reticulum adaptor that facilitates innate immune signalling. Nature 455, 674–678.

Ishov, A.M., Sotnikov, A.G., Negorev, D., Vladimirova, O.V., Neff, N., Kamitani, T., Yeh, E.T., Strauss, J.F., 3rd, and Maul, G.G. (1999). PML is critical for ND10 formation and recruits the PML-interacting protein daxx to this nuclear structure when modified by SUMO-1. J Cell Biol 147, 221–234.

Jarmuz, A., Chester, A., Bayliss, J., Gisbourne, J., Dunham, I., Scott, J., and Navaratnam, N. (2002). An anthropoid-specific locus of orphan C to U RNA-editing enzymes on chromosome 22. Genomics 79, 285–296.

Javanbakht, H., Diaz-Griffero, F., Stremlau, M., Si, Z., and Sodroski, J. (2005). The contribution of RING and

B-Box 2 domains to retroviral restriction mediated by monkey TRIM5alpha. J. Biol. Chem. *280*, 26933–26940.

Javanbakht, H., Yuan, W., Yeung, D.F., Song, B., Diaz-Griffero, F., Li, Y., Li, X., Stremlau, M., and Sodroski, J. (2006). Characterization of TRIM5alpha trimerization and its contribution to human immunodeficiency virus capsid binding. Virology *353*, 234–246.

Jayan, G.C., and Casey, J.L. (2002a). Increased RNA editing and inhibition of hepatitis delta virus replication by high-level expression of ADAR1 and ADAR2. J. Virol. *76*, 3819–3827.

Jayan, G.C., and Casey, J.L. (2002b). Inhibition of hepatitis delta virus RNA editing by short inhibitory RNA-mediated knockdown of ADAR1 but not ADAR2 expression. J. Virol. *76*, 12399–12404.

Jensen, K., Shiels, C., and Freemont, P.S. (2001). PML protein isoforms and the RBCC/TRIM motif. Oncogene *20*, 7223–7233.

Jeon, Y.J., Choi, J.S., Lee, J.Y., Yu, K.R., Ka, S.H., Cho, Y., Choi, E.J., Baek, S.H., Seol, J.H., Park, D., *et al.* (2008). Filamin B serves as a molecular scaffold for type I interferon-induced c-Jun NH2-terminal kinase signaling pathway. Mol. Biol. Cell *19*, 5116–5130.

Jeon, Y.J., Choi, J.S., Lee, J.Y., Yu, K.R., Kim, S.M., Ka, S.H., Oh, K.H., Kim, K.I., Zhang, D.E., Bang, O.S., *et al.* (2009). ISG15 modification of filamin B negatively regulates the type I interferon-induced JNK signalling pathway. EMBO Rep. *10*, 374–380.

Jia, B., Serra-Moreno, R., Neidermyer, W., Rahmberg, A., Mackey, J., Fofana, I.B., Johnson, W.E., Westmoreland, S., and Evans, D.T. (2009). Species-specific activity of SIV Nef and HIV-1 Vpu in overcoming restriction by tetherin/BST2. PLoS Pathog. *5*, e1000429.

Jiang, D., Guo, H., Xu, C., Chang, J., Gu, B., Wang, L., Block, T.M., and Guo, J.T. (2008). Identification of three interferon-inducible cellular enzymes that inhibit the replication of hepatitis C virus. J. Virol. *82*, 1665–1678.

Jin, H.K., Takada, A., Kon, Y., Haller, O., and Watanabe, T. (1999). Identification of the murine Mx2 gene: interferon-induced expression of the Mx2 protein from the feral mouse gene confers resistance to vesicular stomatitis virus. J. Virol. *73*, 4925–4930.

Jouvenet, N., Neil, S.J., Zhadina, M., Zang, T., Kratovac, Z., Lee, Y., McNatt, M., Hatziioannou, T., and Bieniasz, P.D. (2009). Broad-spectrum inhibition of retroviral and filoviral particle release by tetherin. J. Virol. *83*, 1837–1844.

Kakuta, S., Shibata, S., and Iwakura, Y. (2002). Genomic structure of the mouse 2′,5′-oligoadenylate synthetase gene family. J. Interferon Cytokine Res. *22*, 981–993.

Kaletsky, R.L., Francica, J.R., Agrawal-Gamse, C., and Bates, P. (2009). Tetherin-mediated restriction of filovirus budding is antagonized by the Ebola glycoprotein. Proc. Natl. Acad. Sci. U.S.A. *106*, 2886–2891.

Kastner, D.L., Aksentijevich, I., and Goldbach-Mansky, R. (2010). Autoinflammatory disease reloaded: a clinical perspective. Cell *140*, 784–790.

Kerr, I.M., and Brown, R.E. (1978). pppA2′p5′A2′p5′A: an inhibitor of protein synthesis synthesized with an enzyme fraction from interferon-treated cells. Proc. Natl. Acad. Sci. U.S.A. *75*, 256–260.

Khabar, K.S., Siddiqui, Y.M., al-Zoghaibi, F., al-Haj, L., Dhalla, M., Zhou, A., Dong, B., Whitmore, M., Paranjape, J., Al-Ahdal, M.N., *et al.* (2003). RNase L mediates transient control of the interferon response through modulation of the double-stranded RNA-dependent protein kinase PKR. J. Biol. Chem. *278*, 20124–20132.

Khan, M.A., Kao, S., Miyagi, E., Takeuchi, H., Goila-Gaur, R., Opi, S., Gipson, C.L., Parslow, T.G., Ly, H., and Strebel, K. (2005). Viral RNA is required for the association of APOBEC3G with human immunodeficiency virus type 1 nucleoprotein complexes. J. Virol. *79*, 5870–5874.

Khanna, K.K., An, S.J., Wu, J.M., Landolfo, S., and Hovanessian, A.G. (1991). Absence of the 40-kDa form of (2′-5′)oligoadenylate synthetase and its corresponding mRNA from skin fibroblasts derived from Alzheimer disease patients. Proc. Natl. Acad. Sci. U.S.A. *88*, 5852–5856.

Kim, K.I., Giannakopoulos, N.V., Virgin, H.W., and Zhang, D.E. (2004). Interferon-inducible ubiquitin E2, Ubc8, is a conjugating enzyme for protein ISGylation. Mol. Cell. Biol. *24*, 9592–9600.

Kim, U., Garner, T.L., Sanford, T., Speicher, D., Murray, J.M., and Nishikura, K. (1994a). Purification and characterization of double-stranded RNA adenosine deaminase from bovine nuclear extracts. J. Biol. Chem. *269*, 13480–13489.

Kim, U., Wang, Y., Sanford, T., Zeng, Y., and Nishikura, K. (1994b). Molecular cloning of cDNA for double-stranded RNA adenosine deaminase, a candidate enzyme for nuclear RNA editing. Proc. Natl. Acad. Sci. U.S.A. *91*, 11457–11461.

Klimkait, T., Strebel, K., Hoggan, M.D., Martin, M.A., and Orenstein, J.M. (1990). The human immunodeficiency virus type 1-specific protein vpu is required for efficient virus maturation and release. J. Virol. *64*, 621–629.

Knight, E., Jr., Fahey, D., Cordova, B., Hillman, M., Kutny, R., Reich, N., and Blomstrom, D. (1988). A 15-kDa interferon-induced protein is derived by COOH-terminal processing of a 17-kDa precursor. J. Biol. Chem. *263*, 4520–4522.

Knight, M., Cayley, P.J., Silverman, R.H., Wreschner, D.H., Gilbert, C.S., Brown, R.E., and Kerr, I.M. (1980). Radioimmune, radiobinding and HPLC analysis of 2–5A and related oligonucleotides from intact cells. Nature *288*, 189–192.

Kochs, G., and Haller, O. (1999a). GTP-bound human MxA protein interacts with the nucleocapsids of Thogoto virus (Orthomyxoviridae). J. Biol. Chem. *274*, 4370–4376.

Kochs, G., and Haller, O. (1999b). Interferon-induced human MxA GTPase blocks nuclear import of Thogoto virus nucleocapsids. Proc. Natl. Acad. Sci. U.S.A. *96*, 2082–2086.

Kochs, G., Janzen, C., Hohenberg, H., and Haller, O. (2002). Antivirally active MxA protein sequesters La

Crosse virus nucleocapsid protein into perinuclear complexes. Proc. Natl. Acad. Sci. U.S.A. *99*, 3153–3158.

Korant, B.D., Blomstrom, D.C., Jonak, G.J., and Knight, E., Jr. (1984). Interferon-induced proteins. Purification and characterization of a 15,000-dalton protein from human and bovine cells induced by interferon. J. Biol. Chem. *259*, 14835–14839.

Krug, R.M., Shaw, M., Broni, B., Shapiro, G., and Haller, O. (1985). Inhibition of influenza viral mRNA synthesis in cells expressing the interferon-induced Mx gene product. J. Virol. *56*, 201–206.

Kuhen, K.L., and Samuel, C.E. (1997). Isolation of the interferon-inducible RNA-dependent protein kinase Pkr promoter and identification of a novel DNA element within the 5′-flanking region of human and mouse Pkr genes. Virology *227*, 119–130.

Kumar, A., Yang, Y.L., Flati, V., Der, S., Kadereit, S., Deb, A., Haque, J., Reis, L., Weissmann, C., and Williams, B.R. (1997). Deficient cytokine signaling in mouse embryo fibroblasts with a targeted deletion in the PKR gene: role of IRF-1 and NF-kappaB. EMBO J. *16*, 406–416.

Kumar, M., and Carmichael, G.G. (1997). Nuclear antisense RNA induces extensive adenosine modifications and nuclear retention of target transcripts. Proc. Natl. Acad. Sci. U.S.A. *94*, 3542–3547.

Kupzig, S., Korolchuk, V., Rollason, R., Sugden, A., Wilde, A., and Banting, G. (2003). Bst-2/HM1.24 is a raft-associated apical membrane protein with an unusual topology. Traffic *4*, 694–709.

Labrada, L., Liang, X.H., Zheng, W., Johnston, C., and Levine, B. (2002). Age-dependent resistance to lethal alphavirus encephalitis in mice: analysis of gene expression in the central nervous system and identification of a novel interferon-inducible protective gene, mouse ISG12. J. Virol. *76*, 11688–11703.

Lai, C., Struckhoff, J.J., Schneider, J., Martinez-Sobrido, L., Wolff, T., Garcia-Sastre, A., Zhang, D.E., and Lenschow, D.J. (2009). Mice lacking the ISG15 E1 enzyme UbE1L demonstrate increased susceptibility to both mouse-adapted and non-mouse-adapted influenza B virus infection. J. Virol. *83*, 1147–1151.

Lai, F., Drakas, R., and Nishikura, K. (1995). Mutagenic analysis of double-stranded RNA adenosine deaminase, a candidate enzyme for RNA editing of glutamate-gated ion channel transcripts. J. Biol. Chem. *270*, 17098–17105.

Lange, U.C., Adams, D.J., Lee, C., Barton, S., Schneider, R., Bradley, A., and Surani, M.A. (2008). Normal germ line establishment in mice carrying a deletion of the Ifitm/Fragilis gene family cluster. Mol. Cell. Biol. *28*, 4688–4696.

Langland, J.O., Kao, P.N., and Jacobs, B.L. (1999). Nuclear factor-90 of activated T-cells: a double-stranded RNA-binding protein and substrate for the double-stranded RNA-dependent protein kinase, PKR. Biochemistry *38*, 6361–6368.

Le Tortorec, A., and Neil, S.J. (2009). Antagonism to and intracellular sequestration of human tetherin by the human immunodeficiency virus type 2 envelope glycoprotein. J. Virol. *83*, 11966–11978.

Lecossier, D., Bouchonnet, F., Clavel, F., and Hance, A.J. (2003). Hypermutation of HIV-1 DNA in the absence of the Vif protein. Science *300*, 1112.

Lee, S.B., and Esteban, M. (1994). The interferon-induced double-stranded RNA-activated protein kinase induces apoptosis. Virology *199*, 491–496.

Lei, M., Liu, Y., and Samuel, C.E. (1998). Adenovirus VAI RNA antagonizes the RNA-editing activity of the ADAR adenosine deaminase. Virology *245*, 188–196.

Leib, D.A., Machalek, M.A., Williams, B.R., Silverman, R.H., and Virgin, H.W. (2000). Specific phenotypic restoration of an attenuated virus by knockout of a host resistance gene. Proc. Natl. Acad. Sci. U.S.A. *97*, 6097–6101.

Lenschow, D.J., Giannakopoulos, N.V., Gunn, L.J., Johnston, C., O'Guin, A.K., Schmidt, R.E., Levine, B., and Virgin, H.W.t. (2005). Identification of interferon-stimulated gene 15 as an antiviral molecule during Sindbis virus infection *in vivo*. J. Virol. *79*, 13974–13983.

Lenschow, D.J., Lai, C., Frias-Staheli, N., Giannakopoulos, N.V., Lutz, A., Wolff, T., Osiak, A., Levine, B., Schmidt, R.E., Garcia-Sastre, A., *et al.* (2007). IFN-stimulated gene 15 functions as a critical antiviral molecule against influenza, herpes, and Sindbis viruses. Proc. Natl. Acad. Sci. U.S.A. *104*, 1371–1376.

Levy, D., Larner, A., Chaudhuri, A., Babiss, L.E., and Darnell, J.E., Jr. (1986). Interferon-stimulated transcription: isolation of an inducible gene and identification of its regulatory region. Proc. Natl. Acad. Sci. U.S.A. *83*, 8929–8933.

Li, X., and Sodroski, J. (2008). The TRIM5alpha B-Box 2 domain promotes cooperative binding to the retroviral capsid by mediating higher-order self-association. J. Virol. *82*, 11495–11502.

Li, X.L., Blackford, J.A., and Hassel, B.A. (1998). RNase L mediates the antiviral effect of interferon through a selective reduction in viral RNA during encephalomyocarditis virus infection. J. Virol. *72*, 2752–2759.

Li, Y., Li, C., Xue, P., Zhong, B., Mao, A.P., Ran, Y., Chen, H., Wang, Y.Y., Yang, F., and Shu, H.B. (2009). ISG56 is a negative-feedback regulator of virus-triggered signaling and cellular antiviral response. Proc. Natl. Acad. Sci. U.S.A. *106*, 7945–7950.

Li, Z., Wolff, K.C., and Samuel, C.E. (2010). RNA adenosine deaminase ADAR1 deficiency leads to increased activation of protein kinase PKR and reduced vesicular stomatitis virus growth following interferon treatment. Virology *396*, 316–322.

Liddament, M.T., Brown, W.L., Schumacher, A.J., and Harris, R.S. (2004). APOBEC3F properties and hypermutation preferences indicate activity against HIV-1 *in vivo*. Curr Biol *14*, 1385–1391.

Lim, J.K., Lisco, A., McDermott, D.H., Huynh, L., Ward, J.M., Johnson, B., Johnson, H., Pape, J., Foster, G.A., Krysztof, D., *et al.* (2009). Genetic variation in OAS1 is a risk factor for initial infection with West Nile virus in man. PLoS Pathog. *5*, e1000321.

Lindenmann, J., Lane, C.A., and Hobson, D. (1963). The resistance of A2g mice to myxoviruses. J. Immunol. *90*, 942–951.

Liu, M., Li, X.L., and Hassel, B.A. (2003). Proteasomes modulate conjugation to the ubiquitin-like protein, ISG15. J. Biol. Chem. *278*, 1594–1602.

Liu, Y., George, C.X., Patterson, J.B., and Samuel, C.E. (1997). Functionally distinct double-stranded RNA-binding domains associated with alternative splice site variants of the interferon-inducible double-stranded RNA-specific adenosine deaminase. J. Biol. Chem. *272*, 4419–4428.

Liu, Y., and Samuel, C.E. (1996). Mechanism of interferon action: functionally distinct RNA-binding and catalytic domains in the interferon-inducible, double-stranded RNA-specific adenosine deaminase. J. Virol. *70*, 1961–1968.

Liu, Y., Wolff, K.C., Jacobs, B.L., and Samuel, C.E. (2001). Vaccinia virus E3L interferon resistance protein inhibits the interferon-induced adenosine deaminase A-to-I editing activity. Virology *289*, 378–387.

Loeb, K.R., and Haas, A.L. (1992). The interferon-inducible 15-kDa ubiquitin homolog conjugates to intracellular proteins. J. Biol. Chem. *267*, 7806–7813.

Lu, J., O'Hara, E.B., Trieselmann, B.A., Romano, P.R., and Dever, T.E. (1999). The interferon-induced double-stranded RNA-activated protein kinase PKR will phosphorylate serine, threonine, or tyrosine at residue 51 in eukaryotic initiation factor 2alpha. J. Biol. Chem. *274*, 32198–32203.

Lund, S., Christensen, K.V., Hedtjarn, M., Mortensen, A.L., Hagberg, H., Falsig, J., Hasseldam, H., Schrattenholz, A., Porzgen, P., and Leist, M. (2006). The dynamics of the LPS triggered inflammatory response of murine microglia under different culture and *in vivo* conditions. J Neuroimmunol *180*, 71–87.

Maas, S., Melcher, T., Herb, A., Seeburg, P.H., Keller, W., Krause, S., Higuchi, M., and O'Connell, M.A. (1996). Structural requirements for RNA editing in glutamate receptor pre-mRNAs by recombinant double-stranded RNA adenosine deaminase. J. Biol. Chem. *271*, 12221–12226.

Maitra, R.K., and Silverman, R.H. (1998). Regulation of human immunodeficiency virus replication by 2′,5′-oligoadenylate-dependent RNase L. J. Virol. *72*, 1146–1152.

Malakhov, M.P., Kim, K.I., Malakhova, O.A., Jacobs, B.S., Borden, E.C., and Zhang, D.E. (2003). High-throughput immunoblotting. Ubiquitiin-like protein ISG15 modifies key regulators of signal transduction. J. Biol. Chem. *278*, 16608–16613.

Malakhov, M.P., Malakhova, O.A., Kim, K.I., Ritchie, K.J., and Zhang, D.E. (2002). UBP43 (USP18) specifically removes ISG15 from conjugated proteins. J. Biol. Chem. *277*, 9976–9981.

Malakhova, O.A., and Zhang, D.E. (2008). ISG15 inhibits Nedd4 ubiquitin E3 activity and enhances the innate antiviral response. J. Biol. Chem. *283*, 8783–8787.

Malathi, K., Dong, B., Gale, M., Jr., and Silverman, R.H. (2007). Small self-RNA generated by RNase L amplifies antiviral innate immunity. Nature *448*, 816–819.

Mangeat, B., Turelli, P., Caron, G., Friedli, M., Perrin, L., and Trono, D. (2003). Broad antiretroviral defence by human APOBEC3G through lethal editing of nascent reverse transcripts. Nature *424*, 99–103.

Maran, A., Maitra, R.K., Kumar, A., Dong, B., Xiao, W., Li, G., Williams, B.R., Torrence, P.F., and Silverman, R.H. (1994). Blockage of NF-kappa B signaling by selective ablation of an mRNA target by 2–5A antisense chimeras. Science *265*, 789–792.

Mariani, R., Chen, D., Schrofelbauer, B., Navarro, F., Konig, R., Bollman, B., Munk, C., Nymark-McMahon, H., and Landau, N.R. (2003). Species-specific exclusion of APOBEC3G from HIV-1 virions by Vif. Cell *114*, 21–31.

Marie, I., Svab, J., Robert, N., Galabru, J., and Hovanessian, A.G. (1990). Differential expression and distinct structure of 69- and 100-kDa forms of 2–5A synthetase in human cells treated with interferon. J. Biol. Chem. *265*, 18601–18607.

Marques, J., Anwar, J., Eskildsen-Larsen, S., Rebouillat, D., Paludan, S.R., Sen, G., Williams, B.R., and Hartmann, R. (2008). The p59 oligoadenylate synthetase-like protein possesses antiviral activity that requires the C-terminal ubiquitin-like domain. J. Gen. Virol. *89*, 2767–2772.

Martensen, P.M., Sogaard, T.M., Gjermandsen, I.M., Buttenschon, H.N., Rossing, A.B., Bonnevie-Nielsen, V., Rosada, C., Simonsen, J.L., and Justesen, J. (2001). The interferon alpha induced protein ISG12 is localized to the nuclear membrane. Eur. J. Biochem. *268*, 5947–5954.

Mashimo, T., Lucas, M., Simon-Chazottes, D., Frenkiel, M.P., Montagutelli, X., Ceccaldi, P.E., Deubel, V., Guenet, J.L., and Despres, P. (2002). A nonsense mutation in the gene encoding 2′-5′-oligoadenylate synthetase/L1 isoform is associated with West Nile virus susceptibility in laboratory mice. Proc. Natl. Acad. Sci. U.S.A. *99*, 11311–11316.

Masuyama, N., Kuronita, T., Tanaka, R., Muto, T., Hirota, Y., Takigawa, A., Fujita, H., Aso, Y., Amano, J., and Tanaka, Y. (2009). HM1.24 is internalized from lipid rafts by clathrin-mediated endocytosis through interaction with alpha-adaptin. J. Biol. Chem. *284*, 15927–15941.

Matsumoto, A.K., Martin, D.R., Carter, R.H., Klickstein, L.B., Ahearn, J.M., and Fearon, D.T. (1993). Functional dissection of the CD21/CD19/TAPA-1/Leu-13 complex of B lymphocytes. J. Exp. Med. *178*, 1407–1417.

McNatt, M.W., Zang, T., Hatziioannou, T., Bartlett, M., Fofana, I.B., Johnson, W.E., Neil, S.J., and Bieniasz, P.D. (2009). Species-specific activity of HIV-1 Vpu and positive selection of tetherin transmembrane domain variants. PLoS Pathog. *5*, e1000300.

Mellor, H., Flowers, K.M., Kimball, S.R., and Jefferson, L.S. (1994). Cloning and characterization of a cDNA encoding rat PKR, the double-stranded RNA-dependent eukaryotic initiation factor-2 kinase. Biochim. Biophys. Acta *1219*, 693–696.

Meurs, E., Chong, K., Galabru, J., Thomas, N.S., Kerr, I.M., Williams, B.R., and Hovanessian, A.G. (1990). Molecular cloning and characterization of the human double-stranded RNA-activated protein kinase induced by interferon. Cell *62*, 379–390.

Mitchell, R.S., Katsura, C., Skasko, M.A., Fitzpatrick, K., Lau, D., Ruiz, A., Stephens, E.B., Margottin-Goguet, F., Benarous, R., and Guatelli, J.C. (2009). Vpu antagonizes BST-2-mediated restriction of HIV-1 release via beta-TrCP and endo-lysosomal trafficking. PLoS Pathog. *5*, e1000450.

Moser, M.J., Holley, W.R., Chatterjee, A., and Mian, I.S. (1997). The proofreading domain of Escherichia coli DNA polymerase I and other DNA and/or RNA exonuclease domains. Nucleic Acids Res. *25*, 5110–5118.

Muller, S., Moller, P., Bick, M.J., Wurr, S., Becker, S., Gunther, S., and Kummerer, B.M. (2007). Inhibition of filovirus replication by the zinc finger antiviral protein. J. Virol. *81*, 2391–2400.

Murphy, D.G., Dimock, K., and Kang, C.Y. (1991). Numerous transitions in human parainfluenza virus 3 RNA recovered from persistently infected cells. Virology *181*, 760–763.

Nakasato, N., Ikeda, K., Urano, T., Horie-Inoue, K., Takeda, S., and Inoue, S. (2006). A ubiquitin E3 ligase Efp is up-regulated by interferons and conjugated with ISG15. Biochem. Biophys. Res. Commun. *351*, 540–546.

Nakayama, E.E., and Shioda, T. (2010). Anti-retroviral activity of TRIM5 alpha. Rev. Med. Virol. *20*, 77–92.

Nakken, B., Jonsson, R., and Bolstad, A.I. (2001). Polymorphisms of the Ro52 gene associated with anti-Ro 52-kd autoantibodies in patients with primary Sjogren's syndrome. Arthritis Rheum. *44*, 638–646.

Nallagatla, S.R., Hwang, J., Toroney, R., Zheng, X., Cameron, C.E., and Bevilacqua, P.C. (2007). 5'-triphosphate-dependent activation of PKR by RNAs with short stem–loops. Science *318*, 1455–1458.

Nanduri, S., Carpick, B.W., Yang, Y., Williams, B.R., and Qin, J. (1998). Structure of the double-stranded RNA-binding domain of the protein kinase PKR reveals the molecular basis of its dsRNA-mediated activation. EMBO J. *17*, 5458–5465.

Nanduri, S., Rahman, F., Williams, B.R., and Qin, J. (2000). A dynamically tuned double-stranded RNA binding mechanism for the activation of antiviral kinase PKR. EMBO J. *19*, 5567–5574.

Narasimhan, J., Potter, J.L., and Haas, A.L. (1996). Conjugation of the 15-kDa interferon-induced ubiquitin homolog is distinct from that of ubiquitin. J. Biol. Chem. *271*, 324–330.

Narasimhan, J., Wang, M., Fu, Z., Klein, J.M., Haas, A.L., and Kim, J.J. (2005). Crystal structure of the interferon-induced ubiquitin-like protein ISG15. J. Biol. Chem. *280*, 27356–27365.

Navaratnam, N., Morrison, J.R., Bhattacharya, S., Patel, D., Funahashi, T., Giannoni, F., Teng, B.B., Davidson, N.O., and Scott, J. (1993). The p27 catalytic subunit of the apolipoprotein B mRNA editing enzyme is a cytidine deaminase. J. Biol. Chem. *268*, 20709–20712.

Navarro, F., Bollman, B., Chen, H., Konig, R., Yu, Q., Chiles, K., and Landau, N.R. (2005). Complementary function of the two catalytic domains of APOBEC3G. Virology *333*, 374–386.

Neil, S., and Bieniasz, P. (2009). Human immunodeficiency virus, restriction factors, and interferon. J. Interferon Cytokine Res. *29*, 569–580.

Neil, S.J., Zang, T., and Bieniasz, P.D. (2008). Tetherin inhibits retrovirus release and is antagonized by HIV-1 Vpu. Nature *451*, 425–430.

Newman, E.N., Holmes, R.K., Craig, H.M., Klein, K.C., Lingappa, J.R., Malim, M.H., and Sheehy, A.M. (2005). Antiviral function of APOBEC3G can be dissociated from cytidine deaminase activity. Curr. Biol. *15*, 166–170.

Nguyen, L.H., Espert, L., Mechti, N., and Wilson, D.M., 3rd (2001). The human interferon- and estrogen-regulated ISG20/HEM45 gene product degrades single-stranded RNA and DNA *in vitro*. Biochemistry *40*, 7174–7179.

Nie, Y., Hammond, G.L., and Yang, J.H. (2007). Double-stranded RNA deaminase ADAR1 increases host susceptibility to virus infection. J. Virol. *81*, 917–923.

Nisole, S., Stoye, J.P., and Saib, A. (2005). TRIM family proteins: retroviral restriction and antiviral defence. Nat. Rev. Microbiol. *3*, 799–808.

O'Hara, P.J., Nichol, S.T., Horodyski, F.M., and Holland, J.J. (1984). Vesicular stomatitis virus defective interfering particles can contain extensive genomic sequence rearrangements and base substitutions. Cell *36*, 915–924.

Ohtomo, T., Sugamata, Y., Ozaki, Y., Ono, K., Yoshimura, Y., Kawai, S., Koishihara, Y., Ozaki, S., Kosaka, M., Hirano, T., *et al.* (1999). Molecular cloning and characterization of a surface antigen preferentially overexpressed on multiple myeloma cells. Biochem. Biophys. Res. Commun. *258*, 583–591.

Okumura, A., Lu, G., Pitha-Rowe, I., and Pitha, P.M. (2006). Innate antiviral response targets HIV-1 release by the induction of ubiquitin-like protein ISG15. Proc. Natl. Acad. Sci. U.S.A. *103*, 1440–1445.

Okumura, A., Pitha, P.M., and Harty, R.N. (2008). ISG15 inhibits Ebola VP40 VLP budding in an L-domain-dependent manner by blocking Nedd4 ligase activity. Proc. Natl. Acad. Sci. U.S.A. *105*, 3974–3979.

Okumura, F., Zou, W., and Zhang, D.E. (2007). ISG15 modification of the eIF4E cognate 4EHP enhances cap structure-binding activity of 4EHP. Genes Dev *21*, 255–260.

Olofsson, P.S., Jatta, K., Wagsater, D., Gredmark, S., Hedin, U., Paulsson-Berne, G., Soderberg-Naucler, C., Hansson, G.K., and Sirsjo, A. (2005). The antiviral cytomegalovirus inducible gene 5/viperin is expressed in atherosclerosis and regulated by proinflammatory agents. Arterioscler Thromb. Vasc. Biol. *25*, e113–116.

Osiak, A., Utermohlen, O., Niendorf, S., Horak, I., and Knobeloch, K.P. (2005). ISG15, an interferon-stimulated ubiquitin-like protein, is not essential for STAT1 signaling and responses against vesicular stomatitis and lymphocytic choriomeningitis virus. Mol. Cell. Biol. *25*, 6338–6345.

Ozato, K., Shin, D.M., Chang, T.H., and Morse, H.C., 3rd (2008). TRIM family proteins and their emerging roles in innate immunity. Nat. Rev. Immunol. *8*, 849–860.

Pampin, M., Simonin, Y., Blondel, B., Percherancier, Y., and Chelbi-Alix, M.K. (2006). Cross talk between PML and p53 during poliovirus infection: implications for antiviral defense. J. Virol. *80*, 8582–8592.

Patel, C.V., Handy, I., Goldsmith, T., and Patel, R.C. (2000). PACT, a stress-modulated cellular activator of interferon-induced double-stranded RNA-activated protein kinase, PKR. J. Biol. Chem. *275*, 37993–37998.

Patel, R.C., and Sen, G.C. (1994). Characterization of the interactions between double-stranded RNA and the double-stranded RNA binding domain of the interferon induced protein kinase. Cell Mol Biol Res *40*, 671–682.

Patel, R.C., and Sen, G.C. (1998). PACT, a protein activator of the interferon-induced protein kinase, PKR. EMBO J. *17*, 4379–4390.

Patel, R.C., Stanton, P., McMillan, N.M., Williams, B.R., and Sen, G.C. (1995). The interferon-inducible double-stranded RNA-activated protein kinase self-associates *in vitro* and *in vivo*. Proc. Natl. Acad. Sci. U.S.A. *92*, 8283–8287.

Patel, R.C., Stanton, P., and Sen, G.C. (1994). Role of the amino-terminal residues of the interferon-induced protein kinase in its activation by double-stranded RNA and heparin. J. Biol. Chem. *269*, 18593–18598.

Patterson, J.B., and Samuel, C.E. (1995). Expression and regulation by interferon of a double-stranded-RNA-specific adenosine deaminase from human cells: evidence for two forms of the deaminase. Mol. Cell. Biol. *15*, 5376–5388.

Patterson, J.B., Thomis, D.C., Hans, S.L., and Samuel, C.E. (1995). Mechanism of interferon action: double-stranded RNA-specific adenosine deaminase from human cells is inducible by alpha and gamma interferons. Virology *210*, 508–511.

Pattyn, E., Verhee, A., Uyttendaele, I., Piessevaux, J., Timmerman, E., Gevaert, K., Vandekerckhove, J., Peelman, F., and Tavernier, J. (2008). HyperISGylation of Old World monkey ISG15 in human cells. PLoS One 3, e2427.

Pavlovic, J., Arzet, H.A., Hefti, H.P., Frese, M., Rost, D., Ernst, B., Kolb, E., Staeheli, P., and Haller, O. (1995). Enhanced virus resistance of transgenic mice expressing the human MxA protein. J. Virol. *69*, 4506–4510.

Pavlovic, J., and Staeheli, P. (1991). The antiviral potentials of Mx proteins. J Interferon Res *11*, 215–219.

Pavlovic, J., Zurcher, T., Haller, O., and Staeheli, P. (1990). Resistance to influenza virus and vesicular stomatitis virus conferred by expression of human MxA protein. J. Virol. *64*, 3370–3375.

Pentecost, B.T. (1998). Expression and estrogen regulation of the HEM45 MRNA in human tumor lines and in the rat uterus. J Steroid Biochem Mol Biol *64*, 25–33.

Perelygin, A.A., Scherbik, S.V., Zhulin, I.B., Stockman, B.M., Li, Y., and Brinton, M.A. (2002). Positional cloning of the murine flavivirus resistance gene. Proc. Natl. Acad. Sci. U.S.A. *99*, 9322–9327.

Perez-Caballero, D., Zang, T., Ebrahimi, A., McNatt, M.W., Gregory, D.A., Johnson, M.C., and Bieniasz, P.D. (2009). Tetherin inhibits HIV-1 release by directly tethering virions to cells. Cell *139*, 499–511.

Phuphuakrat, A., Kraiwong, R., Boonarkart, C., Lauhakirti, D., Lee, T.H., and Auewarakul, P. (2008). Double-stranded RNA adenosine deaminases enhance expression of human immunodeficiency virus type 1 proteins. J. Virol. *82*, 10864–10872.

Pitossi, F., Blank, A., Schroder, A., Schwarz, A., Hussi, P., Schwemmle, M., Pavlovic, J., and Staeheli, P. (1993). A functional GTP-binding motif is necessary for antiviral activity of Mx proteins. J. Virol. *67*, 6726–6732.

Ponten, A., Sick, C., Weeber, M., Haller, O., and Kochs, G. (1997). Dominant-negative mutants of human MxA protein: domains in the carboxy-terminal moiety are important for oligomerization and antiviral activity. J. Virol. *71*, 2591–2599.

Qiu, L.Q., Cresswell, P., and Chin, K.C. (2009). Viperin is required for optimal Th2 responses and T-cell receptor-mediated activation of NF-kappaB and AP-1. Blood *113*, 3520–3529.

Rajsbaum, R., Stoye, J.P., and O'Garra, A. (2008). Type I interferon-dependent and -independent expression of tripartite motif proteins in immune cells. Eur J. Immunol. *38*, 619–630.

Rasmussen, U.B., Wolf, C., Mattei, M.G., Chenard, M.P., Bellocq, J.P., Chambon, P., Rio, M.C., and Basset, P. (1993). Identification of a new interferon-alpha-inducible gene (p27) on human chromosome 14q32 and its expression in breast carcinoma. Cancer Res *53*, 4096–4101.

Rebagliati, M.R., and Melton, D.A. (1987). Antisense RNA injections in fertilized frog eggs reveal an RNA duplex unwinding activity. Cell *48*, 599–605.

Rebouillat, D., and Hovanessian, A.G. (1999). The human 2',5'-oligoadenylate synthetase family: interferon-induced proteins with unique enzymatic properties. J Interferon Cytokine Res. *19*, 295–308.

Rebouillat, D., Marie, I., and Hovanessian, A.G. (1998). Molecular cloning and characterization of two related and interferon-induced 56-kDa and 30-kDa proteins highly similar to 2'-5' oligoadenylate synthetase. Eur. J. Biochem. *257*, 319–330.

Regad, T., Saib, A., Lallemand-Breitenbach, V., Pandolfi, P.P., de The, H., and Chelbi-Alix, M.K. (2001). PML mediates the interferon-induced antiviral state against a complex retrovirus via its association with the viral transactivator. EMBO J. *20*, 3495–3505.

Richards, N., Schaner, P., Diaz, A., Stuckey, J., Shelden, E., Wadhwa, A., and Gumucio, D.L. (2001). Interaction between pyrin and the apoptotic speck protein (ASC) modulates ASC-induced apoptosis. J. Biol. Chem. *276*, 39320–39329.

Rivieccio, M.A., Suh, H.S., Zhao, Y., Zhao, M.L., Chin, K.C., Lee, S.C., and Brosnan, C.F. (2006). TLR3 ligation activates an antiviral response in human fetal astrocytes: a role for viperin/cig5. J. Immunol. *177*, 4735–4741.

Rold, C.J., and Aiken, C. (2008). Proteasomal degradation of TRIM5alpha during retrovirus restriction. PLoS Pathog. *4*, e1000074.

Rollason, R., Korolchuk, V., Hamilton, C., Schu, P., and Banting, G. (2007). Clathrin-mediated endocytosis of a lipid-raft-associated protein is mediated through a dual tyrosine motif. J. Cell Sci. *120*, 3850–3858.

Romani, B., Engelbrecht, S., and Glashoff, R.H. (2009). Antiviral roles of APOBEC proteins against HIV-1 and suppression by Vif. Arch. Virol. *154*, 1579–1588.

Romano, P.R., Garcia-Barrio, M.T., Zhang, X., Wang, Q., Taylor, D.R., Zhang, F., Herring, C., Mathews, M.B., Qin, J., and Hinnebusch, A.G. (1998). Autophosphorylation in the activation loop is required for full kinase activity *in vivo* of human and yeast eukaryotic initiation factor 2alpha kinases PKR and GCN2. Mol. Cell Biol. *18*, 2282–2297.

Russell, R.A., Wiegand, H.L., Moore, M.D., Schafer, A., McClure, M.O., and Cullen, B.R. (2005). Foamy virus Bet proteins function as novel inhibitors of the APOBEC3 family of innate antiretroviral defense factors. J. Virol. *79*, 8724–8731.

Sadler, A.J., and Williams, B.R. (2008). Interferon-inducible antiviral effectors. Nat. Rev. Immunol. *8*, 559–568.

Sakuma, R., Mael, A.A., and Ikeda, Y. (2007). Alpha interferon enhances TRIM5alpha-mediated antiviral activities in human and rhesus monkey cells. J. Virol. *81*, 10201–10206.

Sakuma, T., Sakurai, A., and Yasuda, J. (2009). Dimerization of tetherin is not essential for its antiviral activity against Lassa and Marburg viruses. PLoS One *4*, e6934.

Samuel, C.E. (2001). Antiviral actions of interferons. Clin. Microbiol. Rev. *14*, 778–809, table of contents.

Samuel, M.A., Whitby, K., Keller, B.C., Marri, A., Barchet, W., Williams, B.R., Silverman, R.H., Gale, M., Jr., and Diamond, M.S. (2006). PKR and RNase L contribute to protection against lethal West Nile Virus infection by controlling early viral spread in the periphery and replication in neurons. J. Virol. *80*, 7009–7019.

Sangster, M.Y., Heliams, D.B., MacKenzie, J.S., and Shellam, G.R. (1993). Genetic studies of flavivirus resistance in inbred strains derived from wild mice: evidence for a new resistance allele at the flavivirus resistance locus (Flv). J. Virol. *67*, 340–347.

Sangster, M.Y., Urosevic, N., Mansfield, J.P., Mackenzie, J.S., and Shellam, G.R. (1994). Mapping the Flv locus controlling resistance to flaviviruses on mouse chromosome 5. J. Virol. *68*, 448–452.

Sarkar, S.N., Bandyopadhyay, S., Ghosh, A., and Sen, G.C. (1999). Enzymatic characteristics of recombinant medium isozyme of 2′-5′ oligoadenylate synthetase. J. Biol. Chem. *274*, 1848–1855.

Sarkis, P.T., Ying, S., Xu, R., and Yu, X.F. (2006). STAT1-independent cell type-specific regulation of antiviral APOBEC3G by IFN-alpha. J. Immunol. *177*, 4530–4540.

Sauter, D., Specht, A., and Kirchhoff, F. (2010). Tetherin: holding on and letting go. Cell *141*, 392–398.

Schafer, A., Bogerd, H.P., and Cullen, B.R. (2004). Specific packaging of APOBEC3G into HIV-1 virions is mediated by the nucleocapsid domain of the gag polyprotein precursor. Virology *328*, 163–168.

Scherbik, S.V., Kluetzman, K., Perelygin, A.A., and Brinton, M.A. (2007). Knock-in of the Oas1b(r) allele into a flavivirus-induced disease susceptible mouse generates the resistant phenotype. Virology *368*, 232–237.

Schneider-Schaulies, S., Schneider-Schaulies, J., Schuster, A., Bayer, M., Pavlovic, J., and ter Meulen, V. (1994). Cell type-specific MxA-mediated inhibition of measles virus transcription in human brain cells. J. Virol. *68*, 6910–6917.

Schnorr, J.J., Schneider-Schaulies, S., Simon-Jodicke, A., Pavlovic, J., Horisberger, M.A., and ter Meulen, V. (1993). MxA-dependent inhibition of measles virus glycoprotein synthesis in a stably transfected human monocytic cell line. J. Virol. *67*, 4760–4768.

Sebastian, S., and Luban, J. (2005). TRIM5alpha selectively binds a restriction-sensitive retroviral capsid. Retrovirology *2*, 40.

Seeler, J.S., and Dejean, A. (2001). SUMO: of branched proteins and nuclear bodies. Oncogene *20*, 7243–7249.

Shaveta, G., Shi, J., Chow, V.T., and Song, J. (2010). Structural characterization reveals that viperin is a radical S-adenosyl-L-methionine (SAM) enzyme. Biochem. Biophys. Res. Commun. *391*, 1390–1395.

Sheehy, A.M., Gaddis, N.C., Choi, J.D., and Malim, M.H. (2002). Isolation of a human gene that inhibits HIV-1 infection and is suppressed by the viral Vif protein. Nature *418*, 646–650.

Sheehy, A.M., Gaddis, N.C., and Malim, M.H. (2003). The antiretroviral enzyme APOBEC3G is degraded by the proteasome in response to HIV-1 Vif. Nat Med *9*, 1404–1407.

Shi, H.X., Yang, K., Liu, X., Liu, X.Y., Wei, B., Shan, Y.F., Zhu, L.H., and Wang, C. (2010). Positive Regulation of IRF3 Activation by Herc5 via ISG15 Modification. Mol Cell Biol. *30*, 2424–2436.

Shi, M., Deng, W., Bi, E., Mao, K., Ji, Y., Lin, G., Wu, X., Tao, Z., Li, Z., Cai, X., *et al.* (2008). TRIM30 alpha negatively regulates TLR-mediated NF-kappa B activation by targeting TAB2 and TAB3 for degradation. .Nat. Immunol. *9*, 369–377.

Shibata, R., Sakai, H., Kawamura, M., Tokunaga, K., and Adachi, A. (1995). Early replication block of human immunodeficiency virus type 1 in monkey cells. J. Gen. Virol. *76*, 2723–2730.

Short, K.M., and Cox, T.C. (2006). Subclassification of the RBCC/TRIM superfamily reveals a novel motif necessary for microtubule binding. J. Biol. Chem. *281*, 8970–8980.

Silverman, R.H., Skehel, J.J., James, T.C., Wreschner, D.H., and Kerr, I.M. (1983). rRNA cleavage as an index of ppp(A2′p)nA activity in interferon-treated encephalomyocarditis virus-infected cells. J. Virol. *46*, 1051–1055.

Smith, R.A., Young, J., Weis, J.J., and Weis, J.H. (2006). Expression of the mouse fragilis gene products in immune cells and association with receptor signaling complexes. Genes Immun 7, 113–121.

Staeheli, P., Haller, O., Boll, W., Lindenmann, J., and Weissmann, C. (1986). Mx protein: constitutive expression in 3T3 cells transformed with cloned Mx cDNA confers selective resistance to influenza virus. Cell *44*, 147–158.

Stirnweiss, A., Ksienzyk, A., Klages, K., Rand, U., Grashoff, M., Hauser, H., and Kroger, A. (2010). IFN Regulatory Factor-1 Bypasses IFN-Mediated Antiviral Effects through Viperin Gene Induction. J. Immunol. *184*, 5179–5185.

Stojdl, D.F., Abraham, N., Knowles, S., Marius, R., Brasey, A., Lichty, B.D., Brown, E.G., Sonenberg, N., and Bell, J.C. (2000). The murine double-stranded RNA-dependent protein kinase PKR is required for resistance to vesicular stomatitis virus. J. Virol. *74*, 9580–9585.

Strebel, K., Daugherty, D., Clouse, K., Cohen, D., Folks, T., and Martin, M.A. (1987). The HIV 'A' (sor) gene product is essential for virus infectivity. Nature *328*, 728–730.

Strebel, K., Klimkait, T., and Martin, M.A. (1988). A novel gene of HIV-1, vpu, and its 16-kilodalton product. Science *241*, 1221–1223.

Stremlau, M., Owens, C.M., Perron, M.J., Kiessling, M., Autissier, P., and Sodroski, J. (2004). The cytoplasmic body component TRIM5alpha restricts HIV-1 infection in Old World monkeys. Nature *427*, 848–853.

Stremlau, M., Perron, M., Lee, M., Li, Y., Song, B., Javanbakht, H., Diaz-Griffero, F., Anderson, D.J., Sundquist, W.I., and Sodroski, J. (2006). Specific recognition and accelerated uncoating of retroviral capsids by the TRIM5alpha restriction factor. Proc. Natl. Acad. Sci. U.S.A. *103*, 5514–5519.

Stremlau, M., Perron, M., Welikala, S., and Sodroski, J. (2005). Species-specific variation in the B30.2(SPRY) domain of TRIM5alpha determines the potency of human immunodeficiency virus restriction. J. Virol. *79*, 3139–3145.

Suspene, R., Guetard, D., Henry, M., Sommer, P., Wain-Hobson, S., and Vartanian, J.P. (2005). Extensive editing of both hepatitis B virus DNA strands by APOBEC3 cytidine deaminases *in vitro* and *in vivo*. Proc. Natl. Acad. Sci. U.S.A. *102*, 8321–8326.

Svarovskaia, E.S., Xu, H., Mbisa, J.L., Barr, R., Gorelick, R.J., Ono, A., Freed, E.O., Hu, W.S., and Pathak, V.K. (2004). Human apolipoprotein B mRNA-editing enzyme-catalytic polypeptide-like 3G (APOBEC3G) is incorporated into HIV-1 virions through interactions with viral and nonviral RNAs. J. Biol. Chem. *279*, 35822–35828.

Taddeo, B., Luo, T.R., Zhang, W., and Roizman, B. (2003). Activation of NF-kappaB in cells productively infected with HSV-1 depends on activated protein kinase R and plays no apparent role in blocking apoptosis. Proc. Natl. Acad. Sci. U.S.A. *100*, 12408–12413.

Takahashi, S., Doss, C., Levy, S., and Levy, R. (1990). TAPA-1, the target of an antiproliferative antibody, is associated on the cell surface with the Leu-13 antigen. J. Immunol. *145*, 2207–2213.

Takeuchi, H., and Matano, T. (2008). Host factors involved in resistance to retroviral infection. Microbiol Immunol 52, 318–325.

Takeuchi, T., Inoue, S., and Yokosawa, H. (2006). Identification and Herc5-mediated ISGylation of novel target proteins. .Biochem. Biophys. Res. Commun. *348*, 473–477.

Takeuchi, T., Iwahara, S., Saeki, Y., Sasajima, H., and Yokosawa, H. (2005). Link between the ubiquitin conjugation system and the ISG15 conjugation system: ISG15 conjugation to the UbcH6 ubiquitin E2 enzyme. J. Biochem. *138*, 711–719.

Tanaka, H., and Samuel, C.E. (1994). Mechanism of interferon action: structure of the mouse PKR gene encoding the interferon-inducible RNA-dependent protein kinase. Proc. Natl. Acad. Sci. U.S.A. *91*, 7995–7999.

Tang, Y., Zhong, G., Zhu, L., Liu, X., Shan, Y., Feng, H., Bu, Z., Chen, H., and Wang, C. (2010). Herc5 attenuates Influenza A virus by catalyzing ISGylation of viral NS1 protein. J. Immunol. 184, 5777–5790.

Tavalai, N., Papior, P., Rechter, S., Leis, M., and Stamminger, T. (2006). Evidence for a role of the cellular ND10 protein PML in mediating intrinsic immunity against human cytomegalovirus infections. J. Virol. *80*, 8006–8018.

Taylor, D.R., Puig, M., Darnell, M.E., Mihalik, K., and Feinstone, S.M. (2005). New antiviral pathway that mediates hepatitis C virus replicon interferon sensitivity through ADAR1. J. Virol. *79*, 6291–6298.

Teng, B., Burant, C.F., and Davidson, N.O. (1993). Molecular cloning of an apolipoprotein B messenger RNA editing protein. Science *260*, 1816–1819.

Terenzi, F., Hui, D.J., Merrick, W.C., and Sen, G.C. (2006). Distinct induction patterns and functions of two closely related interferon-inducible human genes, ISG54 and ISG56. J. Biol. Chem. *281*, 34064–34071.

Terenzi, F., Saikia, P., and Sen, G.C. (2008). Interferon-inducible protein, P56, inhibits HPV DNA replication by binding to the viral protein E1. EMBO J. *27*, 3311–3321.

Terenzi, F., White, C., Pal, S., Williams, B.R., and Sen, G.C. (2007). Tissue-specific and inducer-specific differential induction of ISG56 and ISG54 in mice. J. Virol. *81*, 8656–8665.

Tokarev, A., Skasko, M., Fitzpatrick, K., and Guatelli, J. (2009). Antiviral activity of the interferon-induced cellular protein BST-2/tetherin. AIDS Res. Hum. Retroviruses 25, 1197–1210.

Toniato, E., Chen, X.P., Losman, J., Flati, V., Donahue, L., and Rothman, P. (2002). TRIM8/GERP RING finger protein interacts with SOCS-1. J. Biol. Chem. *277*, 37315–37322.

Toth, A.M., Li, Z., Cattaneo, R., and Samuel, C.E. (2009). RNA-specific adenosine deaminase ADAR1 suppresses measles virus-induced apoptosis and activation of protein kinase PKR. J. Biol. Chem. *284*, 29350–29356.

Turan, K., Mibayashi, M., Sugiyama, K., Saito, S., Numajiri, A., and Nagata, K. (2004). Nuclear MxA proteins form a complex with influenza virus NP and inhibit the transcription of the engineered influenza virus genome. Nucleic Acids Res. *32*, 643–652.

Turelli, P., Mangeat, B., Jost, S., Vianin, S., and Trono, D. (2004). Inhibition of hepatitis B virus replication by APOBEC3G. Science *303*, 1829.

Uchil, P.D., Quinlan, B.D., Chan, W.T., Luna, J.M., and Mothes, W. (2008). TRIM E3 ligases interfere with early and late stages of the retroviral life cycle. PLoS Pathog. *4*, e16.

Van Damme, N., Goff, D., Katsura, C., Jorgenson, R.L., Mitchell, R., Johnson, M.C., Stephens, E.B., and Guatelli, J. (2008). The interferon-induced protein BST-2 restricts HIV-1 release and is downregulated from the cell surface by the viral Vpu protein. Cell Host Microbe *3*, 245–252.

Versteeg, G.A., Hale, B.G., van Boheemen, S., Wolff, T., Lenschow, D.J., and Garcia-Sastre, A. (2010). Species-specific antagonism of host ISGylation by the influenza B virus NS1 protein. J. Virol. *84*, 5423–5430.

Vijay-Kumar, S., Bugg, C.E., and Cook, W.J. (1987). Structure of ubiquitin refined at 1.8 A resolution. J. Mol. Biol. *194*, 531–544.

Wacher, C., Muller, M., Hofer, M.J., Getts, D.R., Zabaras, R., Ousman, S.S., Terenzi, F., Sen, G.C., King, N.J., and Campbell, I.L. (2007). Coordinated regulation and widespread cellular expression of interferon-stimulated genes (ISG) ISG-49, ISG-54, and ISG-56 in the central nervous system after infection with distinct viruses. J. Virol. *81*, 860–871.

Wagner, R.W., Smith, J.E., Cooperman, B.S., and Nishikura, K. (1989). A double-stranded RNA unwinding activity introduces structural alterations by means of adenosine to inosine conversions in mammalian cells and Xenopus eggs. Proc. Natl. Acad. Sci. U.S.A. *86*, 2647–2651.

Wang, X., Hinson, E.R., and Cresswell, P. (2007). The interferon-inducible protein viperin inhibits influenza virus release by perturbing lipid rafts. Cell Host Microbe *2*, 96–105.

Ward, S.V., Markle, D., Das, S., and Samuel, C.E. (2002). The promoter-proximal KCS element of the PKR kinase gene enhances transcription irrespective of orientation and position relative to the ISRE element and is functionally distinct from the KCS-like element of the ADAR deaminase Promoter. J. Interferon Cytokine Res. *22*, 891–898.

Ward, S.V., and Samuel, C.E. (2002). Regulation of the interferon-inducible PKR kinase gene: the KCS element is a constitutive promoter element that functions in concert with the interferon-stimulated response element. Virology *296*, 136–146.

Wathelet, M., Moutschen, S., Defilippi, P., Cravador, A., Collet, M., Huez, G., and Content, J. (1986). Molecular cloning, full-length sequence and preliminary characterization of a 56-kDa protein induced by human interferons. Eur. J. Biochem. *155*, 11–17.

Williams, B.R., Golgher, R.R., and Kerr, I.M. (1979). Activation of a nuclease by pppA2′p5′A2′p5′A in intact cells. FEBS Lett. *105*, 47–52.

Wong, J.J., Pung, Y.F., Sze, N.S., and Chin, K.C. (2006). HERC5 is an IFN-induced HECT-type E3 protein ligase that mediates type I IFN-induced ISGylation of protein targets. Proc. Natl. Acad. Sci. U.S.A. *103*, 10735–10740.

Wreschner, D.H., James, T.C., Silverman, R.H., and Kerr, I.M. (1981a). Ribosomal RNA cleavage, nuclease activation and 2–5A(ppp(A2′p)nA) in interferon-treated cells. Nucleic Acids Res. *9*, 1571–1581.

Wreschner, D.H., McCauley, J.W., Skehel, J.J., and Kerr, I.M. (1981b). Interferon action--sequence specificity of the ppp(A2′p)nA-dependent ribonuclease. Nature *289*, 414–417.

Xiang, Y., Condit, R.C., Vijaysri, S., Jacobs, B., Williams, B.R., and Silverman, R.H. (2002). Blockade of interferon induction and action by the E3L double-stranded RNA binding proteins of vaccinia virus. J. Virol. *76*, 5251–5259.

Yang, K., Shi, H.X., Liu, X.Y., Shan, Y.F., Wei, B., Chen, S., and Wang, C. (2009). TRIM21 is essential to sustain IFN regulatory factor 3 activation during antiviral response. J. Immunol. *182*, 3782–3792.

Yang, Y.L., Reis, L.F., Pavlovic, J., Aguzzi, A., Schafer, R., Kumar, A., Williams, B.R., Aguet, M., and Weissmann, C. (1995). Deficient signaling in mice devoid of double-stranded RNA-dependent protein kinase. EMBO J. *14*, 6095–6106.

Yap, M.W., Nisole, S., and Stoye, J.P. (2005). A single amino acid change in the SPRY domain of human Trim5alpha leads to HIV-1 restriction. Curr. Biol. *15*, 73–78.

Yoon, C.H., Lee, E.S., Lim, D.S., and Bae, Y.S. (2009). PKR, a p53 target gene, plays a crucial role in the tumor-suppressor function of p53. Proc. Natl. Acad. Sci. U.S.A. *106*, 7852–7857.

Yoshimi, R., Chang, T.H., Wang, H., Atsumi, T., Morse, H.C., 3rd, and Ozato, K. (2009). Gene disruption study reveals a nonredundant role for TRIM21/Ro52 in NF-kappaB-dependent cytokine expression in fibroblasts. J. Immunol. *182*, 7527–7538.

Yuan, W., and Krug, R.M. (2001). Influenza B virus NS1 protein inhibits conjugation of the interferon (IFN)-induced ubiquitin-like ISG15 protein. EMBO J. *20*, 362–371.

Zahn, R.C., Schelp, I., Utermohlen, O., and von Laer, D. (2007). A-to-G hypermutation in the genome of lymphocytic choriomeningitis virus. J. Virol. *81*, 457–464.

Zennou, V., Perez-Caballero, D., Gottlinger, H., and Bieniasz, P.D. (2004). APOBEC3G incorporation into human immunodeficiency virus type 1 particles. J. Virol. *78*, 12058–12061.

Zhang, F., Hatziioannou, T., Perez-Caballero, D., Derse, D., and Bieniasz, P.D. (2006). Antiretroviral potential of human tripartite motif-5 and related proteins. Virology *353*, 396–409.

Zhang, F., Wilson, S.J., Landford, W.C., Virgen, B., Gregory, D., Johnson, M.C., Munch, J., Kirchhoff, F., Bieniasz, P.D., and Hatziioannou, T. (2009). Nef proteins from simian immunodeficiency viruses are tetherin antagonists. Cell Host Microbe *6*, 54–67.

Zhang, H., Yang, B., Pomerantz, R.J., Zhang, C., Arunachalam, S.C., and Gao, L. (2003). The cytidine deaminase CEM15 induces hypermutation in newly synthesized HIV-1 DNA. Nature *424*, 94–98.

Zhang, Y., Burke, C.W., Ryman, K.D., and Klimstra, W.B. (2007). Identification and characterization of interferon-induced proteins that inhibit alphavirus replication. J. Virol. *81*, 11246–11255.

Zhao, C., Beaudenon, S.L., Kelley, M.L., Waddell, M.B., Yuan, W., Schulman, B.A., Huibregtse, J.M., and Krug, R.M. (2004). The UbcH8 ubiquitin E2 enzyme is also the E2 enzyme for ISG15, an IFN-alpha/beta-induced ubiquitin-like protein. Proc. Natl. Acad. Sci. U.S.A. *101*, 7578–7582.

Zhao, C., Denison, C., Huibregtse, J.M., Gygi, S., and Krug, R.M. (2005). Human ISG15 conjugation targets both IFN-induced and constitutively expressed proteins functioning in diverse cellular pathways. Proc. Natl. Acad. Sci. U.S.A. *102*, 10200–10205.

Zhao, C., Hsiang, T.Y., Kuo, R.L., and Krug, R.M. (2010). ISG15 conjugation system targets the viral NS1 protein in influenza A virus-infected cells. Proc. Natl. Acad. Sci. U.S.A. *107*, 2253–2258.

Zheng, X., Silverman, R.H., Zhou, A., Goto, T., Kwon, B.S., Kaufman, H.E., and Hill, J.M. (2001). Increased severity of HSV-1 keratitis and mortality in mice lacking the 2–5A-dependent RNase L gene. Invest. Ophthalmol. Vis. Sci. *42*, 120–126.

Zheng, Y.H., Irwin, D., Kurosu, T., Tokunaga, K., Sata, T., and Peterlin, B.M. (2004). Human APOBEC3F is another host factor that blocks human immunodeficiency virus type 1 replication. J. Virol. *78*, 6073–6076.

Zhong, B., Yang, Y., Li, S., Wang, Y.Y., Li, Y., Diao, F., Lei, C., He, X., Zhang, L., Tien, P., *et al.* (2008). The adaptor protein MITA links virus-sensing receptors to IRF3 transcription factor activation. Immunity *29*, 538–550.

Zhong, S., Salomoni, P., and Pandolfi, P.P. (2000). The transcriptional role of PML and the nuclear body. Nat. Cell Biol. 2, E85–90.

Zhou, A., Hassel, B.A., and Silverman, R.H. (1993). Expression cloning of 2–5A-dependent RNAase: a uniquely regulated mediator of interferon action. Cell *72*, 753–765.

Zhou, A., Paranjape, J., Brown, T.L., Nie, H., Naik, S., Dong, B., Chang, A., Trapp, B., Fairchild, R., Colmenares, C., *et al.* (1997). Interferon action and apoptosis are defective in mice devoid of 2′,5′-oligoadenylate-dependent RNase L. EMBO J. *16*, 6355–6363.

Zou, W., Papov, V., Malakhova, O., Kim, K.I., Dao, C., Li, J., and Zhang, D.E. (2005). ISG15 modification of ubiquitin E2 Ubc13 disrupts its ability to form thioester bond with ubiquitin. Biochem. Biophys. Res. Commun. *336*, 61–68.

Zou, W., and Zhang, D.E. (2006). The interferon-inducible ubiquitin-protein isopeptide ligase (E3) EFP also functions as an ISG15 E3 ligase. J. Biol. Chem. *281*, 3989–3994.

Host Interferon: a Silent Partner in the Regulation of Herpes Simplex Virus Latency

5

William P. Halford and Bryan M. Gebhardt

Abstract

Herpes simplex virus (HSV) establishes latent infections as a consequence of a non-cytolytic immune response that represses HSV replication, but fails to destroy neurons that harbour HSV's genetic material. It has become increasingly evident that, in both mice and men, the host interferon system plays a critical role in tipping HSV's latency-replication balance in favour of latency. HSV can resist interferon-induced repression provided that HSV's two interferon antagonists, ICP0 and ICP34.5, are synthesized. Failure to synthesize either protein renders HSV interferon-sensitive and prone to establishing latent infections. Intriguingly, ICP0 and ICP34.5 are encoded within HSV's latency-regulating R_L regions. We propose that differential synthesis of ICP0 and ICP34.5 may endow HSV with the capacity to 'choose' between latency and replication *in vivo*. HSV may choose to establish a latent infection by downregulating ICP0 or ICP34.5, hence rendering itself sensitive to the interferon-induced antiviral state. Conversely, synthesis of ICP0 and ICP34.5 may ensure that HSV resists interferon-induced repression and completes another cycle of replication.

Prologue

Most virus families such as the picornaviruses are similar enough in genetic content to be meaningfully discussed as a group. This is not the case with the herpesviruses. The >130 known herpesviruses employ a conserved set of ~40 enzymes and structural proteins to synthesize virions that are indistinguishable from one another (Davison, 2002; McGeoch *et al.*, 2006). All herpesviruses appear to establish life-long infections in animals that oscillate between relative states of latent versus productive infection. Beyond this, the similarities quickly disappear. For example, herpes simplex virus establishes life-long infections in neurons using a complement of ~75 genes, whereas cytomegalovirus relies on ~200 genes to establish life-long infections in monocytes (Murphy *et al.*, 2003). This review will focus on herpes simplex virus such that the biology of a single herpesvirus may be considered in some depth. In particular, herpes simplex virus 1 (HSV-1) is one of the easiest human herpesviruses to study because it replicates to high titres, is amenable to genetic manipulation and recapitulates most aspects of human infection in animal models. Thus, HSV-1 provides a powerful system to explore the many gaps in knowledge that exist at the interface between virology and immunology.

In recent years, it has become apparent that host interferons play a more pivotal role in the biology of HSV infections than was previously recognized. The central tenet of the current review is that HSV exploits the host interferon system as a negative-feedback mechanism that controls the timing of viral exit from the productive cycle of replication. The intensity of interferon-inducible repression of HSV replication varies over several orders of magnitude (Halford *et al.*, 2005a). Thus, when host interferon signalling exceeds a critical threshold, it is possible that HSV's interferon antagonists (ICP0 and ICP34.5) may fail to accumulate such that HSV is rendered sensitive to repression by the interferon response. Such an interferon-dependent shutoff mechanism might endow HSV with the capacity to cease all viral protein synthesis just as CD8+ T-cells are preparing to destroy HSV antigen+ cells. Unlike cytolysis by T-cells, interferon-inducible repression may be reversible. Once the local immune response wanes, a subset of HSV-infected neurons may re-enter the productive cycle of HSV replication weeks to months later.

It is unlikely that the principles discussed herein are unique to HSV. Rather, the opposing forces provided by interferon-dependent repression and virus-encoded interferon antagonists may be exploited by many herpesviruses to regulate their oscillations between latent and reactivated infections. Likewise, most persistent viral infections may be described as a balance, or equilibrium, that is achieved between the host immune response and a virus that avoids clearance by this host response.

Students are generally familiar with the concept that a process that is static to the eye may, in fact, be controlled by two equal and opposing forces. Therefore, we wish to suggest that students might grasp the biology of persistent viral infections more readily if these processes were described in terms of an equilibrium between virus and host. Aside from the simplicity of the approach, it accurately reflects the weight of evidence that the host immune system plays an integral role in restricting the spread of most, if not all, persistent viral infections. Below, we describe the *in vivo* biology of HSV infections in such terms.

Introduction

Herpes simplex virus types 1 and 2 (HSV-1 and HSV-2) are similar viruses that share a nearly identical set of ~75 co-linear genes. HSV-1 and HSV-2 produce a similar spectrum of disease. In this review, the terms HSV-1 and HSV-2 will be used to denote specific viruses when summarizing the results of a particular study. However, because the biology of HSV-1 and HSV-2 infections are similar in principle, the term 'HSV' will be used wherever possible to indicate that the principles under discussion are probably relevant to both HSV-1 and HSV-2.

In cultured cells infected with HSV, ~75 viral proteins are synthesized in a temporal cascade during productive replication (Honess and Roizman, 1974). Virion protein 16 (VP16) in the tegument of HSV virions forms a complex with the cellular transcription factor Oct 1 to initiate a cascade of viral gene expression (Triezenberg *et al.*, 1988). Only five immediate-early (IE) genes are initially induced based on the presence of VP16-responsive elements in their promoters. Viral IE proteins such as infected cell protein 0 (ICP0) and 4 (ICP4) are thought to activate viral genes, which encode ~70 early (E) and late (L) viral proteins that form the machinery necessary for replicating and packaging HSV genomes into new virions. The IE → E → L model of HSV gene expression summarizes the series of events that lead to HSV virion synthesis (Honess and Roizman, 1974). However, this linear depiction of HSV gene expression does not address one fundamental question that has intrigued scientists for decades, 'How does HSV 'choose' whether infection of a cell will be productive or latent?'

Most reviews of HSV latency tend to focus on the molecular biology of HSV gene expression, and often overlook the fact that latent HSV infections are the consequence of a host immune response to HSV. This is not to say that considerations of HSV gene regulation are irrelevant. Rather, it is important to appreciate that entry of HSV into a latent state occurs in at least two steps: (1) generation of a repressive stimulus that *influences* whether or not HSV replication will stall, and (2) silencing of HSV gene expression. The available evidence indicates that the host immune system is the primary source of the 'repressive stimulus.' In contrast, the effector mechanisms that silence HSV during latency remain unclear, and thus remain an ongoing area of research (Knipe and Cliffe, 2008; Bloom *et al.*, 2010; Perng and Jones, 2010).

Latent HSV infections appear to exist as a dynamic equilibrium that is controlled by the opposing forces of (1) viral activators and (2) host immune repressors of HSV infection (Fig. 5.1A). HSV virology and HSV immunology have largely been studied in isolation from one another for the past 40 years. However, a holistic explanation of HSV latency will require the balanced

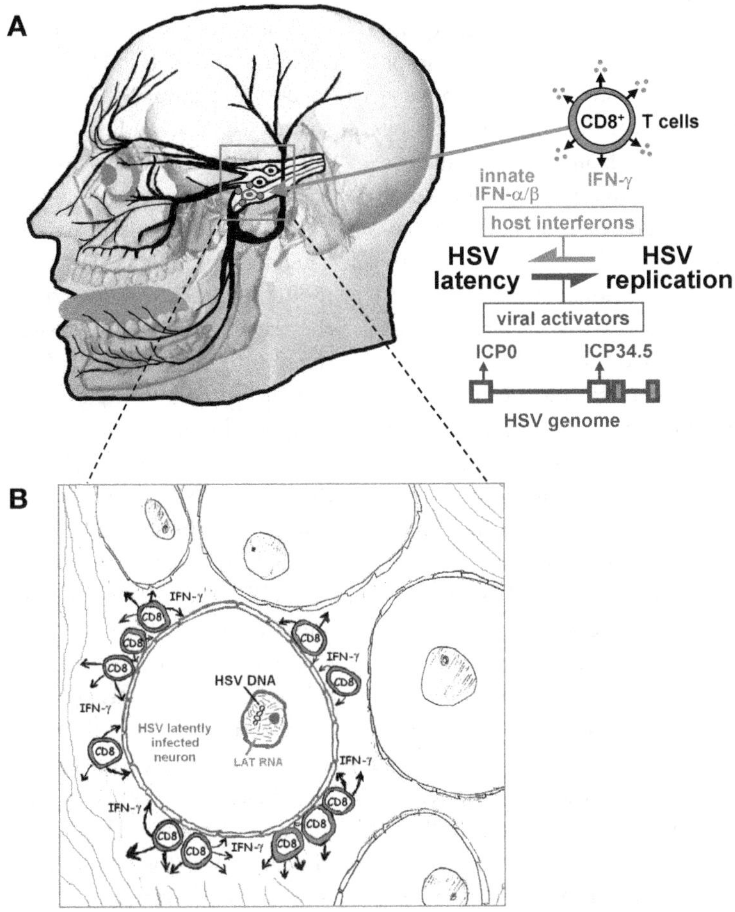

Figure 5.1 An equilibrium model of the latency-replication balance of HSV. (A) Schematic of a human trigeminal ganglion. The ophthalmic and maxillary branches collect nerve fibres from the eye and face, respectively. The lower, mandibular branch collects nerve fibres from the mouth and is the predominant site of latent HSV-1 infection in humans. Neuronal cell bodies collect sensory inputs from the epithelium and relay signals to the brain. A single HSV latently infected neuron is illustrated as a large blue cell surrounded by smaller, red CD8+ T-cells. In the schematic to the right, it is proposed that HSV infections of neurons are controlled by the opposing forces of host interferons (red) and viral activator proteins (blue). Innate interferons (α/β) and CD8+ T-cell-secreted interferon-γ are proposed to favour the establishment of latent HSV infections. Viral activator proteins ICP0 and ICP34.5 are proposed to favour viral spread by dismantling two interferon-inducible blocks to HSV replication. Adapted from the Netter Anatomy Illustration Collection© (Netter, 1997) with permission from Elsevier Inc. All rights reserved. (B) Schematic cross-section of a trigeminal ganglion showing a single, HSV latently infected neuron (blue cell) surrounded by CD8+ T-cells (red cells) that have infiltrated the tissue as part of the host response to viral infection. Green lines in the nucleus of the HSV latently infected neuron represent the HSV latency-associated transcript (LAT) RNAs that may be detected in the HSV latently infected ganglia of experimental animals and humans. A colour version of this figure is located in the plate section at the back of the book.

consideration of both the viral activators and immunological repressors that control how HSV alternates between a silent and reactivated state *in vivo* (Fig. 5.1A).

Reviews of HSV latency that consider only the virus are akin to teaching someone to drive an automobile while focusing solely on the gas pedal (blue arrow in Fig. 5.1A). The accelerator is important because it explains how HSV-encoded activators allow the virus to undergo rapid amplification and spread in an infected animal. However, such virus-focused explanations tend to imply that withdrawal of pressure on HSV's accelerator (i.e. lack of viral activators) is adequate to explain how the spread of HSV infection stops *in vivo*. This is analogous to suggesting that a commuter may safely stop a car during rush hour by simply removing their foot from the gas pedal. As with the commuter who attempts this method of deceleration, such explanations of HSV latency crash upon scrutiny.

Immune-based explanations of HSV latency are akin to teaching someone to drive an automobile while focusing solely on the brake pedal (red arrow in Fig. 5.1A). The braking mechanism is important in explaining how HSV replication stops *in vivo* when T-cells infiltrate an HSV-infected tissue (Fig. 5.1B). If the mechanisms that controlled HSV latency simply involved the degree of pressure on the brake pedal (i.e. magnitude of the T-cell response), then latent HSV infections would never re-enter the productive cycle of replication. However, 2% of the human population is shedding HSV at any given time (Buddingh *et al.*, 1953; Wald *et al.*, 1997; Miller and Danaher, 2008).

Both an accelerator and braking mechanism are required to explain how HSV oscillates between latent and active infections over the lifetime of a human host. Against this background, it is our goal to review the literature that indicates (1) host interferons are key components in the braking mechanism, and (2) HSV-encoded interferon antagonists, ICP0 and ICP34.5, are key components in the accelerator that allows HSV to periodically overpower the immune system and re-enter the productive cycle of replication (Fig. 5.1).

The biology of latent HSV infections

HSV infections in humans and experimental animals

Human carriers of HSV-1 or HSV-2 frequently shed infectious virus in their mouth or genital region, and this commonly occurs in the absence of symptoms (Buddingh *et al.*, 1953; Wald *et al.*, 1997; Miller and Danaher, 2008). When transmitted to a naïve person, HSV replicates in the epithelium at the point of entry, which is typically the mouth or genitals (Fig. 5.1A). Over a 1- to 3-day period, HSV may undergo a 100- to 10,000-fold amplification in the epithelium, and spread to nerve endings. Upon HSV infection of nerve endings, nucleocapsids are carried by retrograde transport through axons to neuronal cell bodies several centimetres away in the innervating ganglia (Fig. 5.1A). Hence, HSV infection of the mouth is quickly followed by HSV infection of the trigeminal ganglia whose neurons innervate the mouth, eyes, and face. In contrast, dorsal root ganglia coming off the lower backbone house peripheral neurons that innervate the anogenital region. Hence, HSV-2 infection of the genital epithelium results in HSV-2 infection of dorsal root ganglia (Stanberry *et al.*, 1982).

Within 3 weeks of infection, sites of productive HSV replication become exceedingly rare and the infection is said to be 'latent.' Empirically, a latent HSV infection simply means that HSV DNA persists in a tissue in the relative absence of infectious HSV virions. The phrase 'relative absence' is important to bear in mind, because the sensitivity of detecting infectious HSV in latently infected animals and humans is poor, and tends to underestimate the actual frequency of spontaneous HSV reactivation (Wald *et al.*, 1997).

During a latent infection, the *LAT* gene is the only locus in the HSV genome that routinely produces high levels of latency-associated transcript (LAT) RNA (Fig. 5.2). LAT-specific riboprobes may be hybridized to nucleic acids present in trigeminal ganglia to reveal the presence of LAT$^+$ neurons, which are a common feature in HSV latently infected ganglia harvested from experimental animals or human cadavers (Stevens *et al.*, 1987, 1988; Fig. 5.1B). LAT RNAs are stable 1.5 and 2.0 kb RNA introns that are derived by differential splicing from a common primary transcript (Farrell *et al.*, 1991). The *LAT* gene lies in the long-repeated (R_L) regions of HSV's genome, which appear to regulate the balance between HSV latency and replication (Fig. 5.2). The LAT introns that accumulate in HSV latently infected neurons share an ~750-bp antisense overlap with the *ICP0* gene (Fig. 5.2) whose IE protein strongly promotes HSV reactivation (Halford and Schaffer, 2001; Halford *et al.*, 2001). Thus, numerous

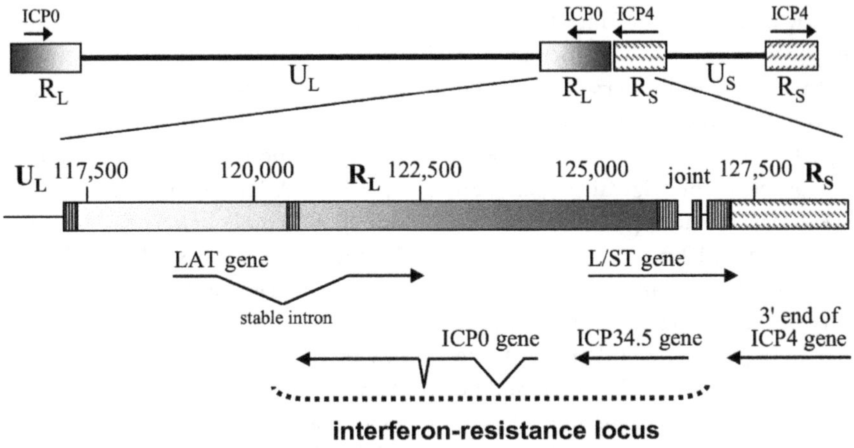

Figure 5.2 The latency-regulating R_L regions of the HSV genome. Numbers refer to base positions in the HSV-1 genome, and arrows denote the *LAT*, *L/ST*, *ICP0*, and *ICP34.5* genes. Microsatellite repeats in the long-repeated (R_L) regions are denoted by small boxes containing vertical bars, and these are abundant in the 'joint' region between the *ICP4* gene in the R_S region and the ICP34.5 gene in the R_L region. The second copy of the R_L region occurs 100 kb away on the opposite end of HSV's long segment (shown on left).

hypotheses including LAT-derived microRNAs have been offered to explain why HSV's *LAT* gene should share >70 turns of a DNA double-helix with the reactivation-promoting *ICP0* gene (Tang *et al.*, 2008; Umbach *et al.*, 2008). While non-coding LAT RNAs provide a useful marker for identifying HSV latently infected neurons (Fig. 5.1B), their function in the HSV life cycle has not been clearly established (Fraser *et al.*, 1992; Bloom, 2004).

How often does HSV spontaneously establish latent infections in neurons?

Stevens and Cook (1971) established that neurons were the long-lived reservoir of latent HSV infections. During the acute phase of a primary HSV infection, homogenization of the innervating ganglia reveals the presence of large quantities of infectious virions. However, one month later, infectious virions are rarely detectable in HSV latently infected ganglia (Stevens and Cook, 1971). When HSV latently infected ganglia are removed from animals and placed in cell culture medium for one week, this physiological change causes latent HSV to reinitiate the *de novo* synthesis of new infectious virions (Stevens and Cook, 1971). The study of Stevens and Cook (1971) was groundbreaking in pinpointing the site of latent HSV infections, and in establishing an experimental system that provided the first means to study how HSV infections may alternate between a latent and reactivated state in *bona fide* latently infected neurons.

Because neurons serve as the reservoir of latent HSV infections, the unique biology of neurons is often suggested to hold the key to explaining how latent HSV infections are established (Roizman *et al.*, 2005; Knipe and Cliffe, 2008). Several neuron-specific repressors including Oct-2 (Lillycrop *et al.*, 1991; Latchman, 1996), Zhangfei (Akhova *et al.*, 2005; Misra *et al.*, 2005), and CoREST (Roizman *et al.*, 2005) have been proposed to explain why HSV establishes latent infections in neurons. Such explanations are based upon the assumption that neurons are inherently refractory to HSV replication, and thus HSV establishes latent infections immediately upon entering neurons. This assumption is inconsistent with the available evidence, which is summarized as follows.

If epithelial cells and neurons intrinsically differed in their propensity to support productive HSV replication, then these differences should be reflected in the quantity of infectious HSV that is recovered from the epithelium versus the innervating ganglia during a primary HSV infection (Fig. 5.1A). However, when animals are inoculated in the cornea with HSV-1, robust viral replication is first observed in the eyes on days 1 and 2 post inoculation (p.i.), and this is followed by equally robust HSV replication in trigeminal ganglion neurons between days 3 and 5 p.i. (Knotts *et al.*, 1974; reviewed in Rawls, 1985).

HSV encodes its own DNA nucleotide biosynthetic enzymes, including thymidine kinase, which are strictly required for HSV to replicate in neurons (Coen *et al.*, 1989). HSV thymidine kinase-null viruses replicate efficiently in cultured cells and in the epithelium of animals, which produce their own deoxynucleotide triphosphates (dNTPs). However, HSV thymidine kinase-null viruses fail to replicate efficiently in terminally differentiated neurons that do not synthesize their own dNTPs. Consequently, HSV thymidine kinase mutants are avirulent in animals because they fail to replicate in neurons, and fail to use peripheral nerve fibres as a conduit of spread (Coen *et al.*, 1989; Valyi-Nagy *et al.*, 1994b).

The relative efficiency with which HSV establishes latent infections in neurons upon its initial entry into the trigeminal ganglia may be determined by comparing the relative abundance of LAT$^+$ neurons versus HSV antigen$^+$ neurons during the acute phase of HSV infection. If HSV established latent infections immediately upon entry into the nuclei of neurons, then a predominance of LAT RNA$^+$ neurons should be observed shortly after inoculation of the eye. Between 3 and 5 days p.i., some HSV LAT RNA$^+$ neurons are indeed observed in the trigeminal ganglia (Table 5.1). However, HSV antigen$^+$ neurons (i.e. sites of viral replication) are ~100-fold more abundant than HSV LAT RNA$^+$ neurons in the trigeminal ganglia at early times after inoculation of mice (Halford *et al.*, 2006; Table 5.1).

The unique biology of neurons is insufficient to explain how latent HSV infections are established *in vivo*. In contrast, the unique biology of neurons appears to be critically important in explaining (1) why LAT RNAs only accumulate in neurons (Stroop *et al.*, 1984; Stevens *et al.*, 1987), (2) why neurons are the only cell type in which latent HSV genomes are maintained over time (Sawtell, 1997), and (3) why neurons are the only cell type from which latent HSV infections are successfully reactivated (Halford *et al.*, 1996a; Halford and Schaffer, 2001).

Table 5.1 Abundance of LAT$^+$ versus HSV antigen$^+$ neurons in trigeminal ganglia during the acute phase of HSV infection

	Day 3		Day 5.5	
Mice[a]	LAT$^+$ neurons[b]	HSV Ag$^+$ neurons[c]	LAT$^+$ neurons	HSV Ag$^+$ neurons
Wild type	0.02–0.04% (12)[d]	7–15% (4399)	0.1–0.2% (45)	9–18% (5300)
rag2$^{-/-}$	0.02–0.04% (12)	10–20% (6122)	0.1–0.3% (82)	18–36% (10897)

[a]Mice were inoculated with 10^5 pfu per eye of wild-type HSV-1 strain KOS, and were sacrificed 72 hours (day 3) or 132 hours (day 5.5) after inoculation.

[b]Neurons in sections of HSV-infected trigeminal ganglia that hybridized with a latency-associated transcript (LAT)-specific riboprobe and which were HSV antigen-negative, as determined under a fluorescent microscope.

[c]Neurons in sections of KOS-infected trigeminal ganglia that were labelled by rabbit anti-HSV antibody and which were LAT-negative, as determined under a fluorescent microscope.

[d]The estimated per cent of neurons that were LAT$^+$, or HSV antigen$^+$, is based on the total number of neurons counted (number shown in parentheses) in a total of 144 TG sections derived from 6 independent TG (i.e. 24 sections per TG were analysed). It was estimated that this number of sections contained at least 30,000 neurons but not more than 60,000 neurons.

This table was adapted from Halford, *et al.* (2006) in accordance with the open-access policy of BioMed Central.

T cells and latent HSV infections: inseparable partners

In animals that lack T-lymphocytes, wild-type HSV fails to exit the productive cycle of replication and HSV infections uniformly spread to the central nervous system, producing a fatal encephalitis (Oakes, 1975; Walz *et al.*, 1976; Nagafuchi *et al.*, 1979). Thus, the establishment of latent HSV infection is experimentally inseparable from the host immune response that represses productive HSV replication in the peripheral nervous system (Fig. 5.1B). This correlation does not prove that immunological repression is the mechanism by which HSV latency is established (Tenser *et al.*, 1993; Gesser *et al.*, 1994). However, the hypothesis that the host immune response is causally related to the establishment of latent HSV infections is based upon more than lethal outcomes of HSV infection in immunodeficient mice.

The strongest support for this hypothesis is derived from the pioneering studies of Simmons and Tscharke (1992) whose work focused attention to the possibility that CD8+ T-cells might be required to suppress HSV replication in neurons (Fig. 5.1B). Simmons and Tscharke asked a simple question that yielded powerful results: 'Do CD8+ T-cells influence outcomes of HSV-1 infection in ganglionic neurons?' In immunocompetent mice inoculated in the left flank, HSV-1 replication peaked in the innervating dorsal root ganglia by day 5 p.i., at which time 13% of neurons were HSV antigen+. On day 8 p.i. in immunocompetent mice, only rare dorsal root ganglion neurons could be found that were HSV antigen+. In contrast, HSV antigen+ neurons were ~50-fold more abundant on day 8 p.i. in the dorsal root ganglia of mice depleted of CD8+ T cells (Simmons and Tscharke, 1992).

The relative frequency of neurons that survived HSV infection was compared in dorsal root ganglia harvested on day 8 p.i. Relative to uninfected ganglia, immunocompetent mice were estimated to have lost no more than 0.5–3.5% of ganglionic neurons, despite the fact that 13% of neurons were HSV antigen+ on day 5 p.i. (Simmons and Tscharke, 1992). In contrast, on day 8 p.i., CD8-depleted mice were estimated to have lost 36% of neurons and gross pathology was observed in their dorsal root ganglia.

The findings of Simmons and Tscharke established the first direct link between CD8+ T-cells and the establishment of latent HSV infection in neurons. Specifically, CD8+ T-cell infiltration of ganglia appeared to be necessary to trigger the cessation of HSV replication in ganglionic neurons (Fig. 5.1B). Moreover, failure to recruit CD8+ T-cells into ganglia in a timely manner resulted in the rapid and widespread destruction of ganglionic neurons, as HSV infection spread unchecked through the peripheral nervous system. Subsequent studies corroborated that T-cell-dependent repression of HSV replication did not compromise the health of ganglionic neurons, but rather precipitated an extraordinarily rapid clearance (turnover) of HSV antigens over a span of just 12–16 hours (Speck and Simmons, 1998).

Host response to HSV infections

Before discussing those aspects of the host immune response most critical to the establishment and maintenance of latent HSV infections, we consider the larger biology of the immune response to HSV infection. This is not a comprehensive treatment of the topic, but rather is meant to provide the non-immunologist a context for appreciating the types of molecules and cells that are called into action when the body responds to an HSV infection. For a more thorough consideration of the host immune response to HSV, we refer the reader to several excellent reviews on the topic (Carr and Tomanek, 2006; Sheridan *et al.*, 2007; Cunningham *et al.*, 2008; Koelle and Corey, 2008).

Innate immune response to HSV infections

HSV infects and destroys epithelial cells. Cell death disrupts the integrity of a tissue such as the epithelium, and this disturbance initiates a healing response. The first phase of healing is the scavenging of cellular debris by acute inflammatory cells such as polymorphonuclear leucocytes (PMNs) and long-lived professional antigen presenting cells (APCs). In the process of phagocytosing dead cells, PMNs and professional APCs secrete an array of chemokines and cytokines, which enhance the inflammatory process causing capillaries to become leaky, and promote the movement of immune mediators out of the vasculature and into the site of tissue damage (Watanabe *et al.*, 1999; Cook *et al.*, 2004; Li *et al.*, 2006). Professional APCs (MHC class II+ cells) circulate through such damaged tissue, gather antigen samples, return to draining lymph nodes, and there present these antigens to T- and B-lymphocytes (Allan *et al.*, 2006; Heath and Carbone, 2009).

During this process of infection and repair, pro-inflammatory cytokines such as interleukin (IL)-1 and tumour necrosis factor (TNF)-α and chemokines such as CXCR-3, CXCL-10, and IL-8 are critical in drawing leucocytes out of the bloodstream to sites of viral replication and tissue damage (Miyazaki *et al.*, 1998; Thapa and Carr, 2008; Wuest and Carr, 2008). IL-1 and TNF-α promote inflammation by causing local capillaries to become leaky and sticky for leucocytes. Leukocyte adhesion molecules upregulated on capillaries cause blood leucocytes to stick to the inner capillary walls, and diapedesis may occur, whereby leucocytes crawl between cells of the blood vessel wall and into the inflamed tissue. Chemokines are critical in establishing the gradient of molecules that leucocytes follow to their source, the site of viral infection and tissue damage.

Much has been claimed about the role of natural resistance and natural killer cells in host defence against HSV infections (Lopez, 1975; Biron *et al.*, 1989). The evidence that supports these claims is equivocal, and is discussed elsewhere (Halford *et al.*, 2004, 2005b). Scores of chemokines and cytokines elicited by the immune system are critical to leucocyte migration, differentiation, and the proliferation of lymphocytes that culminate in a robust adaptive immune response to viral infection (Salazar-Mather and Hokeness, 2003).

Aside from their role as inhibitors of viral replication, host interferons play an immunoregulatory role, which promotes the proliferation of CD4+ T-cells that promote cell-mediated immune responses. Moreover, the second immunoregulatory function of interferons lies in the potent upregulation (10- to 20-fold increase) of MHC class II molecules on macrophage, dendritic cells, and B-lymphocytes. This enhances the vigour of the anti-viral immune response, as MHC class II molecules are the vehicle by which 15–30 amino acid peptide antigens are presented to CD4+ T-helper lymphocytes. CD4+ T-cells act in turn to promote the proliferation and differentiation of effector CD8+ T-lymphocytes and B-lymphocytes that generate the adaptive immune responses that control HSV infection.

Adaptive immune response to HSV infections

B-lymphocytes

Each bone marrow-derived lymphocyte (B-cell) in a draining lymph node is developmentally committed to produce a particular class (or isotype) of antibody, and this antibody can only bind a single, unique antigenic epitope. In terms of protein antigens, the antigen-binding site of an antibody typically accommodates 5 or 6 amino acids derived from a single protein. These amino acids may be continuous in the polypeptide chain of the antigen, but are often discontinuous amino acids derived from 2 distinct polypeptide loops that converge at a single point on the surface of the protein antigen.

Temporally, naïve B-cells always produce IgM isotype antibody first because generation of other antibody isotypes requires (1) cytokine help from CD4+ T-lymphocytes (e.g. IL-4), and (2) the downstream event of genetic rearrangement of the B-cell receptor heavy chain gene. Antibody class switching involves deletion of a 5′ exon that encodes the IgM heavy chain, and this deletion event commits the B-cell to usage of an alternative 3′ exon that encodes the IgA, IgE, or IgG heavy chains. Thus, HSV-specific IgM antibody is measurable in serum 3–5 days into a primary HSV infection, whereas at least 7 days are required before HSV-specific IgG antibody begins to accumulate.

The IgM antibody is a 750 kDa pentameric molecule, which is largely confined to the vascular compartment because of its large size. IgM antibody is critical in host defence against many bacteria, particularly bacteria capable of colonizing the bloodstream. HSV-specific IgM antibodies are a useful diagnostic indicator of primary HSV infections. HSV-specific IgM antibodies do not play a major role in host defence against HSV infections, as IgM antibodies fail to diffuse into virus-infected tissues unless there is a severe loss of vascular integrity. Given the short serum half-life of IgM antibodies (~10 days), and the fact that B-cells undergo isotype class-switching as an obligate step in their differentiation into long-lived memory cells, IgM antibody rapidly declines in titre following the resolution of a primary HSV infection.

The IgG isotype antibody response follows but overlaps the IgM response (Kurtz, 1974). Owing to its lower molecular weight, 150 kDa, IgG antibody is found both intravascularly and extravascularly. Some of the extravascular IgG escapes through leaky capillary walls or is transported across cells into tissues by means of IgG-specific receptors. HSV-specific IgG reduces the severity and spread of HSV infections, but is insufficient to prevent recurrent herpetic infections. A subset of B-cells differentiate into short-lived, IgG antibody-secreting plasma cells, and secrete IgG antibodies that have a serum half-life of ~21 days. Asymptomatic carriers of HSV infection maintain high serum titres of HSV-specific IgG antibody over their lifetime. This raises questions about the source of HSV antigen that drives the continued proliferation and differentiation of HSV-specific B-cells into short-lived plasma cells years after a primary HSV infection, hence maintaining HSV-specific IgG titres over time.

T-lymphocytes

Thymus-derived lymphocytes (T-cells) are the antigen-specific cells of the cell-mediated immune system. Within this category of lymphocytes, there are two subpopulations that are easily identified by plasma membrane markers: Cluster of Differentiation molecule 4 (CD4) and Cluster of Differentiation molecule 8 (CD8). These lymphocytes circulate from blood to lymph and back, and are found in all lymphoid tissues.

CD4+ T-cells have acquired the name 'helper T-cells' because they assist and promote the maturation of antibody responses and cellular immune responses to foreign antigens. During an active HSV infection, HSV-specific CD4+ T-cells promote antibody class-switching and the subsequent differentiation of HSV-specific B-cells into long-lived memory cells. Likewise, HSV-specific CD4+ T-cells promote the clonal expansion of HSV-specific CD8+ T-lymphocytes, and their differentiation into autonomous effector cells capable of (1) destroying HSV-infected cells or (2) secreting cytokines that interfere with HSV replication.

CD8+ T-cells are best known as 'cytotoxic T-lymphocytes' (CTLs) that kill virus-infected target cells. Cytolysis of virus-infected target cells is achieved when a CD8+ T-cell secretes granules containing perforin, granzymes, and fas ligand while directly apposed to a target cell. Internalization of fas ligand by the target cells leads to the initiation of a programmed cell death response, and consequently the virus-infected target cell is destroyed. Because CD8+ T-cells are capable of renewing their cytotoxic granules, a single CTL may kill scores of virus-infected target cells. CD8+ T-cells may also control viral infections via the secretion of interferon-γ, which interferes with viral replication without compromising the health of the host cell. Such non-cytolytic mechanisms are probably critical in explaining how CD8+ T-cells restrict the spread of HSV infection without destroying the neurons in which HSV persists (Fig. 5.1B).

Secondary immune responses to HSV: memory response or chronic response?

Lymphocytes are capable of remembering a prior encounter with antigen and responding with heightened vigour upon a second encounter with their cognate antigen. The primary immune response to acute respiratory viruses such as influenza virus leads to CTL-dependent destruction of virus-infected cells, viral clearance, and subsequent contraction of effector T-cells and IgG-producing plasma cells (Woodland *et al.*, 2001; Hikono *et al.*, 2006).

It is not apparent that the term 'memory immune response' applies to persistent viral infections in which the spread of viral infection is restricted, but the viral infection is never cleared. In such cases, it may be more accurate to think of the adaptive immune response as a chronic immune response to an ongoing infection. Consistent with this viewpoint, experimental animals and humans maintain high and constant serum titres of HSV-specific IgG antibodies over their lifetimes (Halford *et al.*, 1996b, 1997). Likewise, T-cells remain clustered around neurons in latently infected ganglia and at the tips of axons in the genital epithelium of human carriers (Derfuss *et al.*, 2007; Verjans *et al.*, 2007; Zhu *et al.*, 2007).

T-lymphocytes are normally resident in blood, lymphatics, lymph nodes and spleen. It is uncommon to see T-cells in non-lymphoid tissues except when there is an active disease process such as in chronic gingivitis, measles encephalitis, and autoimmune conditions (Liebert, 1997; Yamazaki *et al.*, 2003; Kunz and Ibrahim, 2009). What then can we conclude about the presence of T-cells in the trigeminal ganglia of HSV latently infected hosts months to years after the primary infection (Fig. 5.1B)? The experimental animals and humans in which these subclinical events are observed appear outwardly healthy and exhibit no evidence of an ongoing viral infection. If these T-cells were relics of the primary infection, then we would expect them to be immunologically silent. However, T-cells in HSV latently infected ganglia are not randomly scattered in the ganglionic parenchyma, but are clustered around neuronal cell bodies. Furthermore, these T-cells are often observed secreting a variety of cytokines including TNF-α, IL-2, and interferon-γ (Fig. 5.1B). T-cells not engaged in an immune response would not be secreting such potent mediators of host immunity. Rather, the T-cells that persist in HSV latently infected ganglia appear to be part of an ongoing immune response that restricts the spread of HSV infection, but which fails to clear the virus from the body (Khanna *et al.*, 2004; Decman *et al.*, 2005; Divito *et al.*, 2006).

Host interferons: critical suppressors of HSV replication

Host interferons

Interferon (IFN)-α, IFN-β and IFN-γ are secreted proteins that play important roles in host resistance to viral infections. The >10 isotypes of IFN-α share 20–30% amino acid homology with IFN-β (Pestka, 2007). IFN-α is secreted by leucocytes as an innate response to viral infection, whereas IFN-β is secreted by virus-infected host structural cells. IFN-α and IFN-β bind to a common IFN-α/β receptor, which is expressed on nearly all cells in the body (Samuel, 1998). Activation of IFN-α/β receptors modifies the transcriptional and translational environment in cells inducing an 'antiviral state' (Pestka, 2007). IFN-γ does not share amino acid homology with IFN-α or IFN-β, and binds a distinct IFN-γ receptor that is expressed on most cells in the body. Production of IFN-γ is far more restricted than IFN-α/β, and results when T-cells, natural killer cells, or professional antigen-presenting cells are activated in response to a foreign antigen or activating stimulus (Farrar and Schreiber, 1993; Frucht *et al.*, 2001; Stober *et al.*, 2001; Schoenborn and Wilson, 2007).

Cellular sources of interferon-γ

All cells in the body may produce some IFN-α or IFN-β as an innate response to viral infection (Toth *et al.*, 1990). IFN-α produced by leucocytes appears to be the predominant source of IFN-α/β that restricts HSV spread *in vivo* (Gary-Gouy *et al.*, 2002; Vollstedt *et al.*, 2003). In the case of IFN-γ, the situation is more complex. T-cell biology and the dynamics of the adaptive immune response must be considered to appreciate why the role of IFN-γ in host immunity to HSV changes over time.

A $CD8^+$ T-cell is activated to secrete IFN-γ as a result of contact with a virus-infected cell. One of the primary activating stimuli is the engagement of the T-cell's antigen receptors by MHC class I molecules bearing an 8–10 amino acid viral peptide. In a naïve individual experiencing a primary HSV infection, HSV-specific T-cells initially occur at a frequency of less than 1 per 100,000. Thus, HSV-specific $CD8^+$ T-cells do not contribute to host control of a primary HSV infection until 5–8 days p.i. (McNally *et al.*, 1999). As an infection resolves, antigen receptors on HSV-specific $CD8^+$ T-cells continue to be engaged in draining lymph nodes, which drives their proliferation. HSV-specific $CD8^+$ T-cells increase in number by 10- to 300-fold by the time a latent HSV infection is firmly established (Posavad *et al.*, 1996). Hence, the capacity of HSV-specific $CD8^+$ T-cells to secrete IFN-γ increases by several orders of magnitude in the weeks following a primary HSV infection.

Interferon-γ: a key mediator by which CD8+ T-cells suppress HSV reactivation

The findings of Simmons and Tscharke (1992) raised the possibility that CD8+ T-cells might control HSV replication in neurons. Several labs explored the validity of this inference, and confirmed that T-cells infiltrated HSV-infected ganglia at times that coincided with the shutoff of productive HSV replication (Cantin *et al.*, 1995; Shimeld *et al.*, 1995; Liu *et al.*, 1996). In each study, CD8+ T-cells were observed in direct contact with neurons, but neurons remained healthy in appearance suggesting that CD8+ T-cells relied on non-cytolytic mechanisms to control HSV replication in neurons. Surprisingly, activated CD8+ T-cells were retained in HSV-infected ganglia for months after the resolution of the acute infection (Cantin *et al.*, 1995; Shimeld *et al.*, 1995; Halford *et al.*, 1996b; Liu *et al.*, 1996). These findings suggested that HSV latency might be less 'latent' than originally envisioned. The persistence of IFN-γ-secreting CD8+ T-cells at sites of latent HSV infection suggested that either (1) low-level synthesis of HSV antigen was occurring, (2) HSV was spontaneously reactivating in neurons or (3) both.

Several investigators questioned whether the IFN-γ that persisted at sites of latent HSV infection might play a role in suppressing HSV reactivation. Edouard Cantin's laboratory tested this possibility by comparing the frequency of stress-induced reactivation in HSV latently infected mice that were wild type, lacked IFN-γ or lacked the IFN-γ receptor. Mice lacking IFN-γ exhibited higher rates of HSV reactivation than wild type control mice (Cantin *et al.*, 1999a), and mice lacking IFN-γ receptors exhibited even higher rates of HSV reactivation (Cantin *et al.*, 1999b). These results raised the possibility that the persistence of IFN-γ at sites of latent HSV infection might contribute to ongoing immunological repression of HSV and/or the maintenance of latent HSV infections.

Bob Hendricks' laboratory used a different approach to address a related question. HSV latently infected ganglia may be dissociated to yield primary cultures of trigeminal ganglion cells that contain viable, HSV latently infected neurons (Halford *et al.*, 1996a). Using this *ex vivo* system, Liu, *et al.* (2000) observed that T-cells blocked HSV reactivation in neurons, but a strict set of immunological requirements had to be met. First, T-cells from naïve mice did not block HSV reactivation, whereas HSV-specific T-cells from latently infected mice blocked HSV reactivation in neurons (Liu *et al.*, 2000). Second, HSV-specific T-cells from BALB/c mice did not block HSV reactivation in C57BL/6 mouse neurons, and reciprocally C57BL/6 T-cells did not block reactivation in BALB/c neurons. Hence, HSV-specific T-cells only blocked HSV reactivation when HSV peptides were presented in the context of a correct MHC class I molecule (i.e. BALB/c mice and C57BL/6 mice encode distinct MHC class I molecules). Finally, the capacity of HSV-specific T-cells to block HSV reactivation in neurons was ablated by the addition of (1) anti-CD8 antibodies (but not anti-CD4 antibodies) or (2) neutralizing antibodies to IFN-γ (Liu *et al.*, 2000). Collectively, these results suggested that IFN-γ secreted by CD8+ T-cells might play a direct role in suppressing HSV reactivation in neurons.

How can CD8+ T-cells directly repress HSV reactivation in neurons?

In principle, HSV reactivation in neurons might be recognized by HSV-specific CD8+ T-cells patrolling the ganglia, IFN-γ secretion could result, and secreted IFN-γ could directly suppress HSV replication in neurons (Fig. 5.1B). The data of Cantin, *et al.* (1999a,b) and Liu, *et al.* (2000) were consistent with this possibility. However, the hypothesis that CD8+ T cells directly suppressed HSV replication in neurons seemed incongruous with other observations, considered as follows.

Do neurons express MHC class I?

In the early 1990s, it was widely held that neurons do not express MHC class I, and thus were incapable of presenting intracellular peptides to CD8+ T-cells (Weinstein *et al.*, 1990; Joly and Oldstone, 1992; Bailey *et al.*, 1994). When more sensitive methods were employed, Tony Simmons' laboratory found that (1) neurons expressed low levels of MHC class I molecules on their cell surface, and (2) MHC class I density increased on neurons during HSV infection of ganglia (Pereira *et al.*, 1994). This upregulation of MHC I molecules was most likely a result of HSV-induced IFN-α/β and/or IFN-γ secretion, as both cytokines induce MHC gene expression (Lampson and George, 1986; Njenga *et al.*, 1997). Numerous studies have since corroborated the finding that neurons express a low density of MHC class I molecules on their cell surface (Redwine *et al.*, 2001; Neumann *et al.*, 2002).

How does interferon-γ inhibit HSV reactivation?

The possibility that IFN-γ secreted by CD8+ T-cells might control HSV reactivation in neurons (Cantin *et al.*, 1999a; Liu *et al.*, 2000) paralleled the findings of Chisari and colleagues, who had previously demonstrated that hepatitis B virus infections were controlled in a non-cytolytic manner that was dependent on CD8+ T-cells and IFN-γ (Guidotti *et al.*, 1994; Guidotti *et al.*, 1996). While Cantin and Hendricks' data were impeccable, it was difficult to reconcile how IFN-γ could suppress HSV reactivation in neurons when IFN-γ had only negligible effects on the replication of wild-type HSV in cultured cells (Harle *et al.*, 2002).

The resolution to this paradox lay in considering the *in vivo* context in which T-cells act to control HSV infections. T-cells never arrive at sites of HSV infection, and secrete IFN-γ into environments that are devoid of other cytokines. At a minimum, secretion of IFN-α/β, proinflammatory cytokines (e.g. TNF-α), and chemoattractant cytokines (e.g. IL-8) precedes the recruitment of T-cells into a site of viral infection. Therefore, a more accurate cell culture test of IFN-γ's capacity to inhibit HSV replication would require treatment of cultured cell with a combination of IFN-γ and at least one other cytokine of the innate immune response.

Testing of this hypothesis revealed that IFN-γ acted in concert with IFNα/β to potently inhibit HSV-1 replication in cultured cells. For example, treatment of Vero cells with IFN-β *or* IFN-γ inhibited HSV-1 replication by less than 20-fold, whereas combinations of IFN-β *and* IFN-γ inhibited HSV-1 replication by nearly 1000-fold (Sainz and Halford, 2002). These results suggested that treatment of cells with ligands of IFN-α/β receptors *or* IFN-γ receptors rendered cells only partly resistant to HSV replication, whereas co-activation of both IFN-α/β receptors *and* IFN-γ receptors rendered cells fully resistant to HSV replication. Hence, IFN-γ secreted by CD8+ T-cells might act in concert with IFN-α/β secreted by virus-infected cells to render neurons fully refractory to HSV replication.

Co-activation of interferon-α/β receptors and interferon-γ receptors is sufficient to repress HSV replication in cultured cells

IFN-α/β- and IFN-γ-signalling pathways are often described in terms that suggest that the receptors function independently of one another. However, Takaoka and colleagues demonstrated that IFN-α/β and IFN-γ receptors co-immunoprecipitate in a 1:1 ratio (Takaoka *et al.*, 2000). If the innate and adaptive interferon responses to viral infections are quite literally linked to one another, then ligand binding to IFN-γ receptors may directly enhance the antiviral effects of IFN-α/β. For example, IFN-γ ligand binding to its receptor might multiply the antiviral effect induced by IFN-β ligand binding to its receptor.

To test this hypothesis, Halford, *et al.* (2005) compared HSV DNA yields at 30 hours p.i. in Vero cells that were inoculated with 0.1 pfu per cell of HSV-1, and which had been pre-treated with 64 unique dose combinations of IFN-β and IFN-γ (Fig. 5.3A). The amount of HSV DNA synthesized in each culture was measured by dot blot analysis, where the blackness of each dot provided a direct measure of HSV DNA abundance. Inhibition of HSV DNA synthesis increased as the dosage of IFN-β was increased, and the dose–response relationship was described by a hyperbolic tangent function that produced an S-shaped curve (blue box in Fig. 5.3A; front edge of graph in Fig. 5.3B). Addition of IFN-γ had no effect on the *shape* of the IFN-β dose–response curve, but rather IFN-γ acted as a multiplier that increased the *amplitude* of the curve (red box in

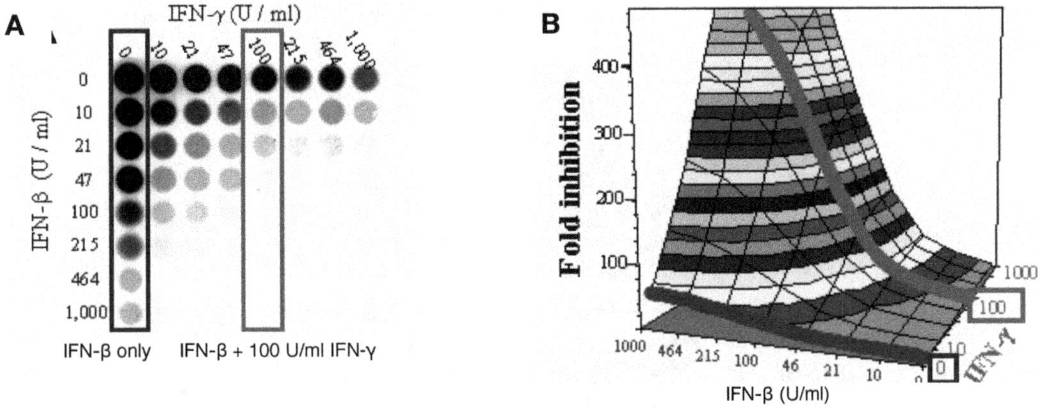

Figure 5.3 Interferon-γ multiplies the potency with which interferon-α/β inhibits HSV-1 replication. (A) Viral DNA harvested 30 h p.i. from Vero cell cultures inoculated with 0.1 pfu per cell of HSV-1 and treated 16 hours earlier with 64 dose combinations of IFN-β and IFN-γ. Viral DNA yields were quantified by the relative abundance of *glycoprotein D* gene-specific oligonucleotide that hybridized to DNA harvested from HSV-1 infected cultures. The blue box denote the effect of IFN-β alone, compared with the red box which denotes the effect of IFN-β when combined with 100 U/ml of IFN-γ. (B) A mathematically modelled response surface that is based on the assumption that combinations of IFN-β and IFN-γ inhibit HSV-1 replication by an amount that is equal to the multiplied product of their individual inhibitory effects. Regression analysis validated that the modelled inhibitory effect (i.e. smoothed) did not significantly differ from the observed effect of combinations of IFN-β and IFN-γ ($P > 0:2$; two-tailed paired *t*-test). The thick blue line superimposed on the response surface denotes the inhibitory effect of IFN-β alone, compared with the thick red line which denotes the effect of IFN-β when combined with 100 U/ml of IFN-γ. Adapted from Halford, *et al.* (2005a) in accordance with the author reproduction policies of Elsevier, Inc. A colour version of this figure is located in the plate section at the back of the book.

Fig. 5.3A; front to back of Fig. 5.3B). Thus, mathematical analysis demonstrated that IFN-γ acted as a multiplier of the potency with which innate IFNs restrict the replication of HSV in otherwise permissive cells (Halford *et al.*, 2005a).

The capacity of IFN-β and IFN-γ to inhibit viral replication synergistically is not unique to HSV-1, but has been observed with cytomegalovirus, SARS virus, and hepatitis C virus replicons (Larkin *et al.*, 2003; Sainz *et al.*, 2004, 2005). The mechanism by which IFN-α/β and IFN-γ synergistically modify the host cell remains obscure (Peng *et al.*, 2008). What is clear is that IFN-α/β and IFN-γ do not synergistically inhibit HSV-1 replication via (1) a cytotoxic or apoptotic effect, (2) a block to HSV entry or (3) an overt suppression of IE and E mRNA or protein synthesis (Pierce *et al.*, 2005). Rather, IFN-α/β and IFN-γ's first, unambiguous synergistic effect on HSV replication is manifest at the level of HSV DNA synthesis, which in turn prevents the synthesis of new HSV virions (Pierce *et al.*, 2005).

Mice that lack IFN-α/β receptors and IFN-γ-receptors fail to control HSV infections

Cell culture studies raised the possibility that IFN-γ might act in concert with IFN-α/β to directly inhibit HSV-1 replication in animals. Luker, *et al.* (2003) put this hypothesis to the test by comparing the spread of a 'glowing' HSV-luciferase virus following rear footpad inoculation of different IFN receptor-knockout mice (Fig. 5.4). In wild type mice, the spread of HSV-luciferase remained largely restricted to the site of inoculation between 1 and 4 days p.i., and all mice survived the infection (top panel in Fig. 5.4; Luker *et al.*, 2003). In IFN-α/β receptor knockout mice, the HSV-luciferase virus transiently spread from rear footpads to the liver on day 3 p.i., but all mice survived the infection (middle panel in Fig. 5.4). In IFN-γ receptor-knockout mice, the spread of HSV-luciferase was indistinguishable from wild-type mice, and all mice survived the infection (Luker *et al.*, 2003).

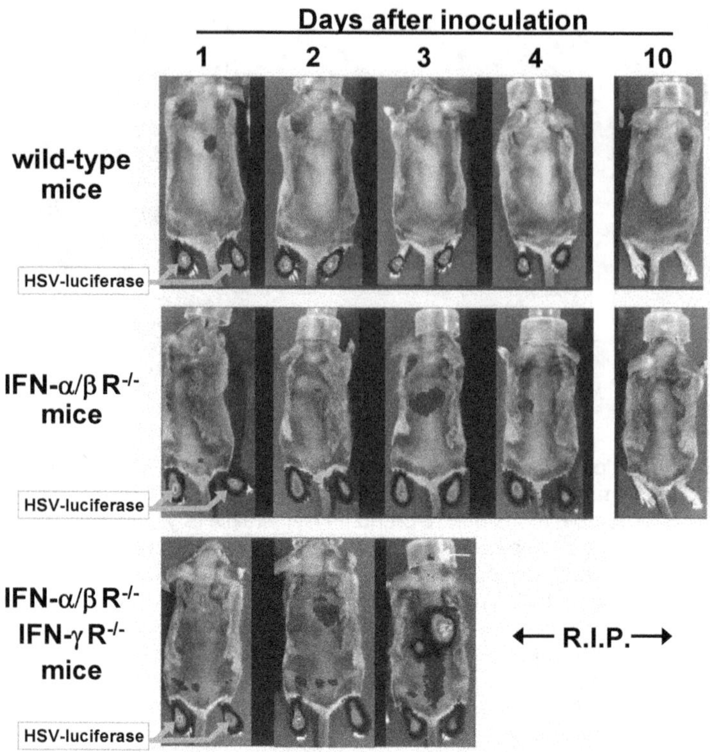

Figure 5.4 Uncontrolled spread of HSV-1 infection in mice lacking both interferon-α/β and interferon-γ signalling pathways. Comparison of the spread of a luciferase-expressing HSV-1 virus in a representative wild-type mouse, IFN-α/β receptor knockout mouse, or a mouse lacking both IFN-α/β receptors and IFN-β receptors. Mice were inoculated in the rear footpads with 2×10^6 pfu of a luciferase-expressing HSV-1 recombinant virus. Bioluminescence imaging was performed every day for 10 days; results of Days 1–4 and 10 are shown. All wild-type and IFN-α/β-receptor knockout mice survived until day 10 p.i.; 0% of double-knockout mice lacking IFN-α/β receptors and IFN-γ receptors survived beyond day 4 p.i. Adapted from Luker *et al.* (2003) with permission from the American Society for Microbiology. A colour version of this figure is located in the plate section at the back of the book.

Mice that lacked both IFN-α/β receptors *and* IFN-γ receptors exhibited a very different pattern of HSV infection (Luker *et al.*, 2003). HSV-luciferase spread in a rapid and uncontrolled manner from the rear footpads to the livers of double-IFN-receptor-knockout mice, and none survived beyond 4 days p.i. (bottom panel in Fig. 5.4). Although animal hosts are intimately dependent on host IFNs to restrict the spread of HSV infection, this fact was overlooked in prior studies that focused on the *partial phenotypes* that result when either IFN-α/β receptors *or* IFN-γ receptors are eliminated. In contrast, Luker, *et al.* (2003) revealed the full phenotype of uncontrolled HSV replication and spread that follows loss of both IFN-α/β and IFN-γ signalling pathways. Hence, IFN signalling is critical to the process by which animals normally restrict the spread of HSV infection to the initial site of infection, and the nerve fibres that innervate this portion of the epithelium (Fig. 5.1A).

Of mice and men

Owing to the wealth of mouse-specific reagents, most advances in HSV immunology have first been made in mice and later tested in humans. Such mouse studies have led us to conclude that T-lymphocytes and the combined activities of the IFN-α/β- and IFN-γ-signalling pathways are required for the process by which wild-type HSV is driven into a latent state in animals. But, do these findings translate to humans? As summarized below, T-cells and host IFN-signalling pathways appear to be highly relevant to the natural process by which latent HSV infections are established and maintained in humans.

CD8+ T cells persist in HSV latently infected ganglia of humans

It was documented in the mid-1990s that CD8+ T-cells persisted at sites of latent HSV infection in mice (Cantin *et al.*, 1995; Shimeld *et al.*, 1995; Halford *et al.*, 1996b; Liu *et al.*, 1996). The relevance of these findings was unclear at the time for two reasons.

First, it was not known if the CD8+ T-cells found in HSV latently infected ganglia represented sentinels that were keeping latent HSV infections in check, or rather were bystanders drawn to the scene of a crime. Bob Hendricks' laboratory addressed this question by determining the fraction of CD8+ T-cells in ganglia that bore antigen receptors specific for HSV peptides. Almost two-thirds of the CD8+ T-cells in HSV latently infected ganglia were specific for amino acids 498–505 of the HSV-1 glycoprotein B, and another 4% were specific for amino acids 822–829 of the HSV-1 ribonucleotide reductase large subunit (Khanna *et al.*, 2003). Therefore, the CD8+ T-cells retained in latently infected ganglia are HSV-specific (Fig. 5.1B), and appear to be actively engaged in the process of responding to subclinical HSV infections (Khanna *et al.*, 2003, 2004; Divito *et al.*, 2006; Knickelbein *et al.*, 2008).

Second, it was possible that the persistence of T-cells in HSV infected ganglia was a bizarre artefact of the mouse model of HSV infection. Using HSV-1 seropositive organ donors as a source of human trigeminal ganglia, Diethilde Theil's laboratory tested this possibility using a combination of *in situ* hybridization for HSV-1 LAT RNA and immunohistochemical staining for T-cells (Theil *et al.*, 2003; Derfuss *et al.*, 2007). Theil found that human ganglia from HSV seropositive donors contained single neurons that were frequently encircled by T-cells (Fig. 5.5A). These neurons were surrounded by T-cells expressing CXCL-3 receptors and other definitive markers of T-cell activation (Derfuss *et al.*, 2007). Yet, neurons remained healthy in appearance, suggesting that T-cells did not unleash their cytotoxic potential on the neurons they encircled (Fig. 5.5A). *In situ* hybridization for HSV-1 LAT RNA confirmed the presence of frequent LAT+ neurons in human trigeminal ganglia that were infiltrated by T-cells (Fig. 5.5B and C). Although T-cell infiltrates were occasionally noted adjacent to HSV LAT+

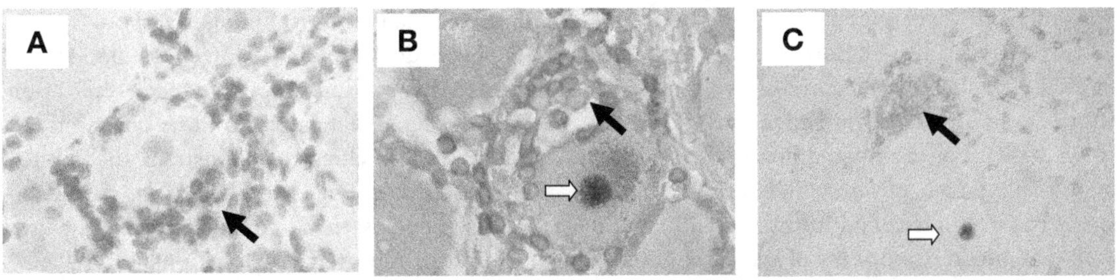

Figure 5.5 T-lymphocytes persist at sites of latent HSV infection in humans. Representative photomicrographs of a tissue section from a human trigeminal ganglion latently infected with HSV-1 stained with anti-CD3 antibodies, which recognize the cell surface CD3 molecule that is associated with the T-cell receptor of both CD4+ T-cells and CD8+ T-cells. (A) A single, large neuronal cell body encircled by an infiltrate of small, brown-stained CD3+ T-cells (denoted by black arrow). Reprinted from *Am. J. Pathol.* 2003, 163: 2179–2184 with permission from the American Society for Investigative Pathology. (B) *In situ* hybridization for HSV-1's LAT RNA introns highlights the nucleus of a single, human neuron that is latently infected with HSV-1 (denoted by white arrow), and immunohistochemical staining reveals the presence of small, brown-stained CD3+ T-cells in an adjacent mononuclear cell infiltrate. Adapted from Derfuss *et al.*, 2007, *Brain Pathol.* 17: 389–398 with permission from John Wiley and Sons. (C) An infiltrate of CD3+ T-cells (denoted by black arrow) drawn to a site several neuronal cell bodies away from an HSV-1 LAT+ neuron (denoted by white arrow). Original images were generously provided by Diethilde Theil. A colour version of this figure is located in the plate section at the back of the book.

neurons (Fig. 5.5B), more often T-cells were noted clustered around neurons that were LAT-negative and which were well separated from HSV LAT^+ neurons (Derfuss *et al.*, 2007; Fig. 5.5C).

Theil's findings were corroborated in an independent study performed on a large cohort of trigeminal ganglia collected from human cadavers (Verjans *et al.*, 2007). Larry Corey and colleagues extended these findings to human carriers of latent HSV-2 infections (Zhu *et al.*, 2007). Biopsies of genital epithelium were taken from the labia of female patients 3 weeks after a recurrence of genital herpes. HSV-specific $CD8^+$ T-cells were found in direct contact with nerve endings at sites of recurrent herpes lesions, but were absent in control biopsies taken from sites where herpes outbreaks did not occur (Zhu *et al.*, 2007).

Deficits in interferon signalling predispose humans to severe HSV infections

Deficits in T-cell-mediated immunity predispose humans to severe herpesviral infections caused by HSV, cytomegalovirus, or varicella-zoster virus. For example, unusually severe recurrences of HSV disease are often noted in patients with AIDS (Posavad *et al.*, 1997), leukaemia (Angarone and Ison, 2008), organ transplants (Miller and Dummer, 2007), severe burns (D'Avignon *et al.*, 2010) or primary immunodeficiencies that prevent T-cell development (Berthet *et al.*, 1994).

Humans may also be born with severe defects in IFN signalling, which predispose these individuals to severe herpesviral infections (Jouanguy *et al.*, 2007). For example, a homozygous defect in the interferon-activated Stat 1 transcription factor has been identified in three children, who each died before 18 months of age (Dupuis *et al.*, 2003; Chapgier *et al.*, 2006). One Stat 1-deficient child died of herpes simplex encephalitis (Dupuis *et al.*, 2003), the second died shortly after a fulminant Epstein–Barr virus infection (Chapgier *et al.*, 2006) and the third child died at home of an undiagnosed, viral-like illness (Dupuis *et al.*, 2003). Likewise, genetic defects in the intracellular UNC-93B protein have been found to compromise the induction of an IFN-induced antiviral state, and have been associated with HSV encephalitis in two small children (Casrouge *et al.*, 2006).

Infiltration of $CD8^+$ T cells in herpes lesions coincides with the cessation of HSV shedding

In humans, the timing of arrival of $CD8^+$ T-cells in recurrent genital lesions correlates strongly with the cessation of HSV-2 replication (Koelle *et al.*, 1998; Posavad *et al.*, 1998). In contrast, immunosuppressed people with deficits in T-cell function present with slowly progressing herpes infections that often fail to resolve without clinical intervention and antiviral drugs (Stanberry *et al.*, 1994; Stewart *et al.*, 1995; Kang *et al.*, 2006; Yudin and Kaul, 2008).

The latency-regulating R_L regions of HSV

Switching gears: from the brakes to the accelerator

The braking mechanism that causes HSV replication to stop *in vivo* is a product of the host immune response that restricts HSV replication, but which fails to clear the virus. While the host immune response explains how the cessation of HSV replication is brought about, it does not address two critical questions: (1) How are latent HSV infections maintained in the human nervous system? (2) How do latent HSV infections periodically re-enter the productive cycle of viral replication? We lack detailed answers to these questions, but possess many clues that constrain the number of possibilities. For example, regulation of HSV latency appears to be controlled from a pair of 10kb long-repeated (R_L) regions in the HSV genome (Fig. 5.2). Synthesis of two R_L-encoded proteins, ICP0 and ICP34.5, promotes HSV replication in neurons, whereas failure to synthesize either protein predisposes HSV to establish a latent infection (Fig. 5.1A). Before discussing the R_L regulatory proteins, we put this discussion of the HSV latency-replication balance into perspective by considering the natural frequency with which HSV infections oscillate between a latent and reactivated state.

How latent is 'latent'?

Approximately 2% of the human population are shedding detectable levels of infectious HSV-1 in their oral cavity on any given day (Miller and Danaher, 2008). Thus, HSV infections are not 'latent' in the absolute sense of the word. Rather, the term 'latent' is relevant in differentiating low-grade HSV infections from viral infections that are truly chronic in nature. For example, humans chronically infected with hepatitis B virus or human immunodeficiency virus may produce *millions* of new infectious virions per day. In contrast, humans who are latently infected with HSV produce only vanishingly small quantities of infectious virus.

The study of Wald, *et al.* (1997) highlighted how the sensitivity of testing for HSV-2 shedding strongly influences our perception of the nature of a latent HSV-2 infection. Over 270 consecutive days, three different measures of HSV-2 reactivation were compared in women with recurrent genital herpes: (1) symptoms of herpetic disease, (2) infectious virus recovered from genital swabs or (3) HSV-2 DNA detected by PCR (Wald *et al.*, 1997). On 3% of days, these women were aware of prodrome (nerve tingling) or active herpes lesions. On 8% of days, infectious HSV-2 was recovered from genital swabs. On 28% of days, genital swabs contained HSV-2 DNA that was detected by PCR, and 60% of HSV-2 DNA shedding events were too brief in duration to produce symptoms (Wald *et al.*, 1997). Thus, our estimates of the frequency of HSV-2 reactivation increase by an order of magnitude when a more sensitive assay is used to detect subclinical HSV-2 reactivation events.

Latent HSV infections are unlikely to be static in nature (Buddingh *et al.*, 1953). Rather, latent HSV infections appear to consist of a population of 1000–10,000 HSV latently infected neurons where, once a week, a single HSV-infected neuron exits the pool of latently infected neurons and re-enters the productive cycle of HSV replication. Hence, HSV virions are exported from the ganglia to the epithelium of human carriers on a regular basis (Buddingh *et al.*, 1953; Wald *et al.*, 1997). Consequently, an immune response persists in human ganglia that harbour these low-grade, persistent HSV infections (Theil *et al.*, 2003; Verjans *et al.*, 2007).

The R_L regions of the HSV genome

The regulation of HSV latency appears to be controlled from a pair of 10 kb, inverted R_L regions in the HSV genome (Fig. 5.2). There are four well-established genes in the R_L regions: *LAT*, *L/ST*, *ICP0* and *ICP34.5*. The *LAT* and *L/ST* genes share extensive antisense overlap with the *ICP0* and *ICP34.5* genes and produce what appear to be non-coding RNAs (Fig. 5.2). It remains unclear what purpose is served by the 1.5 and 2.0 kb LAT RNA-derived introns that accumulate in HSV latently infected neurons (Fig. 5.1B). Likewise, the function of the long/short spanning transcript (L/ST) RNA is unclear. Thus, LAT and L/ST RNAs are enigmatic and do not yet clarify how the HSV R_L regions influence the outcome of an HSV infection.

In contrast, the *ICP0* and *ICP34.5* genes produce two proteins, ICP0 and ICP34.5, which regulate the probability of HSV replication *in vivo*. Importantly, HSV *ICP34.5* null viruses replicate like wild type virus in cultured cells, and HSV *ICP0* null viruses replicate like wild type virus at high multiplicities of infection. Thus, ICP0 and ICP34.5 are not essential cogs in the machinery of viral replication, but rather are regulatory proteins that allow HSV to 'choose' whether its replication machinery will start (ICP0) or keep functioning (ICP34.5) for long enough to produce new infectious virions. Synthesis of ICP0 and ICP34.5 strongly promote HSV replication *in vivo*. It has been postulated that the antisense *LAT* and *L/ST* genes may contribute to the establishment and/or maintenance of latent HSV infections by antagonizing *ICP0* and *ICP34.5* gene expression (Tang *et al.*, 2008; Umbach *et al.*, 2008).

ICP0 functions as an activator/anti-repressor of HSV replication

ICP0 was identified as a protein that transformed the HSV major transcriptional regulator, ICP4, from a weak transcriptional activator to a potent activator of mRNA synthesis; specifically, combinations of ICP0 and ICP4 are 20-fold more potent at driving HSV mRNA synthesis than either ICP0 or ICP4 alone (Everett, 1984; Gelman and Silverstein, 1986). Functionally, synthesis of ICP0 causes the HSV's equilibrium to abruptly tip towards productive replication, whereas absence of ICP0 produces the opposite effect. Synthesis of ICP0 is sufficient to trigger HSV reactivation in trigeminal ganglion neurons and other models of latent HSV infection (Halford *et al.*, 2001; Terry-Allison *et al.*, 2007). While HSV *ICP0*-deficient viruses replicate to nearly wild type levels at high multiplicities of infection, at low multiplicities of infection the same *ICP0*-deficient viruses establish quiescent infections in 99% of the cells they infect (Sacks and Schaffer, 1987; Everett, 1989). Such observations suggest that ICP0 antagonizes a repressor of HSV replication, whose repressive capacity can be saturated with high numbers of HSV genomes (Everett, 1989).

Because ICP0 is an E3 ubiquitin ligase (Boutell *et al.*, 2002), it has been suggested that a repressor may silence HSV mRNA synthesis in the absence of ICP0, whereas accumulation of ICP0 may neutralize this repressor and allow HSV mRNA synthesis to proceed unhindered (Everett, 2000; Preston, 2000). The details remain an area of debate, but the big picture is clear; accumulation of ICP0 promotes productive HSV replication, whereas absence of ICP0 strongly favours the repression of HSV gene expression (Liu *et al.*, 2010).

ICP34.5: a regulator of the progression of HSV replication in neurons

In the late 1980s, Rick Thompson's laboratory identified a neurovirulence factor encoded from the R_L1 gene, which came to be known as the *ICP34.5* gene (Fig. 5.2). Deletion of the R_L1 gene had no effect on the capacity of HSV to replicate in cultured cells, but prevented HSV from completing its replication cycle in neurons *in vivo* (Javier *et al.*, 1987; Thompson *et al.*, 1989; Bolovan *et al.*, 1994). Chou, *et al.* (1995) discovered that the R_L1 gene product, ICP34.5, interacted with cellular protein phosphatase 1 and stimulated the dephosphorylation of eIF-2α (Chou *et al.*, 1995). What is most relevant to initially recognize about ICP34.5 is that absence of this protein strongly predisposes HSV to establish latent infections in neurons (Thompson *et al.*, 1989). In fact, HSV *ICP34.5*-deficient viruses are so predisposed to establish latent infections in neurons, that high doses of HSV *ICP34.5*-deficient viruses may be injected into brain tumours of human patients. Owing to the failure of HSV *ICP34.5*-deficient viruses to replicate in healthy human neurons, these oncolytic viruses can selectively destroy cancer cells in the brain without harming healthy brain tissue in glioma patients (Markert *et al.*, 2000; Todo, 2008).

The R_L regions regulate the sensitivity of HSV to interferon

Hypothesis: host interferons versus viral interferon antagonists

HSV infections oscillate between a state of latency and active replication in a subset of neurons *in vivo* (Fig. 5.1A). The host IFN response serves as a repressive stimulus that pushes this equilibrium towards latency (Fig. 5.1A). The R_L-coded activators, ICP0 and ICP34.5, push this equilibrium in the opposite direction (Fig. 5.1A). As reviewed below, the available evidence indicates that ICP0 and ICP34.5's capacity to promote HSV replication *in vivo* is linked to their function as viral IFN antagonists.

ICP0 and ICP34.5 function as viral IFN antagonists. When synthesized, ICP0 and ICP34.5 endow HSV with the capacity to sustain replication in the face of the antiviral state induced by IFN-α/β (Leib *et al.*, 1999; Leib *et al.*, 2000). Specifically, the resistance of HSV to IFN-α/β-inducible repression increases by ~100-fold when both ICP0 and ICP34.5 are expressed. In contrast, failure to synthesize either ICP0 or ICP34.5 ensures that HSV will succumb to the IFN-α/β-induced antiviral state and will predominantly establish silent infections (Mossman *et al.*, 2000; Harle *et al.*, 2002; Mossman and Smiley, 2002).

The HSV R_L regions function as an IFN resistance locus (Fig. 5.2). When this fact is integrated into an equilibrium model of HSV latency (Fig. 5.1A), three insights emerge. First, synthesis of ICP0 and ICP34.5 is necessary for the 'IFN-resistant' phenotype that allows HSV to sustain replication *in vivo*. Second, failure to express ICP0 or ICP34.5 renders HSV 'IFN-sensitive' and prone to establish latent infections. Finally, the collective gene activity in the HSV R_L regions appears to imbue HSV with the capacity to 'choose' whether viral replication will proceed, or halt, at one of two IFN-inducible restrictions to HSV gene expression (Mossman and Smiley, 2002; Halford *et al.*, 2006). This hypothesis is attractive because it assigns a tangible function to the latency-regulating R_L regions of HSV (Fig. 5.2). Moreover, this hypothesis implies that failure to synthesize ICP0 or ICP34.5 should promote the establishment of a latent HSV infection if, and only if, the host IFN response is intact and available to restrict the progression of HSV replication. We discuss below the evidence that the balance between latency and replication is tipped in opposite directions by ICP0 and ICP34.5 versus the host IFN-α/β response (Fig. 5.1A).

HSV must synthesize ICP0 and ICP34.5 to resist interferon-induced repression *in vitro*

Mossman and colleagues found that the resistance of HSV to IFN-α/β was an active process, and was dependent on the synthesis of both ICP0 and ICP34.5 (Mossman *et al.*, 2000; Mossman and Smiley, 2002). Wild-type HSV replicates ~3-fold less efficiently in IFN-α/β -activated cells relative to untreated cells. In contrast, HSV *ICP0* null viruses replicate ~300-fold less efficiently in IFN-α/β-activated cells relative to untreated cells (Harle *et al.*, 2002; Halford *et al.*, 2006). Likewise, HSV *ICP34.5* null viruses exhibit a similar IFN-sensitive phenotype (Mossman and Smiley, 2002). Importantly, the IFN-sensitive phenotype is not a non-specific phenotype that results when any of the ~30 accessory genes of HSV are mutated; mutations in the *VP16*, *ICP22*, *UL13* or *vhs* genes do not affect the resistance of HSV to IFN-α/β-inducible-repression in cultured cells (Mossman *et al.*, 2000). Therefore, the 'IFN-sensitive' phenotype of HSV may be defined as an ~100-fold increase in sensitivity to IFN-α/β-induced repression *in vitro* that results from failure to synthesize a single HSV protein. To date, ICP0 and ICP34.5 are the only HSV proteins that satisfy this stringent, functional definition of an interferon antagonist.

How do ICP0 and ICP34.5 antagonize interferon-induced repression of HSV?

ICP34.5

IFN-α/β often inhibits the replication of viruses via a shutoff of viral protein translation that is based on the IFN-inducible, dsRNA-activated protein kinase (PKR) and its substrate eIF-2α. When PKR phosphorylates eIF-2α, the eIF2 translation factor is unable to assemble 80S ribosome complexes on mRNA, and protein translation is inhibited (Gebauer and Hentze, 2004). ICP34.5 was first identified as a neurovirulence factor essential for HSV replication in neurons (Thompson *et al.*, 1989). It was later discovered that ICP34.5 stimulated dephosphorylation of eIF-2α by binding protein phosphatase 1 (Chou *et al.*, 1995; Mohr and Gluzman, 1996). Thus, ICP34.5 appeared to be a PKR antagonist, and the replication defect of HSV *ICP34.5* null viruses in neurons was potentially the result of an IFN-inducible block to HSV protein synthesis.

Leib, *et al.* (2000) reasoned that that if ICP34.5 antagonized an IFN-α/β-inducible, PKR-dependent block to HSV replication, then HSV *ICP34.5* null viruses should re-acquire their virulence in knockout mice deficient in IFN-α/β receptors or PKR. Both predictions proved to be correct. High doses of HSV *ICP34.5* null viruses produced little disease when injected into the brains of normal mice. However, the same HSV *ICP34.5* null viruses were uniformly lethal when injected into the brains of PKR-knockout mice or IFN-α/β receptor-knockout mice. Hence, the failure of HSV *ICP34.5* null viruses to replicate in neurons was the result of an *active host response* that was dependent on PKR and IFN-α/β receptors (Leib *et al.*, 2000).

ICP0

Unlike ICP34.5, we lack a crisp, biochemical explanation for the capacity of ICP0 to overcome IFN-inducible repression of HSV. When ICP0 *is* synthesized, presence or absence of IFN-α/β has negligible effects on the efficiency of HSV IE and E mRNA synthesis (Mossman *et al.*, 2000). In contrast, when ICP0 *is not* synthesized, pre-treatment of cells with IFN-α/β prevents the accumulation of HSV IE and E mRNAs (Mossman *et al.*, 2000). Thus, in IFN-α/β -activated cells, synthesis of ICP0 is required if HSV IE and E proteins are to accumulate to levels that support HSV replication and virion synthesis (Mossman *et al.*, 2000; Harle *et al.*, 2002).

HSV must synthesize ICP0 and ICP34.5 to resist innate immune repression in rag2-/- mice

The HSV latency-regulating R_L regions (Fig. 5.2) may control the balance between latency and replication by allowing HSV to alternate between relative states of IFN-sensitivity (latency-prone) versus IFN-resistance (replication-prone). For example, down-regulation of ICP0 or ICP34.5 synthesis may predispose HSV to be repressed by the IFN-α/β -induced antiviral state, hence favouring the establishment of a latent infection. Conversely, efficient synthesis of ICP0 and ICP34.5 may render HSV resistant to repression by the IFN-α/β -induced antiviral state.

Severe-combined immunodeficient (SCID) mice and *rag2-/-* mice lack mature B and T-lymphocytes, and represent a powerful system to test the putative role of ICP0 and ICP34.5 in rendering HSV resistant to repressive effects of the innate immune system. In the absence of T-lymphocytes, wild type HSV-1 infection of SCID mice or *rag2-/-* mice is invariably fatal and culminates in encephalitis. Of course, the arrangement of genes in the R_L regions favours the synthesis of ICP0 and ICP34.5 in the absence of T-lymphocytes, and thus wild-type HSV is genetically predisposed to resist repression by IFN-α/β (Fig. 5.2). But, what happens if this predisposition is eliminated by mutations that prevent the synthesis of ICP0 or ICP34.5?

HSV-1 *ICP0* null viruses and HSV *ICP34.5* null viruses are avirulent in SCID mice and *rag2-/-* mice (Valyi-Nagy *et al.*, 1994a; Halford *et al.*, 2006). HSV-1 *ICP0* null viruses replicate efficiently in the eyes of wild-type mice and *rag2-/-* mice for the first 48 hours p.i. (Fig. 5.6A), and then appear to be driven into a latent state based on the relative absence of viral shedding and disease (Fig. 5.6A,B). When *rag2-/-* mice are sacrificed 60 days p.i., infectious virus is undetectable in homogenates of the eyes or trigeminal ganglia but HSV-1 *ICP0* null viral DNA is readily detected in the trigeminal ganglia by PCR (Halford and Schaffer, 2000; Halford *et al.*, 2006). Likewise, HSV-1 *ICP34.5* null viruses are avirulent in lymphocyte-deficient SCID mice

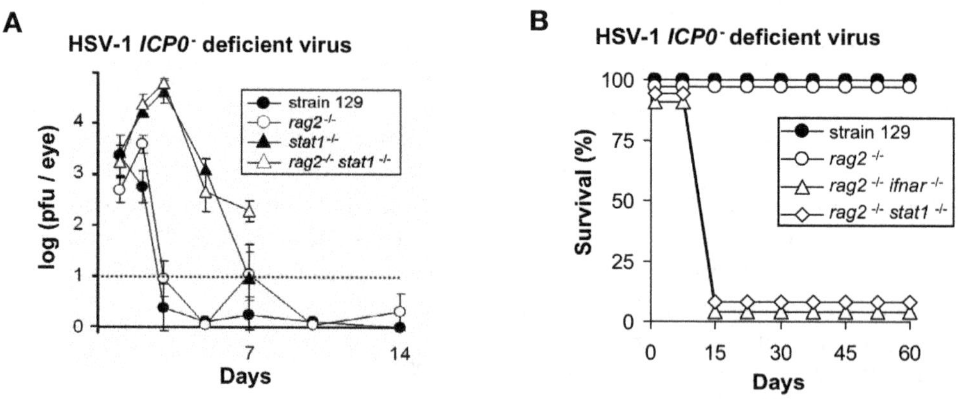

Figure 5.6 HSV *ICP0* null viruses are repressed in *rag2-/-* mice in a manner that is dependent on IFN-α/β receptors and Stat 1. (A) HSV-1 shedding in eyes of strain 129 mice, *rag2-/-* mice, *stat1-/-* mice, or *rag2-/- stat1-/-* mice inoculated with 2×10⁵ pfu per eye of the *ICP0* virus, HSV-1 0-GFP (*n* = 4 mice per group). The dashed line indicates the lower limit of detection of the plaque assay. (B) Results of an independent experiment comparing duration of survival of strain 129 mice, *rag2-/-* mice, *rag2-/- ifnar-/-* mice, or *rag2-/- stat1-/-* mice that were inoculated with 2×10⁵ pfu per eye of the *ICP0-* virus, HSV-1 0-GFP (*n* = 8 mice per group). Adapted from Halford, *et al.* (2006) in accordance with the open-access policy of BioMed Central.

(Valyi-Nagy *et al.*, 1994a), and similar analyses suggest that HSV-1 *ICP34.5* null viruses efficiently establish latent infections in *rag2*-/- mice (W.P. Halford, unpublished).

Repression of HSV ICP0– or ICP34.5– null viruses in rag2–/– mice is IFN-α/β dependent

The fact that HSV *ICP0* null or *ICP34.5* null viruses are avirulent in *rag2*-/- mice does not prove that these viruses are driven into a latent state. Thus, it is critical to distinguish between two equally viable explanations for these observations. Like all of the essential proteins of HSV (e.g. ICP4, ICP27, ICP8), synthesis of ICP0 and ICP34.5 may be required for HSV to sustain productive replication in any experimental animal including a *rag2*-/- mice. Alternatively, the IFN-sensitive phenotype of HSV *ICP0* null or *ICP34.5* null viruses may render these viruses prone to establish latent infections in *rag2*-/- mice that possess an intact IFN-α/β response.

To differentiate between these possibilities, replication of an HSV-1 *ICP0* null virus was compared in lymphocyte-deficient *rag2*-/- mice versus *rag2*-/- mice that possessed secondary defects in their IFN-α/β receptors (*rag2*-/- *ifnar*-/-) or the downstream Stat 1 transcription factor (*rag2*-/- *stat1*-/-). On day 3 p.i., replication of an HSV-1 *ICP0* null virus was low to undetectable in the eyes of *rag2*-/- mice (Fig. 5.6A). In contrast, an HSV-1 *ICP0* null virus replicated to ~1000-fold higher levels in *rag2*-/- *stat1*-/- mice (Fig. 5.6A) or *rag2*-/- *ifnar*-/- mice (Table 3 of Halford *et al.*, 2006). While wild-type mice and *rag2*-/- mice remained disease-free for 60 days p.i. (Fig. 5.6B), the same HSV-1 *ICP0* null virus produced lethal disease within 11 ± 1 days p.i. of *rag2*-/- *ifnar*-/- mice and *rag2*-/- *stat1*-/- mice (Halford *et al.*, 2006; Fig. 5.6B). Similar comparisons of replication and pathogenesis revealed that HSV-1 *ICP34.5* null viruses rapidly produce lethal infections in *rag2*-/- *stat1*-/- mice, despite their avirulent phenotype in *rag2*-/- mice (W.P. Halford, unpublished).

Based on these considerations, neither ICP0 nor ICP34.5 is strictly required for HSV-1 to replicate in animals (Halford and Schaffer, 2000; Leib *et al.*, 2000; Halford *et al.*, 2006). Rather, failure to synthesize ICP0 or ICP34.5 strongly predisposes HSV to establish what appear to be latent infections in animals that retain a functional IFN-α/β-signalling pathway.

Conclusion

Knowledge of HSV latency has grown tremendously in the past 30 years. With the advent of recombinant DNA technology, HSV virology exploded as a science in the 1980s, and yielded new insights into the genetic code of HSV, LAT RNAs, and IE regulatory proteins. In the 1990s, it became clear that T-cells provided the stimuli that triggered the establishment of latent HSV infections in neurons. In the 2000s, host IFN-signalling pathways emerged as a critical connection that linked the repressive stimuli provided by leucocytes to the intracellular restriction of HSV replication in host cells including neurons.

The phenomenon of HSV latency continues to be studied and discussed in predominantly virological terms. Such discussions tend to be confusing because the 'repressive stimulus' that drives HSV into a latent state emanates from a source that lies outside the traditional scope of virology. Recent data illuminate the path out of this quandary: it is time to broaden our scope! Latent HSV infections are the result of an equilibrium between a virus and the host immune response to that virus (Khanna *et al.*, 2004; Decman *et al.*, 2005; Divito *et al.*, 2006; Fig. 5.1). In light of this new principle, understanding of HSV latency is poised to advance to a more mechanistic level as the details emerge regarding how (1) the host IFN response drives HSV into a latent state, and how (2) differential regulation of the IFN antagonists of HSV is achieved.

Acknowledgements

The authors thank Robert Jordan and Bill Sugden for critical review of this manuscript, and their many helpful suggestions. The authors are indebted to the National Institutes of Health for supporting efforts to understand the *Role of the LAT-ICP0 locus in regulating HSV latency* (R01 AI51414), and supporting the *Development of an Effective Genital Herpes Vaccine* (R21 AI81072).

References

Akhova, O., Bainbridge, M., and Misra, V. (2005). The neuronal host cell factor-binding protein Zhangfei inhibits herpes simplex virus replication. J. Virol. *79*, 14708–14718.

Allan, R.S., Waithman, J., Bedoui, S., Jones, C.M., Villadangos, J.A., Zhan, Y., Lew, A.M., Shortman, K., Heath, W.R., and Carbone, F.R. (2006). Migratory dendritic cells transfer antigen to a lymph node-resident dendritic cell population for efficient CTL priming. Immunity *25*, 153–162.

Angarone, M., and Ison, M.G. (2008). Prevention and early treatment of opportunistic viral infections in patients with leukemia and allogeneic stem cell transplantation recipients. J Natl Compr Canc Netw *6*, 191–201.

Bailey, K.A., Drago, J., and Bartlett, P.F. (1994). Neuronal progenitors identified by their inability to express class I histocompatibility antigens in response to interferon-gamma. J. Neurosci. Res. *39*, 166–177.

Berthet, F., Le Deist, F., Duliege, A.M., Griscelli, C., and Fischer, A. (1994). Clinical consequences and treatment of primary immunodeficiency syndromes characterized by functional T and B lymphocyte anomalies (combined immune deficiency). Pediatrics *93*, 265–270.

Biron, C.A., Byron, K.S., and Sullivan, J.L. (1989). Severe herpesvirus infections in an adolescent without natural killer cells. N. Engl. J. Med. *320*, 1731–1735.

Bloom, D.C. (2004). HSV LAT and neuronal survival. Int. Rev. Immunol. *23*, 187–198.

Bloom, D.C., Giordani, N.V., and Kwiatkowski, D.L. (2010). Epigenetic regulation of latent HSV-1 gene expression. Biochim. Biophys. Acta *1799*, 246–256.

Bolovan, C.A., Sawtell, N.M., and Thompson, R.L. (1994). ICP34.5 mutants of herpes simplex virus type 1 strain 17syn+ are attenuated for neurovirulence in mice and for replication in confluent primary mouse embryo cell cultures. J. Virol. *68*, 48–55.

Boutell, C., Sadis, S., and Everett, R.D. (2002). Herpes simplex virus type 1 immediate-early protein ICP0 and is isolated RING finger domain act as ubiquitin E3 ligases *in vitro*. J. Virol. *76*, 841–850.

Buddingh, G.J., Schrum, D.I., Lanier, J.C., and Guidry, D.J. (1953). Studies of the natural history of herpes simplex virus infections. Pediatrics *11*, 593–605.

Cantin, E., Tanamachi, B., and Openshaw, H. (1999a). Role for gamma interferon in control of herpes simplex virus type 1 reactivation. J. Virol. *73*, 3418–3423.

Cantin, E., Tanamachi, B., Openshaw, H., Mann, J., and Clarke, K. (1999b). Gamma interferon (IFN-gamma) receptor null-mutant mice are more susceptible to herpes simplex virus type 1 infection than IFN-gamma ligand null-mutant mice. J. Virol. *73*, 5196–5200.

Cantin, E.M., Hinton, D.R., Chen, J., and Openshaw, H. (1995). Gamma interferon expression during acute and latent nervous system infection by herpes simplex virus type 1. J. Virol. *69*, 4898–4905.

Carr, D.J., and Tomanek, L. (2006). Herpes simplex virus and the chemokines that mediate the inflammation. Curr. Top. Microbiol. Immunol. *303*, 47–65.

Casrouge, A., Zhang, S.Y., Eidenschenk, C., Jouanguy, E., Puel, A., Yang, K., Alcais, A., Picard, C., Mahfoufi, N., Nicolas, N., *et al.* (2006). Herpes simplex virus encephalitis in human UNC-93B deficiency. Science *314*, 308–312.

Chapgier, A., Wynn, R.F., Jouanguy, E., Filipe-Santos, O., Zhang, S., Feinberg, J., Hawkins, K., Casanova, J.L., and Arkwright, P.D. (2006). Human complete Stat-1 deficiency is associated with defective type I and II IFN responses *in vitro* but immunity to some low virulence viruses *in vivo*. J. Immunol. *176*, 5078–5083.

Chou, J., Chen, J.J., Gross, M., and Roizman, B. (1995). Association of a M(r) 90,000 phosphoprotein with protein kinase PKR in cells exhibiting enhanced phosphorylation of translation initiation factor eIF-2 alpha and premature shutoff of protein synthesis after infection with gamma 134.5- mutants of herpes simplex virus 1. Proc. Natl. Acad. Sci. U.S.A. *92*, 10516–10520.

Coen, D.M., Kosz-Vnenchak, M., Jacobson, J.G., Leib, D.A., Bogard, C. L., Schaffer, P.A., Tyler, K.L., and Knipe, D.M. (1989). Thymidine kinase-negative herpes simplex virus mutants establish latency in mouse trigeminal ganglia but do not reactivate. Proc. Natl. Acad. Sci. U.S.A. *86*, 4736–4740.

Cook, W. J., Kramer, M.F., Walker, R.M., Burwell, T.J., Holman, H.A., Coen, D.M., and Knipe, D.M. (2004). Persistent expression of chemokine and chemokine receptor RNAs at primary and latent sites of herpes simplex virus 1 infection. Virol. J. *1*, 5.

Cunningham, A.L., Carbone, F., and Geijtenbeek, T.B. (2008). Langerhans cells and viral immunity. Eur J. Immunol. *38*, 2377–2385.

D'Avignon, L.C., Hogan, B.K., Murray, C.K., Loo, F.L., Hospenthal, D.R., Cancio, L.C., Kim, S.H., Renz, E.M., Barillo, D., Holcomb, J.B., *et al.* (2010). Contribution of bacterial and viral infections to attributable mortality in patients with severe burns: an autopsy series. Burns *36*, 773–779.

Davison, A.J. (2002). Evolution of the herpesviruses. Vet. Microbiol. *86*, 69–88.

Decman, V., Freeman, M.L., Kinchington, P.R., and Hendricks, R.L. (2005). Immune control of HSV-1 latency. Viral Immunol. *18*, 466–473.

Derfuss, T., Segerer, S., Herberger, S., Sinicina, I., Hufner, K., Ebelt, K., Knaus, H.G., Steiner, I., Meinl, E., Dornmair, K., *et al.* (2007). Presence of HSV-1 immediate early genes and clonally expanded T-cells with a memory effector phenotype in human trigeminal ganglia. Brain Pathol. *17*, 389–398.

Divito, S., Cherpes, T.L., and Hendricks, R.L. (2006). A triple entente: virus, neurons, and CD8+ T cells maintain HSV-1 latency. Immunol. Res. *36*, 119–126.

Dupuis, S., Jouanguy, E., Al-Hajjar, S., Fieschi, C., Al-Mohsen, I.Z., Al-Jumaah, S., Yang, K., Chapgier, A., Eidenschenk, C., Eid, P., *et al.* (2003). Impaired response to interferon-alpha/beta and lethal viral disease in human STAT1 deficiency. Nat. Genet. 33, 388–391.

Everett, R.D. (1984). Trans activation of transcription by herpes virus products: requirement for two HSV-1 immediate-early polypeptides for maximum activity. EMBO J. *3*, 3135–3141.

Everett, R.D. (1989). Construction and characterization of herpes simplex virus type 1 mutants with defined lesions in immediate early gene 1. J. Gen. Virol. *70 (Pt 5)*, 1185–1202.

Everett, R.D. (2000). ICP0, a regulator of herpes simplex virus during lytic and latent infection. Bioessays *22*, 761–770.

Farrar, M.A., and Schreiber, R.D. (1993). The molecular cell biology of interferon-gamma and its receptor. Annu. Rev. Immunol. *11*, 571–611.

Farrell, M.J., Dobson, A.T., and Feldman, L.T. (1991). Herpes simplex virus latency-associated transcript is a stable intron. Proc. Natl. Acad. Sci. U.S.A. *88*, 790–794.

Fraser, N.W., Block, T., and Spivack, J.G. (1992). The latency-associated transcripts of herpes simplex virus: RNA in search of function. Virology *191*, 1–8.

Frucht, D.M., Fukao, T., Bogdan, C., Schindler, H., O'Shea, J.J., and Koyasu, S. (2001). IFN-gamma production by antigen-presenting cells: mechanisms emerge. Trends Immunol *22*, 556–560.

Gary-Gouy, H., Lebon, P., and Dalloul, A.H. (2002). Type I interferon production by plasmacytoid dendritic cells and monocytes is triggered by viruses, but the level of production is controlled by distinct cytokines. J. Interferon Cytokine Res. *22*, 653–659.

Gebauer, F., and Hentze, M.W. (2004). Molecular mechanisms of translational control. Nat. Rev. Mol. Cell Biol. *5*, 827–835.

Gelman, I.H., and Silverstein, S. (1986). Co-ordinate regulation of herpes simplex virus gene expression is mediated by the functional interaction of two immediate early gene products. J. Mol. Biol. *191*, 395–409.

Gesser, R.M., Valyi-Nagy, T., and Fraser, N.W. (1994). Restricted herpes simplex virus type 1 gene expression within sensory neurons in the absence of functional B and T lymphocytes. Virology *200*, 791–795.

Guidotti, L.G., Ando, K., Hobbs, M.V., Ishikawa, T., Runkel, L., Schreiber, R.D., and Chisari, F.V. (1994). Cytotoxic T lymphocytes inhibit hepatitis B virus gene expression by a noncytolytic mechanism in transgenic mice. Proc. Natl. Acad. Sci. U.S.A. *91*, 3764–3768.

Guidotti, L.G., Ishikawa, T., Hobbs, M.V., Matzke, B., Schreiber, R., and Chisari, F.V. (1996). Intracellular inactivation of the hepatitis B virus by cytotoxic T lymphocytes. Immunity *4*, 25–36.

Halford, W.P., and Schaffer, P.A. (2000). Optimized viral dose and transient immunosuppression enable herpes simplex virus ICP0-null mutants to establish wild-type levels of latency *in vivo*. J. Virol. *74*, 5957–5967.

Halford, W.P., and Schaffer, P.A. (2001). ICP0 is required for efficient reactivation of herpes simplex virus type 1 from neuronal latency. J. Virol. *75*, 3240–3249.

Halford, W.P., Gebhardt, B.M., and Carr, D.J. (1996a). Mechanisms of herpes simplex virus type 1 reactivation. J. Virol. *70*, 5051–5060.

Halford, W.P., Gebhardt, B.M., and Carr, D.J. (1996b). Persistent cytokine expression in trigeminal ganglion latently infected with herpes simplex virus type 1. J. Immunol. *157*, 3542–3549.

Halford, W.P., Gebhardt, B.M., and Carr, D. J.J. (1997). Acyclovir blocks cytokine gene expression in trigeminal ganglia latently infected with herpes simplex virus type 1. Virology *238*, 53–63.

Halford, W.P., Kemp, C.D., Isler, J.A., Davido, D.J., and Schaffer, P.A. (2001). ICP0, ICP4, or VP16 expressed from adenovirus vectors induces reactivation of latent herpes simplex virus type 1 in primary cultures of latently infected trigeminal ganglion cells. J. Virol. *75*, 6143–6153.

Halford, W.P., Balliet, J.D., and Gebhardt, B.M. (2004). Re-evaluating natural resistance to herpes simplex virus type 1. J. Virol. *78*, 10086–10095.

Halford, W.P., Halford, K.J., and Pierce, A.T. (2005a). Mathematical analysis demonstrates that interferons-beta and -gamma interact in a multiplicative manner to disrupt herpes simplex virus replication. J. Theor. Biol. *235*, 439–454.

Halford, W.P., Maender, J.L., and Gebhardt, B.M. (2005b). Re-evaluating the role of natural killer cells in innate resistance to herpes simplex virus type 1. Virol. J. *2*, 56.

Halford, W.P., Weisend, C., Grace, J., Soboleski, M., Carr, D.J., Balliet, J.W., Imai, Y., Margolis, T.P., and Gebhardt, B.M. (2006). ICP0 antagonizes Stat 1-dependent repression of herpes simplex virus: implications for the regulation of viral latency. Virol. J. *3*, 44.

Harle, P., Sainz, B., Jr., Carr, D.J., and Halford, W.P. (2002). The immediate-early protein, ICP0, is essential for the resistance of herpes simplex virus to interferon-alpha/beta. Virology *293*, 295–304.

Heath, W.R., and Carbone, F.R. (2009). Dendritic cell subsets in primary and secondary T cell responses at body surfaces. Nat. Immunol. *10*, 1237–1244.

Hikono, H., Kohlmeier, J.E., Ely, K.H., Scott, I., Roberts, A.D., Blackman, M.A., and Woodland, D.L. (2006). T-cell memory and recall responses to respiratory virus infections. Immunol. Rev. *211*, 119–132.

Honess, R.W., and Roizman, B. (1974). Regulation of herpesvirus macromolecular synthesis. I. Cascade regulation of the synthesis of three groups of viral proteins. J. Virol. *14*, 8–19.

Javier, R.T., Thompson, R.L., and Stevens, J.G. (1987). Genetic and biological analyses of a herpes simplex virus intertypic recombinant reduced specifically for neurovirulence. J. Virol. *61*, 1978–1984.

Joly, E., and Oldstone, M.B. (1992). Neuronal cells are deficient in loading peptides onto MHC class I molecules. Neuron *8*, 1185–1190.

Jouanguy, E., Zhang, S.Y., Chapgier, A., Sancho-Shimizu, V., Puel, A., Picard, C., Boisson-Dupuis, S., Abel, L., and Casanova, J.L. (2007). Human primary immunodeficiencies of type I interferons. Biochimie *89*, 878–883.

Kang, Y.N., Oh, H.K., Chang, Y.C., Kim, H.C., Lee, S.L., Hwang, M., and Park, K.K. (2006). Systemic herpes simplex virus infection following cadaveric renal transplantation: a case report. Transplant. Proc. *38*, 1346–1347.

Khanna, K., Bonneau, R., Kinchington, P., and Hendricks, R. (2003). Herpes simplex virus-specific memory CD8(+) T cells are selectively activated and retained in latently infected sensory ganglia. Immunity *18*, 593–603.

Khanna, K.M., Lepisto, A.J., and Hendricks, R.L. (2004). Immunity to latent viral infection: many skirmishes but few fatalities. Trends Immunol. *25*, 230–234.

Knickelbein, J.E., Khanna, K.M., Yee, M.B., Baty, C.J., Kinchington, P.R., and Hendricks, R.L. (2008). Noncytotoxic lytic granule-mediated CD8+ T cell inhibition of HSV-1 reactivation from neuronal latency. Science *322*, 268–271.

Knipe, D.M., and Cliffe, A. (2008). Chromatin control of herpes simplex virus lytic and latent infection. Nat. Rev. Microbiol. *6*, 211–221.

Knotts, F.B., Cook, M.L., and Stevens, J.G. (1974). Pathogenesis of herpetic encephalitis in mice after ophthalmic inoculation. J. Infect. Dis. *130*, 16–27.

Koelle, D.M., and Corey, L. (2008). Herpes simplex: insights on pathogenesis and possible vaccines. Annu. Rev. Med. *59*, 381–395.

Koelle, D.M., Posavad, C.M., Barnum, G.R., Johnson, M.L., Frank, J.M., and Corey, L. (1998). Clearance of HSV-2 from recurrent genital lesions correlates with infiltration of HSV-specific cytotoxic T lymphocytes. J. Clin. Invest. *101*, 1500–1508.

Kunz, M., and Ibrahim, S.M. (2009). Cytokines and cytokine profiles in human autoimmune diseases and animal models of autoimmunity. Mediators Inflamm. *2009*, 979258.

Kurtz, J.B. (1974). Specific IgG and IgM antibody responses in herpes-simplex-virus infections. J. Med. Microbiol. *7*, 333–341.

Lampson, L.A., and George, D.L. (1986). Interferon-mediated induction of class I MHC products in human neuronal cell lines: analysis of HLA and beta 2-m RNA, and HLA-A and HLA-B proteins and polymorphic specificities. J. Interferon Res. *6*, 257–265.

Larkin, J., Jin, L., Farmen, M., Venable, D., Huang, Y., Tan, S.L., and Glass, J.I. (2003). Synergistic antiviral activity of human interferon combinations in the hepatitis C virus replicon system. J. Interferon Cytokine Res. *23*, 247–257.

Latchman, D.S. (1996). The Oct-2 transcription factor. Int. J. Biochem. Cell Biol. *28*, 1081–1083.

Leib, D.A., Harrison, T.E., Laslo, K.M., Machalek, M.A., Moorman, N.J., and Virgin, H.W. (1999). Interferons regulate the phenotype of wild-type and mutant herpes simplex viruses *in vivo*. J. Exp. Med. *189*, 663–672.

Leib, D.A., Machalek, M.A., Williams, B.R., Silverman, R.H., and Virgin, H.W. (2000). Specific phenotypic restoration of an attenuated virus by knockout of a host resistance gene. Proc. Natl. Acad. Sci. U.S.A. *97*, 6097–6101.

Li, H., Zhang, J., Kumar, A., Zheng, M., Atherton, S.S., and Yu, F.S. (2006). Herpes simplex virus 1 infection induces the expression of proinflammatory cytokines, interferons and TLR7 in human corneal epithelial cells. Immunology *117*, 167–176.

Liebert, U.G. (1997). Measles virus infections of the central nervous system. Intervirology *40*, 176–184.

Lillycrop, K.A., Dent, C.L., Wheatley, S.C., Beech, M.N., Ninkina, N.N., Wood, J.N., and Latchman, D.S. (1991). The octamer-binding protein Oct-2 represses HSV immediate-early genes in cell lines derived from latently infectable sensory neurons. Neuron *7*, 381–390.

Liu, M., Rakowski, B., Gershburg, E., Weisend, C.M., Lucas, O., Schmidt, E.E., and Halford, W.P. (2010). ICP0 antagonizes ICP4-dependent silencing of the herpes simplex virus ICP0 gene. PLoS One *5*, e8837.

Liu, T., Khanna, K.M., Chen, X., Fink, D.J., and Hendricks, R.L. (2000). CD8(+) T cells can block herpes simplex virus type 1 (HSV-1) reactivation from latency in sensory neurons. J. Exp. Med. *191*, 1459–1466.

Liu, T., Tang, Q., and Hendricks, R.L. (1996). Inflammatory infiltration of the trigeminal ganglion after herpes simplex virus type 1 corneal infection. J. Virol. *70*, 264–271.

Lopez, C. (1975). Genetics of natural resistance to herpesvirus infections in mice. Nature *258*, 152–153.

Luker, G.D., Prior, J.L., Song, J., Pica, C.M., and Leib, D.A. (2003). Bioluminescence imaging reveals systemic dissemination of herpes simplex virus type 1 in the absence of interferon receptors. J. Virol. *77*, 11082–11093.

Markert, J.M., Medlock, M.D., Rabkin, S.D., Gillespie, G.Y., Todo, T., Hunter, W.D., Palmer, C.A., Feigenbaum, F., Tornatore, C., Tufaro, F., and Martuza, R.L. (2000). Conditionally replicating herpes simplex virus mutant, G207 for the treatment of malignant glioma: results of a phase I trial. Gene Ther *7*, 867–874.

McGeoch, D.J., Rixon, F.J., and Davison, A.J. (2006). Topics in herpesvirus genomics and evolution. Virus Res. *117*, 90–104.

McNally, J.M., Dempsey, D., Wolcott, R.M., Chervenak, R., and Jennings, S.R. (1999). Phenotypic identification of antigen-dependent and antigen-independent CD8 CTL precursors in the draining lymph node during acute cutaneous herpes simplex virus type 1 infection. J. Immunol. *163*, 675–681.

Miller, C.S., and Danaher, R.J. (2008). Asymptomatic shedding of herpes simplex virus (HSV) in the oral cavity. Oral Surg. Oral Med. Oral Pathol. Oral Radiol. Endod. *105*, 43–50.

Miller, G.G., and Dummer, J.S. (2007). Herpes simplex and varicella zoster viruses: forgotten but not gone. Am. J. Transplant. *7*, 741–747.

Misra, V., Rapin, N., Akhova, O., Bainbridge, M., and Korchinski, P. (2005). Zhangfei is a potent and specific inhibitor of the host cell factor-binding transcription factor Luman. J. Biol. Chem. *280*, 15257–15266.

Miyazaki, D., Inoue, Y., Araki-Sasaki, K., Shimomura, Y., Tano, Y., and Hayashi, K. (1998). Neutrophil chemotaxis induced by corneal epithelial cells after herpes simplex virus type 1 infection. Curr. Eye Res. *17*, 687–693.

Mohr, I., and Gluzman, Y. (1996). A herpesvirus genetic element which affects translation in the absence of the viral GADD34 function. EMBO J. *15*, 4759–4766.

Mossman, K.L., Saffran, H.A., and Smiley, J.R. (2000). Herpes simplex virus ICP0 mutants are hypersensitive to interferon. J. Virol. *74*, 2052–2056.

Mossman, K.L., and Smiley, J.R. (2002). Herpes simplex virus ICP0 and ICP34.5 counteract distinct interferon-induced barriers to virus replication. J. Virol. *76*, 1995–1998.

Murphy, E., Rigoutsos, I., Shibuya, T., and Shenk, T.E. (2003). Reevaluation of human cytomegalovirus coding potential. Proc. Natl. Acad. Sci. U.S.A. *100*, 13585–13590.

Nagafuchi, S., Oda, H., Mori, R., and Taniguchi, T. (1979). Mechanism of acquired resistance to herpes simplex virus infection as studied in nude mice.J. Gen. Virol. *44*, 715–723.

Netter, F.H. (1997). Atlas of Human Anatomy, 2nd edn (Elsevier, Inc., Philadelphia).

Neumann, H., Medana, I.M., Bauer, J., and Lassmann, H. (2002). Cytotoxic T lymphocytes in autoimmune and degenerative CNS diseases. Trends Neurosci. *25*, 313–319.

Njenga, M.K., Pease, L.R., Wettstein, P., Mak, T., and Rodriguez, M. (1997). Interferon alpha/beta mediates early virus-induced expression of H-2D and H-2K in the central nervous system. Lab. Invest. *77*, 71–84.

Oakes, J.E. (1975). Role for cell-mediated immunity in the resistance of mice to subcutaneous herpes simplex virus infection. Infect. Immun. *12*, 166–172.

Peng, T., Zhu, J., Hwangbo, Y., Corey, L., and Bumgarner, R.E. (2008). Independent and cooperative antiviral actions of beta interferon and gamma interferon against herpes simplex virus replication in primary human fibroblasts. J. Virol. *82*, 1934–1945.

Pereira, R.A., Tscharke, D.C., and Simmons, A. (1994). Upregulation of class I major histocompatibility complex gene expression in primary sensory neurons, satellite cells, and Schwann cells of mice in response to acute but not latent herpes simplex virus infection *in vivo*. J. Exp. Med. *180*, 841–850.

Perng, G.C., and Jones, C. (2010). Towards an understanding of the herpes simplex virus type 1 latency-reactivation cycle. Interdiscip. Perspect. Infect. Dis. *2010*, 262415.

Pestka, S. (2007). The interferons: 50 years after their discovery, there is much more to learn. J. Biol. Chem. *282*, 20047–20051.

Pierce, A.T., DeSalvo, J., Foster, T.P., Kosinski, A., Weller, S.K., and Halford, W.P. (2005). Interferon-beta and interferon-gamma synergize to block viral DNA and virion synthesis in herpes simplex virus-infected cells.J. Gen. Virol. *86*, 2421–2432.

Posavad, C.M., Koelle, D.M., and Corey, L. (1996). High frequency of CD8+ cytotoxic T-lymphocyte precursors specific for herpes simplex viruses in persons with genital herpes. J. Virol. *70*, 8165–8168.

Posavad, C.M., Koelle, D.M., and Corey, L. (1998). Tipping the scales of herpes simplex virus reactivation: the important responses are local. Nat. Med. *4*, 381–382.

Posavad, C.M., Koelle, D.M., Shaughnessy, M.F., and Corey, L. (1997). Severe genital herpes infections in HIV-infected individuals with impaired herpes simplex virus-specific CD8+ cytotoxic T lymphocyte responses. Proc. Natl. Acad. Sci. U.S.A. *94*, 10289–10294.

Preston, C.M. (2000). Repression of viral transcription during herpes simplex virus latency. J. Gen. Virol. *81*, 1–19.

Rawls, W.E. (1985). Herpes simplex virus infections, In Virology, Fields B.N., ed. (Raven Press, New York), pp. 527–561.

Redwine, J.M., Buchmeier, M.J., and Evans, C.F. (2001). *In vivo* expression of major histocompatibility complex molecules on oligodendrocytes and neurons during viral infection. Am. J. Pathol. *159*, 1219–1224.

Roizman, B., Gu, H., and Mandel, G. (2005). The first 30 minutes in the life of a virus: unREST in the nucleus. Cell Cycle *4*, 1019–1021.

Sacks, W.R., and Schaffer, P.A. (1987). Deletion mutants in the gene encoding the herpes simplex virus type 1 immediate-early protein ICP0 exhibit impaired growth in cell culture. J. Virol. *61*, 829–839.

Sainz, B., Jr. and Halford, W.P. (2002). Alpha/Beta interferon and gamma interferon synergize to inhibit the replication of herpes simplex virus type 1. J. Virol. *76*, 11541–11550.

Sainz, B., Jr., Lamarca, H.L., Garry, R.F., and Morris, C.A. (2005). Synergistic inhibition of human cytomegalovirus replication by interferon-alpha/beta and interferon-gamma. Virol. J. *2*, 14.

Sainz, B., Jr., Mossel, E.C., Peters, C.J., and Garry, R.F. (2004). Interferon-beta and interferon-gamma synergistically inhibit the replication of severe acute respiratory syndrome-associated coronavirus (SARS-CoV). Virology *329*, 11–17.

Salazar-Mather, T.P., and Hokeness, K.L. (2003). Calling in the troops: regulation of inflammatory cell trafficking through innate cytokine/chemokine networks. Viral Immunol. *16*, 291–306.

Samuel, C.E. (1998). Reoviruses and the interferon system. Curr. Top. Microbiol. Immunol. *233*, 125–145.

Sawtell, N.M. (1997). Comprehensive quantification of herpes simplex virus latency at the single-cell level. J. Virol. *71*, 5423–5431.

Schoenborn, J.R., and Wilson, C.B. (2007). Regulation of interferon-gamma during innate and adaptive immune responses. Adv. Immunol. *96*, 41–101.

Sheridan, B.S., Knickelbein, J.E., and Hendricks, R.L. (2007). CD8 T cells and latent herpes simplex virus type 1: keeping the peace in sensory ganglia. Expert Opin. Biol. Ther. *7*, 1323–1331.

Shimeld, C., Whiteland, J.L., Nicholls, S.M., Grinfeld, E., Easty, D.L., Gao, H., and Hill, T.J. (1995). Immune cell infiltration and persistence in the mouse trigeminal ganglion after infection of the cornea with herpes simplex virus type 1. J. Neuroimmunol. *61*, 7–16.

Simmons, A., and Tscharke, D.C. (1992). Anti-CD8 impairs clearance of herpes simplex virus from the nervous system: implications for the fate of virally infected neurons. J. Exp. Med. *175*, 1337–1344.

Speck, P., and Simmons, A. (1998). Precipitous clearance of herpes simplex virus antigens from the peripheral nervous systems of experimentally infected C57BL/10 mice. J. Gen. Virol. *79*, 561–564.

Stanberry, L.R., Floyd-Reising, S.A., Connelly, B.L., Alter, S.J., Gilchrist, M.J., Rubio, C., and Myers, M.G. (1994). Herpes simplex viremia: report of eight pediatric cases and review of the literature. Clin. Infect. Dis. *18*, 401–407.

Stanberry, L.R., Kern, E.R., Richards, J.T., Abbott, T.M., and Overall, J.C., Jr. (1982). Genital herpes in guinea pigs: pathogenesis of the primary infection and description of recurrent disease. J. Infect. Dis. *146*, 397–404.

Stevens, J.G., and Cook, M.L. (1971). Latent herpes simplex virus in spinal ganglia of mice. Science *173*, 843–845.

Stevens, J.G., Haar, L., Porter, D., Cook, M.L., and Wagner, E.K. (1988). Prominence of the herpes simplex virus latency associated transcript in trigeminal ganglia from seropositive humans. J. Infect. Dis. *158*, 117–123.

Stevens, J.G., Wagner, E.K., Devi-Rao, G.B., Cook, M.L., and Feldman, L.T. (1987). RNA complementary to a herpesvirus a gene mRNA is prominent in latently infected neurons. Science *235*, 1056–1059.

Stewart, J.A., Reef, S.E., Pellett, P.E., Corey, L., and Whitley, R.J. (1995). Herpesvirus infections in persons infected with human immunodeficiency virus. Clin. Infect. Dis. *21*, S114–120.

Stober, D., Schirmbeck, R., and Reimann, J. (2001). IL-12/IL-18-dependent IFN-gamma release by murine dendritic cells. J. Immunol. *167*, 957–965.

Stroop, W. G., Rock, D. L., and Fraser, N.W. (1984). Localization of herpes simplex virus in the trigeminal and olfactory systems of the mouse central nervous system during acute and latent infections by in situ hybridization. Lab. Invest. *51*, 27–38.

Takaoka, A., Mitani, Y., Suemori, H., Sato, M., Yokochi, T., Noguchi, S., Tanaka, N., and Taniguchi, T. (2000). Cross talk between interferon-gamma and -alpha/beta signaling components in caveolar membrane domains. Science *288*, 2357–2360.

Tang, S., Patel, A., and Krause, P.R. (2008). Novel less-abundant viral miRNAs encoded by herpes simplex virus 2 latency-associated transcript and their roles in regulating ICP34.5 and ICP0 mRNAs. J. Virol. *83*, 1433–1442.

Tenser, R.B., Edris, W.A., and Hay, K.A. (1993). Neuronal control of herpes simplex virus latency. Virology *195*, 337–347.

Terry-Allison, T., Smith, C.A., and DeLuca, N.A. (2007). Relaxed repression of herpes simplex virus type 1 genomes in Murine trigeminal neurons. J. Virol. *81*, 12394–12405.

Thapa, M., and Carr, D.J. (2008). Chemokines and chemokine receptors critical to host resistance following genital herpes simplex virus Type 2 (HSV-2) infection. Open Immunol. J. *1*, 33–41.

Theil, D., Derfuss, T., Paripovic, I., Herberger, S., Meinl, E., Schueler, O., Strupp, M., Arbusow, V., and Brandt, T. (2003). Latent herpesvirus infection in human trigeminal ganglia causes chronic immune response. Am. J. Pathol. *163*, 2179–2184.

Thompson, R.L., Rogers, S.K., and Zerhusen, M.A. (1989). Herpes simplex virus neurovirulence and productive infection of neural cells is associated with a function which maps between 0.82 and 0.832 map units on the HSV genome. Virology *172*, 435–450.

Todo, T. (2008). Oncolytic virus therapy using genetically engineered herpes simplex viruses. Front. Biosci. *13*, 2060–2064.

Toth, F.D., Norskov-Lauritsen, N., Juhl, C.B., Aboagye-Mathiesen, G., and Ebbesen, P. (1990). Interferon production by cultured human trophoblasts and choriocarcinoma cell lines induced by Sendai virus. J. Gen. Virol. *71 (Pt 12)*, 3067–3069.

Triezenberg, S.J., Kingsbury, R.C., and McKnight, S.L. (1988). Functional dissection of VP16, the trans-activator of herpes simplex virus immediate early gene expression. Genes Dev. *2*, 718–729.

Umbach, J.L., Kramer, M.F., Jurak, I., Karnowski, H.W., Coen, D.M., and Cullen, B.R. (2008). MicroRNAs expressed by herpes simplex virus 1 during latent infection regulate viral mRNAs. Nature *454*, 780–783.

Valyi-Nagy, T., Fareed, M.U., O'Keefe, J.S., Gesser, R.M., MacLean, A.R., Brown, S.M., Spivack, J.G., and Fraser, N.W. (1994a). The herpes simplex virus type 1 strain 17+ gamma 34.5 deletion mutant 1716 is avirulent in SCID mice. J. Gen. Virol. *75*, 2059–2063.

Valyi-Nagy, T., Gesser, R.M., Raengsakulrach, B., Deshmane, S.L., Randazzo, B.P., Dillner, A. J., and Fraser, N.W. (1994b). A thymidine kinase-negative HSV-1 strain establishes a persistent infection in SCID mice that features uncontrolled peripheral replication but only marginal nervous system involvement. Virology *199*, 484–490.

Verjans, G.M., Hintzen, R.Q., van Dun, J.M., Poot, A., Milikan, J.C., Laman, J.D., Langerak, A.W., Kinchington, P.R., and Osterhaus, A.D. (2007). Selective retention of herpes simplex virus-specific T cells in latently infected human trigeminal ganglia. Proc. Natl. Acad. Sci. U.S.A. *104*, 3496–3501.

Vollstedt, S., Franchini, M., Hefti, H.P., Odermatt, B., O'Keeffe, M., Alber, G., Glanzmann, B., Riesen, M., Ackermann, M., and Suter, M. (2003). Flt3 ligand-treated neonatal mice have increased innate immunity against intracellular pathogens and efficiently control virus infections. J. Exp. Med. *197*, 575–584.

Wald, A., Corey, L., Cone, R., Hobson, A., Davis, G., and Zeh, J. (1997). Frequent genital herpes simplex virus 2 shedding in immunocompetent women. Effect of acyclovir treatment. J. Clin. Invest. *99*, 1092–1097.

Walz, M.A., Yamamoto, H., and Notkins, A.L. (1976). Immunological response restricts number of cells in sensory ganglia infected with herpes simplex virus. Nature *264*, 554–556.

Watanabe, D., Adachi, A., Tomita, Y., Yamamoto, M., Kobayashi, M., and Nishiyama, Y. (1999). The role of polymorphonuclear leukocyte infiltration in herpes simplex virus infection of murine skin. Arch. Dermatol. Res. *291*, 28–36.

Weinstein, D.L., Walker, D.G., Akiyama, H., and McGeer, P.L. (1990). Herpes simplex virus type I infection of the CNS induces major histocompatibility complex antigen expression on rat microglia. J. Neurosci. Res. *26*, 55–65.

Woodland, D.L., Hogan, R.J., and Zhong, W. (2001). Cellular immunity and memory to respiratory virus infections. Immunol. Res. *24*, 53–67.

Wuest, T.R., and Carr, D.J. (2008). Dysregulation of CXCR3 signaling due to CXCL10 deficiency impairs the antiviral response to herpes simplex virus 1 infection. J. Immunol. *181*, 7985–7993.

Yamazaki, K., Yoshie, H., and Seymour, G.J. (2003). T cell regulation of the immune response to infection in periodontal diseases. Histol. Histopathol. *18*, 889–896.

Yudin, M.H., and Kaul, R. (2008). Progressive hypertrophic genital herpes in an HIV-infected woman despite immune recovery on antiretroviral therapy. Infect. Dis. Obstet. Gynecol. *2008*, 592532.

Zhu, J., Koelle, D.M., Cao, J., Vazquez, J., Huang, M.L., Hladik, F., Wald, A., and Corey, L. (2007). Virus-specific CD8+ T cells accumulate near sensory nerve endings in genital skin during subclinical HSV-2 reactivation. J. Exp. Med. *204*, 595–603.

Poxviruses and Interferons

6

Beatriz Perdiguero and Mariano Esteban

Abstract

Since the discovery of interferons (IFNs) more than half a century ago, these molecules have become key players of many cellular processes, particularly in the control of viral infections, cell growth and immune regulation. How the cells respond to IFN and identification of the molecular signals involved is a major goal in research. In this chapter we address these issues through the current understanding of the interaction between poxviruses and the IFN system. These large DNA containing viruses have acquired during their evolution an array of genes that counteract the IFN pathways at multiple levels. How the viral genes act is presented.

Introduction

The co-evolution of viruses with their hosts is a dynamic process involving the generation of host immune responses and their evasion by viruses using an impressive array of immunomodulatory mechanisms (Spriggs, 1996; Alcamí and Koszinowski, 2000; Loo and Gale, 2007; Schroder and Bowie, 2007; Unterholzner and Bowie, 2008). Viral infection of host cells leads to the initiation of antiviral antigen non-specific innate immune responses, which results in the expression of type I interferons (IFNs) and pro-inflammatory cytokines. The acquisition of viral proteins by dendritic cells in the tissues and their migration to local lymph nodes to interact with antigen specific-lymphocytes initiate the adaptive immune response to infection.

Poxviruses (family Poxviridae) are complex viruses that infect vertebrates (subfamily Chordopoxvirinae) or insects (subfamily Entomopoxvirinae; Moss, 2007). Members of this family are related genetically and are characterized by having a large linear double-stranded DNA genome, a complex enveloped brick-shaped virion, a cytoplasmic site of replication, their own enzymes for transcription (Moss, 1990) and DNA replication and a complex morphogenic pathway that produces distinct forms of infectious virions from each infected cell (Moss, 2007). Despite their common features, poxviruses evolved to infect a wide variety of hosts and some cause acute and others persistent infections (Table 6.1). Of the known poxviruses, only two exclusively infect humans: variola virus (VARV), the causative agent of smallpox, which is now eradicated, and molluscum contagiosum virus (MoCV), which is still prevalent throughout the world. Vaccinia virus (VACV), the prototype orthopoxvirus used as a vaccine against smallpox, is of unknown origin and has no natural host at this time. Chordopoxvirinae genomes range in size from 135,000 to 289,000 basepairs and encode between 136 and 260 open reading frames, respectively. The genes involved in viral replication are highly conserved among most poxviruses and are generally located in the central portion of the genome (~100 kbp; Gubser *et al.*, 2004). In contrast, genes located towards the termini are more variable and, although non-essential for virus replication, affect virulence, host range and modulation of the host immune system (Moss, 2007).

Table 6.1 Members of the family *Poxviridae*

Family/subfamily/genus	Members	Description
Poxviridae		Replicate in the host cell cytoplasm. Linear dsDNA genome, 130–300 kbp
Chordopoxvirinae		Infection of vertebrates
Orthopoxvirus	VACV (vaccinia virus); VARV (variola virus); CPXV (cowpox virus); ECTV (ectromelia virus); CMLV (camelpox virus); MPXV (monkeypox virus)	Acute and chronic infections. Multiple tissues + skin
Parapoxvirus	ORFV (orf virus); bovine papular stomatitis virus; pseudocowpox virus; parapoxviruses of squirrels, seals and deer (DPXV; deerpox virus)	Narrow host range that includes man (except squirrel and deer parapoxviruses-not known). Acute and chronic infections of the skin
Suipoxvirus	SPXV (swinepox virus)	Arthropod borne: skin lesions, mild
Capripoxvirus	CaPV (goat pox, sheep pox viruses)	Lumpy skin disease of cattle. Arthropod borne. High mortality in sheep and goats
Leporipoxvirus	MYXV (myxoma virus); SFV (Shope fibroma virus)	Arthropod borne. Severe generalized disease (MYXV), local benign tumours (SFV)
Molluscipoxvirus	MoCV (molluscum contagiosum virus)	Human pathogen. Localized tumours
Yatapoxvirus	TPV (tanapox virus); YLDV (Yaba-like disease virus)	Primate pathogen. Arthropod borne. Skin lesions in humans
Avipoxvirus	Canarypox virus; FWPV (fowlpox virus)	Arthropod transmission to birds
Entomopoxvirinae		Infection of insects

Poxviruses have evolved strategies to inhibit key host antiviral responses including IFN production, inflammatory cytokine production, chemokine production, apoptosis and the activity of cytotoxic T cells, NK cells and antibodies (Smith *et al.*, 1997b; McFadden and Murphy, 2000; Barrett *et al.*, 2001; Dunlop *et al.*, 2003; Seet *et al.*, 2003; Haga and Bowie, 2005; Alcami, 2007; Table 6.2). The most extraordinary evasion strategy is the acquisition and subsequent modification of immune response genes from the host. In this review, we provide an overview of the multiple strategies used by poxviruses to systematically modulate key components of the innate and adaptive immune responses, and in particular the IFN system. We describe current knowledge on these pathways and the poxvirus proteins involved. Immune evasion mechanisms used by insect poxviruses (family Entomopoxvirinae) are not covered in this review. For a more detailed description of the immune evasion mechanisms used by VACV the reader is referred to (Perdiguero and Esteban, 2009).

Interferon signalling pathways and inhibition by poxvirus proteins

IFNs are a group of secreted cytokines that constitute the first line of defence against viral infection and can induce direct antiviral effects as well as promote Th1 responses (Samuel, 2001; Sen, 2001; Honda *et al.*, 2006; Stetson and Medzhitov, 2006). The antiviral activity of IFNs was demonstrated half a century ago as an activity secreted from cells which renders cells resistant to subsequent virus infection (Isaacs and Lindenmann, 1957), and later confirmed in IFNα/β receptor (IFNAR)-deficient mice, who showed an extreme susceptibility to different viral infections (Muller *et al.*, 1994).

The IFN system is regulated by two signalling cascades: a virus-induced IFN-producing signal and an IFN receptor-mediated secondary signal (Fig. 6.1). The first signal is initiated by the detection of viral pathogen-associated molecular patterns (PAMPs) or endogenous molecules released from damaged cells, termed

Table 6.2 Poxvirus immune evasion proteins

Function	Virus/protein (gene)	Mechanism	References
Poxvirus immunomodulatory proteins interfering with PRRs signalling pathways			
Inhibition of TIR signalling and NF-κB activation	VACV A46 (A46R)	Sequesters adaptor proteins MyD88, TRIF, MAL and TRAM interfering with downstream activation of MAPKs, NF-κB and IRFs	Bowie *et al.* (2000), Stack *et al.* (2005)
	VACV A52 (A52R)	Inhibits TLR-induced NF-κB activation by binding IRAK2 and TRAF6. Enhances TLR-induced IL-10 production by binding TRAF6	Bowie *et al.* (2000), Harte *et al.* (2003), Maloney *et al.* (2005), Keating *et al.* (2007)
	VACV K7 (K7R)	Inhibits PRR-mediated induction of IFNβ by preventing TBK1/Ikki-mediated IRF activation owing to its ability to target human DEAD box protein 3 (DDX3)	Schroder *et al.*, 2008
	VACV N1 (N1L)	Inhibits TRAF6-induced NF-κB activation at the level of TRAF6 or downstream proteins preceding the IKK complex	Bartlett *et al.* (2002), DiPerna *et al.* (2004), Cooray *et al.* (2007), Chen *et al.* (2008), Tscharke and Smith (1999), Kotwal *et al.* (1989), Jacobs *et al.* (2008)
	VACV B14 (B14R)[a]	Inhibits the phosphorylation and subsequent degradation of IkBα, the inhibitor of NF-κB, by its interaction with IKKβ	Chen *et al.* (2006), Chen *et al.* (2008)
	VACV M2 (M2L)	Inhibits NF-κB activation by preventing ERK2 phosphorylation	Gedey *et al.* (2006), Hinthong *et al.* (2008)
	MoCV MC159, MC160	Inhibit NF-κB activation	Gil *et al.* (2001), Murao and Shisler (2005), Nichols and Shisler (2009)
	ORFV 024	Inhibits activation of NF-κB by inhibiting IKKs activation	Diel *et al.* (2010)
	CPXV 016	NF-κB regulator	Oie and Pickup (2001)
	VACV K1 (K1L)	Inhibits NF-κB activation by preventing the degradation of IkBα	Shisler and Jin (2004), Aravalli *et al.* (2008)
	MYXV MNF (M150R)	Inhibits NF-κB activation	Camus-Bouclainville *et al.* (2004)
	CPXV CP77	Inhibits TNFα-induced NF-κB activation by direct binding to the NF-κB/p65 subunit	Chang *et al.* (2009)
	MYXV M013 (M013L)	Inhibitor of the NF-κB/p65 signalling pathway	Rahman *et al.* (2009)

Table 6.2 continued

Function	Virus/protein (gene)	Mechanism	References
Inhibition of NLR signalling pathways	MYXV M013 (M013L)	Disrupts NLRP–ASC interaction and caspase-1 activation by direct association with ASC	Johnston *et al.* (2005)
	SFV gp013	Inhibits PYD-mediated activation of caspase-1 by direct association with ASC	Dorfleutner *et al.* (2007)
	CPXV CrmA	Caspase-1 inhibitor	Komiyama *et al.* (1994), Ray *et al.* (1992), Pickup *et al.* (1986)
	VACV B13 (B13R)	Caspase-1 inhibitor	Kettle *et al.* (1997)
Poxvirus immunomodulatory proteins interfering with IFN signalling pathways			
Inhibition of IFN binding	VACV B19 (B19R) (vIFN-α/βR)	Secreted and cell surface protein that binds type I IFN from various species (the affinity for mouse type I IFN is considerably lower than for the other species tested) inhibiting binding of type I IFNs to its receptor	Symons *et al.* (1995), Colamonici *et al.* (1995), Alcami *et al.* (2000), Waibler *et al.* (2009), Morikawa and Ueda (1993), Ueda *et al.* (1990), Liptakova *et al.* (1997)
	YLDV 136	Binds and inhibits both type I and type III IFNs	Huang *et al.* (2007)
	VACV B8 (B8R) (vIFN-γR)	Secreted protein that binds IFNγ from various species but not mouse inhibiting binding of IFNγ to its receptor	Alcami and Smith (2002), Alcami and Smith (1995), Alcami and Smith (1996b), Mossman *et al.* (1995a), Symons *et al.* (2002b), Puehler *et al.* (1998), Sroller *et al.* (2001), Verardi *et al.* (2001)
	MYXV M-T7	Binds and inhibits binding of IFNγ to its receptor	Upton *et al.* (1992), Mossman *et al.* (1995b)
Dephosphorylation of STAT	VACV H1 (H1L) (VH1)	Binds and dephosphorylates STAT1 inhibiting IFNγ signalling	Najarro *et al.* (2001), Liu *et al.* (1995)
Poxvirus immunomodulatory proteins interfering with IFN effector pathways			
Inhibition of PKR activity and other ISGs	VACV E3 (E3L) (vdsRNA-BP)	Bind dsRNA and prevent PKR activation	Davies *et al.* (1993), Chang *et al.* (1992), Romano *et al.* (1998)
		Inhibits PKR activity by direct binding	Sharp *et al.* (1998)
		Reduces the adenosine-to-inosine editing activity of IFN-induced ADAR	Liu *et al.* (2001)
		Inhibits IRF3/7 activation	Smith *et al.* (2001), Xiang *et al.* (2002)
		Binds to and disables ISG15 function	Guerra *et al.* (2008)
		Binds dsRNA and prevent activation of 2′5′OAS/RNaseL	Rivas *et al.* (1998)

Inhibition of eIF-2α phosphorylation and translational arrest	VACV K3 (K3L)	Homologue of eIF2α, inhibits PKR preventing eIF2α phosphorylation	Beattie *et al.* (1991), Davies *et al.* (1992), Davies *et al.* (1993)
	VACV C7 (C7L)	Inhibits phosphorylation of eIF2α and apoptosis induction	Najera *et al.* (2006)
	MYXV M-62 (M62R)	VACV C7L homologue. Equivalent function when expressed by recombinant VACV	Meng *et al.* (2008)
	YLDV 67 (67R)	VACV C7L homologue. Equivalent function when expressed by recombinant VACV	Meng *et al.* (2008)
Poxvirus immunomodulatory proteins intercepting cytokines			
Viral TNF receptors	MYXV M-T2 SFV S-T2	Virus-encoded TNFRs (vTNFRs). Bind extracellular TNF inhibiting its binding to the receptor	Upton *et al.* (1991), Xu *et al.* (2000), Smith *et al.* (1991)
	CPXV CrmB, C, D and E	Soluble vTNFRs	Hu *et al.* (1994), Loparev *et al.* (1998), Saraiva and Alcami (2001), Smith *et al.* (1996)
	VACV A53 (A53R)	Functional, soluble TNF receptor homologue that binds TNF inhibiting binding of TNF to receptor (strain Lister)	Alcami *et al.* (1999)
	VACV CrmE gene	Soluble and membrane-bound TNFR (strain USSR)	Reading *et al.* (2002)
TNF-binding proteins	TPV 2L	Viral TNF-binding protein (vTNF-BP)	Brunetti *et al.* (2003), Rahman *et al.* (2006)
Viral IL-1β receptors	VACV B16 (B16R)	Viral soluble IL-1βR	Alcami and Smith (1992), Spriggs *et al.* (1992)
	CPXV B14 (B14R)	Viral soluble IL-1βR	Spriggs *et al.* (1992)
	ECTV 191	Viral soluble IL-1βR	Smith and Alcami (2000)
Viral IL-18-binding proteins	MoCV MC54 (MC54L)	Viral IL-18-binding protein	Xiang and Moss (1999), Xiang and Moss (2003)
	ECTV p13	Viral IL-18-binding protein	Born *et al.* (2000)
	VACV C12 (C12L)[b]	Secreted protein that binds IL-18 inhibiting binding of IL-18 to receptor. Inhibits IL-18-induced IFNγproduction and NK response	Symons *et al.* (2002a), Reading and Smith (2003)

Table 6.2 continued

Function	Virus/protein (gene)	Mechanism	References
Poxvirus immunomodulatory proteins capturing chemokines			
Viral chemokine-binding proteins	MYXV M-T7	Type I CBP. Binds a broad spectrum of CXC, CC and C-chemokines with low affinity	Lalani *et al.* (1997)
	MYXV M-T1	Type II CBP. Binds to CC-chemokines with high affinity	Graham *et al.* (1997), Lalani *et al.* (1999)
	VACV B29/C23 (B29R/C23L)	Type II CBP. Binds to CC-chemokines with high affinity (strains Lister and Evans)	Smith *et al.* (1997b), Lalani *et al.* (2000), Alcami *et al.* (1998)
	ECTV 163	Binds a limited number of CC and CXC chemokines with high affinity	Ruiz-Arguello *et al.* (2008)
	ORFV CBP	Secreted high affinity CC-chemokine-binding protein	Lateef *et al.* (2009), Lateef *et al.* (2010)
	MPXV vCCI	Secreted chemokine-binding protein. Binds rhesus MIP-1α	Jones *et al.* (2008)
	VARV CBP-II	Secreted CBP-II	Dunlop *et al.* (2003)
Viral chemokine receptor homologues	YLDV 7 (7L)	Chemokine receptor with 53% amino acid identity to human CC chemokine receptor 8 (CCR8)	Najarro *et al.* (2006)
Viral chemokine homologues	MoCV MC148 (MC148R)	Related to the CC-chemokine family but with a deletion in the N-terminal region, necessary for activation of receptors	Bugert *et al.* (1998), Damon *et al.* (1998), Krathwohl *et al.* (1997)
	FWPV Fpv060, Fpv061, Fpv116 and Fpv121	CC-chemokine-like proteins	Jeshtadi *et al.* (2005)
Poxvirus immunomodulatory proteins inhibiting apoptosis			
Viral caspase inhibitors	CPXV CrmA	Caspase-8 inhibitor	Pickup *et al.* (1986), Zhou *et al.* (1997), Tewari and Dixit (1995)
		Granzyme B inhibitor	Quan *et al.* (1995)
	MYXV Serp-2	SPI-2-like serpin	Petit *et al.* (1996), Messud-Petit *et al.* (1998)
	MoCV MC159, MC160	Caspase-8 inhibitor	Shisler and Moss (2001b), Garvey *et al.* (2002), Hu *et al.* (1997), Bertin *et al.* (1997), Garvey *et al.* (2002)

Other anti-apoptotic functions	MoCV 066	Prevents cell death triggered by stress catalyzing the conversion of toxic reactive oxygen species (H_2O_2) to water	Shisler *et al.* (1998)
	MYXV M11 (M11L)	Inhibits transduction of death signals via the mitochondrial checkpoint	Macen *et al.* (1996), Opgenorth *et al.* (1992), Everett *et al.* (2000)
	VACV F1 (F1L)	Mitochondrial-associated inhibitor of apoptosis	Wasilenko *et al.* (2003), Wasilenko *et al.* (2005), Stewart *et al.* (2005), Fischer *et al.* (2006)
	FWPV 039	Bcl-2-like inhibitor	Banadyga *et al.* (2007)
	ORFV 125	Bcl-2-like inhibitor	Westphal *et al.* (2007)
	MYXV M-T4	Inhibits ER stress response to infection	Barry *et al.* (1997), Hnatiuk *et al.* (1999)
	MYXV M-T5 CPXV CHOhr	Can prevent apoptosis associated with virus infection	Mossman *et al.* (1996a), Ink *et al.* (1995)
	MYXV M-T2	Intracellular inhibitor of apoptosis in virus-infected lymphocytes	Macen *et al.* (1996), Schreiber *et al.* (1997), Sedger and McFadden (1996), Xu *et al.* (2000)
	VACV N1 (N1L)	Inhibits staurosporine-induced apoptosis in transfected and infected cells	Cooray *et al.* (2007)
		Binds BH3 peptides from Bid, Bim and Bak *in vitro*	Aoyagi *et al.* (2007)
		Co-precipitates Bid, Bad and Bax from VACV-infected cells	Cooray *et al.* (2007)
	VACV C7 (C7L)	Inhibits phosphorylation of eIF2α and apoptosis induction	Najera *et al.* (2006)

VACV, vaccinia virus; MoCV, Molluscum contagiosum virus; ORFV, orf virus; CPXV, cowpox virus; MYXV, myxoma virus; SFV, Shope fibroma virus; YLDV, Yaba-like disease virus; TPV, tanapox virus; ECTV, ectromelia (mousepox) virus; VARV, variola (smallpox) virus; FWPV, fowlpox virus.

The gene nomenclature corresponds to Copenhagen strain for the case of VACV.

[a]B14R: VACV WR nomenclature. Homologue to fragmented VACV Cop B15R.

[b]C12L: VACV WR nomenclature. Absent in Copenhagen strain.

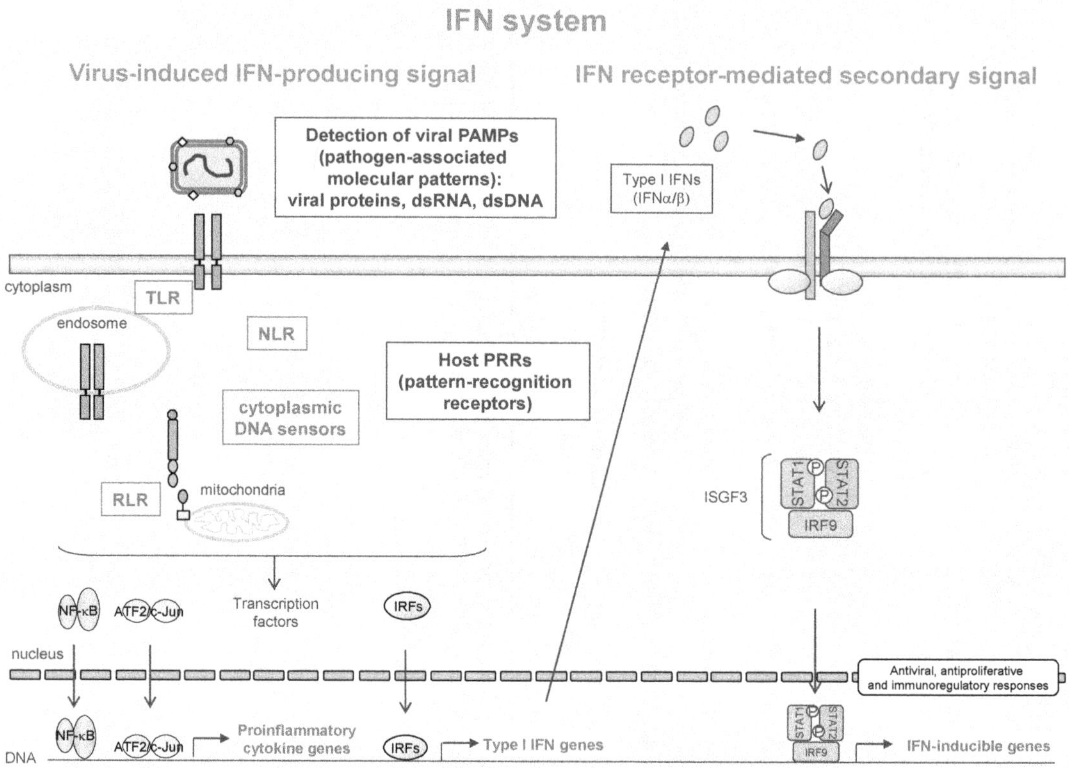

Figure 6.1 Scheme of the different components involved in the IFN signalling pathway. The IFN system is regulated by two signalling cascades: a virus-induced IFN-producing signal (left) and an IFN receptor-mediated secondary signal (right). The first signal is initiated by the detection of viral PAMPs by different host PRRs (TLRs, RLRs, NLRs and cytoplasmic DNA sensors). The activation of these PRRs leads to the recruitment of several adaptor proteins and kinases and the subsequent activation of transcription factors (NF-κB, ATF2/c-Jun and IRFs) that are translocated into the nucleus and activate the transcription of proinflammatory cytokine and type I IFN genes. IFNs expressed interact with the IFN receptors (IFNRs) triggering a secondary signal involving Janus kinases (JAKs) and STAT factors, leading to the transcription of IFN-inducible genes. These IFN-stimulated genes are involved in the generation of an antiviral response.

damage-associated molecular patterns (DAMPs), by host pattern-recognition receptors (PRRs). The subsequent triggering of a signalling cascade results in the activation of a series of transcription factors, such as IFN regulatory factors (IRFs), activating transcription factor 2 (ATF2)/c-Jun and nuclear factor κB (NF-κB). These activated factors are translocated into the nucleus where they bind to the promoter region of IFN genes (IRFs) or proinflammatory cytokine genes (ATF2/c-Jun and NF-κB; Miyamoto *et al.*, 1988; Fujita *et al.*, 1989; Lenardo *et al.*, 1989; Du and Maniatis, 1992) activating the transcription of these genes (Carey, 1998). IRF3 and IRF7 are key regulators of IFN gene activation (Lin *et al.*, 1998; Sato *et al.*, 1998b; Wathelet *et al.*, 1998; Weaver *et al.*, 1998; Yoneyama *et al.*, 1998). IRF3 is a potent activator of the *IFNB* gene but not the *IFNA* genes whereas IRF7 efficiently activates both *IFNA* and *IFNB* genes (Marie *et al.*, 1998; Sato *et al.*, 1998a, 2000; Romieu-Mourez *et al.*, 2006).

IFNs interact with the broadly expressed IFN receptors (IFNRs), in an autocrine or paracrine manner (Cutrone and Langer, 2001), activating a signal transduction cascade involving Janus kinases (JAKs) and signal transducers and activators of transcription (STAT) factors, which leads to the transcription of IFN-stimulated genes (ISGs). These ISGs are involved in eliminating viral components from infected cells, inducing apoptosis of infected cells and conferring resistance to viral infection in uninfected cells. In addition to these cell-intrinsic effects in the establishment of the antiviral state, type I IFNs, especially TLR-induced IFNs in immunospecific cells, are known to function in the regulation of subsequent innate and adaptive immunity. For example, it has been demonstrated that IFNs are involved in the maturation of DCs, including the cross-presentation of viral antigens for CD8+ T cells and the activation of NK cells (Stetson and Medzhitov, 2006). Proinflammatory cytokines are also critical for eliminating virus infection by inducing inflammation and recruiting innate and adaptive immune cells.

PRR (pattern-recognition receptor) signalling

The initial sensing of infection by the host is mediated by innate pattern-recognition receptors (PRRs), which detect poxvirus components such as viral envelope glycoproteins, intermediate double-stranded RNA (dsRNA) products or genomic dsDNA. Currently, four classes of PRRs have been identified: Toll-like receptors (TLRs), which are transmembrane proteins, retinoic acid-inducible gene I (RIG-I)-like receptors (RLRs), nucleotide-binding oligomerization domain (NOD)-like receptors (NLRs) and cytoplasmic DNA sensors, which comprise cytoplasmic proteins (Kumagai and Akira, 2010; Takeuchi and Akira, 2010; Yoneyama and Fujita, 2010). The sensing of viral PAMPs or DAMPs by TLRs, RLRs or cytoplasmic DNA sensors upregulates the transcription of type I IFN and proinflammatory cytokine genes, whereas NLRs regulate interleukin 1β (IL-1β) maturation through activation of IL-1β-converting enzyme (ICE; also known as caspase-1; Kanneganti *et al.*, 2007; Petrilli *et al.*, 2007).

TLR signalling

The TLR family is one of the best characterized PRR families and is responsible for sensing invading pathogens outside of the cell and in intracellular endosomes and lysosomes (Akira, 2006; Akira *et al.*, 2006). Ten TLRs have been identified in humans and 12 in mice (O'Neill, 2006). All TLRs are type I transmembrane glycoprotein receptors comprised of an N-terminal leucine-rich repeat (LRR) domain involved in ligand binding, a transmembrane domain and a cytoplasmic Toll/IL-1R homology (TIR) domain, which mediates interactions between TLRs and downstream signalling molecules (Bowie and O'Neill, 2000; Akira *et al.*, 2006; Beutler *et al.*, 2006). TLRs are expressed not only on a range of immune cells (macrophages, dendritic cells, B cells and certain types of T cells) but also on certain non immune cells, such as epithelial cells, which lie at potential sites of entry. TLR2, TLR3, TLR4, TLR6, TLR7 (mice)/TLR8 (humans) and TLR9 are involved in the recognition of viral components: TLR2, TLR4 and TLR6, present on the plasma membrane, are involved in the recognition of viral envelope glycoproteins, while TLR3, TLR7/8 and TLR9 are localized on cytoplasmic vesicles, such as endosomes and lysosomes, and recognize viral nucleic acids. Specifically, TLR3 recognizes dsRNA, while TLR7/8 and TLR9 recognize ssRNA and DNA with CpG motifs, respectively. While TLR3 recognizes dsRNA in conventional DCs (cDCs) and possibly epithelial cells, TLR7/8 and TLR9 are highly expressed in plasmacytoid DCs (pDCs), a cell type that produces extremely high levels of type I IFNs in response to virus infection (Kato *et al.*, 2005).

Stimulation with TLR2 ligands induces the production of various proinflammatory cytokines (but not type I IFNs) in macrophages and DCs. However, another report showed that TLR2 in inflammatory monocytes induced type I IFNs in response to viral infection, suggesting that cellular responses to TLR2 ligands differ depending on the cell types involved (Barbalat *et al.*, 2009). TLR4, together with myeloid differentiation factor 2 (MD-2), is involved in the recognition of viral envelope proteins. Activation of TLRs implicated in the sensing of nucleic acids derived from viruses (TLR3, 7/8, 9) leads to the production of type I IFNs in addition to proinflammatory cytokines.

After ligand binding, TLRs dimerize and undergo conformational changes, which are required for the recruitment of adaptor molecules, via their TIR domains. There are five TIR domain-containing adaptor proteins: MyD88 (myeloid differentiation factor 88), MAL (MyD88 adaptor-like; also known as TIRAP), TRIF (TIR domain-containing adaptor inducing IFNβ, also known as TICAM-1), TRAM (TRIF-related adaptor molecule) and SARM (sterile-α and armadillo motif-containing protein; O'Neill and Bowie, 2007). These adaptor molecules contribute to the specificity of individual responses to viruses. Each TLR can mediate a tailored response in association with different combinations of these adaptors. TLR signalling can be roughly divided into two distinct pathways depending on the usage of MyD88 or TRIF adaptor molecules: the MyD88-dependent pathway, used by all TLRs except TLR3, and the TRIF-dependent pathway, used by TLR3 and also by TLR4 (Fig. 6.2).

MyD88-dependent signalling pathway

MyD88 is composed of a death domain (DD) in addition to a TIR domain. TLR2 and TLR4 signalling requires MAL/TIRAP for bridging between TLR and MyD88 (Fitzgerald *et al.*, 2001). Signalling is initiated by recruitment of MyD88 to TLRs, via TIR–TIR interactions (O'Neill, 2006). MyD88 interacts with IL-1R-associated kinase (IRAK)-4, a serine/threonine kinase with an N-terminal death domain. IRAK-4 activates IRAK-1 and IRAK-2 (Kawagoe *et al.*, 2008). The IRAKs then dissociate from MyD88 and interact

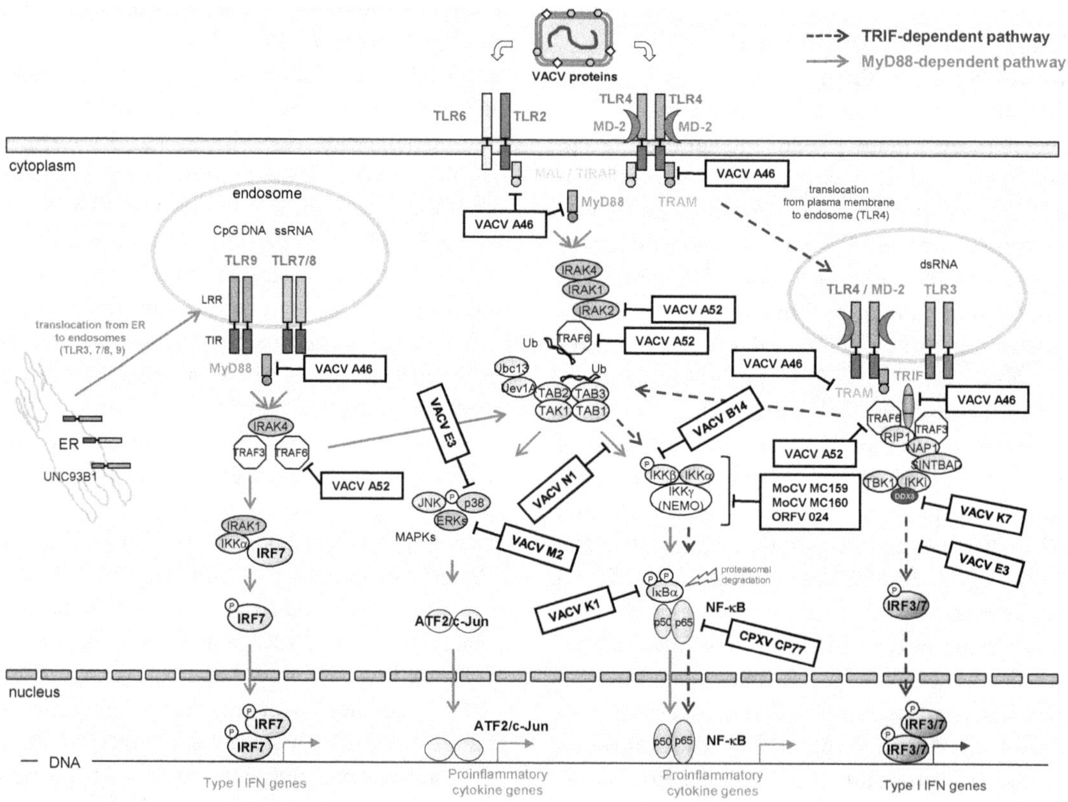

Figure 6.2 Interplay between toll-like receptor (TLR) signalling pathways and poxvirus proteins. Viral ligands are recognized by different TLRs: TLR2 and TLR4, localized on the plasma membrane, and TLR3, TLR7/8 and TLR9 which are translocated from the ER to endosomes. Only TLR3 exclusively uses the TRIF-dependent signalling pathway (dotted line), while other TLRs use the MyD88-dependent pathway (solid line). TLR4 uses both pathways. Through these pathways, transcription factors (IRFs, ATF2/c-Jun and NF-κB) are activated and translocated into the nucleus where they induce the transcription of type I IFNs and proinflammatory cytokine genes. Poxvirus-encoded immunomodulatory proteins interfering with these signalling pathways at different levels are indicated.

with TNFR-associated factor 6 (TRAF6), which acts as an E3 ubiquitin protein ligase. Together with an E2 ubiquitin-conjugating enzyme complex comprising Ubc13 and Uev1A, TRAF6 catalyses the formation of a lysine 63-linked polyubiquitin chain on TRAF6 itself as well as the generation of an unconjugated free polyubiquitin chain (Xia *et al.*, 2009). A complex of TGFβ-activated kinase 1 (TAK1) and TAK1-binding (TAB) proteins 1, 2 and 3 (TAB1, TAB2 and TAB3) is activated by the unconjugated free lysine 63 polyubiquitin chain and phosphorylates IκB kinase (IKK)β and members of the mitogen-activated protein kinase (MAPK) family JNK, p38 and extracellular signal-related kinases (ERKs). Subsequently, the IKK complex, composed of IKKα, IKKβ and the regulatory subunit IKKγ (also known as NEMO), phosphorylates IκBα, an NF-κB inhibitory protein. Phosphorylated IκBα undergoes degradation by the ubiquitin-proteasome system, thereby freeing NF-κB to translocate into the nucleus and activate expression of proinflammatory cytokine genes (Akira, 2006). Activation of the MAP kinase cascade is responsible for the formation of another transcription factor complex, ATF2/c-Jun, that targets cytokine genes.

TLR7/8 and TLR9 signalling induces the production of type I IFNs in addition to other NF-κB-dependent cytokines in a MyD88-dependent manner. In pDCs, MyD88 forms a complex with IRAK-4, TRAF6, TRAF3, IRAK-1, IKKα and IRF7. Phosphorylated IRF7 translocates into the nucleus to activate the expression of genes encoding type I IFNs. In cDCs, IRF3, but not IRF7, is activated downstream of TLR7 and TLR9, resulting in the activation of IFNβ gene expression (Negishi *et al.*, 2006; Schmitz *et al.*, 2007).

TRIF-dependent signalling pathway

TLR3 and TLR4 make use of this pathway, in which the adaptor molecule TRIF is essential (Yamamoto *et al.*, 2003). Additionally, the adaptor molecule TRAM is necessary for the TLR4-mediated TRIF-dependent pathway and appears to act as a bridging adaptor between TLR4 and TRIF (Oshiumi *et al.*, 2003). TRIF associates with TRAF3 and TRAF6 through TRAF-binding motifs present in its N-terminal portion. In humans, SARM has been described as a specific inhibitor of TRIF-dependent signalling (Carty *et al.*, 2006). TRAF6 and RIP1 activate NF-κB whereas TRAF3 interacts with and activates two IKK-related kinases, designated TRAF family member-associated NF-κB activator (TANK)-binding kinase 1 (TBK1) and inducible IκB kinase (IKKi; also known as IKKε; Oganesyan *et al.*, 2006). NAP1 and SINTBAD are required for the activation of the TBK1/IKKi complex, which then phosphorylates IRF3 and IRF7. Phosphorylated IRF3 and IRF7 dimers are then translocated into the nucleus, resulting in the induction of type I IFN genes.

Spatial regulation of TLRs appears to be important for ligand recognition and proper signalling (Barton and Kagan, 2009). While TLR2, TLR4 and TLR6 are present on the plasma membrane, nucleic acid-sensing TLRs (TLR3, TLR7/8 and TLR9) are mainly present on the endoplasmic reticulum (ER) membrane and are recruited to endolysosomes following stimulation by their ligands. An ER membrane protein, UNC93B1, was identified as an essential molecule for the translocation of TLR3, TLR7/8 and TLR9 from the ER to endosomes (Tabeta *et al.*, 2006; Kim *et al.*, 2008). When TLR9 is recruited from the ER to the endolysosome, it undergoes processing by proteases, such as cathepsins, in the endolysosome (Ewald *et al.*, 2008; Park *et al.*, 2008). The processed form of TLR9 is responsible for CpG-DNA recognition. Another ER protein, PRAT4A, also controls TLR9 trafficking from the ER to endosomes/lysosomes (Takahashi *et al.*, 2007). TLR4 is also translocated from the plasma membrane to endosomes together with TRAM in the TRIF-dependent signalling pathway.

Some of the factors involved in TLR signalling are shared with other signalling pathways. More specifically, it has been reported that TNFR-associated death domain protein (TRADD), an essential adaptor for TNFR signalling, FAS-associated death domain-containing protein (FADD) and RIP1 are required for TRIF-dependent signalling (Ermolaeva *et al.*, 2008; Pobezinskaya *et al.*, 2008) and for both RIG-I and MDA5 signalling pathways (Michallet *et al.*, 2008). This might provide the opportunity for cross-talk between the different innate immune signalling events.

Poxvirus evasion of TLR signalling pathways

In order to counteract host innate immunity, poxviruses encode immunomodulatory proteins that antagonize important components of the TLR signalling pathways (Fig. 6.2). The first demonstration of viral proteins that inhibit TLRs was reported for the case of VACV A46 and A52 proteins (Bowie *et al.*, 2000). Several members of the VACV A46 and N1 families, with a Bcl-2-like structure, are involved in the regulation of the host innate immune response where they act non-redundantly at different levels of the TLR signalling pathway. Based on sequence and structure similarity, it has been proposed that other VACV members like N2, C1, C6 and C16/B22 might have a similar role in suppression of host immune responses as A46, A52, B15 and K7, by antagonizing at different levels with the TLR signalling pathway (Gonzalez and Esteban, 2010).

VACV A46 protein is the only reported viral protein containing a TIR domain (Bowie *et al.*, 2000). Through this domain A46 binds directly to the TIR-domain-containing adaptors MyD88, MAL, TRIF and TRAM, interfering with the downstream activation of MAPKs, NF-κB and IRFs triggered by multiple TLR pathways (Bowie *et al.*, 2000; Stack *et al.*, 2005). However, A46 does not interact with SARM, which is a negative regulator of TLR signalling. When the gene that encodes A46 was deleted from the VACV genome, the virus was attenuated in a murine intranasal model, demonstrating the importance of A46 for VACV virulence (Stack *et al.*, 2005).

The VACV A52 protein interacts with both IRAK2 and TRAF6, two signalling molecules that are downstream of the TIR-domain-containing adaptors (Harte *et al.*, 2003). Although A52 inhibits TLR-mediated NF-κB activation, it can simultaneously activate MAPK p38 and enhance the production of IL-10 (Maloney *et al.*, 2005). Interaction with IRAK2 is sufficient for the ability of A52 to inhibit NF-κB, whereas interaction with TRAF6 facilitates A52-mediated induction of MAPK p38 (Keating *et al.*, 2007). Thus, under conditions of TLR stimulation, A52 inhibits the expression of NF-κB-dependent genes, such as IL-8 and RANTES, and enhances the production of IL-10, a cytokine required for viral persistence (Brooks *et al.*, 2006; Ejrnaes *et al.*, 2006). This dual activity might contribute to the role of A52 in VACV virulence as a virus deletion mutant lacking the A52R gene was attenuated in a murine intranasal model of infection (Harte *et al.*, 2003).

VACV also encodes the K7 protein that shares significant sequence similarity to A52. K7 inhibits PRR-mediated induction of IFNβ by preventing TBK1/IKKi-mediated IRF activation owing to its ability to target human DEAD box protein 3 (DDX3; Schroder *et al.*, 2008; Soulat *et al.*, 2008). DDX3 is part of the TBK1/IKKi complex downstream of TLRs, RLRs and cytoplasmic DNA receptors. Furthermore, virus infection induced an association between DDX3 and IKKi.

Poxvirus NF-κB inhibitors (Mohamed and McFadden, 2009)

VACV N1 protein, an intracellular homodimeric virulence factor (Bartlett *et al.*, 2002) which shares significant similarity to A52, has been reported to inhibit TRAF- and IKKα/IKKβ-dependent signalling to NF-κB owing to its ability to associate with several components of the multisubunit IKK complex and with TBK1 (DiPerna *et al.*, 2004). However, Cooray *et al.* reported that N1 does not inhibit IL-1-induced NF-κB-dependent gene expression in VACV-infected cells, probably because of the presence of other VACV NF-κB signalling inhibitors (Cooray *et al.*, 2007). Moreover, a recent study showed that N1 did not co-precipitate with the IKK complex, unlike another VACV protein, B14 (Chen *et al.*, 2008; see further). The observation that N1 inhibits TRAF6-induced NF-κB activation but does not seem to interact with the IKK complex (Chen *et al.*, 2008), in contrast to previous observations (DiPerna *et al.*, 2004), suggests that N1 acts at the level of TRAF6 or on downstream proteins in the pathway that precede the IKK complex. In addition, the N1 crystal structure reveals that N1 is a Bcl-2-like protein (Aoyagi *et al.*, 2007; Cooray *et al.*, 2007) and its role in the inhibition of apoptosis will be discussed below. Regarding virulence, VACV lacking the N1L gene has been reported to be attenuated in mouse intradermal (Tscharke and Smith, 1999; Bartlett *et al.*, 2002), intracranial (Kotwal *et al.*, 1989) and intranasal (Bartlett *et al.*, 2002; Jacobs *et al.*, 2008) models of infection.

Another VACV NF-κB inhibitor is B14, a 17-kDa cytosolic protein that inhibits the IKK complex (Chen *et al.*, 2008) and contributes to VACV virulence (Chen *et al.*, 2006). B14 specifically interacts with IKKβ and prevents its activation by inhibiting the phosphorylation of its activation loop (Chen *et al.*, 2008). Consequently, B14 inhibits the subsequent phosphorylation and degradation of IκBα, the inhibitor of NF-κB (Chen *et al.*, 2008). A52 and B14 are Bcl-2 family members but, unlike N1, lack a surface groove for binding BH3 peptides and therefore they do not modulate apoptosis (Graham *et al.*, 2008). B14 has been shown to contribute to VACV virulence in a mouse intradermal model of infection (Chen *et al.*, 2006).

Another example is VACV M2, an early protein that localizes to the endoplasmic reticulum and inhibits NF-κB activation by preventing phosphorylation of the ERK2 protein (Gedey *et al.*, 2006). Introduction of the M2L gene from VACV strain WR into the MVA genome re-establishes the ability of MVA to inhibit NF-κB (Gedey *et al.*, 2006). The M2 protein contains motifs characteristics of ER-localized proteins and this organelle location affects M2 protein function since elimination of the N-terminal leader sequence or the putative ER retention and retrieval motifs from M2 compromises both its ER location and its ability to inhibit virus-induced NF-κB activation (Hinthong *et al.*, 2008).

In addition to VACV, other orthopoxviruses encode NF-κB inhibitors. MoCV encodes two proteins, MC159 and MC160, which contain death effector domains (DEDs) and are involved in the inhibition of apoptosis (see below). MC159 was also reported to inhibit NF-κB activation induced by the overexpression of TNFR1 molecules (Chaudhary *et al.*, 1999) and by the dsRNA-dependent protein kinase R (PKR; Gil *et al.*, 2001). MC159 prevents TNF-induced IκBβ degradation but not IκBα degradation and interaction with TRAF2 was shown to be critical for the inhibitory function of MC159 (Murao and Shisler, 2005). MC160 was reported to induce IKKα degradation by competitively interacting with the cellular heat shock protein 90 (HSP90), necessary for IKKα stabilization (Nichols and Shisler, 2009).

The parapoxvirus orf virus (ORFV) ORFV024 protein has been recently shown to inhibit activation of the NF-κB signalling pathway (Diel *et al.*, 2010). Infection of primary ovine cells with an ORFV024 deletion mutant virus resulted in a marked increase in expression of NF-κB-regulated chemokines and other proinflammatory host genes. Expression of ORFV024 decreased phosphorylation of IKKα and IKKβ, degradation of IκBα and phosphorylation and nuclear translocation of NF-κB/p65 induced by TNFα, indicating that ORFV024 functions by inhibiting activation of IKKs (Diel *et al.*, 2010). Finally, the cowpox virus (CPXV) CPXV016 gene encodes an NF-κB regulator (Oie and Pickup, 2001).

Ankyrin repeat (ANK) proteins comprise the largest family of poxvirus proteins and are encoded by almost all chordopoxviruses (Sonnberg *et al.*, 2008). Several studies have reported the ability of some of these ANK-repeat proteins to inhibit NF-κB activation. For example, VACV K1 protein has been shown to inhibit NF-κB activation in RK-13 cells by preventing IκBα degradation (Shisler and Jin, 2004) and was recently shown to inhibit signalling from TLR2, TLR4 and TLR9 in primary microglial cells (Aravalli *et al.*, 2008). K1 may inhibit IκBα degradation by interfering either directly with IKK or indirectly with kinases that act upstream of IKK to prevent IκBα phosphorylation (Shisler and Jin, 2004). This K1 inhibition of NF-κB activation does not seem to be related to the host range function of K1 (Werden *et al.*, 2008) since this inhibition was observed both in permissive and non-permissive cell lines (Shisler and Jin, 2004).

Another ANK-containing protein from myxoma virus (MYXV), MNF (M150R gene), has been shown to co-localize with NF-κB following TNF treatment of cells (Camus-Bouclainville *et al.*, 2004). *In vivo* experiments using an M150R deletion mutant virus showed that MNF is a critical virulence factor, with its deletion generating an increase in the inflammatory response, consistent with the interference of MNF with the NF-κB-induced proinflammatory pathway (Camus-Bouclainville *et al.*, 2004).

Recently, additional ANK-containing NF-κB inhibitors have been reported from several orthopoxviruses including CPXV, monkeypox virus

(MPXV), ectromelia virus (ECTV) and VARV (Chang *et al.*, 2009; Mohamed *et al.*, 2009a,b). The CP77 host range protein from CPXV was reported to block TNFα-induced NF-κB activation by direct binding to the NF-κB/p65 subunit, independent of its host range function (Chang *et al.*, 2009). Another gene from CPXV, CPXV-006, encodes an inhibitor of the atypical NF-κB activation pathway involving the p105 subunit. *In vitro* infection of human myeloid (THP1) cells with CPXV-006 deletion mutant rapidly induces high levels of NF-κB-controlled proinflammatory cytokines (Mohamed *et al.*, 2009b). *In vivo*, this deletion mutant was attenuated in a murine intranasal model and induced a significantly elevated cellular inflammatory response (Mohamed *et al.*, 2009b). G1 protein from VARV (gene G1R) also interacts with NF-κB/p105 to inhibit NF-κB activation (Mohamed *et al.*, 2009a).

RLR signalling and inhibition by poxviruses

RLR signalling pathways play essential roles in the recognition of viral RNA in various cell types, such as fibroblasts and cDCs, with the exception of pDCs (Kato *et al.*, 2005). The RLR family has three members: retinoic acid-inducible gene I (RIG-I), melanoma differentiation-associated gene 5 (MDA5) and laboratory of genetics and physiology 2 (LGP2; Yoneyama and Fujita, 2008; Takeuchi and Akira, 2009). RIG-I and MDA5 consist of two N-terminal caspase recruitment domains (CARDs), a central DExD/H box RNA helicase domain and a C-terminal repressor domain (RD), whereas LGP2 lacks a CARD. The RLRs recognize viral RNAs in the cytoplasm (short dsRNA with 5′ triphosphate ends for RIG-I and long dsRNA for MDA5). The helicase domain and RD are important for the recognition of these RNAs, while the CARDs are essential for triggering intracellular signalling cascades (Yoneyama *et al.*, 2004; Saito *et al.*, 2007). LGP2 is a positive regulator of the MDA5-mediated immune response (Satoh *et al.*, 2010) but the role of LGP2 in the RIG-I-mediated response is controversial.

The CARDs of activated RIG-I or MDA5 are responsible for triggering signalling cascades by interacting with the N-terminal CARD-containing adaptor IFNβ-promoter stimulator 1 (IPS-1; also known as MAVS, VISA or CARDIF; Kawai and Akira, 2006; Potter *et al.*, 2008), which localizes to the outer mitochondrial membrane through a C-terminal transmembrane domain, suggesting that mitochondria may be important for IFN responses, in addition to their roles in metabolism and cell death (Seth *et al.*, 2005). NLRX1 (also known as NOD9), an NLR family member, is localized on the mitochondrial membrane and acts as an inhibitor of IPS-1 signalling (Moore *et al.*, 2008). The downstream signalling molecules are shared between the IPS-1 and TRIF signalling pathways. IPS-1 activates signalling cascades leading to the expression of type I IFN genes via TRAF3, NAP1/SINTBAD, TBK1/IKKi and IRF3/7 (Fitzgerald *et al.*, 2003; Sharma *et al.*, 2003; Hemmi *et al.*, 2004). Simultaneously, IPS-1 signalling induces nuclear translocation of NF-κB via TRADD, FADD and caspase-8/–10. A cleaved fragment of caspase-8/-10 is responsible for the activation of NF-κB (Takahashi *et al.*, 2006).

A recent study identified a novel protein named stimulator of IFN genes (STING) localized in the endoplasmic reticulum membrane as an important molecule for RIG-I/MDA5 signalling (Ishikawa and Barber, 2008). Additionally, an IPS-1-independent function of RIG-I has been recently described: RIG-I directly activates the inflammasome, a protein complex that cleaves pro-IL-1β into mature IL-1β (Poeck *et al.*, 2010; Fig. 6.3).

Poxviruses produce considerable amounts of dsRNA owing to convergent transcription [that is, when two adjacent open reading frames are transcribed at opposite directions, resulting in a partial 3′ overlap between their transcripts (Weber *et al.*, 2006)]. It has been reported that VACV generates an MDA5-stimulatory RNA during infection (Pichlmair *et al.*, 2009). To avoid the antiviral responses initiated by RLRs (and by the dsRNA-activated effector proteins PKR and 2′,5′-oligoadenylate synthase [2′5′OAS]), some poxviruses encode cytoplasmic dsRNA-binding proteins, including VACV E3 (see below). In addition to sequestering viral RNA, some of these proteins also bind to PKR and/or RIG-I directly, thereby inhibiting the antiviral response at many levels.

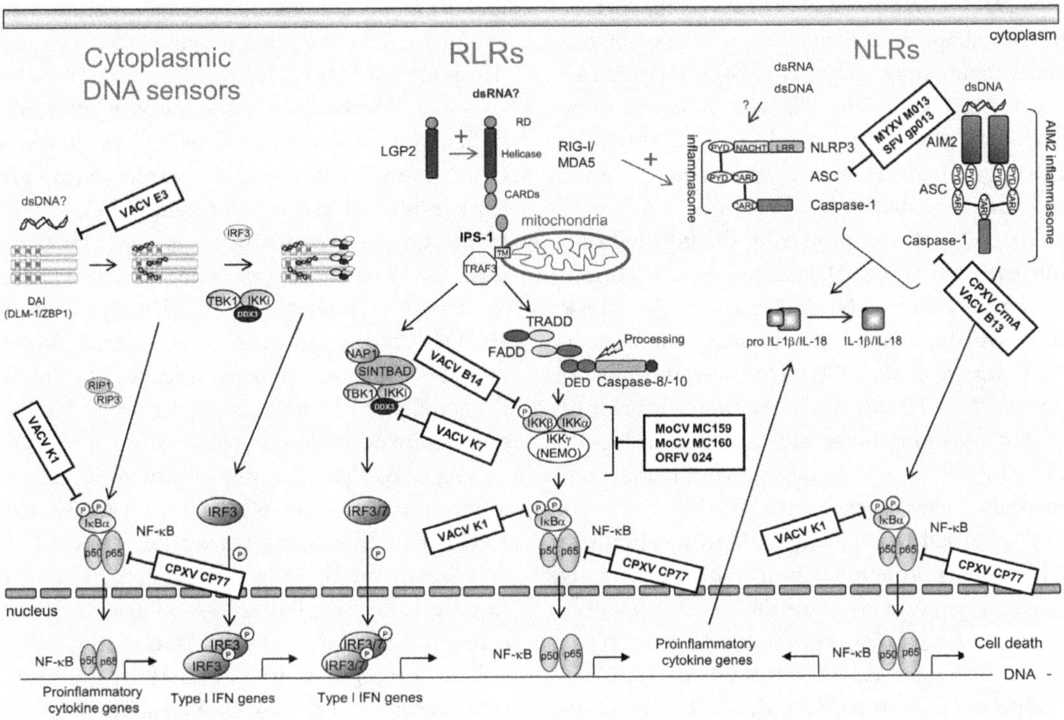

Figure 6.3 Interplay between RLR, NLR and cytoplasmic DNA sensor signalling pathways and poxvirus proteins. Activated RIG-I promotes the induction of IFNs and other proinflammatory cytokines via the mitochondrial adaptor IPS-1. IPS-1-dependent induction of pro-IL-1β allows it to be processed into mature IL-1β by the inflammasome, which can be directly activated by RIG-I in an IPS-1-independent manner. Viral DNA delivered into the cytoplasm binds to AIM2, which, in turn, binds to ASC and assembles an inflammasome. Formation of the AIM2 inflammasome leads to the activation of caspase-1 from its pro-form. Active caspase-1 mediates cleavage of pro-IL-1β and pro-IL-18, which are then released to the extracellular space. In parallel, AIM2 inflammasome formation leads to the activation of NF-κB and the induction of cell death. Poxvirus immunomodulatory proteins interfering with these signalling pathways are indicated.

NLR signalling

The NLR family is a group of cytoplasmic PRRs involved in the recognition of cytoplasmic microbial PAMPs and/or endogenous danger signals and in the initiation of proinflammatory responses (Muruve *et al.*, 2008). The NLR family members have similar structures: an effector N-terminal domain consisting of either a pyrin domain (PYD), a caspase activation and recruitment domain (CARD) or a baculovirus inhibitor of apoptosis protein repeat (BIR), a central nucleotide binding and oligomerization domain (NACHT) and a C-terminal ligand binding leucine-rich repeat domain (LRR). Some NLRs, such as NLRP1, NLRP3, IPAF and NAIP5, are components of cytosolic protein complexes called inflammasomes (Martinon *et al.*, 2009; Pedra *et al.*, 2009). The inflammasome complex comprises one or some of the NLR proteins and caspase-1, which is activated in this complex by self-cleavage and cleaves substrates, such as pro-IL-1β and pro-IL-18, to produce mature proteins for secretion, thus driving the inflammatory response that serves to recruit immune cells to the site of infection (Bryant and Fitzgerald, 2009). The key role of caspase-1 in the processing of IL-1β and, surprisingly, also in the production of IL-1α, is evidenced by the phenotype of mice in which the caspase-1 gene has been disrupted (Li *et al.*, 1995). These mice are defective for mature IL-1β production, have diminished IL-1α compared with littermate controls and are resistant to the induction of endotoxic shock, a condition believed to result in large part from excessive IL-1 production.

NLRP3 (also known as NALP3 or cryopyrin), with its adaptor molecule, apoptotic speck-like protein containing a CARD (ASC), recruits caspase-1 making up the NLRP3 inflammasome (Petrilli *et al.*, 2007; Cassel *et al.*, 2009). The NLRP3 inflammasome is stimulated during acute infection by both adenovirus and VACV, leading to IL-1β secretion from the infected cells (Muruve *et al.*, 2008; Delaloye *et al.*, 2009), supporting the idea that NLRP3 may serve as a PRR for DNA-based pathogens to trigger inflammation (Muruve *et al.*, 2008). However, it is not yet known if NLRP3 directly detects nucleic acids or if these inflammasomes are activated indirectly following nucleic acid recognition by an unknown sensor that may associate with NLRP3.

As opposed to NLRP3, the AIM2 (absent in melanoma 2) protein has been shown to directly bind and respond to cytosolic dsDNA (Burckstummer *et al.*, 2009; Fernandes-Alnemri *et al.*, 2009; Hornung *et al.*, 2009; Roberts *et al.*, 2009). Binding of DNA to AIM2 induces its oligomerization and recruitment of ASC through its pyrin domain, leading to caspase-1 activation as in the case of NLRs (Fig. 6.3). VACV, which can trigger inflammasome assembly in an NLRP3-independent but ASC-dependent manner, requires AIM2 for this activity (Hornung and Latz, 2010).

Poxvirus evasion of NLR signalling

It has been reported that a PYD-containing protein encoded by the MYXV M013L gene disrupts NLRP–ASC interaction and caspase-1 activation (Johnston *et al.*, 2005). The M013 protein was shown to directly associate with ASC and inhibit activation of pro-caspase-1 and subsequent processing of pro-IL-1β and pro-IL-18 (Johnston *et al.*, 2005). Recently, the M013 protein has also been described as an inhibitor of the NF-κB/p65 signalling pathway (Rahman *et al.*, 2009). Thus, the M013 inhibitor exerts a dual immune-subversive role in the simultaneous co-regulation of both the cellular inflammasome complex and NF-κB-mediated proinflammatory responses.

Another PYD-containing protein from Shope fibroma virus (SFV), gp013, was also reported to directly associate with ASC and inhibit PYD-mediated activation of pro-caspase-1 (Dorfleutner *et al.*, 2007). Additional PYD-only proteins were identified in the genomes of swinepox virus (SPXV), Yaba-like disease virus (YLDV) and deerpox virus (DPXV).

Some poxviruses encode caspase inhibitors, such as CrmA from CPXV or B13 from VACV. CrmA protein belongs to the serpin (serine protease inhibitor) family of proteins (Pickup *et al.*, 1986). One of two identified substrates for CrmA is the IL-1β converting enzyme [ICE, also known as caspase-1 (Ray *et al.*, 1992; Komiyama *et al.*, 1994)]. Thus, expression of the CrmA protein probably functions during infection to inhibit the activity of ICE and, subsequently, to diminish production of mature IL-1β.

VACV B13 protein, also known as SPI-2, is an intracellular inhibitor of caspase-1 (Kettle *et al.*, 1997). B13 blocks the conversion of pro-IL-1β into the mature IL-1β by inhibiting the activity of caspase-1. Because B13 is intracellular, its activity is restricted to infected cells; thus it is unable to influence the production of IL-1β by uninfected cells recruited to the infected area. The B13 protein is also predicted to inhibit the formation of IL-18 (IFNγ-inducing factor) from infected cells because, like IL-1β, IL-18 is synthesised as a precursor that needs to be proteolytically cleaved by caspase-1 (Ghayur *et al.*, 1997). Similarly, other caspase-1 inhibitors have been reported from MYXV (Petit *et al.*, 1996) and SFV (Moss and Shisler, 2001).

Cytoplasmic DNA sensors and their inhibition by poxviruses

DNA-dependent activator of IRFs (DAI; previously also known as Z-DNA binding protein 1 (DLM-1) or ZBP1) was the first molecule identified as a possible candidate for a cytoplasmic DNA sensor (Takaoka *et al.*, 2007). DAI can directly interact with B-DNA/Z-DNA and with TBK1 and IRF3, and activate IRF3 leading to the production of type I IFN genes (Takaoka *et al.*, 2007). Furthermore, it was demonstrated that RIP1 and RIP3 are involved in the DAI-mediated activation of NF-κB (Kaiser *et al.*, 2008; Rebsamen *et al.*, 2009; Fig. 6.3). However, gene knockout analysis demonstrated that DAI deficiency had no effect on DNA-induced IFN production (Ishii *et al.*, 2008), suggesting that cytoplasmic DNA is redundantly recognized by other dsDNA sensors

(Ishii *et al.*, 2008). This hypothesis has been confirmed by the recent identification of two other dsDNA sensors: AIM2 (previously described in the NLR signalling pathway; Burckstummer *et al.*, 2009; Fernandes-Alnemri *et al.*, 2009; Hornung *et al.*, 2009; Roberts *et al.*, 2009) and RNA polymerase III (Ablasser *et al.*, 2009; Chiu *et al.*, 2009).

DAI has an N-terminal Z-DNA-binding domain that is homologous to the Zα and Zβ domains of adenosine deaminase RNA-specific 1 (ADAR 1) and also to the VACV virulence factor E3. In this context, data from Takaoka *et al.* indicated that IFNβ induction by VACV genomic DNA is mediated by DAI (Takaoka *et al.*, 2007). As the VACV E3 protein has a Z-DNA-binding domain, it is presumed that DAI function is targeted by E3 (Takaoka and Taniguchi, 2008; Wang *et al.*, 2008), protecting the VACV DNA genome from recognition by DNA receptors. However, detailed studies with VACV lacking only the Z-DNA-binding domain of E3 will further clarify the involvement of DAI and of the other dsDNA sensors in VACV infection.

The different PRR signalling pathways described above, specific sensing of poxviruses by these PRRs and poxviruses proteins interfering with these signalling pathways are depicted schematically in Figs 6.2 and 6.3.

IFN receptor signalling

IFNs are grouped into three classes, designated IFNs types I to III, according to the receptor complex they signal through. Type I IFNs, which in humans comprise 13 IFNα subtypes, IFNβ, IFNκ, IFNε, IFNω, IFNτ and IFNδ, are known to be essential for mounting a robust host response against viral infection and mediate their antiviral activities through the induction of cellular proteins that in turn participate in RNA degradation, the inhibition of translation and MHC antigen regulation (Samuel, 1991). Type II IFN comprises a single member called IFNγ (Bach *et al.*, 1997; Ikeda *et al.*, 2002) and is secreted by activated T cells and natural killer (NK) cells; rather than in direct response to viral infection, IFNγ is a potent inducer of the cell-mediated (Th1) immune response (Pestka *et al.*, 1987; Farrar and Schreiber, 1993). Besides possessing autocrine functions, IFNγ activates macrophages by upregulating their MHC class II expression and regulates B-lymphocyte differentiation (Farrar and Schreiber, 1993). Like type I IFNs, IFNγ also induces cellular proteins that ultimately lead to the inhibition of translation and RNA degradation associated with 'antiviral' activity. The more recently described type III IFNs include IFNλ1, IFNλ2 and IFNλ3, also referred as IL-29, IL-28A and IL-28B, respectively (Ank *et al.*, 2006; Uze and Monneron, 2007). The type III IFN family elicits an antiviral response that is nearly identical to that evoked by IFNα/β and has been proposed to be the ancestral type I IFNs (Levraud *et al.*, 2007).

The action of the IFNs is mediated through three receptor complexes: a heterodimer of IFNAR1 and IFNAR2 binds type I IFNs (Uze *et al.*, 2007); the IL-10R2 associates with IFNLR1 to bind the three IFNλ subtypes; and a tetramer consisting of two IFNGR2 chains and two IFNGR1 chains binds dimers of the type II IFNγ (Pestka, 1997). Following binding of IFNs to specific receptors, signal transduction is initiated by pre-associated tyrosine Janus kinases JAK1, JAK2 and TYK2, which Tyr phosphorylate the corresponding IFN receptor leading to the recruitment and Tyr phosphorylation of the signal transducers and activators of transcription (STAT) proteins (Schindler *et al.*, 2007). Phosphorylation of these STATs leads to the formation of two transcriptional activator complexes: STAT1–STAT2 heterodimers associate with IRF9 to form IFN-stimulated gene factor 3 (ISGF3; type I and III IFN signalling) and STAT1–STAT1 homodimers form the IFNγ activation factor (GAF; type II IFN signalling; Platanias, 2005). These complexes translocate into the nucleus to induce more than 300 ISGs (Der *et al.*, 1998) from IFN-stimulated response elements (ISREs; type I and type III IFN responses) or GAS promoter elements (type II IFN response; Fig. 6.4). Divergence from this signalling pathways can occur: type I and type III IFNs are reported to elicit STAT1 homodimers and type II IFNs STAT1–STAT2 heterodimers; STAT3 or STAT5 homodimers and STAT1-STAT3 or STAT5-CrkL (v-crk sarcoma virus CT10 oncogene homologue avian-like) heterodimers have also been reported to be elicited by type I IFNs (Fish *et al.*, 1999).

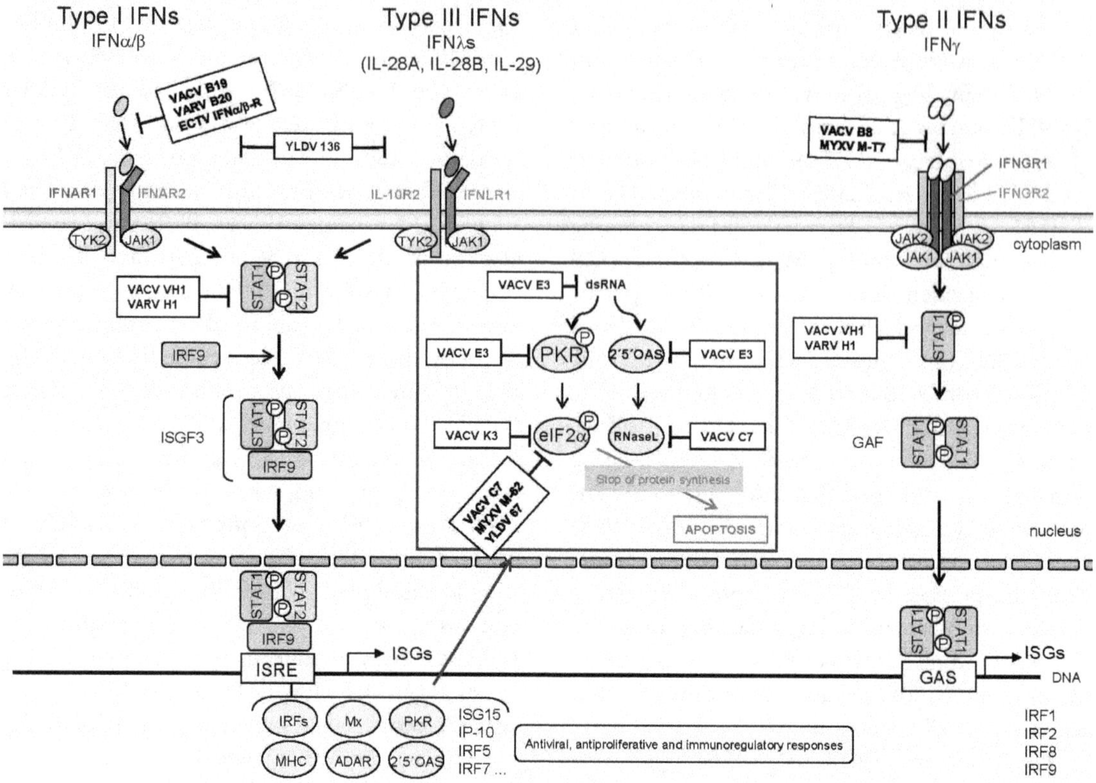

Figure 6.4 Interplay between interferon (IFN) signalling and effector pathways and poxvirus proteins. IFNs produced interact with specific IFN receptors (IFNRs), in an autocrine or paracrine manner, activating a signal transduction cascade involving JAKs and STAT factors, which leads to the transcription of IFN-stimulated genes (ISGs). These ISGs are involved in eliminating viral components from infected cells, inducing apoptosis of infected cells and conferring resistance to viral infection on uninfected cells. Poxvirus immunomodulatory proteins interfering with these signalling pathways are indicated.

ISGs are involved in the amplification and regulation of the IFN response (RIG-I, MDA5, DAI, some microRNAs, TRIM) as well as in antiviral mechanisms that interfere with the viral life cycle, such as 2′5′OAS (Clemens, 2005; Silverman, 2007), RNaseL (Silverman, 2007), PKR (Garcia *et al.*, 2006), myxovirus resistance GTPase (Mx; Martens and Howard, 2006; Haller *et al.*, 2007), RNA-specific adenosine deaminase (ADAR), apolipoprotein B mRNA-editing enzyme-catalytic polypeptide-like 3 (APOBEC3), IFNAR1, IPS1, IFN-stimulated protein of 15 kDa (ISG15), myeloid differentiation protein-2 (MD-2) and STATs (Sadler and Williams, 2008). Furthermore, type I IFNs secreted by immune cells (macrophages or DCs) regulate the activation of adaptive immune responses, including the maturation of DCs, activation of NK cells and differentiation of cytotoxic T cells (Stetson and Medzhitov, 2006).

Poxvirus evasion of IFN signalling pathways

Poxviruses encode several proteins that inhibit the action of IFN either by preventing IFN binding to its natural receptor (soluble IFN binding proteins) or within the infected cell by blocking IFN signalling cascades or the IFN-induced antiviral state.

Viral soluble IFN-binding proteins

VACV encodes a soluble IFNα/β receptor that belongs to the immunoglobulin superfamily and has limited homology to cellular IFNα/βRs (Colamonici *et al.*, 1995). It is encoded by the

B18R gene in WR strain or the B19R gene in Copenhagen strain and is both secreted and at the surface of VACV-infected cells by adopting a type II membrane topology (Ueda *et al.*, 1990; Morikawa and Ueda, 1993). Secreted B19 can bind to the cell surface via an unidentified cellular molecule (Colamonici *et al.*, 1995; Alcami *et al.*, 2000). B19 binds and inhibits type I IFNs from human (Liptakova *et al.*, 1997), cow, rabbit, rat and mouse, although the affinity for mouse type I IFN is considerably lower than for the other species tested (Symons *et al.*, 1995). A virus lacking B19R was attenuated in both intranasally and intracranially infected mice (Colamonici *et al.*, 1995; Symons *et al.*, 1995) despite the relatively low affinity of B19 for mouse type I IFNs. B19 has been recently involved in the multistep process of VACV-mediated inhibition of type I IFN responses by professional IFN-producing pDC (Waibler *et al.*, 2009).

Most other orthopoxviruses also encode soluble IFNα/β-binding proteins: VARV B20R gene is the homologue of VACV B19R gene and the ectromelia virus-encoded IFNα/β-receptor has been reported to potently inhibit human and mouse IFNα and human, but not mouse, IFNβ (Smith and Alcami, 2002). YLDV has a protein (YLDV 136) related to VACV B19, but Y136 protein binds and inhibits both type I and type III IFNs (Huang *et al.*, 2007).

VACV also encodes a soluble IFNγ receptor (gene B8R) with amino acid similarity to the extracellular domain of the type II IFNR. This protein is secreted from infected cells early in infection as a homodimer (Alcami and Smith, 2002), probably assisting in the binding of the homodimeric IFNγ molecules. B8 binds and inhibits IFNγ from human, cow, rat, rabbit, horse and chicken but not mouse (Alcami and Smith, 1995; Mossman *et al.*, 1995a; Puehler *et al.*, 1998; Symons *et al.*, 2002b). Deletion of the B8R gene from the VACV genome originally showed no difference in virulence compared with wild-type virus in a mouse model, but the knockout virus was attenuated in rabbits (Sroller *et al.*, 2001; Symons *et al.*, 2002b). This result was anticipated given that B8 does not bind or inhibit mouse IFNγ with high affinity but inhibits the rabbit ligand. However, contrasting results reported from another group demonstrated that deletion of B8R in VACV results in a significant decrease in weight loss and mortality in normal mice following intranasal infection (Verardi *et al.*, 2001). This suggests that B8 may recognize additional ligands or possess other activities not related to its binding of IFNγ.

Other viral IFNγ-binding proteins are secreted by cells infected with MYXV (M-T7 protein; Upton *et al.*, 1992; Mossman *et al.*, 1995b), ECTV (Mossman *et al.*, 1995a; Smith and Alcami, 2002), CPXV, camelpox virus (CMLV; Alcami and Smith, 1995), VARV (Massung *et al.*, 1993; Shchelkunov *et al.*, 1993) and SFV (Alcami and Smith, 1996b). European rabbits infected with a recombinant MYXV lacking M-T7 showed a dramatic attenuation in lethality, disease symptoms and viral dissemination to secondary sites (Mossman *et al.*, 1996b) and an alteration in the infiltrating leucocyte profile at sites of infection. However, the subsequent observation that M-T7 is also a chemokine-binding protein (Lalani *et al.*, 1997; see below) makes this phenotype difficult to attribute solely to IFNγ inhibition.

Blocking of IFN signalling cascades

Earlier studies indicated a role of a viral phosphatase and of an ATPase in the resistance of VACV to IFN (Paez and Esteban, 1984; Diaz Guerra *et al.*, 1993). VACV encodes a dual-specificity phosphatase, VH1 (encoded by H1L gene), which has been shown to dephosphorylate STAT1 after IFNγ stimulation of epithelial cells, greatly attenuating IFN-induced biological functions (Najarro *et al.*, 2001). VH1 is required for viral replication as mutant VACV lacking H1L has greatly reduced virus production (Najarro *et al.*, 2001). Although the H1L gene is expressed late, VH1 is able to have an immediate effect on a new host cell because each infecting virion contains approximately 200 molecules of VH1 (Liu *et al.*, 1995). More recently, it has been reported that VACV infection is capable of inhibiting the phosphorylation of STAT1 and STAT2 after stimulation of fibroblasts or bone marrow-derived macrophages with either type I or type II IFNs, but did not inhibit the activation of STAT3 or STAT5 in either cell type (Mann *et al.*, 2008). By using recombinant proteins for *in vitro* assays, it was observed that VARV H1 is more active than VACV H1, although it has similar selectivity for STAT targets (Mann *et al.*, 2008).

Blocking of the IFN-induced antiviral state

VACV E3 and K3 proteins block the action of PKR and/or 2′5′OAS. Activation of these enzymes by dsRNA produced during VACV convergent transcription inhibits the translation and synthesis of viral and cellular proteins in infected cells, and frequently leads to the induction of apoptosis (Barber, 2001; Gil and Esteban, 2000; Jacobs and Langland, 1996; Samuel, 1991). Once activated by binding dsRNA, PKR phosphorylates first itself and then the Ser51 residue of eIF2α. Phosphorylated eIF2α is inactive so protein synthesis stops (Garcia *et al.*, 2006). After activation of 2′5′OAS by binding dsRNA, it synthesises 2′,5′oligoadenylates which activate RNaseL which in turn degrades RNA resulting in the inhibition of protein synthesis. The block in protein synthesis by both pathways stops virus replication.

VACV E3 is a 25 kDa dsRNA-binding protein (encoded by gene E3L) that binds and sequesters dsRNA to prevent the activation of PKR and 2′5′OAS (Chang *et al.*, 1992; Davies *et al.*, 1993; Rivas *et al.*, 1998; Romano *et al.*, 1998). E3 can also bind directly to PKR and inhibits its activity, resulting in reduced phosphorylation of eIF2α (Sharp *et al.*, 1998). Additional functions of VACV E3 protein have been reported: E3 was shown to reduce adenosine deaminase editing activity of ADAR (Liu *et al.*, 2001) and to bind to SUMO-1 (Rogan and Heaphy, 2000). E3 also blocks IRF3/7 activation and prevents upregulation of the host cell IFN genes (Smith *et al.*, 2001; Xiang *et al.*, 2002) and it inhibits p38 and NF-κB (Deng *et al.*, 2006) as well as cytokine expression (Langland *et al.*, 2006). E3 has recently been shown to directly target and disable ISG15 function (Guerra *et al.*, 2008) and to inhibit the IFNλ-mediated antiviral response through a PKR-dependent pathway (Bandi *et al.*, 2009). The relevance of E3 as an inhibitor of the IFN system was shown with E3-transgenic mice that showed reduced sensitivity to viruses and parasites (Domingo-Gil *et al.*, 2008).

A second VACV gene, K3L, encodes a 10.5 kDa homologue of the eIF2α subunit that acts as a non-phosphorylatable pseudosubstrate of PKR and competitively inhibits phosphorylation of eIF2α (Beattie *et al.*, 1991; Davies *et al.*, 1992, 1993; Carroll *et al.*, 1993). Deletion of E3L or K3L VACV genes renders the virus sensitive to IFN and severely limits host range and disease progression (Beattie *et al.*, 1995a,b, 1996). Therefore, E3 and K3 have different but overlapping roles in blocking IFN action. It is possible that these cytoplasmic proteins may also prevent the redistribution of dsRNA to endosomes, where it would be recognized by TLR3.

E3L and K3L orthologues have been identified in myxoma virus (Cameron *et al.*, 1999), Yaba-like disease virus, variola virus, Shope fibroma virus (Willer *et al.*, 1999), swinepox virus (C8L gene; Kawagishi-Kobayashi *et al.*, 2000) and orf virus (McInnes *et al.*, 1998).

Finally, VACV C7 protein, a viral determinant for multiplication in human cells (Oguiura *et al.*, 1993), has been shown to be involved in the inhibition of eIF2α phosphorylation and apoptosis (Najera *et al.*, 2006). These results were obtained when the C7L gene was re-introduced into the NYVAC genome, an attenuated VACV derived from Copenhagen strain (Najera *et al.*, 2006). It has also been demonstrated that recombinant NYVAC-C7L maintains the attenuated phenotype in mice (Najera *et al.*, 2006). Additionally, it has been reported that myxoma virus M62R gene and Yaba-like disease virus 67R gene, two different C7L homologues, function equivalently to VACV C7 protein when expressed by recombinant VACV (Meng *et al.*, 2008).

Poxviruses immunomodulatory proteins interfering with IFN signalling pathways described above are represented in Fig 6.4.

Virus infection is manifested by the production of type I IFNs and inflammatory cytokines (IL-1, IL-18, TNF-α), as previously described, but also of chemokines (IL-8, MIP-1α) and vasoactive amines. These act together to stimulate vascular permeability for effector blood proteins, increased blood flow and the recruitment and activation of inflammatory cells (monocytes/macrophages) and lymphocytes (including NK cells) to the site of infection. These cells can also produce cytokines, chemokines and vasoactive amines to maintain or regulate the inflammation and signal a particular type of adaptive immune response. Virus infection can also lead to apoptosis or programmed cell death as a potent antiviral response in which infected cells commit suicide preventing virus replication (Haig, 2001). In this context, poxviruses encode proteins targeting these components of the antiviral response: cytokines, chemokines and apoptosis.

Poxvirus proteins intercepting cytokines

Cytokines are secreted proteins with growth, differentiation and activation functions that regulate and determine the nature of immune responses (Commins *et al.*, 2010). Many poxviruses express soluble, secreted versions of innate immune receptors or ligand-binding proteins for numerous cytokines, such as TNF, lymphotoxin-α (LTα), IL-1β, IL-18 and CD153 (Seet *et al.*, 2003; Alcami, 2007). Expression of these soluble viral receptors or cytokine binding proteins can intercept cellular ligand–receptor interactions, blocking downstream signalling pathways.

Viral TNF receptors

TNF is a potent proinflammatory cytokine secreted primarily by macrophages and activated T cells (Smith *et al.*, 1994). The pleiotropic effects of TNF on the host response to infection, which include promoting an antiviral state and mediating the cytolysis of infected cells (Locksley *et al.*, 2001), has provided powerful selection pressure for viruses to evolve strategies to circumvent the TNF-mediated response to infection. The best characterized anti-TNF strategy used by poxviruses involves encoded homologues of TNFR, termed vTNFRs, that function by binding and sequestering extracellular TNF prior to cellular TNFR engagement (Cunnion, 1999).

The best studied vTNFRs are the T2-like family members encoded by leporipoxviruses and the cytokine response modifier (Crm)-like orthologues encoded by orthopoxviruses. T2 proteins from the leporipoxviruses SFV and MYXV were the first vTNFRs to be characterized. T2 protein is expressed as a secreted glycoprotein that binds and inhibits rabbit TNF (Smith *et al.*, 1991; Xu *et al.*, 2000) and was reported to be important for MYXV virulence in the infected rabbit host (Upton *et al.*, 1991). T2 protein also acts as an intracellular inhibitor of apoptosis (see below; Xu *et al.*, 2000).

In addition, the related CPXV proteins CrmB, C, D and E were also shown to act as soluble vTNFRs (Hu *et al.*, 1994; Smith *et al.*, 1996; Loparev *et al.*, 1998; Saraiva and Alcami, 2001) that vary in ligand specificity and patterns of expression. In addition, CPXV also encodes an additional member of the TNFR family, a soluble, secreted form of CD30, the receptor for CD153 (Panus *et al.*, 2002). CrmB and CrmD possess sequence similarity to T2-like vTNFRs. These two genes encode proteins that bind and inhibit both TNF and LTα (Loparev *et al.*, 1998). Orthologues of these CPXV vTNFRs were also found in other orthopoxviruses, including VACV, ECTV, MPXV, VARV and CMLV (Shchelkunov, 2003). Variola and camelpox viruses possess a single predicted CrmB-like protein whereas ectromelia virus encodes a functional CrmD variant that binds rodent and human TNF (Smith and Alcami, 2000). Although CrmE genes have been identified in several orthopoxviruses, only the cowpox virus and VACV strain USSR versions have shown to encode functional TNF inhibitors (Saraiva and Alcami, 2001). Three strains of VACV (Lister, USSR and Evans) have been shown to possess both soluble and cell-associated vTNFR activity (Alcami *et al.*, 1999). These activities have been mapped to an intact A53R gene in the Lister strain that encodes a functional, soluble vTNFR (Alcami *et al.*, 1999) and to the CrmE gene in the USSR strain, a 18 kDa protein associated with soluble and membrane-bound TNF-binding activity (Reading *et al.*, 2002).

More recently, another class of secreted poxvirus-encoded TNF inhibitors was described for members of the genus *Yatapoxvirus* (Lee *et al.*, 2001; Brunetti *et al.*, 2003). For example, the 2L protein of tanapox virus (TPV-2L) is a potent inhibitor of human TNF, but more closely resembles a secreted version of the MHC-I heavy chain, rather than any known TNF receptor species (Brunetti *et al.*, 2003). Thus, these poxviral proteins are referred to as viral TNF-binding proteins (vTNF-BPs) rather than vTNFRs. Secreted

TPV-2L was reported to inhibit TNF-mediated NF-κB signalling as measured by IκBα degradation (Rahman *et al.*, 2006).

Viral IL-1β receptors

IL-1, an important proinflammatory cytokine produced in response to infection, comprises three forms: IL-1α, IL-1β and the IL-1 receptor antagonist (IL-1ra) that negatively regulates IL-1 activity by binding to IL-1 receptors without inducing signal transduction (Dinarello, 1996). IL-1β is expressed in many cell types and mediates a broad spectrum response to tissue injury and infection (Dinarello, 1996). It is synthesised as a precursor which is proteolytically cleaved into the mature form by caspase-1, which also cleaves the pro-form of IL-18 but not IL-1α. Once released, IL-1β binds to IL-1 receptors to trigger signal transduction and the production of other mediators of inflammation, such as IL-2.

VACV B16R gene and CPXV B14R gene encode an abundant glycoprotein that functions as a soluble IL-1R, which in contrast with cellular counterparts, binds only IL-1β and not IL-1α or the natural competitor IL-1ra (Alcami and Smith, 1992; Spriggs *et al.*, 1992). VACV soluble IL-1βR (vIL-1βR) is encoded by the B15R gene in WR strain or the B16R gene in Copenhagen strain. Animal studies with knockout VACV demonstrated that B16R suppresses the febrile response of infected mice, being the first demonstration that IL-1β functions as the major endogenous pyrogen during poxvirus infection (Alcami and Smith, 1996a). Deletion of B16R reduced virulence when administered to mice by an intracranial route (Spriggs *et al.*, 1992) but enhanced virulence when administered intranasally (Alcami and Smith, 1992). An IL-1βR homologue in ectromelia virus (E191) has also been reported (Smith and Alcami, 2000).

Viral IL-18-binding proteins

IL-18, first known as the IFNγ-inducing factor, is a potent, pleiotropic cytokine that can stimulate the synthesis of various cytokines and chemokines, regulate Th1 and Th2 cell responses and activate NK and cytotoxic T cells (Nakanishi *et al.*, 2001). The antiviral properties of IL-18 have been demonstrated by the ability of this cytokine to protect mice from pathological effects of herpesvirus (Fujioka *et al.*, 1999) and VACV infections (Tanaka-Kataoka *et al.*, 1999). IL-18 is a member of the IL-1 family and exhibits a structural resemblance to IL-1β. Similar to IL-1β, IL-18 is cleaved from a precursor form into its active, secreted form by caspase-1 (Gu *et al.*, 1997) and this cleavage renders IL-18 susceptible to the action of poxvirus inhibitors of caspase-1. One mechanism that modulates the effects of IL-18 is the mammalian IL-18-binding protein (IL-18BP), a naturally occurring secreted antagonist that blocks IL-18 from binding its receptor (Aizawa *et al.*, 1999; Novick *et al.*, 1999) but has no sequence similarity to membrane IL-18 receptors.

The identification of the human IL-18BP prompted the immediate identification of poxvirus-encoded IL-18BP that possesses sequence similarities to the immunoglobulin domain of the human IL-18BP (Novick *et al.*, 1999). The MoCV MC54L gene has been reported to bind IL-18 (Xiang and Moss, 1999, 2003). The *in vivo* contribution of the virus-encoded IL-18BP during infection has been assessed in mice infected with ECTV containing a disrupted IL-18BP (ECTV-p13⁻; Born *et al.*, 2000). Following peritoneal infection with ECTV-p13⁻, mice exhibited an increase in cytotoxic activity mediated primarily by NK cells that correlated with an increase in the levels of IFNγ. VACV also expresses a functional vIL-18BP encoded by the C12L gene, which has been reported to inhibit mouse IL-18 and promote virus virulence in the murine intranasal model (Symons *et al.*, 2002a). It has also been reported that C12 promotes virulence by reducing IFNγ production and NK and T cell activity in a murine intranasal model of infection (Reading and Smith, 2003). Other poxviruses containing functional IL-18BPs include CPXV (Smith *et al.*, 2000), VARV (Massung *et al.*, 1993), YLDV (Lee *et al.*, 2001), MPXV (Shchelkunov *et al.*, 2001) and SPXV (Afonso *et al.*, 2002).

Poxvirus proteins capturing chemokines

Chemokines (chemoattractant cytokines) are a family of soluble proteins that regulate numerous inflammatory responses, such as the recruitment of specific leucocyte subsets and antigen-presenting cells to areas of inflammation, and are also important in defence against pathogens (Rollins, 1997; Baggiolini, 1998). Chemokines are classified into four families (CC, CXC, C and CX3C) based on their structure, of which CC (MIP-1α, MIP-1β, MCP-1 and RANTES) and CXC (IL-8, GRO and NAP-2) are predominant. In general, CC chemokines attract macrophages and T cell subsets, while CXC chemokines attract neutrophils.

Poxviruses encode a variety of proteins that interfere with chemokine function, including viral chemokine-binding proteins (vCBPs), homologues of chemokine receptors and chemokine ligand mimics (Lalani and McFadden, 1997; Kotwal, 2000; Lalani *et al.*, 2000; Mahalingam and Karupiah, 2000; Murphy, 2001; Rosenkilde *et al.*, 2001; Seet and McFadden, 2002; Boomker *et al.*, 2005; Alcami, 2007; Alcami and Saraiva, 2009).

Viral chemokine-binding proteins

The myxoma virus IFNγR homologue, termed M-T7, binds not only IFNγ but also a broad spectrum of CXC, CC and C-chemokines with no apparent species restriction (Lalani *et al.*, 1997). Designated a type I CBP (CBP-I), M-T7 binds chemokines through a low affinity interaction with the C-terminal heparin-binding domain present on many chemokines. The binding of rabbit IFNγ and chemokines by M-T7 is mutually exclusive, suggesting that the chemokine-binding domain and the IFNγ-binding domain on M-T7 may overlap (Lalani *et al.*, 1997). A study to evaluate M-T7 inhibition of inflammatory cells has been recently reported (Bartee *et al.*, 2009).

Members of the genera *Leporipoxvirus* and *Orthopoxvirus* produce a mechanistically different class of secreted CBP (CBP-II; also called vCCI) that binds to CC-chemokines with high affinity and prevents the binding of these CC-chemokines, but not C, CXC or CX3C chemokines, to their receptors (Graham *et al.*, 1997; Smith *et al.*, 1997a; Alcami *et al.*, 1998; Smith and Alcami, 2000). Remarkably, these related poxvirus proteins lack recognizable sequence similarity to any known chemokine receptor. Disruption of the CBP-II/M-T1 gene in myxoma virus or CBP-II/35-kDa gene in rabbitpox virus results in an increase in early leucocyte infiltration into tissues at sites of infection (Martinez Pomares *et al.*, 1995; Graham *et al.*, 1997; Lalani *et al.*, 1999). However, unlike CBP-I, no effect on virus lethality was observed in either of the deletion viruses.

VACV strains Lister and Evans also encode a secreted type II CBP (B29R/C23L gene) but the commonly used VACV strains WR and Copenhagen lacked this activity (Graham *et al.*, 1997; Smith *et al.*, 1997a; Alcami *et al.*, 1998). Ectromelia virus E163 is a secreted 31 kDa glycoprotein that selectively binds a limited number of CC and CXC chemokines with high affinity (Ruiz-Arguello *et al.*, 2008). Orf virus also encodes a secreted CBP shown to bind murine chemokines CCL2, CCL3 and CCL5 with high affinity by surface plasmon resonance (Lateef *et al.*, 2009) but also constitutive CCL19 and CCL21 (Lateef *et al.*, 2010). CBP inhibits chemokine-induced recruitment of monocytes (Lateef *et al.*, 2009) and the recruitment of DCs from sites of skin inflammation and migration to peripheral lymph nodes in a mouse skin model (Lateef *et al.*, 2010). Monkeypox virus encodes a secreted chemokine-binding protein (vCCI) that binds rhesus MIP-1α, completely inhibiting MIP-1α-mediated chemotaxis (Jones *et al.*, 2008). Finally, VARV also encodes a secreted CBP-II (Dunlop *et al.*, 2003).

More recently, a chemokine-binding domain (named SECRET domain) has been identified and defines a new family of five vCBP in poxviruses that are either fused to the viral TNFR or expressed independently (Alejo *et al.*, 2006). Examples of these proteins are expressed by VACV WR (gene B7R) and MVA (gene 175R) strains.

Viral chemokine receptor homologues

Yaba-like disease virus encodes a chemokine receptor, termed 7L, with 53% amino acid identity to human CC chemokine receptor 8 (CCR8). The contribution of 7L to poxvirus pathogenesis was investigated by infection of mice with a recombinant VACV expressing YLDV-7L. In both intranasal and intradermal models, expression of 7L caused attenuation of VACV (Najarro *et al.*, 2006). Other poxviruses, such as swinepox virus and capripox virus, encode putative

CC-chemokine receptors (Cao *et al.*, 1995; Lalani *et al.*, 2000) but these proteins remain to be functionally characterized.

Viral chemokine homologues

Only two poxviruses have been reported to encode chemokine-mimics: MoCV (Senkevich *et al.*, 1996; Bugert *et al.*, 1998; Ishikawa-Mochizuki *et al.*, 1999) and fowlpox virus (FWPV; Afonso *et al.*, 2000). MoCV chemokine-mimic, termed MC148R, is a 104-amino-acid protein structurally related to the CC-chemokine family of chemokines but contains a deletion in the N-terminal region preceding the dicysteine motif, a region necessary for activation of receptors. MC148R is detected as a secreted protein (Bugert *et al.*, 1998) and was initially shown to potently inhibit binding, signalling and chemotaxis of various leucocytes in response to a number of CC and CXC chemokines (Krathwohl *et al.*, 1997; Damon *et al.*, 1998). Fowlpox virus encodes four CC chemokine-like proteins termed Fpv060, Fpv061, Fpv116 and Fpv121 (Jeshtadi *et al.*, 2005).

Poxvirus proteins inhibiting apoptosis

Another important function of IFNs is to establish a pro-apoptotic state in target cells (Clemens, 2003; Maher *et al.*, 2007), which helps to remove infected cells. There are several pathways by which IFNs can act. For example, procaspase genes are IFN-inducible (Tamura *et al.*, 1996; Chin *et al.*, 1997; Dai and Krantz, 1999) and both PKR (Lee and Esteban, 1994; Garcia *et al.*, 2006) and 2′5′OAS (Diaz-Guerra *et al.*, 1997; Castelli *et al.*, 1998) affect apoptosis and are activated by viral dsRNA. Finally, the apoptosis-activating gene p53 is also IFN-inducible and can be activated indirectly in response to dsRNA to trigger an apoptotic response (Takaoka *et al.*, 2003). The tumour suppressive and antiviral functions of p53 can be explained by being directly phosphorylated by PKR, its activity enhanced by PKR (Cuddihy *et al.*, 1999a,b) and PKR being a direct target gene of p53, resulting in translational inhibition and cell apoptosis (Yoon *et al.*, 2009).

Poxviruses have evolved different anti-apoptotic strategies by producing viral proteins that are rapidly expressed at the early stages of replication (Roulston *et al.*, 1999; Shisler and Moss, 2001a; Everett and McFadden, 2002). These anti-apoptotic effectors can be secreted and neutralize signals from the extracellular environment (for example, viral TNF receptors described previously) or they can modulate cell death transduction pathways within the cell, as described below.

Viral caspase inhibitors

A key event in the induction of apoptosis is the activation of a family of pro-apoptotic proteases termed caspases (Hengartner, 2000). Consequently, these apoptotic proteases are frequently targeted for inhibition by different poxviruses.

CPXV encodes the CrmA protein, which functions to inhibit caspase activity. CrmA is synthesized early during infection and displays significant homology to members of the serine protease inhibitor (serpin) superfamily (Pickup *et al.*, 1986). Serpins function as pseudosubstrates that form stable inhibitory complexes with active proteases (Silverman *et al.*, 2001). CrmA is unique in that it inhibits both serine and cysteine proteases (Komiyama *et al.*, 1994; Quan *et al.*, 1995; Zhou *et al.*, 1997). CrmA is a potent inhibitor of both caspase-1 (IL-1β-converting enzyme), which converts pro-IL-1β to its active form (Ray *et al.*, 1992), and caspase-8 (Zhou *et al.*, 1997), the apical caspase that is crucial for death receptor mediated apoptosis. As such, CrmA expression inhibits apoptosis induced by TNF and triggering of the Fas receptor (Tewari and Dixit, 1995) but fails to inhibit apoptosis triggered through the intrinsic mitochondrial route (Datta *et al.*, 1997). Additionally, CrmA has been reported to inhibit granzyme B, a serine protease released from the granules of cytotoxic T lymphocytes (Quan *et al.*, 1995). The delivery of granzyme B initiates apoptosis by the direct activation of caspases and activation of the mitochondrial death pathway (Lord *et al.*, 2003; Trapani and Sutton, 2003). Although CrmA inhibits granzyme B *in vitro* it is unable to inhibit granule-mediated CTL death (Tewari *et al.*, 1995; Barry *et al.*, 2000). Therefore, CrmA inhibits two important aspects of the host immune response: apoptosis and the processing of proinflammatory cytokines.

The SPI-2 family of poxvirus serpins, such as B13 from VACV, can protect cells from Fas- and TNF-mediated apoptosis (Dobbelstein and Shenk, 1996; Kettle *et al.*, 1997) but, in comparison to CPXV CrmA, are generally less potent apoptosis inhibitors. Serp-2, a SPI-2-like serpin expressed by MYXV that was first described as a caspase-1 inhibitor, plays an important role in pathogenesis by preventing apoptosis of lymphoid cells in infected rabbits (Petit *et al.*, 1996; Messud-Petit *et al.*, 1998). However, Serp-2 is considerably less effective at inhibiting caspase-1 and granzyme B than CrmA.

Finally, the strategy employed by MoCV is directed towards preventing activation of the initiator caspase, caspase-8 (Bertin *et al.*, 1997; Hu *et al.*, 1997; Thome *et al.*, 1997). Activation of caspase-8 is triggered by the ligation of cell surface Fas, TNF or TRAIL death receptors. MoCV prevents this caspase-8 activation by producing two DED-containing proteins: MC159 and MC160 (Bertin *et al.*, 1997; Hu *et al.*, 1997; Thome *et al.*, 1997; Shisler and Moss, 2001b; Garvey *et al.*, 2002).

Inhibition of apoptosis induced by PKR

Activation of PKR results in the inhibition of translation and, frequently, in the induction of apoptosis (Gil and Esteban, 2000). VACV encodes two PKR inhibitors, E3 and K3 (as discussed previously), that can be categorized as anti-apoptotic proteins given their target pathway (Davies *et al.*, 1993). It has also been reported that PKR triggers apoptosis by activating caspase-8. Consequently, the poxvirus caspase-8 inhibitors, MC159 (MoCV; Gil *et al.*, 2001) and CrmA (CPXV; Ezelle *et al.*, 2001), have also been shown to counteract PKR-mediated apoptosis.

Preventing oxidative stress as an anti-apoptotic strategy

MoCV MC066 protein prevents cell death triggered by stress catalyzing the conversion of toxic reactive oxygen species, such as H_2O_2, to water (Shisler *et al.*, 1998). As H_2O_2 can be produced following UV irradiation, MC066 could be important for virus replication at the primary site of infection in basal keratinocytes of the skin. Two unrelated poxvirus RING finger proteins, p28 of ectromelia virus (Brick *et al.*, 2000) and N1R of Shope fibroma virus (Brick *et al.*, 1998), are able to inhibit UV-light-induced but not TNF- or Fas-induced apoptosis.

Other viral anti-apoptotic proteins

Sensors of cellular stress are localized within specific organelles, of which mitochondria play a key role in the coordination of apoptotic signalling pathways (Everett and McFadden, 2001). Because of the key role of this organelle in apoptosis induction, viruses have developed multiple strategies to counteract mitochondrial-mediated apoptosis (Galluzzi *et al.*, 2008). Myxoma virus M11L protein is able to inhibit transduction of death signals via the mitochondrial checkpoint. M11L was first identified as a virulence factor and the knockout virus also displayed a pro-apoptotic phenotype following infection of a rabbit lymphocyte cell line (Opgenorth *et al.*, 1992; Macen *et al.*, 1996). M11L is targeted to mitochondria and prevents mitochondrial changes associated with apoptosis including the loss of the inner mitochondrial membrane potential (Everett *et al.*, 2000). VACV F1L gene encodes a mitochondrial-associated inhibitor of apoptosis (Wasilenko *et al.*, 2003; Wasilenko *et al.*, 2005). Expression of F1L inhibits loss of the inner mitochondrial membrane potential and cytochrome *c* release induced by a variety of stimuli, including UV-light, staurosporine, TNF and anti-Fas (Wasilenko *et al.*, 2003; Stewart *et al.*, 2005). F1L associates with the pro-apoptotic Bcl-2 family member Bak and inhibits its activation (Wasilenko *et al.*, 2005). F1L also has the ability to interact with the BH3 peptide of Bim, and to a more limited degree, the BH3 peptides of Bak and Bax (Fischer *et al.*, 2006).

Recently, two novel Bcl-2-like inhibitors of apoptosis encoded by poxviruses have been discovered: FPV039 from fowlpox virus (Banadyga *et al.*, 2007) and ORFV125 from orf virus (Westphal *et al.*, 2007). In human and chicken cells, FPV039 localizes at mitochondria and constitutively interacts with Bak, thereby suppressing apoptosis induced by TNFα and Bak overexpression. Moreover, FPV039 is able to substitute for F1L in inhibiting Bak activation and apoptosis triggered by staurosporine and VACV infection, confirming that FPV039 is a functional homologue of F1L (Banadyga *et al.*, 2007). In UV-irradiated HeLa cells, ORFV125 fully inhibits DNA fragmentation, caspase activation and cytochrome *c* release. The mitochondrial localization of ORFV125 is required for its anti-apoptotic function (Westphal *et al.*, 2007).

Other apoptotic modulators from myxoma virus have been identified: the M-T4 protein, an ER-resident protein, may inhibit an ER stress response to infection (Barry *et al.*, 1997; Hnatiuk *et al.*, 1999); and the M-T5 protein, an ankyrin repeat protein with homology to the Chinese hamster ovary host range (CHOhr) gene of cowpox virus (Mossman *et al.*, 1996a). M-T5 and CHOhr proteins can prevent apoptosis associated with virus infection (Ink *et al.*, 1995; Mossman *et al.*, 1996a). M-T2, the myxoma TNFR homologue, also functions as an intracellular inhibitor of apoptosis in virus-infected lymphocytes (Macen *et al.*, 1996; Sedger and McFadden, 1996; Schreiber *et al.*, 1997; Xu *et al.*, 2000).

VACV N1L and C7L genes are also involved in the inhibition of apoptosis: N1 was shown to inhibit staurosporine-induced apoptosis in transfected and infected cells (Cooray *et al.*, 2007), to bind BH3 peptides from pro-apoptotic Bcl-2 family proteins Bid, Bim and Bak *in vitro* (Aoyagi *et al.*, 2007) and to co-precipitate Bid, Bad and Bax from VACV-infected cells (Cooray *et al.*, 2007). C7 has been reported to be involved in the inhibition of eIF2α phosphorylation and apoptosis (Najera *et al.*, 2006).

The specific anti-apoptotic functions of these poxviral proteins suggest that the inhibition of apoptosis at multiple points is an important survival strategy for poxviruses.

Concluding remarks

Poxviruses have developed multiple strategies to counteract, either directly or indirectly, IFN signalling pathways as well as IFN effects. The numerous studies regarding immunological targets and strategies used by poxviruses to disrupt these host signalling pathways have provided important insights into diverse aspects of immunology, virology and inflammation. While a variety of viral genes have assigned specific interfering actions, however, it is probable that more viral interfering genes will be found. In addition, it is likely that more new functions will also be discovered in the already known genes. The elucidation of the function of these virus molecules will help in our understanding of viral pathogenesis, provide tools with which to study and manipulate the normal host immune response and allow the design of safer virus vaccines and virus-based drug delivery systems. Furthermore, because of their anti-inflammatory nature, many of these poxvirus proteins hold promise as potential therapeutic agents for acute or chronic inflammatory conditions.

Acknowledgements

Research was conducted as part of the Poxvirus T Cell Vaccine Discovery Consortium (PTVDC) as part of the Collaboration for AIDS Vaccine Discovery with support from the Bill and Melinda Gates Foundation. This work was also supported by the Spanish Foundation for AIDS Research (FIPSE) and Foundation Botín. We would like to thank Carmen E. Gómez and José Luis Nájera for thoughtful suggestions.

References

Ablasser, A., Bauernfeind, F., Hartmann, G., Latz, E., Fitzgerald, K.A., and Hornung, V. (2009). RIG-I-dependent sensing of poly(dA:dT) through the induction of an RNA polymerase III-transcribed RNA intermediate. Nat. Immunol. *10*, 1065–1072.

Afonso, C.L., Tulman, E.R., Lu, Z., Zsak, L., Kutish, G.F., and Rock, D.L. (2000). The genome of fowlpox virus. J. Virol. *74*, 3815–3831.

Afonso, C.L., Tulman, E.R., Lu, Z., Zsak, L., Osorio, F.A., Balinsky, C., Kutish, G.F., and Rock, D.L. (2002). The genome of swinepox virus. J. Virol. *76*, 783–790.

Aizawa, Y., Akita, K., Taniai, M., Torigoe, K., Mori, T., Nishida, Y., Ushio, S., Nukada, Y., Tanimoto, T., Ikegami, H., *et al.* (1999). Cloning and expression of interleukin-18 binding protein. FEBS Lett. *445*, 338–342.

Akira, S. (2006). TLR signaling. Curr. Top. Microbiol. Immunol. *311*, 1–16.

Akira, S., Uematsu, S., and Takeuchi, O. (2006). Pathogen recognition and innate immunity. Cell *124*, 783–801.

Alcami, A. (2007). New insights into the subversion of the chemokine system by poxviruses. Eur. J. Immunol. *37*, 880–883.

Alcami, A., Khanna, A., Paul, N.L., and Smith, G.L. (1999). Vaccinia virus strains Lister, USSR and Evans express soluble and cell-surface tumour necrosis factor receptors. J. Gen. Virol. *80*, 949–959.

Alcamí, A., and Koszinowski, U.H. (2000). Viral mechanisms of immune evasion. Trends Microbiol. *8*, 410–418.

Alcami, A., and Saraiva, M. (2009). Chemokine binding proteins encoded by pathogens. Adv. Exp. Med. Biol. *666*, 167–179.

Alcami, A., and Smith, G.L. (1992). A soluble receptor for interleukin-1 beta encoded by vaccinia virus: a novel mechanism of virus modulation of the host response to infection. Cell *71*, 153–167.

Alcami, A., and Smith, G.L. (1995). Vaccinia, cowpox, and camelpox viruses encode soluble gamma interferon receptors with novel broad species specificity. J. Virol. *69*, 4633–4639.

Alcami, A., and Smith, G.L. (1996a). A mechanism for the inhibition of fever by a virus. Proc. Natl. Acad. Sci. U.S.A. *93*, 11029–11034.

Alcami, A., and Smith, G.L. (1996b). Receptors for gamma-interferon encoded by poxviruses: implications for the unknown origin of vaccinia virus. Trends Microbiol. *4*, 321–326.

Alcami, A., and Smith, G.L. (2002). The vaccinia virus soluble interferon-gamma receptor is a homodimer. J. Gen. Virol. *83*, 545–549.

Alcami, A., Symons, J.A., Collins, P.D., Williams, T.J., and Smith, G.L. (1998). Blockade of chemokine activity by a soluble chemokine binding protein from vaccinia virus. J. Immunol. *160*, 624–633.

Alcami, A., Symons, J.A., and Smith, G.L. (2000). The vaccinia virus soluble alpha/beta interferon (IFN) receptor binds to the cell surface and protects cells from the antiviral effects of IFN. J. Virol. *74*, 11230–11239.

Alejo, A., Ruiz-Arguello, M.B., Ho, Y., Smith, V.P., Saraiva, M., and Alcami, A. (2006). A chemokine-binding domain in the tumor necrosis factor receptor from variola (smallpox) virus. Proc. Natl. Acad. Sci. U.S.A. *103*, 5995–6000.

Ank, N., West, H., and Paludan, S.R. (2006). IFN-lambda: novel antiviral cytokines. J. Interferon Cytokine Res. *26*, 373–379.

Aoyagi, M., Zhai, D., Jin, C., Aleshin, A.E., Stec, B., Reed, J.C., and Liddington, R.C. (2007). Vaccinia virus N1L protein resembles a B cell lymphoma-2 (Bcl-2) family protein. Protein Sci. *16*, 118–124.

Aravalli, R.N., Hu, S., and Lokensgard, J.R. (2008). Inhibition of toll-like receptor signaling in primary murine microglia. J. Neuroimmune Pharmacol. *3*, 5–11.

Bach, E.A., Aguet, M., and Schreiber, R.D. (1997). The IFN gamma receptor: a paradigm for cytokine receptor signaling. Annu. Rev. Immunol. *15*, 563–591.

Baggiolini, M. (1998). Chemokines and leukocyte traffic. Nature *392*, 565–568.

Banadyga, L., Gerig, J., Stewart, T., and Barry, M. (2007). Fowlpox virus encodes a Bcl-2 homologue that protects cells from apoptotic death through interaction with the proapoptotic protein Bak. J. Virol. *81*, 11032–11045.

Bandi, P., Pagliaccetti, N.E., and Robek, M.D. (2009). Inhibition of type III interferon activity by orthopoxvirus immunomodulatory proteins. J. Interferon Cytokine Res. *30*, 123–134.

Barbalat, R., Lau, L., Locksley, R.M., and Barton, G.M. (2009). Toll-like receptor 2 on inflammatory monocytes induces type I interferon in response to viral but not bacterial ligands. Nat. Immunol. *10*, 1200–1207.

Barber, G.N. (2001). Host defense, viruses and apoptosis. Cell Death Differ. *8*, 113–126.

Barrett, J.W., Cao, J.X., Hota-Mitchell, S., and McFadden, G. (2001). Immunomodulatory proteins of myxoma virus. Semin. Immunol. *13*, 73–84.

Barry, M., Heibein, J.A., Pinkoski, M.J., Lee, S.F., Moyer, R.W., Green, D.R., and Bleackley, R.C. (2000). Granzyme B short-circuits the need for caspase 8 activity during granule-mediated cytotoxic T-lymphocyte killing by directly cleaving Bid. Mol. Cell. Biol. *20*, 3781–3794.

Barry, M., Hnatiuk, S., Mossman, K., Lee, S.F., Boshkov, L., and McFadden, G. (1997). The myxoma virus M-T4 gene encodes a novel RDEL-containing protein that is retained within the endoplasmic reticulum and is important for the productive infection of lymphocytes. Virology *239*, 360–377.

Bartee, M.Y., Dai, E., Liu, L., Munuswamy-Ramanujam, G., Macaulay, C., McIvor, D., McFadden, G., and Lucas, A.R. (2009). 10 M-T7: measuring chemokine-modulating activity. Methods Enzymol. *460*, 209–228.

Bartlett, N., Symons, J.A., Tscharke, D.C., and Smith, G.L. (2002). The vaccinia virus N1L protein is an intracellular homodimer that promotes virulence. J. Gen. Virol. *83*, 1965–1976.

Barton, G.M., and Kagan, J.C. (2009). A cell biological view of Toll-like receptor function: regulation through compartmentalization. Nat. Rev. Immunol. *9*, 535–542.

Beattie, E., Denzler, K.L., Tartaglia, J., Perkus, M.E., Paoletti, E., and Jacobs, B.L. (1995a). Reversal of the interferon-sensitive phenotype of a vaccinia virus lacking E3L by expression of the reovirus S4 gene. J. Virol. *69*, 499–505.

Beattie, E., Kauffman, E.B., Martinez, H., Perkus, M.E., Jacobs, B.L., Paoletti, E., and Tartaglia, J. (1996). Host-range restriction of vaccinia virus E3L-specific deletion mutants. Virus Genes *12*, 89–94.

Beattie, E., Paoletti, E., and Tartaglia, J. (1995b). Distinct patterns of IFN sensitivity observed in cells infected with vaccinia K3L- and E3L-mutant viruses. Virology *210*, 254–263.

Beattie, E., Tartaglia, J., and Paoletti, E. (1991). Vaccinia virus-encoded eIF-2 alpha homolog abrogates the antiviral effect of interferon. Virology *183*, 419–422.

Bertin, J., Armstrong, R.C., Ottilie, S., Martin, D.A., Wang, Y., Banks, S., Wang, G.H., Senkevich, T.G., Alnemri, E.S., Moss, B., *et al.* (1997). Death effector domain-containing herpesvirus and poxvirus proteins inhibit both Fas- and TNFR1-induced apoptosis. Proc. Natl. Acad. Sci. U.S.A. *94*, 1172–1176.

Beutler, B., Jiang, Z., Georgel, P., Crozat, K., Croker, B., Rutschmann, S., Du, X., and Hoebe, K. (2006). Genetic analysis of host resistance: Toll-like receptor signaling and immunity at large. Annu. Rev. Immunol. *24*, 353–389.

Boomker, J.M., de Leij, L.F., The, T.H., and Harmsen, M.C. (2005). Viral chemokine-modulatory proteins: tools and targets. Cytokine Growth Factor Rev *16*, 91–103.

Born, T.L., Morrison, L.A., Esteban, D.J., VandenBos, T., Thebeau, L.G., Chen, N., Spriggs, M.K., Sims, J.E., and Buller, R.M. (2000). A poxvirus protein that binds to and inactivates IL-18, and inhibits NK cell response. J. Immunol. *164*, 3246–3254.

Bowie, A., Kiss-Toth, E., Symons, J.A., Smith, G.L., Dower, S.K., and O'Neill, L.A. (2000). A46R and A52R from vaccinia virus are antagonists of host IL-1 and toll-like receptor signaling. Proc. Natl. Acad. Sci. U.S.A. *97*, 10162–10167.

Bowie, A., and O'Neill, L.A. (2000). The interleukin-1 receptor/Toll-like receptor superfamily: signal generators for pro-inflammatory interleukins and microbial products. J. Leukoc. Biol. *67*, 508–514.

Brick, D.J., Burke, R.D., Minkley, A.A., and Upton, C. (2000). Ectromelia virus virulence factor p28 acts upstream of caspase-3 in response to UV light-induced apoptosis. J. Gen. Virol. *81*, 1087–1097.

Brick, D.J., Burke, R.D., Schiff, L., and Upton, C. (1998). Shope fibroma virus RING finger protein N1R binds DNA and inhibits apoptosis. Virology *249*, 42–51.

Brooks, D.G., Trifilo, M.J., Edelmann, K.H., Teyton, L., McGavern, D.B., and Oldstone, M.B. (2006). Interleukin-10 determines viral clearance or persistence *in vivo*. Nat. Med. *12*, 1301–1309.

Brunetti, C.R., Paulose-Murphy, M., Singh, R., Qin, J., Barrett, J.W., Tardivel, A., Schneider, P., Essani, K., and McFadden, G. (2003). A secreted high-affinity inhibitor of human TNF from Tanapox virus. Proc. Natl. Acad. Sci. U.S.A. *100*, 4831–4836.

Bryant, C., and Fitzgerald, K.A. (2009). Molecular mechanisms involved in inflammasome activation. Trends Cell. Biol. *19*, 455–464.

Bugert, J.J., Lohmuller, C., Damon, I., Moss, B., and Darai, G. (1998). Chemokine homolog of molluscum contagiosum virus: sequence conservation and expression. Virology *242*, 51–59.

Burckstummer, T., Baumann, C., Bluml, S., Dixit, E., Durnberger, G., Jahn, H., Planyavsky, M., Bilban, M., Colinge, J., Bennett, K.L., *et al.* (2009). An orthogonal proteomic-genomic screen identifies AIM2 as a cytoplasmic DNA sensor for the inflammasome. Nat. Immunol. *10*, 266–272.

Cameron, C., Hota-Mitchell, S., Chen, L., Barrett, J., Cao, J.X., Macaulay, C., Willer, D., Evans, D., and McFadden, G. (1999). The complete DNA sequence of myxoma virus. Virology *264*, 298–318.

Camus-Bouclainville, C., Fiette, L., Bouchiha, S., Pignolet, B., Counor, D., Filipe, C., Gelfi, J., and Messud-Petit, F. (2004). A virulence factor of myxoma virus colocalizes with NF-kappaB in the nucleus and interferes with inflammation. J. Virol. *78*, 2510–2516.

Cao, J.X., Gershon, P.D., and Black, D.N. (1995). Sequence analysis of HindIII Q2 fragment of capripoxvirus reveals a putative gene encoding a G-protein-coupled chemokine receptor homologue. Virology *209*, 207–212.

Carey, M. (1998). The enhanceosome and transcriptional synergy. Cell *92*, 5–8.

Carroll, K., Elroy Stein, O., Moss, B., and Jagus, R. (1993). Recombinant vaccinia virus K3L gene product prevents activation of double-stranded RNA-dependent, initiation factor 2 alpha-specific protein kinase. J. Biol. Chem. *268*, 12837–12842.

Carty, M., Goodbody, R., Schroder, M., Stack, J., Moynagh, P.N., and Bowie, A.G. (2006). The human adaptor SARM negatively regulates adaptor protein TRIF-dependent Toll-like receptor signaling. Nat. Immunol. *7*, 1074–1081.

Cassel, S.L., Joly, S., and Sutterwala, F.S. (2009). The NLRP3 inflammasome: a sensor of immune danger signals. Semin. Immunol. *21*, 194–198.

Castelli, J.C., Hassel, B.A., Maran, A., Paranjape, J., Hewitt, J.A., Li, X.L., Hsu, Y.T., Silverman, R.H., and Youle, R.J. (1998). The role of 2′-5′ oligoadenylate-activated ribonuclease L in apoptosis. Cell Death Differ. *5*, 313–320.

Clemens, M.J. (2003). Interferons and apoptosis. J. Interferon Cytokine Res. *23*, 277–292.

Clemens, M.J. (2005). Translational control in virus-infected cells: models for cellular stress responses. Semin. Cell Dev. Biol. *16*, 13–20.

Colamonici, O.R., Domanski, P., Sweitzer, S.M., Larner, A., and Buller, R.M. (1995). Vaccinia virus B18R gene encodes a type I interferon-binding protein that blocks interferon alpha transmembrane signaling. J. Biol. Chem. *270*, 15974–15978.

Commins, S.P., Borish, L., and Steinke, J.W. (2010). Immunologic messenger molecules: cytokines, interferons, and chemokines. J. Allergy Clin. Immunol. *125*, S53–72.

Cooray, S., Bahar, M.W., Abrescia, N.G., McVey, C.E., Bartlett, N.W., Chen, R.A., Stuart, D.I., Grimes, J.M., and Smith, G.L. (2007). Functional and structural studies of the vaccinia virus virulence factor N1 reveal a Bcl-2-like anti-apoptotic protein. J. Gen. Virol. *88*, 1656–1666.

Cuddihy, A.R., Li, S., Tam, N.W., Wong, A.H., Taya, Y., Abraham, N., Bell, J.C., and Koromilas, A.E. (1999a). Double-stranded-RNA-activated protein kinase PKR enhances transcriptional activation by tumor suppressor p53. Mol. Cell. Biol. *19*, 2475–2484.

Cuddihy, A.R., Wong, A.H., Tam, N.W., Li, S., and Koromilas, A.E. (1999b). The double-stranded RNA activated protein kinase PKR physically associates with the tumor suppressor p53 protein and phosphorylates human p53 on serine 392 *in vitro*. Oncogene *18*, 2690–2702.

Cunnion, K.M. (1999). Tumor necrosis factor receptors encoded by poxviruses. Mol. Genet. Metab. *67*, 278–282.

Cutrone, E.C., and Langer, J.A. (2001). Identification of critical residues in bovine IFNAR-1 responsible for interferon binding. J. Biol. Chem. *276*, 17140–17148.

Chang, H.W., Watson, J.C., and Jacobs, B.L. (1992). The E3L gene of vaccinia virus encodes an inhibitor of the interferon-induced, double-stranded RNA-dependent protein kinase. Proc. Natl. Acad. Sci. U.S.A. *89*, 4825–4829.

Chang, S.J., Hsiao, J.C., Sonnberg, S., Chiang, C.T., Yang, M.H., Tzou, D.L., Mercer, A.A., and Chang, W. (2009). Poxvirus host range protein CP77 contains an F-box-like domain that is necessary to suppress NF-kappaB activation by tumor necrosis factor alpha but is independent of its host range function. J. Virol. *83*, 4140–4152.

Chaudhary, P.M., Jasmin, A., Eby, M.T., and Hood, L. (1999). Modulation of the NF-kappa B pathway by virally encoded death effector domains-containing proteins. Oncogene *18*, 5738–5746.

Chen, R.A., Jacobs, N., and Smith, G.L. (2006). Vaccinia virus strain Western Reserve protein B14 is an intracellular virulence factor. J. Gen. Virol. *87*, 1451–1458.

Chen, R.A., Ryzhakov, G., Cooray, S., Randow, F., and Smith, G.L. (2008). Inhibition of IkappaB kinase by vaccinia virus virulence factor B14. PLoS Pathog. *4*, e22.

Chin, Y.E., Kitagawa, M., Kuida, K., Flavell, R.A., and Fu, X.Y. (1997). Activation of the STAT signaling pathway can cause expression of caspase 1 and apoptosis. Mol. Cell. Biol. *17*, 5328–5337.

Chiu, Y.H., Macmillan, J.B., and Chen, Z.J. (2009). RNA polymerase III detects cytosolic DNA and induces type I interferons through the RIG-I pathway. Cell *138*, 576–591.

Dai, C., and Krantz, S.B. (1999). Interferon gamma induces upregulation and activation of caspases 1, 3, and 8 to produce apoptosis in human erythroid progenitor cells. Blood *93*, 3309–3316.

Damon, I., Murphy, P.M., and Moss, B. (1998). Broad spectrum chemokine antagonistic activity of a human poxvirus chemokine homolog. Proc. Natl. Acad. Sci. U.S.A. *95*, 6403–6407.

Datta, R., Kojima, H., Banach, D., Bump, N.J., Talanian, R.V., Alnemri, E.S., Weichselbaum, R.R., Wong, W.W., and Kufe, D.W. (1997). Activation of a CrmA-insensitive, p35-sensitive pathway in ionizing radiation-induced apoptosis. J. Biol. Chem. *272*, 1965–1969.

Davies, M.V., Chang, H.W., Jacobs, B.L., and Kaufman, R.J. (1993). The E3L and K3L vaccinia virus gene products stimulate translation through inhibition of the double-stranded RNA-dependent protein kinase by different mechanisms. J. Virol. *67*, 1688–1692.

Davies, M.V., Elroy Stein, O., Jagus, R., Moss, B., and Kaufman, R.J. (1992). The vaccinia virus K3L gene product potentiates translation by inhibiting double-stranded-RNA-activated protein kinase and phosphorylation of the alpha subunit of eukaryotic initiation factor 2. J. Virol. *66*, 1943–1950.

Delaloye, J., Roger, T., Steiner-Tardivel, Q.G., Le Roy, D., Knaup Reymond, M., Akira, S., Petrilli, V., Gomez, C.E., Perdiguero, B., Tschopp, J., *et al.* (2009). Innate immune sensing of modified vaccinia virus Ankara (MVA) is mediated by TLR2-TLR6, MDA-5 and the NALP3 inflammasome. PLoS Pathog. *5*, e1000480.

Deng, L., Dai, P., Ding, W., Granstein, R.D., and Shuman, S. (2006). Vaccinia virus infection attenuates innate immune responses and antigen presentation by epidermal dendritic cells. J. Virol. *80*, 9977–9987.

Der, S.D., Zhou, A., Williams, B.R., and Silverman, R.H. (1998). Identification of genes differentially regulated by interferon alpha, beta, or gamma using oligonucleotide arrays. Proc. Natl. Acad. Sci. U.S.A. *95*, 15623–15628.

Diaz-Guerra, M., Rivas, C., and Esteban, M. (1997). Activation of the IFN-inducible enzyme RNase L causes apoptosis of animal cells. Virology *236*, 354–363.

Diaz Guerra, M., Kahn, J.S., and Esteban, M. (1993). A mutation of the nucleoside triphosphate phosphohydrolase I (NPH-I) gene confers sensitivity of vaccinia virus to interferon. Virology *197*, 485–491.

Diel, D.G., Delhon, G., Luo, S., Flores, E.F., and Rock, D.L. (2010). A novel inhibitor of the NF-{kappa}B signaling pathway encoded by the parapoxvirus orf virus. J. Virol. *84*, 3962–3973.

Dinarello, C.A. (1996). Biologic basis for interleukin-1 in disease. Blood *87*, 2095–2147.

DiPerna, G., Stack, J., Bowie, A.G., Boyd, A., Kotwal, G., Zhang, Z., Arvikar, S., Latz, E., Fitzgerald, K.A., and Marshall, W.L. (2004). Poxvirus protein N1L targets the I-kappaB kinase complex, inhibits signaling to NF-kappaB by the tumor necrosis factor superfamily of receptors, and inhibits NF-kappaB and IRF3 signaling by toll-like receptors. J. Biol. Chem. *279*, 36570–36578.

Dobbelstein, M., and Shenk, T. (1996). Protection against apoptosis by the vaccinia virus SPI-2 (B13R) gene product. J. Virol. *70*, 6479–6485.

Domingo-Gil, E., Perez-Jimenez, E., Ventoso, I., Najera, J.L., and Esteban, M. (2008). Expression of the E3L gene of vaccinia virus in transgenic mice decreases host resistance to vaccinia virus and Leishmania major infections. J. Virol. *82*, 254–267.

Dorfleutner, A., Talbott, S.J., Bryan, N.B., Funya, K.N., Rellick, S.L., Reed, J.C., Shi, X., Rojanasakul, Y., Flynn, D.C., and Stehlik, C. (2007). A Shope Fibroma virus PYRIN-only protein modulates the host immune response. Virus Genes *35*, 685–694.

Du, W., and Maniatis, T. (1992). An ATF/CREB binding site is required for virus induction of the human interferon beta gene [corrected]. Proc. Natl. Acad. Sci. U.S.A. *89*, 2150–2154.

Dunlop, L.R., Oehlberg, K.A., Reid, J.J., Avci, D., and Rosengard, A.M. (2003). Variola virus immune evasion proteins. Microbes Infect *5*, 1049–1056.

Ejrnaes, M., Filippi, C.M., Martinic, M.M., Ling, E.M., Togher, L.M., Crotty, S., and von Herrath, M.G. (2006). Resolution of a chronic viral infection after interleukin-10 receptor blockade. J. Exp. Med. *203*, 2461–2472.

Ermolaeva, M.A., Michallet, M.C., Papadopoulou, N., Utermohlen, O., Kranidioti, K., Kollias, G., Tschopp, J., and Pasparakis, M. (2008). Function of TRADD in tumor necrosis factor receptor 1 signaling and in TRIF-dependent inflammatory responses. Nat. Immunol. *9*, 1037–1046.

Everett, H., Barry, M., Lee, S.F., Sun, X., Graham, K., Stone, J., Bleackley, R.C., and McFadden, G. (2000). M11L: a novel mitochondria-localized protein of myxoma virus that blocks apoptosis of infected leukocytes. J. Exp. Med. *191*, 1487–1498.

Everett, H., and McFadden, G. (2001). Viruses and apoptosis: meddling with mitochondria. Virology *288*, 1–7.

Everett, H., and McFadden, G. (2002). Poxviruses and apoptosis: a time to die. Curr. Opin. Microbiol. *5*, 395–402.

Ewald, S.E., Lee, B.L., Lau, L., Wickliffe, K.E., Shi, G.P., Chapman, H.A., and Barton, G.M. (2008). The ectodomain of Toll-like receptor 9 is cleaved to generate a functional receptor. Nature *456*, 658–662.

Ezelle, H.J., Balachandran, S., Sicheri, F., Polyak, S.J., and Barber, G.N. (2001). Analyzing the mechanisms of interferon-induced apoptosis using CrmA and hepatitis C virus NS5A. Virology *281*, 124–137.

Farrar, M.A., and Schreiber, R.D. (1993). The molecular cell biology of interferon-gamma and its receptor. Annu. Rev. Immunol. *11*, 571–611.

Fernandes-Alnemri, T., Yu, J.W., Datta, P., Wu, J., and Alnemri, E.S. (2009). AIM2 activates the inflammasome and cell death in response to cytoplasmic DNA. Nature *458*, 509–513.

Fischer, S.F., Ludwig, H., Holzapfel, J., Kvansakul, M., Chen, L., Huang, D.C., Sutter, G., Knese, M., and Hacker, G. (2006). Modified vaccinia virus Ankara protein F1L is a novel BH3-domain-binding protein and acts together with the early viral protein E3L to block virus-associated apoptosis. Cell Death Differ. *13*, 109–118.

Fish, E.N., Uddin, S., Korkmaz, M., Majchrzak, B., Druker, B.J., and Platanias, L.C. (1999). Activation of a CrkL-stat5 signaling complex by type I interferons. J. Biol. Chem. *274*, 571–573.

Fitzgerald, K.A., McWhirter, S.M., Faia, K.L., Rowe, D.C., Latz, E., Golenbock, D.T., Coyle, A.J., Liao, S.M., and Maniatis, T. (2003). IKKepsilon and TBK1 are essential components of the IRF3 signaling pathway. Nat. Immunol. *4*, 491–496.

Fitzgerald, K.A., Palsson-McDermott, E.M., Bowie, A.G., Jefferies, C.A., Mansell, A.S., Brady, G., Brint, E., Dunne, A., Gray, P., Harte, M.T., *et al.* (2001). Mal (MyD88-adapter-like) is required for Toll-like receptor-4 signal transduction. Nature *413*, 78–83.

Fujioka, N., Akazawa, R., Ohashi, K., Fujii, M., Ikeda, M., and Kurimoto, M. (1999). Interleukin-18 protects mice against acute herpes simplex virus type 1 infection. J. Virol. 73, 2401–2409.

Fujita, T., Miyamoto, M., Kimura, Y., Hammer, J., and Taniguchi, T. (1989). Involvement of a cis-element that binds an H2TF-1/NF kappa B like factor(s) in the virus-induced interferon-beta gene expression. Nucleic Acids Res. *17*, 3335–3346.

Galluzzi, L., Brenner, C., Morselli, E., Touat, Z., and Kroemer, G. (2008). Viral control of mitochondrial apoptosis. PLoS Pathog. *4*, e1000018.

Garcia, M.A., Gil, J., Ventoso, I., Guerra, S., Domingo, E., Rivas, C., and Esteban, M. (2006). Impact of protein kinase PKR in cell biology: from antiviral to antiproliferative action. Microbiol. Mol. Biol. Rev. *70*, 1032–1060.

Garvey, T.L., Bertin, J., Siegel, R.M., Wang, G.H., Lenardo, M.J., and Cohen, J.I. (2002). Binding of FADD and caspase-8 to molluscum contagiosum virus MC159 v-FLIP is not sufficient for its antiapoptotic function. J. Virol. *76*, 697–706.

Gedey, R., Jin, X.L., Hinthong, O., and Shisler, J.L. (2006). Poxviral regulation of the host NF-kappaB response: the vaccinia virus M2L protein inhibits induction of NF-kappaB activation via an ERK2 pathway in virus-infected human embryonic kidney cells. J. Virol. *80*, 8676–8685.

Ghayur, T., Banerjee, S., Hugunin, M., Butler, D., Herzog, L., Carter, A., Quintal, L., Sekut, L., Talanian, R., Paskind, M., *et al.* (1997). Caspase-1 processes IFN-gamma-inducing factor and regulates LPS-induced IFN-gamma production. Nature *386*, 619–623.

Gil, J., and Esteban, M. (2000). Induction of apoptosis by the dsRNA-dependent protein kinase (PKR): mechanism of action. Apoptosis 5, 107–114.

Gil, J., Rullas, J., Alcami, J., and Esteban, M. (2001). MC159L protein from the poxvirus molluscum contagiosum virus inhibits NF-kappaB activation and apoptosis induced by PKR. J. Gen. Virol. *82*, 3027–3034.

Gonzalez, J.M., and Esteban, M. (2010). A poxvirus Bcl-2-like gene family involved in regulation of host immune response: sequence similarity and evolutionary history. Virol. J. 7, 59.

Graham, K.A., Lalani, A.S., Macen, J.L., Ness, T.L., Barry, M., Liu, L.Y., Lucas, A., Clark-Lewis, I., Moyer, R.W., and McFadden, G. (1997). The T1/35kDa family of poxvirus-secreted proteins bind chemokines and modulate leukocyte influx into virus-infected tissues. Virology *229*, 12–24.

Graham, S.C., Bahar, M.W., Cooray, S., Chen, R.A., Whalen, D.M., Abrescia, N.G., Alderton, D., Owens, R.J., Stuart, D.I., Smith, G.L., *et al.* (2008). Vaccinia virus proteins A52 and B14 Share a Bcl-2-like fold but have evolved to inhibit NF-kappaB rather than apoptosis. PLoS Pathog. *4*, e1000128.

Gu, Y., Kuida, K., Tsutsui, H., Ku, G., Hsiao, K., Fleming, M.A., Hayashi, N., Higashino, K., Okamura, H., Nakanishi, K., *et al.* (1997). Activation of interferon-gamma inducing factor mediated by interleukin-1beta converting enzyme. Science *275*, 206–209.

Gubser, C., Hue, S., Kellam, P., and Smith, G.L. (2004). Poxvirus genomes: a phylogenetic analysis. J. Gen. Virol. *85*, 105–117.

Guerra, S., Caceres, A., Knobeloch, K.P., Horak, I., and Esteban, M. (2008). Vaccinia virus E3 protein prevents the antiviral action of ISG15. PLoS Pathog. *4*, e1000096.

Haga, I.R., and Bowie, A.G. (2005). Evasion of innate immunity by vaccinia virus. Parasitology *130* Suppl, S11–25.

Haig, D.M. (2001). Subversion and piracy: DNA viruses and immune evasion. Res. Vet. Sci. *70*, 205–219.

Haller, O., Staeheli, P., and Kochs, G. (2007). Interferon-induced Mx proteins in antiviral host defense. Biochimie *89*, 812–818.

Harte, M.T., Haga, I.R., Maloney, G., Gray, P., Reading, P.C., Bartlett, N.W., Smith, G.L., Bowie, A., and O'Neill, L.A. (2003). The poxvirus protein A52R targets Toll-like receptor signaling complexes to suppress host defense. J. Exp. Med. *197*, 343–351.

Hemmi, H., Takeuchi, O., Sato, S., Yamamoto, M., Kaisho, T., Sanjo, H., Kawai, T., Hoshino, K., Takeda, K., and Akira, S. (2004). The roles of two IkappaB kinase-related kinases in lipopolysaccharide and double-stranded RNA signaling and viral infection. J. Exp. Med. *199*, 1641–1650.

Hengartner, M.O. (2000). The biochemistry of apoptosis. Nature *407*, 770–776.

Hinthong, O., Jin, X.L., and Shisler, J.L. (2008). Characterization of wild-type and mutant vaccinia virus M2L proteins' abilities to localize to the endoplasmic reticulum and to inhibit NF-kappaB activation during infection. Virology *373*, 248–262.

Hnatiuk, S., Barry, M., Zeng, W., Liu, L., Lucas, A., Percy, D., and McFadden, G. (1999). Role of the C-terminal RDEL motif of the myxoma virus M-T4 protein in terms of apoptosis regulation and viral pathogenesis. Virology *263*, 290–306.

Honda, K., Takaoka, A., and Taniguchi, T. (2006). Type I interferon gene induction by the interferon regulatory factor family of transcription factors. Immunity *25*, 349–360.

Hornung, V., Ablasser, A., Charrel-Dennis, M., Bauernfeind, F., Horvath, G., Caffrey, D.R., Latz, E., and Fitzgerald, K.A. (2009). AIM2 recognizes cytosolic dsDNA and forms a caspase-1-activating inflammasome with ASC. Nature *458*, 514–518.

Hornung, V., and Latz, E. (2010). Intracellular DNA recognition. Nat. Rev. Immunol. *10*, 123–130.

Hu, F.Q., Smith, C.A., and Pickup, D.J. (1994). Cowpox virus contains two copies of an early gene encoding a soluble secreted form of the type II TNF receptor. Virology *204*, 343–356.

Hu, S., Vincenz, C., Buller, M., and Dixit, V.M. (1997). A novel family of viral death effector domain-containing molecules that inhibit both CD-95- and tumor necrosis factor receptor-1-induced apoptosis. J. Biol. Chem. *272*, 9621–9624.

Huang, J., Smirnov, S.V., Lewis-Antes, A., Balan, M., Li, W., Tang, S., Silke, G.V., Putz, M.M., Smith, G.L., and Kotenko, S.V. (2007). Inhibition of type I and type III interferons by a secreted glycoprotein from Yaba-like disease virus. Proc. Natl. Acad. Sci. U.S.A. *104*, 9822–9827.

Ikeda, H., Old, L.J., and Schreiber, R.D. (2002). The roles of IFN gamma in protection against tumor development and cancer immunoediting. Cytokine Growth Factor Rev. *13*, 95–109.

Ink, B.S., Gilbert, C.S., and Evan, G.I. (1995). Delay of vaccinia virus-induced apoptosis in nonpermissive Chinese hamster ovary cells by the cowpox virus CHOhr and adenovirus E1B 19K genes. J. Virol. *69*, 661–668.

Isaacs, A., and Lindenmann, J. (1957). Virus interference. I. The interferon. Proc. R. Soc. Lond. B Biol. Sci. *147*, 258–267.

Ishii, K.J., Kawagoe, T., Koyama, S., Matsui, K., Kumar, H., Kawai, T., Uematsu, S., Takeuchi, O., Takeshita, F., Coban, C., *et al.* (2008). TANK-binding kinase-1 delineates innate and adaptive immune responses to DNA vaccines. Nature *451*, 725–729.

Ishikawa-Mochizuki, I., Kitaura, M., Baba, M., Nakayama, T., Izawa, D., Imai, T., Yamada, H., Hieshima, K., Suzuki, R., Nomiyama, H., *et al.* (1999). Molecular cloning of a novel CC chemokine, interleukin-11 receptor alpha-locus chemokine (ILC), which is located on chromosome 9p13 and a potential homologue of a CC chemokine encoded by molluscum contagiosum virus. FEBS Lett. *460*, 544–548.

Ishikawa, H., and Barber, G.N. (2008). STING is an endoplasmic reticulum adaptor that facilitates innate immune signalling. Nature *455*, 674–678.

Jacobs, B.L., and Langland, J.O. (1996). When two strands are better than one: the mediators and modulators of the cellular responses to double-stranded RNA. Virology *219*, 339–349.

Jacobs, N., Bartlett, N.W., Clark, R.H., and Smith, G.L. (2008). Vaccinia virus lacking the Bcl-2-like protein N1 induces a stronger natural killer cell response to infection. J. Gen. Virol. *89*, 2877–2881.

Jeshtadi, A., Henriquet, G., Laidlaw, S.M., Hot, D., Zhang, Y., and Skinner, M.A. (2005). *In vitro* expression and analysis of secreted fowlpox virus CC chemokine-like proteins Fpv060, Fpv061, Fpv116 and Fpv121. Arch. Virol. *150*, 1745–1762.

Johnston, J.B., Barrett, J.W., Nazarian, S.H., Goodwin, M., Ricciuto, D., Wang, G., and McFadden, G. (2005). A poxvirus-encoded pyrin domain protein interacts with ASC-1 to inhibit host inflammatory and apoptotic responses to infection. Immunity *23*, 587–598.

Jones, J.M., Messauodi, I., Estep, R.D., Orzechowska, B., and Wong, S.W. (2008). Monkeypox virus viral chemokine inhibitor (MPV vCCI), a potent inhibitor of rhesus macrophage inflammatory protein-1. Cytokine *43*, 220–228.

Kaiser, W.J., Upton, J.W., and Mocarski, E.S. (2008). Receptor-interacting protein homotypic interaction motif-dependent control of NF-kappa B activation via the DNA-dependent activator of IFN regulatory factors. J. Immunol. *181*, 6427–6434.

Kanneganti, T.D., Lamkanfi, M., and Núñez, G. (2007). Intracellular NOD-like receptors in host defense and disease. Immunity *27*, 549–559.

Kato, H., Sato, S., Yoneyama, M., Yamamoto, M., Uematsu, S., Matsui, K., Tsujimura, T., Takeda, K., Fujita, T., Takeuchi, O., *et al.* (2005). Cell type-specific involvement of RIG-I in antiviral response. Immunity *23*, 19–28.

Kawagishi-Kobayashi, M., Cao, C., Lu, J., Ozato, K., and Dever, T.E. (2000). Pseudosubstrate inhibition of protein kinase PKR by swine pox virus C8L gene product. Virology *276*, 424–434.

Kawagoe, T., Sato, S., Matsushita, K., Kato, H., Matsui, K., Kumagai, Y., Saitoh, T., Kawai, T., Takeuchi, O., and Akira, S. (2008). Sequential control of Toll-like receptor-dependent responses by IRAK1 and IRAK2. Nat. Immunol. *9*, 684–691.

Kawai, T., and Akira, S. (2006). Innate immune recognition of viral infection. Nat. Immunol. *7*, 131–137.

Keating, S.E., Maloney, G.M., Moran, E.M., and Bowie, A.G. (2007). IRAK-2 participates in multiple toll-like receptor signaling pathways to NFkappaB via activation of TRAF6 ubiquitination. J. Biol. Chem. *282*, 33435–33443.

Kettle, S., Alcami, A., Khanna, A., Ehret, R., Jassoy, C., and Smith, G.L. (1997). Vaccinia virus serpin B13R (SPI-2) inhibits interleukin-1beta-converting enzyme and protects virus-infected cells from TNF- and Fas- mediated apoptosis, but does not prevent IL-1beta-induced fever. J. Gen. Virol. *78*, 677–685.

Kim, Y.M., Brinkmann, M.M., Paquet, M.E., and Ploegh, H.L. (2008). UNC93B1 delivers nucleotide-sensing toll-like receptors to endolysosomes. Nature *452*, 234–238.

Komiyama, T., Ray, C.A., Pickup, D.J., Howard, A.D., Thornberry, N.A., Peterson, E.P., and Salvesen, G. (1994). Inhibition of interleukin-1 beta converting enzyme by the cowpox virus serpin CrmA. An example of cross-class inhibition. J. Biol. Chem. *269*, 19331–19337.

Kotwal, G.J. (2000). Poxviral mimicry of complement and chemokine system components: what's the end game? Immunol. Today *21*, 242–248.

Kotwal, G.J., Hügin, A.W., and Moss, B. (1989). Mapping and insertional mutagenesis of a vaccinia virus gene encoding a 13,800-Da secreted protein. Virology *171*, 579–587.

Krathwohl, M.D., Hromas, R., Brown, D.R., Broxmeyer, H.E., and Fife, K.H. (1997). Functional characterization of the C---C chemokine-like molecules encoded by molluscum contagiosum virus types 1 and 2. Proc. Natl. Acad. Sci. U.S.A. *94*, 9875–9880.

Kumagai, Y., and Akira, S. (2010). Identification and functions of pattern-recognition receptors. J. Allergy Clin. Immunol. *125*, 985–992.

Lalani, A.S., Barrett, J.W., and McFadden, G. (2000). Modulating chemokines: more lessons from viruses. Immunol. Today *21*, 100–106.

Lalani, A.S., Graham, K., Mossman, K., Rajarathnam, K., Clark-Lewis, I., Kelvin, D., and McFadden, G. (1997). The purified myxoma virus gamma interferon receptor homolog M-T7 interacts with the heparin-binding domains of chemokines. J. Virol. *71*, 4356–4363.

Lalani, A.S., Masters, J., Graham, K., Liu, L., Lucas, A., and McFadden, G. (1999). Role of the myxoma virus soluble CC-chemokine inhibitor glycoprotein, M-T1, during myxoma virus pathogenesis. Virology *256*, 233–245.

Lalani, A.S., and McFadden, G. (1997). Secreted poxvirus chemokine binding proteins. J. Leukoc. Biol. *62*, 570–576.

Langland, J.O., Kash, J.C., Carter, V., Thomas, M.J., Katze, M.G., and Jacobs, B.L. (2006). Suppression of proinflammatory signal transduction and gene expression by the dual nucleic acid binding domains of the vaccinia virus E3L proteins. J. Virol. *80*, 10083–10095.

Lateef, Z., Baird, M.A., Wise, L.M., Mercer, A.A., and Fleming, S.B. (2009). Orf virus-encoded chemokine-binding protein is a potent inhibitor of inflammatory monocyte recruitment in a mouse skin model. J. Gen. Virol. *90*, 1477–1482.

Lateef, Z., Baird, M.A., Wise, L.M., Young, S., Mercer, A.A., and Fleming, S.B. (2010). The chemokine-binding protein encoded by the poxvirus orf virus inhibits recruitment of dendritic cells to sites of skin inflammation and migration to peripheral lymph nodes. Cell. Microbiol. *12*, 665–676.

Lee, H.J., Essani, K., and Smith, G.L. (2001). The genome sequence of Yaba-like disease virus, a yatapoxvirus. Virology *281*, 170–192.

Lee, S.B., and Esteban, M. (1994). The interferon-induced double-stranded RNA-activated protein kinase induces apoptosis. Virology *199*, 491–496.

Lenardo, M.J., Fan, C.M., Maniatis, T., and Baltimore, D. (1989). The involvement of NF-kappa B in beta-interferon gene regulation reveals its role as widely inducible mediator of signal transduction. Cell *57*, 287–294.

Levraud, J.P., Boudinot, P., Colin, I., Benmansour, A., Peyrieras, N., Herbomel, P., and Lutfalla, G. (2007). Identification of the zebrafish IFN receptor: implications for the origin of the vertebrate IFN system. J. Immunol. *178*, 4385–4394.

Li, P., Allen, H., Banerjee, S., Franklin, S., Herzog, L., Johnston, C., McDowell, J., Paskind, M., Rodman, L., Salfeld, J., *et al.* (1995). Mice deficient in IL-1 beta-converting enzyme are defective in production of mature IL-1 beta and resistant to endotoxic shock. Cell *80*, 401–411.

Lin, R., Heylbroeck, C., Pitha, P.M., and Hiscott, J. (1998). Virus-dependent phosphorylation of the IRF-3 transcription factor regulates nuclear translocation, transactivation potential, and proteasome-mediated degradation. Mol. Cell. Biol. *18*, 2986–2996.

Liptakova, H., Kontsekova, E., Alcami, A., Smith, G.L., and Kontsek, P. (1997). Analysis of an interaction between the soluble vaccinia virus-coded type I interferon (IFN)-receptor and human IFN-alpha1 and IFN-alpha2. Virology *232*, 86–90.

Liu, K., Lemon, B., and Traktman, P. (1995). The dual-specificity phosphatase encoded by vaccinia virus, VH1, is essential for viral transcription *in vivo* and *in vitro*. J. Virol. *69*, 7823–7834.

Liu, Y., Wolff, K.C., Jacobs, B.L., and Samuel, C.E. (2001). Vaccinia virus E3L interferon resistance protein inhibits the interferon-induced adenosine deaminase A-to-I editing activity. Virology *289*, 378–387.

Locksley, R.M., Killeen, N., and Lenardo, M.J. (2001). The TNF and TNF receptor superfamilies: integrating mammalian biology. Cell *104*, 487–501.

Loo, Y.M., and Gale, M., Jr. (2007). Viral regulation and evasion of the host response. Curr. Top. Microbiol. Immunol. *316*, 295–313.

Loparev, V.N., Parsons, J.M., Knight, J.C., Panus, J.F., Ray, C.A., Buller, R.M., Pickup, D.J., and Esposito, J.J. (1998). A third distinct tumor necrosis factor receptor of orthopoxviruses. Proc. Natl. Acad. Sci. U.S.A. *95*, 3786–3791.

Lord, S.J., Rajotte, R.V., Korbutt, G.S., and Bleackley, R.C. (2003). Granzyme B: a natural born killer. Immunol. Rev. *193*, 31–38.

Macen, J.L., Graham, K.A., Lee, S.F., Schreiber, M., Boshkov, L.K., and McFadden, G. (1996). Expression of the myxoma virus tumor necrosis factor receptor homologue and M11L genes is required to prevent virus-induced apoptosis in infected rabbit T lymphocytes. Virology *218*, 232–237.

Mahalingam, S., and Karupiah, G. (2000). Modulation of chemokines by poxvirus infections. Curr. Opin. Immunol. *12*, 409–412.

Maher, S.G., Romero-Weaver, A.L., Scarzello, A.J., and Gamero, A.M. (2007). Interferon: cellular executioner or white knight? Curr. Med. Chem. *14*, 1279–1289.

Maloney, G., Schroder, M., and Bowie, A.G. (2005). Vaccinia virus protein A52R activates p38 mitogen-activated protein kinase and potentiates lipopolysaccharide-induced interleukin-10. J. Biol. Chem. *280*, 30838–30844.

Mann, B.A., Huang, J.H., Li, P., Chang, H.C., Slee, R.B., O'Sullivan, A., Anita, M., Yeh, N., Klemsz, M.J., Brutkiewicz, R.R., *et al.* (2008). Vaccinia virus blocks Stat1-dependent and Stat1-independent gene expression induced by type I and type II interferons. J. Interferon Cytokine Res. *28*, 367–380.

Marie, I., Durbin, J.E., and Levy, D.E. (1998). Differential viral induction of distinct interferon-alpha genes by positive feedback through interferon regulatory factor-7. EMBO J. *17*, 6660–6669.

Martens, S., and Howard, J. (2006). The interferon-inducible GTPases. Annu. Rev. Cell Dev. Biol. *22*, 559–589.

Martinez Pomares, L., Thompson, J.P., and Moyer, R.W. (1995). Mapping and investigation of the role in pathogenesis of the major unique secreted 35-kDa protein of rabbitpox virus. Virology *206*, 591–600.

Martinon, F., Mayor, A., and Tschopp, J. (2009). The inflammasomes: guardians of the body. Annu. Rev. Immunol. *27*, 229–265.

Massung, R.F., Esposito, J.J., Liu, L.I., Qi, J., Utterback, T.R., Knight, J.C., Aubin, L., Yuran, T.E., Parsons, J.M., Loparev, V.N., *et al.* (1993). Potential virulence determinants in terminal regions of variola smallpox virus genome [see comments]. Nature *366*, 748–751.

McFadden, G., and Murphy, P.M. (2000). Host-related immunomodulators encoded by poxviruses and herpesviruses. Curr. Opin. Microbiol. *3*, 371–378.

McInnes, C.J., Wood, A.R., and Mercer, A.A. (1998). Orf virus encodes a homolog of the vaccinia virus interferon-resistance gene E3L. Virus Genes *17*, 107–115.

Meng, X., Chao, J., and Xiang, Y. (2008). Identification from diverse mammalian poxviruses of host-range regulatory genes functioning equivalently to vaccinia virus C7L. Virology *372*, 372–383.

Messud-Petit, F., Gelfi, J., Delverdier, M., Amardeilh, M.F., Py, R., Sutter, G., and Bertagnoli, S. (1998). Serp2, an inhibitor of the interleukin-1beta-converting enzyme, is critical in the pathobiology of myxoma virus. J. Virol. *72*, 7830–7839.

Michallet, M.C., Meylan, E., Ermolaeva, M.A., Vazquez, J., Rebsamen, M., Curran, J., Poeck, H., Bscheider, M., Hartmann, G., Konig, M., *et al.* (2008). TRADD protein is an essential component of the RIG-like helicase antiviral pathway. Immunity *28*, 651–661.

Miyamoto, M., Fujita, T., Kimura, Y., Maruyama, M., Harada, H., Sudo, Y., Miyata, T., and Taniguchi, T. (1988). Regulated expression of a gene encoding a nuclear factor, IRF-1, that specifically binds to IFN-beta gene regulatory elements. Cell *54*, 903–913.

Mohamed, M.R., and McFadden, G. (2009). NFkB inhibitors: strategies from poxviruses. Cell Cycle *8*, 3125–3132.

Mohamed, M.R., Rahman, M.M., Lanchbury, J.S., Shattuck, D., Neff, C., Dufford, M., van Buuren, N., Fagan, K., Barry, M., Smith, S., *et al.* (2009a). Proteomic screening of variola virus reveals a unique NF-kappaB inhibitor that is highly conserved among pathogenic orthopoxviruses. Proc. Natl. Acad. Sci. U.S.A. *106*, 9045–9050.

Mohamed, M.R., Rahman, M.M., Rice, A., Moyer, R.W., Werden, S.J., and McFadden, G. (2009b). Cowpox virus expresses a novel ankyrin repeat NF-kappaB inhibitor that controls inflammatory cell influx into virus-infected tissues and is critical for virus pathogenesis. J. Virol. *83*, 9223–9236.

Moore, C.B., Bergstralh, D.T., Duncan, J.A., Lei, Y., Morrison, T.E., Zimmermann, A.G., Accavitti-Loper, M.A., Madden, V.J., Sun, L., Ye, Z., *et al.* (2008). NLRX1 is a regulator of mitochondrial antiviral immunity. Nature *451*, 573–577.

Morikawa, S., and Ueda, Y. (1993). Characterization of vaccinia surface antigen expressed by recombinant baculovirus. Virology *193*, 753–761.

Moss, B. (1990). Regulation of vaccinia virus transcription. Annu. Rev. Biochem. *59*, 661–688.

Moss, B. (2007). Poxviridae: the viruses and their replication. In Fields Virology, 5th edn, Knipe DM and Howley PM eds (Wolters Kluwer Health/Lippincott Williams & Wilkins, Philadelphia), pp. 2905–2946.

Moss, B., and Shisler, J.L. (2001). Immunology 101 at poxvirus U: immune evasion genes. Semin. Immunol. *13*, 59–66.

Mossman, K., Lee, S.F., Barry, M., Boshkov, L., and McFadden, G. (1996a). Disruption of M-T5, a novel myxoma virus gene member of poxvirus host range superfamily, results in dramatic attenuation of myxomatosis in infected European rabbits. J. Virol. *70*, 4394–4410.

Mossman, K., Nation, P., Macen, J., Garbutt, M., Lucas, A., and McFadden, G. (1996b). Myxoma virus M-T7, a secreted homolog of the interferon-gamma receptor, is a critical virulence factor for the development of myxomatosis in European rabbits. Virology *215*, 17–30.

Mossman, K., Upton, C., Buller, R.M., and McFadden, G. (1995a). Species specificity of ectromelia virus and vaccinia virus interferon-gamma binding proteins. Virology *208*, 762–769.

Mossman, K., Upton, C., and McFadden, G. (1995b). The myxoma virus-soluble interferon-gamma receptor homolog, M-T7, inhibits interferon-gamma in a species-specific manner. J. Biol. Chem. *270*, 3031–3038.

Muller, U., Steinhoff, U., Reis, L.F., Hemmi, S., Pavlovic, J., Zinkernagel, R.M., and Aguet, M. (1994). Functional role of type I and type II interferons in antiviral defense. Science *264*, 1918–1921.

Murao, L.E., and Shisler, J.L. (2005). The MCV MC159 protein inhibits late, but not early, events of TNF-alpha-induced NF-kappaB activation. Virology *340*, 255–264.

Murphy, P.M. (2001). Viral exploitation and subversion of the immune system through chemokine mimicry. Nat. Immunol. *2*, 116–122.

Muruve, D.A., Petrilli, V., Zaiss, A.K., White, L.R., Clark, S.A., Ross, P.J., Parks, R.J., and Tschopp, J. (2008). The inflammasome recognizes cytosolic microbial and host DNA and triggers an innate immune response. Nature *452*, 103–107.

Najarro, P., Gubser, C., Hollinshead, M., Fox, J., Pease, J., and Smith, G.L. (2006). Yaba-like disease virus chemokine receptor 7L, a CCR8 orthologue. J. Gen. Virol. *87*, 809–816.

Najarro, P., Traktman, P., and Lewis, J.A. (2001). Vaccinia virus blocks gamma interferon signal transduction: viral VH1 phosphatase reverses Stat1 activation. J. Virol. *75*, 3185–3196.

Najera, J.L., Gomez, C.E., Domingo-Gil, E., Gherardi, M.M., and Esteban, M. (2006). Cellular and biochemical differences between two attenuated poxvirus vaccine candidates (MVA and NYVAC) and role of the C7L gene. J. Virol. *80*, 6033–6047.

Nakanishi, K., Yoshimoto, T., Tsutsui, H., and Okamura, H. (2001). Interleukin-18 regulates both Th1 and Th2 responses. Annu. Rev. Immunol. *19*, 423–474.

Negishi, H., Fujita, Y., Yanai, H., Sakaguchi, S., Ouyang, X., Shinohara, M., Takayanagi, H., Ohba, Y., Taniguchi, T., and Honda, K. (2006). Evidence for licensing of IFN-gamma-induced IFN regulatory factor 1 transcription factor by MyD88 in Toll-like receptor-dependent gene induction program. Proc. Natl. Acad. Sci. U.S.A. *103*, 15136–15141.

Nichols, D.B., and Shisler, J.L. (2009). Poxvirus MC160 protein utilizes multiple mechanisms to inhibit NF-kappaB activation mediated via components of the tumor necrosis factor receptor 1 signal transduction pathway. J. Virol. *83*, 3162–3174.

Novick, D., Kim, S.H., Fantuzzi, G., Reznikov, L.L., Dinarello, C.A., and Rubinstein, M. (1999). Interleukin-18 binding protein: a novel modulator of the Th1 cytokine response. Immunity *10*, 127–136.

O'Neill, L.A. (2006). How Toll-like receptors signal: what we know and what we don't know. Curr. Opin. Immunol. *18*, 3–9.

O'Neill, L.A., and Bowie, A.G. (2007). The family of five: TIR-domain-containing adaptors in Toll-like receptor signalling. Nat. Rev. Immunol. *7*, 353–364.

Oganesyan, G., Saha, S.K., Guo, B., He, J.Q., Shahangian, A., Zarnegar, B., Perry, A., and Cheng, G. (2006). Critical role of TRAF3 in the Toll-like receptor-dependent and -independent antiviral response. Nature *439*, 208–211.

Oguiura, N., Spehner, D., and Drillien, R. (1993). Detection of a protein encoded by the vaccinia virus C7L open reading frame and study of its effect on virus multiplication in different cell lines. J. Gen. Virol. *74*, 1409–1413.

Oie, K.L., and Pickup, D.J. (2001). Cowpox virus and other members of the orthopoxvirus genus interfere with the regulation of NF-kappaB activation. Virology *288*, 175–187.

Opgenorth, A., Graham, K., Nation, N., Strayer, D., and McFadden, G. (1992). Deletion analysis of two tandemly arranged virulence genes in myxoma virus, M11L and myxoma growth factor. J. Virol. *66*, 4720–4731.

Oshiumi, H., Sasai, M., Shida, K., Fujita, T., Matsumoto, M., and Seya, T. (2003). TIR-containing adapter molecule (TICAM)-2, a bridging adapter recruiting to toll-like receptor 4 TICAM-1 that induces interferon-beta. J. Biol. Chem. *278*, 49751–49762.

Paez, E., and Esteban, M. (1984). Resistance of vaccinia virus to interferon is related to an interference phenomenon between the virus and the interferon system. Virology *134*, 12–28.

Panus, J.F., Smith, C.A., Ray, C.A., Smith, T.D., Patel, D.D., and Pickup, D.J. (2002). Cowpox virus encodes a fifth member of the tumor necrosis factor receptor family: a soluble, secreted CD30 homologue. Proc. Natl. Acad. Sci. U.S.A. *99*, 8348–8353.

Park, B., Brinkmann, M.M., Spooner, E., Lee, C.C., Kim, Y.M., and Ploegh, H.L. (2008). Proteolytic cleavage in an endolysosomal compartment is required for activation of Toll-like receptor 9. Nat. Immunol. *9*, 1407–1414.

Pedra, J.H., Cassel, S.L., and Sutterwala, F.S. (2009). Sensing pathogens and danger signals by the inflammasome. Curr. Opin. Immunol. *21*, 10–16.

Perdiguero, B., and Esteban, M. (2009). The interferon system and vaccinia virus evasion mechanisms. J. Interferon Cytokine Res. *29*, 581–598.

Pestka, S. (1997). The interferon receptors. Semin. Oncol. *24*, S9–18, S19–40.

Pestka, S., Langer, J.A., Zoon, K.C., and Samuel, C.E. (1987). Interferons and their actions. Annu. Rev. Biochem. *56*, 727–777.

Petit, F., Bertagnoli, S., Gelfi, J., Fassy, F., Boucraut-Baralon, C., and Milon, A. (1996). Characterization of a myxoma virus-encoded serpin-like protein with activity against interleukin-1 beta-converting enzyme. J. Virol. *70*, 5860–5866.

Petrilli, V., Dostert, C., Muruve, D.A., and Tschopp, J. (2007). The inflammasome: a danger sensing complex triggering innate immunity. Curr. Opin. Immunol. *19*, 615–622.

Pickup, D.J., Ink, B.S., Hu, W., Ray, C.A., and Joklik, W.K. (1986). Hemorrhage in lesions caused by cowpox virus is induced by a viral protein that is related to plasma protein inhibitors of serine proteases. Proc. Natl. Acad. Sci. U.S.A. *83*, 7698–7702.

Pichlmair, A., Schulz, O., Tan, C.P., Rehwinkel, J., Kato, H., Takeuchi, O., Akira, S., Way, M., Schiavo, G., and Reis e Sousa, C. (2009). Activation of MDA5 requires higher-order RNA structures generated during virus infection. J. Virol. *83*, 10761–10769.

Platanias, L.C. (2005). Mechanisms of type-I- and type-II-interferon-mediated signalling. Nat. Rev. Immunol. *5*, 375–386.

Pobezinskaya, Y.L., Kim, Y.S., Choksi, S., Morgan, M.J., Li, T., Liu, C., and Liu, Z. (2008). The function of TRADD in signaling through tumor necrosis factor receptor 1 and TRIF-dependent Toll-like receptors. Nat. Immunol. *9*, 1047–1054.

Poeck, H., Bscheider, M., Gross, O., Finger, K., Roth, S., Rebsamen, M., Hannesschlager, N., Schlee, M., Rothenfusser, S., Barchet, W., *et al.* (2010). Recognition of RNA virus by RIG-I results in activation of CARD9 and inflammasome signaling for interleukin 1 beta production. Nat. Immunol. *11*, 63–69.

Potter, J.A., Randall, R.E., and Taylor, G.L. (2008). Crystal structure of human IPS-1/MAVS/VISA/Cardif caspase activation recruitment domain. BMC Struct. Biol. *8*, 11.

Puehler, F., Weining, K.C., Symons, J.A., Smith, G.L., and Staeheli, P. (1998). Vaccinia virus-encoded cytokine receptor binds and neutralizes chicken interferon-gamma. Virology *248*, 231–240.

Quan, L.T., Caputo, A., Bleackley, R.C., Pickup, D.J., and Salvesen, G.S. (1995). Granzyme B is inhibited by the cowpox virus serpin cytokine response modifier A. J. Biol. Chem. *270*, 10377–10379.

Rahman, M.M., Barrett, J.W., Brouckaert, P., and McFadden, G. (2006). Variation in ligand binding specificities of a novel class of poxvirus-encoded tumor necrosis factor-binding protein. J. Biol. Chem. *281*, 22517–22526.

Rahman, M.M., Mohamed, M.R., Kim, M., Smallwood, S., and McFadden, G. (2009). Co-regulation of NF-kappaB and inflammasome-mediated inflammatory responses by myxoma virus pyrin domain-containing protein M013. PLoS Pathog. *5*, e1000635.

Ray, C.A., Black, R.A., Kronheim, S.R., Greenstreet, T.A., Sleath, P.R., Salvesen, G.S., and Pickup, D.J. (1992). Viral inhibition of inflammation: cowpox virus encodes an inhibitor of the interleukin-1 beta converting enzyme. Cell *69*, 597–604.

Reading, P.C., Khanna, A., and Smith, G.L. (2002). Vaccinia virus CrmE encodes a soluble and cell surface tumor necrosis factor receptor that contributes to virus virulence. Virology *292*, 285–298.

Reading, P.C., and Smith, G.L. (2003). Vaccinia virus interleukin-18-binding protein promotes virulence by reducing gamma interferon production and natural killer and T-cell activity. J. Virol. *77*, 9960–9968.

Rebsamen, M., Heinz, L.X., Meylan, E., Michallet, M.C., Schroder, K., Hofmann, K., Vazquez, J., Benedict, C.A., and Tschopp, J. (2009). DAI/ZBP1 recruits RIP1 and RIP3 through RIP homotypic interaction motifs to activate NF-kappaB. EMBO Rep. *10*, 916–922.

Rivas, C., Gil, J., Melkova, Z., Esteban, M., and Diaz-Guerra, M. (1998). Vaccinia virus E3L protein is an inhibitor of the interferon (i.f.n.)-induced 2–5A synthetase enzyme. Virology *243*, 406–414.

Roberts, T.L., Idris, A., Dunn, J.A., Kelly, G.M., Burnton, C.M., Hodgson, S., Hardy, L.L., Garceau, V., Sweet, M.J., Ross, I.L., *et al.* (2009). HIN-200 proteins regulate caspase activation in response to foreign cytoplasmic DNA. Science *323*, 1057–1060.

Rogan, S., and Heaphy, S. (2000). The vaccinia virus E3L protein interacts with SUMO-1 and ribosomal protein L23a in a yeast two hybrid assay. Virus Genes *21*, 193–195.

Rollins, B.J. (1997). Chemokines. Blood *90*, 909–928.

Romano, P.R., Zhang, F., Tan, S.L., Garcia-Barrio, M.T., Katze, M.G., Dever, T.E., and Hinnebusch, A.G. (1998). Inhibition of double-stranded RNA-dependent protein kinase PKR by vaccinia virus E3: role of complex formation and the E3 N-terminal domain. Mol. Cell. Biol. *18*, 7304–7316.

Romieu-Mourez, R., Solis, M., Nardin, A., Goubau, D., Baron-Bodo, V., Lin, R., Massie, B., Salcedo, M., and Hiscott, J. (2006). Distinct roles for IFN regulatory factor (IRF)-3 and IRF-7 in the activation of antitumor properties of human macrophages. Cancer Res *66*, 10576–10585.

Rosenkilde, M.M., Waldhoer, M., Luttichau, H.R., and Schwartz, T.W. (2001). Virally encoded 7TM receptors. Oncogene *20*, 1582–1593.

Roulston, A., Marcellus, R.C., and Branton, P.E. (1999). Viruses and apoptosis. Annu. Rev. Microbiol. *53*, 577–628.

Ruiz-Arguello, M.B., Smith, V.P., Campanella, G.S., Baleux, F., Arenzana-Seisdedos, F., Luster, A.D., and Alcami, A. (2008). An ectromelia virus protein that interacts with chemokines through their glycosaminoglycan binding domain. J. Virol. *82*, 917–926.

Sadler, A.J., and Williams, B.R. (2008). Interferon-inducible antiviral effectors. Nat. Rev. Immunol. *8*, 559–568.

Saito, T., Hirai, R., Loo, Y.M., Owen, D., Johnson, C.L., Sinha, S.C., Akira, S., Fujita, T., and Gale, M., Jr. (2007). Regulation of innate antiviral defenses through a shared repressor domain in RIG-I and LGP2. Proc. Natl. Acad. Sci. U.S.A. *104*, 582–587.

Samuel, C.E. (1991). Antiviral actions of interferon. Interferon-regulated cellular proteins and their surprisingly selective antiviral activities. Virology *183*, 1–11.

Samuel, C.E. (2001). Antiviral actions of interferons. Clin. Microbiol. Rev. *14*, 778–809.

Saraiva, M., and Alcami, A. (2001). CrmE, a novel soluble tumor necrosis factor receptor encoded by poxviruses. J. Virol. *75*, 226–233.

Sato, M., Hata, N., Asagiri, M., Nakaya, T., Taniguchi, T., and Tanaka, N. (1998a). Positive feedback regulation of type I IFN genes by the IFN-inducible transcription factor IRF-7. FEBS Lett. *441*, 106–110.

Sato, M., Suemori, H., Hata, N., Asagiri, M., Ogasawara, K., Nakao, K., Nakaya, T., Katsuki, M., Noguchi, S., Tanaka, N., *et al.* (2000). Distinct and essential roles of transcription factors IRF-3 and IRF-7 in response to viruses for IFN-alpha/beta gene induction. Immunity *13*, 539–548.

Sato, M., Tanaka, N., Hata, N., Oda, E., and Taniguchi, T. (1998b). Involvement of the IRF family transcription factor IRF-3 in virus-induced activation of the IFN-beta gene. FEBS Lett. *425*, 112–116.

Satoh, T., Kato, H., Kumagai, Y., Yoneyama, M., Sato, S., Matsushita, K., Tsujimura, T., Fujita, T., Akira, S., and Takeuchi, O. (2010). LGP2 is a positive regulator of RIG-I- and MDA5-mediated antiviral responses. Proc. Natl. Acad. Sci. U.S.A. *107*, 1512–1517.

Schindler, C., Levy, D.E., and Decker, T. (2007). JAK-STAT signaling: from interferons to cytokines. J. Biol. Chem. *282*, 20059–20063.

Schmitz, F., Heit, A., Guggemoos, S., Krug, A., Mages, J., Schiemann, M., Adler, H., Drexler, I., Haas, T., Lang, R., *et al.* (2007). Interferon-regulatory-factor 1 controls Toll-like receptor 9-mediated IFN-beta production in myeloid dendritic cells. Eur. J. Immunol. *37*, 315–327.

Schreiber, M., Sedger, L., and McFadden, G. (1997). Distinct domains of M-T2, the myxoma virus tumor necrosis factor (TNF) receptor homolog, mediate extracellular TNF binding and intracellular apoptosis inhibition. J. Virol. *71*, 2171–2181.

Schroder, M., Baran, M., and Bowie, A.G. (2008). Viral targeting of DEAD box protein 3 reveals its role in TBK1/IKKepsilon-mediated IRF activation. EMBO J. *27*, 2147–2157.

Schroder, M., and Bowie, A.G. (2007). An arms race: innate antiviral responses and counteracting viral strategies. Biochem. Soc. Trans. *35*, 1512–1514.

Sedger, L., and McFadden, G. (1996). M-T2: a poxvirus TNF receptor homologue with dual activities. Immunol. Cell Biol. *74*, 538–545.

Seet, B.T., Johnston, J.B., Brunetti, C.R., Barrett, J.W., Everett, H., Cameron, C., Sypula, J., Nazarian, S.H., Lucas, A., and McFadden, G. (2003). Poxviruses and immune evasion. Annu. Rev. Immunol. *21*, 377–423.

Seet, B.T., and McFadden, G. (2002). Viral chemokine-binding proteins. J. Leukoc. Biol. *72*, 24–34.

Sen, G.C. (2001). Viruses and interferons. Annu. Rev. Microbiol. *55*, 255–281.

Senkevich, T.G., Bugert, J.J., Sisler, J.R., Koonin, E.V., Darai, G., and Moss, B. (1996). Genome sequence of a human tumorigenic poxvirus: prediction of specific host response-evasion genes. Science *273*, 813–816.

Seth, R.B., Sun, L., Ea, C.K., and Chen, Z.J. (2005). Identification and characterization of MAVS, a mitochondrial antiviral signaling protein that activates NF-kappaB and IRF 3. Cell *122*, 669–682.

Sharma, S., tenOever, B.R., Grandvaux, N., Zhou, G.P., Lin, R., and Hiscott, J. (2003). Triggering the interferon antiviral response through an IKK-related pathway. Science *300*, 1148–1151.

Sharp, T.V., Moonan, F., Romashko, A., Joshi, B., Barber, G.N., and Jagus, R. (1998). The vaccinia virus E3L gene product interacts with both the regulatory and the substrate binding regions of PKR: implications for PKR autoregulation. Virology *250*, 302–315.

Shchelkunov, S.N. (2003). [Immunomodulatory proteins of orthopoxviruses]. Mol. Biol. (Mosk.) *37*, 41–53.

Shchelkunov, S.N., Blinov, V.M., and Sandakhchiev, L.S. (1993). Genes of variola and vaccinia viruses necessary to overcome the host protective mechanisms. FEBS Lett. *319*, 80–83.

Shchelkunov, S.N., Totmenin, A.V., Babkin, I.V., Safronov, P.F., Ryazankina, O.I., Petrov, N.A., Gutorov, V.V., Uvarova, E.A., Mikheev, M.V., Sisler, J.R., *et al.* (2001). Human monkeypox and smallpox viruses: genomic comparison. FEBS Lett. *509*, 66–70.

Shisler, J.L., and Jin, X.L. (2004). The vaccinia virus K1L gene product inhibits host NF-kappaB activation by preventing IkappaBalpha degradation. J. Virol. *78*, 3553–3560.

Shisler, J.L., and Moss, B. (2001a). Immunology 102 at poxvirus U: avoiding apoptosis. Semin. Immunol. *13*, 67–72.

Shisler, J.L., and Moss, B. (2001b). Molluscum contagiosum virus inhibitors of apoptosis: The MC159 v-FLIP protein blocks Fas-induced activation of procaspases and degradation of the related MC160 protein. Virology *282*, 14–25.

Shisler, J.L., Senkevich, T.G., Berry, M.J., and Moss, B. (1998). Ultraviolet-induced cell death blocked by a selenoprotein from a human dermatotropic poxvirus. Science *279*, 102–105.

Silverman, G.A., Bird, P.I., Carrell, R.W., Church, F.C., Coughlin, P.B., Gettins, P.G., Irving, J.A., Lomas, D.A., Luke, C.J., Moyer, R.W., *et al.* (2001). The serpins are an expanding superfamily of structurally similar but functionally diverse proteins. Evolution, mechanism of inhibition, novel functions, and a revised nomenclature. J. Biol. Chem. *276*, 33293–33296.

Silverman, R.H. (2007). Viral encounters with 2′,5′-oligoadenylate synthetase and RNase L during the interferon antiviral response. J. Virol. *81*, 12720–12729.

Smith, C.A., Davis, T., Wignall, J.M., Din, W.S., Farrah, T., Upton, C., McFadden, G., and Goodwin, R.G. (1991). T2 open reading frame from the Shope fibroma virus encodes a soluble form of the TNF receptor. Biochem. Biophys. Res. Commun. *176*, 335–342.

Smith, C.A., Farrah, T., and Goodwin, R.G. (1994). The TNF receptor superfamily of cellular and viral proteins: activation, costimulation, and death. Cell *76*, 959–962.

Smith, C.A., Hu, F.Q., Smith, T.D., Richards, C.L., Smolak, P., Goodwin, R.G., and Pickup, D.J. (1996). Cowpox virus genome encodes a second soluble homologue of cellular TNF receptors, distinct from CrmB, that binds TNF but not LT alpha. Virology *223*, 132–147.

Smith, C.A., Smith, T.D., Smolak, P.J., Friend, D., Hagen, H., Gerhart, M., Park, L., Pickup, D.J., Torrance, D., Mohler, K., *et al.* (1997a). Poxvirus genomes encode a secreted, soluble protein that preferentially inhibits beta chemokine activity yet lacks sequence homology to known chemokine receptors. Virology *236*, 316–327.

Smith, E.J., Marie, I., Prakash, A., Garcia-Sastre, A., and Levy, D.E. (2001). IRF3 and IRF7 phosphorylation in virus-infected cells does not require double-stranded RNA-dependent protein kinase R or Ikappa B kinase but is blocked by Vaccinia virus E3L protein. J. Biol. Chem. *276*, 8951–8957.

Smith, G.L., Symons, J.A., Khanna, A., Vanderplasschen, A., and Alcami, A. (1997b). Vaccinia virus immune evasion. Immunol. Rev. *159*, 137–154.

Smith, V.P., and Alcami, A. (2000). Expression of secreted cytokine and chemokine inhibitors by ectromelia virus. J. Virol. *74*, 8460–8471.

Smith, V.P., and Alcami, A. (2002). Inhibition of interferons by ectromelia virus. J. Virol. *76*, 1124–1134.

Smith, V.P., Bryant, N.A., and Alcami, A. (2000). Ectromelia, vaccinia and cowpox viruses encode secreted interleukin-18-binding proteins. J. Gen. Virol. *81*, 1223–1230.

Sonnberg, S., Seet, B.T., Pawson, T., Fleming, S.B., and Mercer, A.A. (2008). Poxvirus ankyrin repeat proteins are a unique class of F-box proteins that associate with cellular SCF1 ubiquitin ligase complexes. Proc. Natl. Acad. Sci. U.S.A. *105*, 10955–10960.

Soulat, D., Burckstummer, T., Westermayer, S., Goncalves, A., Bauch, A., Stefanovic, A., Hantschel, O., Bennett, K.L., Decker, T., and Superti-Furga, G. (2008). The DEAD-box helicase DDX3X is a critical component of the TANK-binding kinase 1-dependent innate immune response. EMBO J. *27*, 2135–2146.

Spriggs, M.K. (1996). One step ahead of the game: viral immunomodulatory molecules. Annu. Rev. Immunol. *14*, 101–130.

Spriggs, M.K., Hruby, D.E., Maliszewski, C.R., Pickup, D.J., Sims, J.E., Buller, R.M., and VanSlyke, J. (1992). Vaccinia and cowpox viruses encode a novel secreted interleukin-1-binding protein. Cell *71*, 145–152.

Sroller, V., Ludvikova, V., Maresova, L., Hainz, P., and Nemeckova, S. (2001). Effect of IFN-gamma receptor gene deletion on vaccinia virus virulence. Arch. Virol. *146*, 239–249.

Stack, J., Haga, I.R., Schroder, M., Bartlett, N.W., Maloney, G., Reading, P.C., Fitzgerald, K.A., Smith, G.L., and Bowie, A.G. (2005). Vaccinia virus protein A46R targets multiple Toll-like-interleukin-1 receptor adaptors and contributes to virulence. J. Exp. Med. *201*, 1007–1018.

Stetson, D.B., and Medzhitov, R. (2006). Type I interferons in host defense. Immunity *25*, 373–381.

Stewart, T.L., Wasilenko, S.T., and Barry, M. (2005). Vaccinia virus F1L protein is a tail-anchored protein that functions at the mitochondria to inhibit apoptosis. J. Virol. *79*, 1084–1098.

Symons, J.A., Adams, E., Tscharke, D.C., Reading, P.C., Waldmann, H., and Smith, G.L. (2002a). The vaccinia virus C12L protein inhibits mouse IL-18 and promotes virus virulence in the murine intranasal model. J. Gen. Virol. *83*, 2833–2844.

Symons, J.A., Alcami, A., and Smith, G.L. (1995). Vaccinia virus encodes a soluble type I interferon receptor of novel structure and broad species specificity. Cell *81*, 551–560.

Symons, J.A., Tscharke, D.C., Price, N., and Smith, G.L. (2002b). A study of the vaccinia virus interferon-gamma receptor and its contribution to virus virulence. J. Gen. Virol. *83*, 1953–1964.

Tabeta, K., Hoebe, K., Janssen, E.M., Du, X., Georgel, P., Crozat, K., Mudd, S., Mann, N., Sovath, S., Goode, J., *et al.* (2006). The Unc93b1 mutation 3d disrupts exogenous antigen presentation and signaling via Toll-like receptors 3, 7 and 9. Nat. Immunol. *7*, 156–164.

Takahashi, K., Kawai, T., Kumar, H., Sato, S., Yonehara, S., and Akira, S. (2006). Roles of caspase-8 and caspase-10 in innate immune responses to double-stranded RNA. J. Immunol. *176*, 4520–4524.

Takahashi, K., Shibata, T., Akashi-Takamura, S., Kiyokawa, T., Wakabayashi, Y., Tanimura, N., Kobayashi, T., Matsumoto, F., Fukui, R., Kouro, T., *et al.* (2007). A protein associated with Toll-like receptor (TLR) 4 (PRAT4A) is required for TLR-dependent immune responses. J. Exp. Med. *204*, 2963–2976.

Takaoka, A., Hayakawa, S., Yanai, H., Stoiber, D., Negishi, H., Kikuchi, H., Sasaki, S., Imai, K., Shibue, T., Honda, K., *et al.* (2003). Integration of interferon-alpha/beta signalling to p53 responses in tumour suppression and antiviral defence. Nature *424*, 516–523.

Takaoka, A., and Taniguchi, T. (2008). Cytosolic DNA recognition for triggering innate immune responses. Adv. Drug Deliv. Rev. *60*, 847–857.

Takaoka, A., Wang, Z., Choi, M.K., Yanai, H., Negishi, H., Ban, T., Lu, Y., Miyagishi, M., Kodama, T., Honda, K., *et al.* (2007). DAI (DLM-1/ZBP1) is a cytosolic DNA sensor and an activator of innate immune response. Nature *448*, 501–505.

Takeuchi, O., and Akira, S. (2009). Innate immunity to virus infection. Immunol. Rev. *227*, 75–86.

Takeuchi, O., and Akira, S. (2010). Pattern recognition receptors and inflammation. Cell *140*, 805–820.

Tamura, T., Ueda, S., Yoshida, M., Matsuzaki, M., Mohri, H., and Okubo, T. (1996). Interferon-gamma induces Ice gene expression and enhances cellular susceptibility to apoptosis in the U937 leukemia cell line. Biochem. Biophys. Res. Commun. *229*, 21–26.

Tanaka-Kataoka, M., Kunikata, T., Takayama, S., Iwaki, K., Ohashi, K., Ikeda, M., and Kurimoto, M. (1999). *In vivo* antiviral effect of interleukin 18 in a mouse model of vaccinia virus infection. Cytokine *11*, 593–599.

Tewari, M., and Dixit, V.M. (1995). Fas- and tumor necrosis factor-induced apoptosis is inhibited by the poxvirus crmA gene product. J. Biol. Chem. *270*, 3255–3260.

Tewari, M., Telford, W.G., Miller, R.A., and Dixit, V.M. (1995). CrmA, a poxvirus-encoded serpin, inhibits cytotoxic T-lymphocyte-mediated apoptosis. J. Biol. Chem. *270*, 22705–22708.

Thome, M., Schneider, P., Hofmann, K., Fickenscher, H., Meinl, E., Neipel, F., Mattmann, C., Burns, K., Bodmer, J.L., Schroter, M., *et al.* (1997). Viral FLICE-inhibitory proteins (FLIPs) prevent apoptosis induced by death receptors. Nature *386*, 517–521.

Trapani, J.A., and Sutton, V.R. (2003). Granzyme B: pro-apoptotic, antiviral and antitumor functions. Curr. Opin. Immunol. *15*, 533–543.

Tscharke, D.C., and Smith, G.L. (1999). A model for vaccinia virus pathogenesis and immunity based on intradermal injection of mouse ear pinnae. J. Gen. Virol. *80*, 2751–2755.

Ueda, Y., Morikawa, S., and Matsuura, Y. (1990). Identification and nucleotide sequence of the gene encoding a surface antigen induced by vaccinia virus. Virology *177*, 588–594.

Unterholzner, L., and Bowie, A.G. (2008). The interplay between viruses and innate immune signaling: recent insights and therapeutic opportunities. Biochem Pharmacol 75, 589–602.

Upton, C., Macen, J.L., Schreiber, M., and McFadden, G. (1991). Myxoma virus expresses a secreted protein with homology to the tumor necrosis factor receptor gene family that contributes to viral virulence. Virology *184*, 370–382.

Upton, C., Mossman, K., and McFadden, G. (1992). Encoding of a homolog of the IFN-gamma receptor by myxoma virus. Science 258, 1369–1372.

Uze, G., and Monneron, D. (2007). IL-28 and IL-29: newcomers to the interferon family. Biochimie *89*, 729–734.

Uze, G., Schreiber, G., Piehler, J., and Pellegrini, S. (2007). The receptor of the type I interferon family. Curr. Top. Microbiol. Immunol. *316*, 71–95.

Verardi, P.H., Jones, L.A., Aziz, F.H., Ahmad, S., and Yilma, T.D. (2001). Vaccinia virus vectors with an inactivated gamma interferon receptor homolog gene (B8R) are attenuated *In vivo* without a concomitant reduction in immunogenicity. J. Virol. 75, 11–18.

Waibler, Z., Anzaghe, M., Frenz, T., Schwantes, A., Pohlmann, C., Ludwig, H., Palomo-Otero, M., Alcami, A., Sutter, G., and Kalinke, U. (2009). Vaccinia virus-mediated inhibition of type I interferon responses is a multifactorial process involving the soluble type I interferon receptor B18 and intracellular components. J. Virol. *83*, 1563–1571.

Wang, Z., Choi, M.K., Ban, T., Yanai, H., Negishi, H., Lu, Y., Tamura, T., Takaoka, A., Nishikura, K., and Taniguchi, T. (2008). Regulation of innate immune responses by DAI (DLM-1/ZBP1) and other DNA-sensing molecules. Proc. Natl. Acad. Sci. U.S.A. *105*, 5477–5482.

Wasilenko, S.T., Banadyga, L., Bond, D., and Barry, M. (2005). The vaccinia virus F1L protein interacts with the proapoptotic protein Bak and inhibits Bak activation. J. Virol. *79*, 14031–14043.

Wasilenko, S.T., Stewart, T.L., Meyers, A.F., and Barry, M. (2003). Vaccinia virus encodes a previously uncharacterized mitochondrial-associated inhibitor of apoptosis. Proc. Natl. Acad. Sci. U.S.A. *100*, 14345–14350.

Wathelet, M.G., Lin, C.H., Parekh, B.S., Ronco, L.V., Howley, P.M., and Maniatis, T. (1998). Virus infection induces the assembly of coordinately activated transcription factors on the IFN-beta enhancer *in vivo*. Mol Cell *1*, 507–518.

Weaver, B.K., Kumar, K.P., and Reich, N.C. (1998). Interferon regulatory factor 3 and CREB-binding protein/p300 are subunits of double-stranded RNA-activated transcription factor DRAF1. Mol. Cell Biol. *18*, 1359–1368.

Weber, F., Wagner, V., Rasmussen, S.B., Hartmann, R., and Paludan, S.R. (2006). Double-stranded RNA is produced by positive-strand RNA viruses and DNA viruses but not in detectable amounts by negative-strand RNA viruses. J. Virol. *80*, 5059–5064.

Werden, S.J., Rahman, M.M., and McFadden, G. (2008). Poxvirus host range genes. Adv. Virus Res. *71*, 135–171.

Westphal, D., Ledgerwood, E.C., Hibma, M.H., Fleming, S.B., Whelan, E.M., and Mercer, A.A. (2007). A novel Bcl-2-like inhibitor of apoptosis is encoded by the parapoxvirus ORF virus. J. Virol. *81*, 7178–7188.

Willer, D.O., McFadden, G., and Evans, D.H. (1999). The complete genome sequence of shope (rabbit) fibroma virus. Virology *264*, 319–343.

Xia, Z.P., Sun, L., Chen, X., Pineda, G., Jiang, X., Adhikari, A., Zeng, W., and Chen, Z. J. (2009). Direct activation of protein kinases by unanchored polyubiquitin chains. Nature *461*, 114–119.

Xiang, Y., Condit, R.C., Vijaysri, S., Jacobs, B., Williams, B.R., and Silverman, R.H. (2002). Blockade of interferon induction and action by the E3L double-stranded RNA binding proteins of vaccinia virus. J. Virol. *76*, 5251–5259.

Xiang, Y., and Moss, B. (1999). IL-18 binding and inhibition of interferon gamma induction by human poxvirus-encoded proteins. Proc. Natl. Acad. Sci. U.S.A. *96*, 11537–11542.

Xiang, Y., and Moss, B. (2003). Molluscum contagiosum virus interleukin-18 (IL-18) binding protein is secreted as a full-length form that binds cell surface glycosaminoglycans through the C-terminal tail and a furin-cleaved form with only the IL-18 binding domain. J. Virol. *77*, 2623–2630.

Xu, X., Nash, P., and McFadden, G. (2000). Myxoma virus expresses a TNF receptor homolog with two distinct functions. Virus Genes *21*, 97–109.

Yamamoto, M., Sato, S., Hemmi, H., Hoshino, K., Kaisho, T., Sanjo, H., Takeuchi, O., Sugiyama, M., Okabe, M., Takeda, K., *et al.* (2003). Role of adaptor TRIF in the MyD88-independent toll-like receptor signaling pathway. Science *301*, 640–643.

Yoneyama, M., and Fujita, T. (2008). Structural mechanism of RNA recognition by the RIG-I-like receptors. Immunity *29*, 178–181.

Yoneyama, M., and Fujita, T. (2010). Recognition of viral nucleic acids in innate immunity. Rev Med Virol *20*, 4–22.

Yoneyama, M., Kikuchi, M., Natsukawa, T., Shinobu, N., Imaizumi, T., Miyagishi, M., Taira, K., Akira, S., and Fujita, T. (2004). The RNA helicase RIG-I has an essential function in double-stranded RNA-induced innate antiviral responses. Nat. Immunol. *5*, 730–737.

Yoneyama, M., Suhara, W., Fukuhara, Y., Fukuda, M., Nishida, E., and Fujita, T. (1998). Direct triggering of the type I interferon system by virus infection: activation of a transcription factor complex containing IRF-3 and CBP/p300. EMBO J. *17*, 1087–1095.

Yoon, C.H., Lee, E.S., Lim, D.S., and Bae, Y.S. (2009). PKR, a p53 target gene, plays a crucial role in the tumor-suppressor function of p53. Proc. Natl. Acad. Sci. U.S.A. *106*, 7852–7857.

Zhou, Q., Snipas, S., Orth, K., Muzio, M., Dixit, V.M., and Salvesen, G.S. (1997). Target protease specificity of the viral serpin CrmA. Analysis of five caspases. J. Biol. Chem. *272*, 7797–7800.

Evasion of Interferon Responses by Haemorrhagic Fever Viruses

7

Christopher F. Basler and Gaya K. Amarasinghe

Abstract

Viral haemorrhagic fever, a clinical syndrome characterized by fever, shock and bleeding, can be caused in humans by members of several RNA virus families, including filoviruses, bunyaviruses, arenaviruses and flaviviruses. None of the haemorrhagic fever viruses uniformly cause haemorrhage in humans. However, some viruses show greater propensity to cause severe, life threatening disease than do others. Because of their potential to cause life threatening disease, these viruses are public health concerns and many of the haemorrhagic fever viruses are considered to be potential weapons of terror. Recent emphasis on these viruses has prompted research into the mechanisms by which they interact with and evade host innate immune responses, particularly antiviral interferon (IFN) responses. Research has identified a variety of mechanisms of innate immune antagonism, and data from filovirus and bunyavirus systems link these functions to virulence. This sets the stage for studies to evaluate how specific mechanisms of IFN evasion contribute to the clinical manifestations of viral haemorrhagic fever.

Introduction

Viral haemorrhagic fever (VHF) is a clinical syndrome that is characterized by general malaise, fever, vascular permeability accompanied by decreased volume of plasma, coagulation abnormalities and manifestations of haemorrhage (Bray, 2005). The viruses that cause VHF are typically transmitted by either an animal or arthropod vector to humans. The frequency with which human infections occur varies among the different pathogens as does the frequency of the most severe forms of disease. However, the lethality of this syndrome has made the viruses that cause VHF of concern as public health threats, in part because several of these pathogens are viewed as biological agents that may be employed by terrorists (Borio *et al.*, 2002). These concerns have prompted research to define mechanisms of pathogenesis and methods for controlling these infections. A major area of interest in this regard is to define how haemorrhagic fever viruses interact with and evade host interferon (IFN) responses, as the systemic infections associated with VHF must evade or counteract the innate antiviral response mounted by the IFN system. This chapter will focus on recent studies that have defined mechanisms of innate immune evasion by these pathogens.

Several families of viruses are associated with VHF, including the filoviruses, as well as select bunyaviruses; arenaviruses and flaviviruses (Table 7.1). Features common to several haemorrhagic fever viruses as well as differentiating features have been reviewed (Bray, 2005). These are all enveloped viruses that replicate in the cell cytoplasm and possess a single-stranded RNA genome, although this can be either a negative-sense RNA, as in filoviruses, bunyaviruses and arenaviruses, or positive-sense, as in the flaviviruses. While some haemorrhagic fever viruses destroy infected cells, others appear to cause little cytopathic effect. However, significant production of pro-inflammatory cytokines is a common manifestation of VHF, and these cytokines may make

Table 7.1 Viruses associated with VHF (Borio *et al.*, 2002)

Family	Genus	Specific viruses
Filoviruses	*Ebolavirus*	Ebolaviruses
	Marburgvirus	Marburgvirus
Bunyaviruses	*Phlebovirus*	Rift Valley fever virus
	Nairovirus	Crimean–Congo haemorrhagic fever virus
	Hantavirus	Hantaan virus and other hantaviruses
Arenaviruses	*Arenavirus*	Lassa virus
		New World arenaviruses
Flaviviruses	*Flavivirus*	Dengue virus
		Yellow fever virus

an important contribution to the pathogenesis of disease (Bray, 2005). Although the haemorrhagic fever viruses may infect a wide variety of cells *in vivo*, macrophages and dendritic cells appear to be common infection targets in VHF, and infection of these cells may be an important source of damaging cytokines (Bray, 2005). The challenge of defining how each virus family causes VHF is hampered, in many cases, by the lack of animal models that closely mimic human disease. However, in the case of ebolaviruses (EBOVs), non-human primate models do exist and closely mimic severe VHF in humans (Geisbert and Hensley, 2004). Therefore, this system holds promise for defining the molecular determinants of VHF pathogenesis, including how evasion of IFN responses contributes to development of VHF. As detailed below, numerous different mechanisms are employed by haemorrhagic fever viruses to evade IFN responses (Table 7.2). This chapter will focus first on how filoviruses evade innate immunity. The mechanisms by which other VHF agents antagonize IFN responses are then presented for comparison.

Ebola haemorrhagic fever

Thanks in large part to the availability of animal models that mimic human disease, detailed descriptions of EBOV-induced haemorrhagic fever are available. These serve as a useful review of the manifestations of a VHF (for review see Bray and Geisbert, 2005). Human infections exhibit an incubation period of up to 2 weeks, followed by fever, myalgias, headache, and gastrointestinal symptoms (reviewed in Mahanty and Bray, 2004). Half of patients also develop a rash, and changes in coagulation occur, although not all patients exhibit bleeding. Fatal infections correlate with increasing viraemia over the course of infection and typically result in shock, convulsions and 'coagulopathy' (Geisbert and Hensley, 2004). Non-human primates experimentally infected with Zaire EBOV exhibit a similar disease although these infections progress more rapidly and are almost invariably lethal (Geisbert and Hensley, 2004). These experimental infections identified macrophages and dendritic cells as early targets of infection, but dissemination to other tissues occurs relatively rapidly (Geisbert *et al.*, 2003). The pathology associated with filovirus infection is thought to be coupled to the induction of inflammatory mediators, such as cytokines and tissue factor, and probably systemic replication to high titres. That substantial replication is probably required for pathological responses to infection suggests that EBOVs must overcome innate IFN responses to cause disease *in vivo*.

Host cell IFN-α/β responses

Type I interferons, IFN-α/β, signal through the IFN-α/β receptor to exert an innate antiviral effect on many cell types. They also modulate adaptive immune responses to infection (Biron, 2001; Hoebe and Beutler, 2004). As is the case for many virus families, those that cause VHF also encode mechanisms to evade IFN responses. Production of IFN-α/β in response to RNA virus infection can be triggered by either of two classes of pattern recognition receptors, the RIG-I-like receptors (RLRs), or by select Toll-like receptors.

Table 7.2 Mechanisms of immune evasion by haemorrhagic fever viruses

Inhibition of IFN-α/β production	
Genome replication avoids detection by RIG-I	Bunyaviruses Arenaviruses
Inhibition of components of the RIG-I signalling pathway	Ebolavirus VP35 Hantavirus glycoprotein arenaviruses NP New World arenavirus Z Dengue virus NS2B3
Inhibition of IFNβ gene transcription	RVFV NSs
Inhibition of IFN-induced signalling pathways	
Inhibition of Jak1	Marburgvirus VP40
Inhibition of STAT1	Flavivirus NS4B
Block to STAT1 nuclear transport	Ebolavirus VP24
Degradation of STAT2	Dengue virus NS5

The RLRs, which include RIG-I, MDA-5 and LGP2, are cytoplasmic RNA helicases that serve as sensors of viral nucleic acids. Among the virus families that cause VHF, RIG-I is most likely the primary sensor of virus infection (e.g. Kato *et al.*, 2006; Habjan *et al.*, 2008; Loo *et al.*, 2008).

RIG-I has homology to RNA helicases at its carboxy-terminal end and has two protein–protein interaction domains of the caspase recruitment domain (CARD) family at its N-terminus (Yoneyama *et al.*, 2004). Current data indicates that RIG-I responds to RNAs possessing 5′ triphosphates (Hornung *et al.*, 2006; Pichlmair *et al.*, 2006). As the products of replication of many RNA viruses would possess 5′ triphosphates, this feature would identify these RNAs as 'foreign' and distinguish them from 5′ capped mRNAs. In addition to the 5′ triphosphate, studies have implicated the AU content of an RNA as well as the presence of double-stranded regions as important for inducing RIG-I signalling (Saito *et al.*, 2008; Schlee *et al.*, 2009; Schmidt *et al.*, 2009). Interaction with RNA ligands is thought to trigger structural changes in RIG-I allowing downstream signalling events. Signalling requires the presence of the ubiquitin ligase TRIM25, which promotes ubiquitination of the CARD domains of RIG-I (Gack *et al.*, 2007; Gack *et al.*, 2008). The CARD domains can also bind to lysine 63-linked polyubiquitin chains, and this binding is sufficient to permit RIG-I activation in the presence of ATP and an activating RNA (Zeng *et al.*). Activation of RIG-I allows it to interact with the CARD domains of the mitochondria-associated adapter protein MAVS (also known as IPS-1, VISA and CARDIF; Kawai *et al.*, 2005; Meylan *et al.*, 2005; Seth *et al.*, 2005; Xu *et al.*, 2005). This results in the activation of either of two cellular kinases, IKKε and TBK-1, which phosphorylate and activate interferon regulatory factor (IRF)–3. Signalling through these IFN-α/β inducing pathways ultimately results in the activation of IKKε and/or TBK-1 (Fitzgerald *et al.*, 2003; Sharma *et al.*, 2003). The subsequent activation of IRF-3 or the related transcription factor, IRF-7, the expression of which is induced by IFN-α/β, contributes to the activation of IFN-α/β gene expression (Hiscott, 2007).

IFNα and IFNβ are released from cells and bind to a single common receptor, the IFN-α/β receptor. This binding triggers signalling of a JAK/STAT pathway (Platanias, 2005). JAK family tyrosine kinases, Jak1 and Tyk2, become tyrosine phosphorylated and phosphorylate tyrosine residues on STAT1 and STAT2. These STAT proteins then heterodimerize via phosphotyrosine-SH2 domain interactions. This dimeric complex further complexes with IRF-9, forming a transcription factor complex, ISGF-3. STAT nuclear accumulation is required for activation of IFN-α/β responsive genes and induction in cells of an antiviral state. Among the best studied IFN–α/β-induced genes whose products exert antiviral effects are the dsRNA-activated protein kinase PKR, 2′,5′-oligoadenylate synthetase (OAS) and the MxA protein (called Mx1 in mouse cells), although other IFN induced genes also contribute to IFN-induced antiviral effects (e.g. Lenschow *et al.*, 2007; Wang *et al.*, 2007).

IFNγ is produced primarily by T cells and natural killer (NK) cells and is relevant to this review because it can also exert antiviral effects, although it is also an important regulator of adaptive immune responses. Signalling by IFNγ is through a distinct receptor, the IFNγ receptor. It also triggers Jak-STAT signalling. However, IFNγ activates Jak1 and Jak2 tyrosine kinases which leads primarily to the tyrosine phosphorylation of STAT1, which homodimerizes and localizes to the nucleus (Stark *et al.*, 1998).

Filoviruses

The filoviruses include the most lethal VHF pathogens, and because of the availability of well-developed animal models of EBOV VHF, this family may provide the best opportunity to determine the links between mechanisms of viral IFN evasion and this disease. The family Filoviridae consists of two genera: *Ebolavirus* and *Marburgvirus* (Sanchez *et al.*, 2007). The genus *Ebolavirus* is further divided into five distinct species: Zaire, Sudan, Cote d'Ivoire, Bundibugyo and Reston, whereas the genus *Marburgvirus* (MARV) contains only one species, Lake Victoria MARV. These viruses are zoonotic pathogens, whose natural reservoirs may be bats (Leroy *et al.*, 2005; Pourrut *et al.*, 2007, 2009; Towner *et al.*, 2007). Filoviruses are well known for their ability to cause outbreaks with very high mortality rates (Sanchez *et al.*, 2007). Zaire EBOV has consistently been the most lethal with fatality rates approaching 90% in some outbreaks. Other filoviruses have been associated with high mortality rates as well. For example, Sudan EBOV outbreaks have exhibited fatality rates in the 40–60% range. Although several MARV outbreaks have been reported to have fatality rates of approximately 30%, a 2005 outbreak in Angola had a fatality rate among 374 cases of 88%. Reston EBOV is unique among filoviruses in that, despite documented exposures of humans to this virus, no human disease has been reported, suggesting that the virus may be avirulent in humans (Sanchez *et al.*, 2007).

Filovirus molecular biology

Filoviruses are enveloped, non-segmented, negative sense RNA viruses with a filamentous shape (Sanchez *et al.*, 2007). The filovirus genome possesses 7 genes. In EBOV-infected cells, these genes produce eight major polypeptides whereas in MBGV-infected cells, 7 major proteins are produced. Similarities in the protein coding strategies between the EBOVs and MARVs include the production of a nucleoprotein (NP), VP35, VP40, glycoprotein (GP), VP30, VP24 and large (L) proteins. Transcription and replication of the viral genome, which occurs in the cell cytoplasm, requires a functional viral polymerase complex comprised of NP, VP35, VP30 and L, which is the enzymatic subunit of the viral RNA-dependent RNA polymerase (Muhlberger *et al.*, 1998; Muhlberger *et al.*, 1999). VP40 serves as the viral matrix protein and drives budding membrane of nascent virions from the plasma (Elliott *et al.*, 1985; Harty *et al.*, 2000; Timmins *et al.*, 2001; Noda *et al.*, 2002). VP24 has been described as a 'minor matrix protein' and promotes virus assembly (Huang *et al.*, 2002; Han *et al.*, 2003; Bamberg *et al.*, 2005; Watanabe *et al.*, 2007). GP, the transmembrane glycoprotein, accumulates at sites of virus budding and is present in the viral envelope where it functions to mediate attachment and entry of new viral particles into host cells (Feldmann *et al.*, 2001). GP may also facilitate immune evasion by masking cell surface proteins, including major histocompatibility antigens (e.g. Volchkov *et al.*, 2001; Simmons *et al.*, 2002; Ray *et al.*, 2004; Sullivan *et al.*, 2005; Wahl-Jensen *et al.*, 2005; Zampieri *et al.*, 2007). GP also appears capable of counteracting the antiviral effect of tetherin, an IFN-induced protein which can inhibit VP40 budding in the absence of GP (Kaletsky *et al.*, 2009; Lopez *et al.*, 2010). The primary difference between the protein coding strategies of EBOVs versus MARV is that EBOVs encode two distinct primary translation products from their glycoprotein gene, the membrane bound glycoprotein (GP) and a secreted glycoprotein (sGP), whereas MARVs only encode the integral membrane GP. In EBOV-infected cells, the integral membrane GP mRNA is translated from an mRNA that has undergone 'editing' by the viral polymerase such that a non-template encoded adenosine is inserted into the mRNA. This results in a frame-shift yielding a protein with a transmembrane domain (Volchkov *et al.*, 1995). sGP, in contrast, is translated from an unedited mRNA (Sanchez *et al.*, 2007). Relative to GP, sGP is shortened because of a stop codon, resulting in a protein product lacking a transmembrane domain which is therefore secreted from the host cell (Volchkov *et al.*, 1995). Although the precise role(s) of sGP remains ambiguous, several functions, such as evasion of immune responses, have been proposed (Yang *et al.*, 1998; Kindzelskii *et al.*, 2000; Wahl-Jensen *et al.*, 2005). In addition to the functions described above, the VP35 proteins of EBOVs and MARVs, the EBOV VP24 and the MARV VP40 proteins have been demonstrated to interfere with the host IFN response (Basler *et al.*, 2000; Basler *et al.*, 2003; Reid *et al.*, 2006; Reid *et al.*, 2007).

EBOVs evade IFN responses

Substantial amounts of IFNα appear in the blood of EBOV-infected humans and non-human primates (Villinger *et al.*, 1999; Gupta *et al.*, 2001; Hensley *et al.*, 2002; Hutchinson and Rollin, 2007). While the source(s) of this systemic IFN response remains to be identified, the response typically fails to control the infection. Why the response is ineffective remains to be elucidated, but treatment of infected non-human primates with IFNα has also proven ineffective (Jahrling *et al.*, 1999). There is evidence from mouse models of infection that induction of an early IFN-α/β response can control EBOV infection (Bray, 2001; Bray *et al.*, 2002; Mahanty *et al.*, 2003). Therefore, by better understanding the mechanisms by which these pathogens evade IFN responses, it may be possible to understand why infections are not better controlled and to devise strategies to overcome viral resistance to innate immunity.

EBOV VP35 inhibits IFNα/β production

The VP35 protein of EBOVs is perhaps the most extensively studied inhibitor of IFN responses among the haemorrhagic fever viruses and is a critical EBOV virulence factor (Prins *et al.*, 2010). The availability of high resolution X-ray crystal structures of a functionally critical region of the protein has provided insight into its mechanisms of action and suggests strategies for antiviral development. The function of VP35 as an inhibitor of IFN-α/β responses has been reviewed (Basler and Amarasinghe, 2009). Transfection into cells of VP35 could functionally replace, *in trans*, the influenza virus NS1 protein to promote influenza virus replication. Because NS1 was known to inhibit the IFNα/β response to influenza viral infections, the complementation suggested that VP35 was also able to block host IFN responses (Basler *et al.*, 2000). The ability of VP35 to functionally substitute for NS1 was further correlated with its ability to suppress production of IFNβ mRNA production in cells and to inhibit activation of the IFNβ or ISG54 promoter induced by transfected dsRNA or by virus-infection (Basler *et al.*, 2000). VP35 can also prevent virus infection-induced IRF-3 dimerization and nuclear accumulation by blocking the phosphorylation of IRF-3 that typically triggers its activation (Basler *et al.*, 2003; Cardenas *et al.*, 2006). Consistent with these observations produced by over-expression studies are observations produced in EBOV-infected cells. For example, little IRF-3 nuclear accumulation is detected in wild-type Zaire EBOV-infected cells (Basler *et al.*, 2003; Hartman *et al.*, 2006), whereas infection with a Zaire EBOV encoding a mutated VP35 results in enhanced nuclear accumulation of IRF-3 (Hartman *et al.*, 2006).

EBOV inhibition of the RIG-I pathway

That filoviruses will also trigger IFNα/β by a RIG-I pathway is suggested by the fact that purified Zaire EBOV RNA activates the IFNβ promoter in a RIG-I-dependent manner upon transfection into 293T cells (Habjan *et al.*, 2008). VP35 is able to block RIG-I-induced expression of IFNα/β (Cardenas *et al.*, 2006). When RIG-I is overexpressed in 293T cells, IFNβ activation is seen; but expression of VP35 blocks this activation. Similarly, VP35 expression can also inhibit IFNβ promoter activation induced by over-expression of MAVS, IKKε or TBK-1 (Cardenas *et al.*, 2006). The inhibition of IFNβ promoter induced by RIG-I correlates with the ability of VP35 to block

IRF-3 phosphorylation and to inhibit production of IFNβ from endogenous cellular genes (Cardenas *et al.*, 2006). However, VP35 does not detectably inhibit induction of gene expression by a constitutively active mutant IRF-3. Cumulatively, these studies indicate that VP35 specifically inhibits RIG-I dependent signalling that triggers phosphorylation of IRF-3 (Basler *et al.*, 2003).

VP35 interacts with the kinase domain of IKKε and TBK-1 to block IRF-3/IRF-7 activation

VP35 can serve as an alternative substrate for IKKε/TBK-1, affecting the ability of these kinases to interact with IRF-3 and IRF-7 (Prins *et al.*, 2009a). Co-immunoprecipitation experiments demonstrated the interaction of VP35 with the amino-terminal kinase domains of either IKKε or TBK-1, which share homology. In *in vitro* kinase reactions, recombinant IKKε was able to phosphorylate VP35 but not a control protein, glutathione S-transferase (Prins *et al.*, 2009b). Mass spectroscopy performed on VP35 immunoprecipitated from both the *in vitro* kinase assays and from cells in which both VP35 and IKKε were overexpressed identified several serine residues in VP35 that are phosphorylated by IKKε (unpublished observation), although the functional significance of these phosphorylations remains to be elucidated. VP35 interaction with IKKε or TBK-1 was also found to affect their interaction with either IRF-3 or IRF-7. In co-immunoprecipitation studies, VP35 blocked the interaction of either IRF-3 or IRF-7 with IKKε and with TBK-1 (Prins *et al.*, 2009b). When *in vitro* kinase assays were performed using IKKε from transfected cells and a GST-IRF-3 as substrate, VP35 inhibited IRF-3 phosphorylation (Prins *et al.*, 2009b). IKKε also interacts with the upstream adaptor molecule MAVS (Hiscott *et al.*, 2006), and expression of increasing amounts of VP35 can inhibit MAVS interaction with IKKε (Prins *et al.*, 2009b). Overall, these data suggest that VP35 interacts with the kinase domains of IKKε and TBK-1 to inhibit activation of IRF-3 and IRF-7.

Other functions of VP35

VP35 also inhibits the activation of PKR, an IFN-induced and dsRNA-activated cellular serine/threonine kinase. PKR exerts an antiviral effect by phosphorylating the α subunit of the translation initiation factor eIF-2, thereby inhibiting translation (Gale and Katze, 1998). It is well established that viruses often encode PKR inhibitors (Gale and Katze, 1998). Examples include the adenovirus VA_I RNA, the vaccinia virus E3L protein and the influenza A virus NS1 protein (Kitajewski *et al.*, 1986; Chang *et al.*, 1992; Lu *et al.*, 1995).

VP35 could complement a recombinant mutant herpes simplex virus deleted of the $\gamma_1$34.5 gene. $\gamma_1$34.5 encodes a viral protein that recruits the cellular protein phosphatase 1 to dephosphorylate eIF-2α (He *et al.*, 1997). HSV-1 mutants lacking this function display increased sensitivity to the antiviral effects of IFN, but the presence of VP35 restores protein expression and replication of HSV-1 in the presence of IFNα. In contrast, the $\gamma_1$34.5 deletion virus without VP35 was inhibited by IFNα (Feng *et al.*, 2007). Rescue of virus replication by VP35 was accompanied by inhibition of PKR activation and reduced eIF-2α phosphorylation (Feng *et al.*, 2007). In this system, an R312A VP35 mutant which lacks dsRNA binding activity still inhibits PKR (Cardenas *et al.*, 2006; Feng *et al.*, 2007; Leung *et al.*, 2009, 2010a). However, other studies suggest a role for dsRNA binding in VP35 inhibition of PKR (Schumann *et al.*, 2009). Therefore, the importance of dsRNA binding in PKR inhibition remains to be established. It also remains to be determined whether PKR mediates an important antiviral function in the context of EBOV infection.

Biochemical and structural studies on VP35

Recently, structural and biochemical studies have begun to shed light on how VP35 binds to dsRNA and how this function contributes to inhibition of IFNα/β production. VP35 possesses an N-terminal oligomerization domain and a C-terminal dsRNA binding domain (Hartman *et al.*, 2004; Reid *et al.*, 2005; Cardenas *et al.*, 2006; Hartman *et al.*, 2006; Hartman *et al.*, 2008b). Mutations designed to disrupt a predicted coiled-coil domain within the N-terminus of the MARV VP35 impair its ability to function as a component of the viral RNA polymerase complex. Similar mutations in the N-terminal region of

Zaire EBOV VP35 diminish the ability of VP35 to antagonize the RIG-I pathway and to inhibit transcription of early IFN genes (Hartman *et al.*, 2004, 2006, 2008b). However, a heterologous trimerization domain could substitute for the coiled-coil domain and restore VP35 antagonism of the RIG-I pathway (Reid *et al.*, 2005). This suggested that the IFN inhibitory domain (IID) maps to the C-terminus of VP35 (Moller *et al.*, 2005; Reid *et al.*, 2005). Further supporting a role for the VP35 C-terminus in suppression of RIG-I signalling, deletion of the C-terminal 40 residues of VP35 eliminates IFN inhibition and leads to loss of dsRNA binding (Hartman *et al.*, 2004). The structure of the EBOV VP35 IID was recently solved, and the structure reveals that the VP35 IID forms a unique fold that binds dsRNA (Fig. 7.1). Comparison of the VP35 IID structure with other IFN antagonists, such as influenza NS1A/NS1B, reveal that their structures are distinct from other canonical dsRNA binding domains (Leung *et al.*, 2009).

The structure of the EBOV VP35 IID consists of two subdomains, both of which are required to maintain the overall fold and, therefore, function (Basler and Amarasinghe, 2009; Leung *et al.*, 2009, 2010a, 2010b). The alpha helical subdomain contains a four helical bundle that spans approximately 70 residues, where the helices are arranged in an anti-parallel manner. The beta sheet subdomain contains a four-stranded mixed beta sheet, an alpha helix, and a left-handed type II polyproline helix. The inter-subdomain interface is largely formed by residues from helices 2

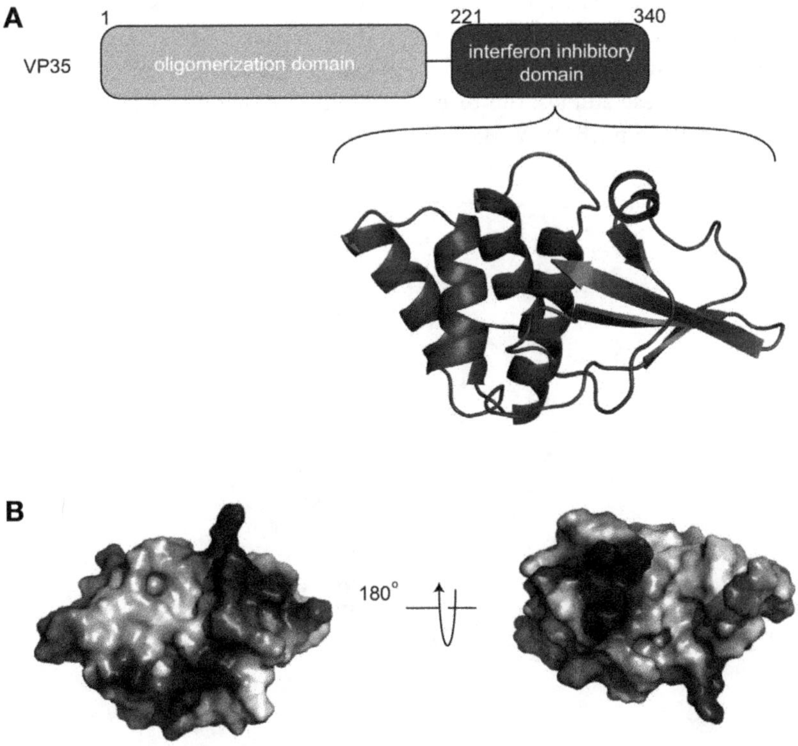

Figure 7.1 Domain organization and structure of the EBOV VP35 interferon inhibitory domain (IID). (A) A diagram representing the organization of the EBOV VP35 protein is presented. A N-terminal domain contains an oligomerization function, while the C-terminal region from amino acids 221–340, comprises a dsRNA-binding domain required for maximum VP35 IFN-antagonist function. The C-terminal 'interferon inhibitory domain' has been crystallized, and the structure of this domain is shown. (B) Two surface representations of the VP35 IID structure shown in A. A colour version of this figure is located in the plate section at the back of the book.

and 4 and beta sheets 3 and 4, and the polyproline region. The residues within these structural elements are nearly identical in all EBOV VP35 isolates sequenced thus far (Leung *et al.*, 2009). Previous studies have shown that mutation of basic residues R305, K309, and R312 leads to loss of dsRNA binding and IFN inhibition (Cardenas *et al.*, 2006; Hartman *et al.*, 2004, 2006). These residues were initially identified through a sequence similarity match with the influenza NS1A protein during a bioinformatics search of viral antagonists of the host IFN system (Hartman *et al.*, 2004). In the VP35 structure, these basic residues form an extended basic patch, which contains several additional basic residues (K319, R322, K339) that are also highly conserved. Unlike the sequences surrounding residues R305, K309 and R312, these additional basic residues do not show sequence similarities to influenza NS1 or other viral antagonists (Leung *et al.*, 2009). However, it appears that the charged patch formed by these residues is important for dsRNA binding and for VP35-mediated functions (Leung *et al.*, 2009). Consistent with these studies, the recent structures of Zaire and Reston EBOV VP35 IIDs bound to dsRNA reveal that these basic residues play important roles in dsRNA recognition (Kimberlin *et al.*, 2010; Leung *et al.*, 2010a; Fig. 7.2). In the Zaire EBOV VP35–dsRNA complex, VP35 IID interacts with the dsRNA backbone and 'end-caps' the blunt end of the dsRNA. These interactions are carried out by two different VP35 IID molecules. While both types of interactions include protein–backbone interactions, only one type of interactions is seen between the protein and the blunt end of the dsRNA. Interestingly, all residues that are important for dsRNA binding are also important for IFN inhibition. Comparison of the VP35 IID–dsRNA complex structure with RIG-I like helicase RNA binding domains suggests that the VP35 IID mimics interactions between viral RNA and RIG-I, suggesting a potential mechanism for RIG-I antagonism (Leung *et al.*, 2010a; Lu *et al.*, 2010; Wang *et al.*, 2010). *In vitro* experiments suggest that VP35 IID can compete with RIG-I dsRNA binding domains, and transfection of dsRNA and RIG-I with and without VP35 also show that VP35 is able to inhibit dsRNA-dependent RIG-I activation (Leung *et al.*, 2010a).

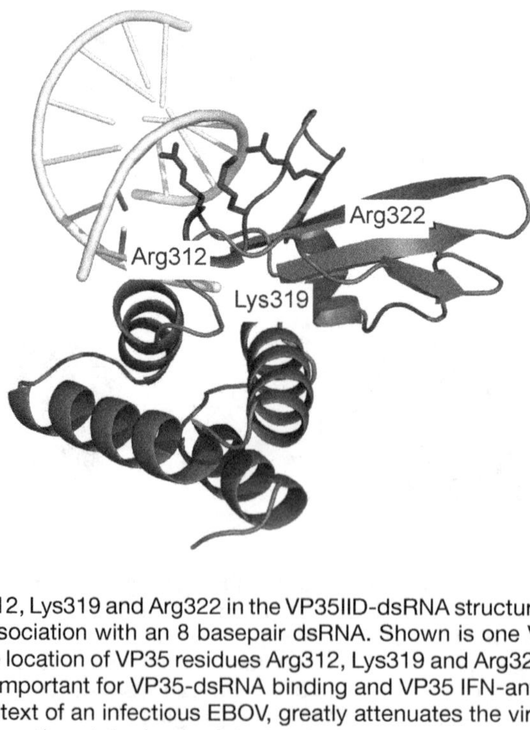

Figure 7.2 Location of Arg312, Lys319 and Arg322 in the VP35IID-dsRNA structure. The VP35 IID crystallized as a pair of IID dimers in association with an 8 basepair dsRNA. Shown is one VP35 IID monomer making contact with the dsRNA. The location of VP35 residues Arg312, Lys319 and Arg322 are shown. As detailed in the text, these residues are important for VP35-dsRNA binding and VP35 IFN-antagonist function. Mutation of these residues, in the context of an infectious EBOV, greatly attenuates the virus. A colour version of this figure is located in the plate section at the back of the book.

While the structure of the EBOV VP35 IID has provided a wealth of structural information and identified several key residues important for dsRNA dependent functions (e.g. the central basic patch), it also raises a series of yet unanswered questions. For example, the C-terminal IID is probably monomeric in solution, but forms an oligomer in the full length form through the N-terminus (Leung *et al.*, 2009). Moreover, the existence of several different oligomerization interfaces during crystallization (e.g. the structure is a dimer in the crystallographic asymmetric unit) suggests that the C-terminus is likely to support additional interactions in the context of the homotrimer. It is also unclear if VP35 recognition of dsRNA is sequence and/or length dependent and if dsRNA interactions with the oligomeric form of the full length VP35 molecule can induce additional protein–protein interactions that play important roles in IFN inhibition and in other VP35-mediated functions such as genome replication and nucleocapsid formation. Given the ability of EBOV VP35 to antagonize IRF3 phosphorylation in a dsRNA-dependent and -independent manner, coupled with its ability to interact with TBK-1 and IKKε kinases, suggests that structural characterization of the full length protein and complexes with nucleic acids will be necessary to fully appreciate the extent of contributions of EBOV VP35 towards immune evasion.

Connecting VP35 IFN-antagonist function to virulence

The filoviruses are of interest primarily because of their extreme virulence. The availability of a well characterized IFN antagonist protein, VP35, coupled with systems that allow the genetic manipulation of these viruses and established animal models have established the role of VP35 in virulence. Specifically, recombinant viruses have been built with mutant VP35 proteins in which central basic patch residues have been changed to alanines. In one case, an R312A mutant virus was found to promote IRF-3 activation and induction of IFN responses and to replicate poorly in mice, relative to a wild-type VP35 virus (Hartman *et al.*, 2006, 2008a,b). In another study, the dual mutations K319A and R322A were tested in a lethal guinea pig model of infection which exhibits some manifestations of VHF following infection with a guinea pig-adapted Zaire EBOV (Connolly *et al.*, 1999). Strikingly, while the parental, wild-type VP35 virus killed 100% of infected animals, the VP35 mutant was avirulent and caused no signs of disease (Prins *et al.*, 2010). When the mutant VP35 animals were challenged, approximately three weeks later, with the parental virus, they again showed no signs of disease. Therefore, the mutations in VP35 that abrogate interaction with dsRNA made the virus avirulent but immunogenic (Prins *et al.*, 2010). These data link VP35 structure to IFN-antagonism which is in turn linked to virulence. The *in vivo* data suggest VP35 as a target for antiviral development, and the availability of structural data on the critical residues should facilitate antiviral design (Fig. 7.2). An important follow-up will be to evaluate VP35 mutants in non-human primates that best reproduce severe VHF.

Filovirus evasion of IFN signalling

Filoviruses not only block the activation phase of the IFN-α/β response, they are also able to impair cellular responses to exogenously added IFN-α/β or IFNγ. It is interesting to note, however, that despite their genetic relatedness, EBOVs and MARVs appear to have evolved different mechanisms to suppress IFN signalling. For example, Zaire EBOV infection of human umbilical vein endothelial cells (HUVECs) impaired the ability of these cells to respond to exogenously added IFNα or IFNγ as evidenced by the suppression of IFN-induced gene expression in infected cells (Harcourt *et al.*, 1999). In a separate study, microarray analyses of Huh7 cells infected with filoviruses revealed inhibition of cellular responses to IFNα by Zaire EBOV- or MARV-infection (Kash *et al.*, 2006). This ability to counter the signalling induced by IFNs may explain, at least in part, why IFNs have proven relatively ineffective as therapeutics in non-human primate models of infection (Jahrling *et al.*, 1999).

In EBOVs, the VP24 protein appears to be a primary factor required to inhibit IFN signalling as this protein inhibits the nuclear accumulation of tyrosine-phosphorylated STAT1. Among EBOV proteins, VP24 had the unique capability to potently inhibit both IFN-α/β and IFNγ induced gene expression. In addition, IFN-α-treatment of cells expressing VP24 did not result in the effective activation of an antiviral state, whereas IFN-treatment of control cells transfected with an empty expression plasmid or expressing other EBOV proteins rendered the cells resistant to infection with Newcastle disease virus (Reid *et al.*, 2006). Although numerous viral inhibitors of STAT1 activation have previously been described, VP24 exerts its effect in a novel fashion. In cells expressing VP24 or in cells infected with Zaire EBOV, IFN treatment results in tyrosine phosphorylation of STAT1, the expected response to IFN. However, in contrast to the typical scenario, rather than resulting in the relocalization of the tyrosine phosphorylated STAT1 from the cytoplasm to the nucleus, VP24 prevents the nuclear accumulation of phospho-STAT1.

Typically, IFNs trigger STAT protein tyrosine phosphorylation and dimerization, via phospho-tyrosine-SH2 interactions. IFNγ treatment triggers formation primarily of STAT1 homodimers while IFN-α/β mainly results in STAT1–STAT2 heterodimer formation. In each case, dimerization reveals an atypical nuclear localization signal (NLS) and subsequent nuclear accumulation mediated by interaction with the nuclear import factors of the karyopherin α (importin α) family. Karyopherin α proteins interact with cargoes bearing NLSs and with importin β. Importin β mediates the nuclear import of the complex. Importin β docks the trimeric complex at the nuclear pore, and the complex is subsequently translocated into the nucleus (Gorlich *et al.*, 1996; Weis *et al.*, 1996).

Human cells encode six karyopherin α proteins which can be assigned, based on sequence similarity, into three subfamilies (Kohler *et al.*, 1997), the Rch1 subfamily (karyopherin α2; Cuomo *et al.*, 1994; Weis *et al.*, 1995), the Qip1 subfamily (karyopherin α3 and karyopherin α4; Kohler *et al.*, 1997; Miyamoto *et al.*, 1997; Seki *et al.*, 1997; Nachury *et al.*, 1998), and the NPI-1 subfamily (karyopherin α1, karyopherin α5 and karyopherin α6; Cortes *et al.*, 1994; Moroianu *et al.*, 1995; O'Neill and Palese, 1995; Kohler *et al.*, 1997; Kohler *et al.*, 1999;). STAT1 nuclear accumulation has been most clearly demonstrated to occur via karyopherin α1 (importin α5). IFNα/β or IFNγ activation leads to STAT1:STAT2 heterodimer or STAT1:STAT1 homodimer interaction with karyopherin α1 (Sekimoto *et al.*, 1997; McBride *et al.*, 2002; Melen *et al.*, 2003), and this interaction mediates the nuclear accumulation of these STAT1-containing complexes (Sekimoto *et al.*, 1997; McBride *et al.*, 2002; Melen *et al.*, 2003).

EBOV VP24 was first demonstrated to interact, via co-immunoprecipitation, with karyopherin α1 and to prevent phospho-STAT1-karyopherin α1 interaction, explaining why phospho-STAT1 fails to enter the nucleus and induce gene expression when VP24 is present. Subsequent studies demonstrated that VP24 and phospho-STAT1 bind to a similar carboxy-terminal domain on karyopherin α1 and that VP24 and phospho-STAT1 could compete with each other for binding to karyopherin α1 in *in vitro* assays (Reid *et al.*, 2007). Further support for this model was provided by the identification of mutations which, when introduced into VP24, eliminate its capacity to bind karyopherin α proteins. These mutants lose the ability to inhibit STAT1 nuclear import and to block IFN-induced gene expression (Mateo *et al.*). Interestingly, VP24 was found to interact with all three members of the NPI-1 karyopherin α subfamily (karyopherins α1, α5 and α6). Phospho-STAT1 co-immunoprecipitated with the same subset of karyopherin αs, suggesting that nuclear import of activated STAT1 may occur through any of the NPI-1 subfamily members and that VP24 is able to block STAT1 import by any of these transport proteins (Reid *et al.*, 2007).

Interestingly, VP24 also influences the capacity of EBOVs to cause lethal disease in different host species. For example, as alluded to above, Zaire EBOV, although very lethal in primates, does not cause disease in guinea pigs or mice. However, infection of mice lacking an intact IFN-α/β receptor also resulted in susceptibility to non-adapted Zaire EBOV. Alternatively, the virus can be adapted by serial passage in these animals resulting in an ability to cause lethality in these species (Bray *et al.*, 1998; Volchkov *et al.*, 2000). Adaptation resulted in the acquisition of several genetic changes in the adapted virus relative to the parental strain (Volchkov *et al.*, 2000; Ebihara *et al.*, 2006). Changes were found in VP24 both in guinea pig and mouse-adapted strains (Ebihara *et al.*, 2006). In the mouse system, only two amino acid changes were found to be necessary for recombinant EBOVs to cause lethal disease in mice, one in NP and one in VP24 (Ebihara *et al.*, 2006). Interestingly, the growth of the lethal viruses in mouse macrophages was less sensitive to the antiviral effects of IFN-α/β than non-lethal control EBOVs (Ebihara *et al.*, 2006). However, a direct connection between VP24-karyopherin α interaction and the mouse adapted phenotype has not been demonstrated (Reid *et al.*, 2007).

In addition to its role as an inhibitor of IFN signalling, VP24 is required, along with the EBOV VP35 and NP proteins, for the assembly of filamentous viral nucleocapsids which are comprised mainly of NP and VP35 (Huang *et al.*, 2002; Noda *et al.*, 2007), and is needed for production of infectious viral particles (Bamberg *et al.*, 2005). VP24 also influences viral budding and modulates the function of the viral polymerase complex (Han *et al.*, 2002; Licata *et al.*, 2004; Watanabe *et al.*, 2007). It remains possible that one or more of these functions is altered following acquisition of mouse- or guineapig-adaptive changes.

The EBOV and MARV VP24s are clearly homologous. For example, the Lake Victoria MARV from a 2005 Angola outbreak is 36% identical and roughly 50% similar, at the amino acid level, to the Zaire EBOV VP24. However, the MARV VP24 does not detectably interact with karyopherin α proteins and does not block IFN signalling (Valmas *et al.*, 2010). MARV infection does, however, result in an inhibition of cellular responses to IFNs, as assessed by microarray analysis (Kash *et al.*, 2006). These observations suggested that MARV uses a protein(s) other than VP24 to suppress IFN responses. The cellular factor(s) targeted by MARV also appear to be different from the karyopherin αs targeted by EBOV. In EBOV-infected cells, IFN treatment results in the tyrosine phosphorylation of STAT1, but STAT1 fails to accumulate in the nucleus. In contrast, in MARV-infected cells, IFN-induced STAT1 and STAT2 phosphorylation is impaired (Kash *et al.*, 2006; Valmas *et al.*, 2010). Among MARV proteins tested, VP40 was able to prevent induction in IFN-treated cells of an antiviral state and could inhibit IFN-α/β and IFNγ-induced gene expression (Valmas *et al.*, 2010). VP40 could further prevent the tyrosine phosphorylation of Jak and STAT proteins in response to IFNs and to IL-6 (Valmas *et al.*, 2010). Specifically, when IFN-α/β was added to cells tyrosine phosphorylation of Jak1, Tyk2, STAT1 and STAT2 was inhibited. When IFNγ was added to VP40-expressing cells, JAK1, JAK2 and STAT1 tyrosine phosphorylation was impaired. Infection of cells with MARV mirrored the phenotype of VP40-expressing cells suggesting that VP40 mediates the inhibition of IFN signalling in infected cells. To assess the impact of VP40 on signalling pathways that are independent of IFN, TNFα and IL-6 were chosen. TNFα activates IL-8 expression in a manner independent of Jak-STAT signalling, and MARV VP40 did not significantly inhibit this response, demonstrating that VP40 is not a global inhibitor of cell signalling pathways (Valmas *et al.*, 2010).

However, the impact of MARV VP40 on host cell signalling is not restricted to IFN signalling. For example, VP40 can inhibit the IL-6-induced tyrosine phosphorylation of STAT1 and STAT3. The IL-6 receptor is comprised of IL-6 receptor α and GP130 subunits. Binding of IL-6 to IL-6 receptor α leads to IL-6 receptor α association with gp130, resulting in activation of several signalling cascades, including a Jak-STAT pathway in which STAT1 and STAT3 become tyrosine phosphorylated. When VP40 was present, no detectable tyrosine phosphorylation of STAT1 was detected and STAT3 phosphorylation was substantially decreased (Valmas *et al.*, 2010). Interestingly, the phenotype of a Jak1 knockout cell mirrors the loss of Jak and STAT tyrosine phosphorylation seen in VP40 expressing cells. Specifically, in Jak1-deficient cells treated with IFN-α/β, Jak1, Tyk2, STAT1 and STAT2 tyrosine phosphorylation is abrogated; while in IFNγ treated cells, Jak1, Jak2 and STAT1 tyrosine phosphorylation is lost. Finally, Jak1 knockout cells treated with IL-6 showed a loss of detectable STAT1 tyrosine phosphorylation and a decrease in phospho-STAT3 levels (Muller *et al.*, 1993; Guschin *et al.*, 1995; Rodig *et al.*, 1998). These data suggest that VP40 specifically targets Jak1. Consistent with this, VP40 was demonstrated to inhibit STAT1, 2 or 3 tyrosine phosphorylation induced by over-expression of Jak1 but not over-expressed Tyk2 (Valmas *et al.*, 2010). Over-expression of these kinases also results in the tyrosine phosphorylation of the over-expressed kinase. VP40 inhibited this apparent autophosphorylation of Jak1 but not Tyk2. Interestingly, while VP40 is a matrix protein which buds from cells in the absence of other viral proteins, the Jak1-inhibitory function appears to be separable from VP40's budding function, as a mutation severely affecting VP40 budding had no detectable impact upon its capacity to inhibit Jak1 signalling (Valmas *et al.*, 2010).

These data demonstrate a role for VP40 in MARV suppression of IFN responses and suggest that VP40 may also significantly impair Jak1-dependent signalling by other cytokines and growth factors. The ability to overcome the antiviral effects of IFNs has obvious implications for sustaining virus replication in the face of a host IFN response. It is also intriguing that when MARV was adapted to the mouse, VP40, among other genes, underwent genetic changes (Warfield *et al.*, 2009). It will be of interest to determine whether these changes affect host specific inhibition of Jak1 and thereby contribute to viral adaptation to a new host, as this would imply that the Jak1-inhibitory function of VP40 influences virulence.

Bunyaviruses causing VHF

The bunyaviruses are enveloped, negative-sense RNA viruses with segmented genomes. The family is comprised of five genera, four of which infect animals: *Orthobunyavirus*, *Phlebovirus*, *Nairovirus*, and *Hantavirus*. Among these, a nairovirus [Crimean-Congo haemorrhagic fever virus (CCHFV)], a phlebovirus [Rift Valley fever virus (RVFV)] and some hantaviruses cause VHF. CCHFV and RVFV are arthropod-transmitted, CCHFV is transmitted by ticks, while RVFV is transmitted mainly by mosquitoes of the species *Aedes*. In contrast, hantaviruses are transmitted to humans by infected rodents that serve as reservoir species.

Bunyavirus replication cycle

The bunyavirus tri-segmented genomes are designated based on size as small (S), medium (M) and large (L). The S segment encodes the viral nucleocapsid (N) protein, and depending on the genus of a given virus, may also encode a non-structural protein designated NSs. The M segment encodes the viral glycoproteins Gn and Gc which bind to the host cell receptor and mediate fusion of viral and host cell membranes following endocytosis of the infecting virion. For some viruses, M also encodes a non-structural protein NSm. The L segment encodes the polymerase or large (L) protein which possesses several enzymatic functions required for viral RNA synthesis. For some nairoviruses, the L segment also contains a protease domain of the ovarian tumour family (OTU domain). The N, Gn, Gc and L proteins are always encoded in the negative-sense by bunyaviruses. However, the NSs protein is present in the phlebovirus genome in the positive-sense, and its expression requires the production of an mRNA that is in the same 'sense' as the viral genomic RNA. For orthobunyaviruses, NSs is encoded in an alternative open reading frame by the same mRNA that encodes N, whereas hantaviruses and nairoviruses do not appear to produce an NSs protein.

Following acid-dependent fusion of viral and endocytic membranes, viral genomic RNAs are released, in the form of ribonucleoproteins, into the host cell cytoplasm. Replication then proceeds within the cytoplasm. Viral mRNA synthesis requires the N and L proteins as well as host cell mRNAs. The host cell mRNAs are cleaved by an L-associated endonuclease activity that allows the viral polymerase to 'snatch' 5′ capped oligonucleotides and use them to prime mRNA synthesis. This results in mRNAs that possess non-template encoded sequence at their 5′ ends. More importantly for host innate immunity, the presence of a 5′ cap is expected to render the mRNAs unable to stimulate signalling through RIG-I.

Genome replication strategies to evade the RIG-I pathway

One genome replication strategy employed by bunyaviruses may prevent or reduce viral detection by RIG-I. It was noted that genomic RNAs from CCHFV and Hantaan virus did not activate RIG-I when transfected into cells, whereas RVFV RNA did (Habjan *et al.*, 2008). The difference appears to be that RVFV RNA retains a 5′ triphosphate whereas CCHF and Hantaan virus RNAs possess a 5′ monophosphate (Habjan *et al.*, 2008). The presence of a 5′ monophosphate at the ends of these genomes probably reflects the fact that the initial product of genome replication undergoes an endonucleolytic cleavage, removing the first nucleotide that was added to the initiated RNA (Habjan *et al.*, 2008). This can occur without loss of viral genetic information because of the 'prime and realign' mechanism used to initiate genome replication. In this model, genome replication is primed by a host cell RNA-derived primer, and the first nucleotide added to the nascent RNA is internal, corresponding to the +3 position on the template. However, after the polymerase begins to extend the new RNA, the product 'realigns' on the template such that the nucleotide that was initially added to correspond to the +4 position is paired with the +1 template nucleotide. The unpaired 5′ sequence is then removed by an endonuclease activity associated with the viral polymerase (Garcin *et al.*, 1995).

Hantaviruses and IFN

Hantaviruses are associated with either of two clinical syndromes, haemorrhagic fever with renal syndrome or hantavirus pulmonary syndrome (Schmaljohn and Hjelle, 1997). The viruses that cause haemorrhagic fever include Hantaan virus, Seoul virus, and Puumala virus. Causing hantavirus pulmonary syndrome are NY-1 virus (NY-1V), sin nombre virus, and Andes virus (ANDV). In addition there are hantaviruses that appear to be non-pathogenic to humans, including Prospect Hill virus (PHV) and Tula virus (Schmaljohn and Hjelle, 1997). Several studies have addressed the potential of IFNs to inhibit hantaviruses. For example, infection with Hantaan virus induces IFN and is inhibited by pretreatment with IFNα, IFNγ and IFNλ (Nam *et al.*, 2003; Oelschlegel *et al.*, 2007; Stoltz *et al.*, 2007). It has also been demonstrated that MxA inhibits replication of representative members of bunyaviruses from the genera *Hantavirus*, *Bunyavirus* and *Phlebovirus*, providing at least one explanation as to why IFN is an inhibitor of these viruses (Frese *et al.*, 1996; Kanerva *et al.*, 1996; Andersson *et al.*, 2006). These observations suggest that it is important for these viruses to counteract host IFN responses *in vivo*.

Connection between *Hantavirus* pathogenesis and innate immune evasion

Endothelial cells are major targets of hantavirus infection, and upon infection of endothelial cells, the pathogenic Hantaan virus appears to suppress early IFNα/β responses more effectively that the non-pathogenic Prospect Hill virus, although at later times in infection, a substantial IFN response, as measured by global gene expression patterns, is detectable in cells infected by both viruses (Geimonen *et al.*, 2002). When human umbilical vein endothelial cell (HUVEC) infection by the pathogenic hantavirus was compared with infection by the non-pathogenic Tula virus by microarray, Hantaan virus induced more IFNβ but Tula virus resulted in a greater induction of MxA expression (Kraus *et al.*, 2004). These differences may thus contribute to the relative virulence of these viruses, and induction of TLR4 expression by Hantaan virus may explain the virus' capacity to induce IFNβ and other cytokines (Jiang *et al.*, 2008).

Differences in IFN induction may also be related to the presence of suppressing functions encoded by pathogenic viruses. Expression of the NY-1 virus Gn (G1) cytoplasmic tail was demonstrated to inhibit induction of IRF-3 responsive genes, including the IFNβ gene, whether it was induced by a constitutively active RIG-I or by over-expression of TBK-1. However, because the Gn cytoplasmic tail did not inhibit gene induction by a constitutively active mutant of IRF-3, Gn appears to act upstream of IRF-3 (Geimonen *et al.*, 2002). The Gn cytoplasmic tail also inhibits NF-κB activation induced by TBK-1 or TRAF2 over-expression. This was correlated with the ability of the cytoplasmic tail to bind to TRAF3 and to prevent TBK-1–TRAF3 interaction. Further supporting this as a mode of IFN evasion by hantaviruses, NY-1 virus infection also impaired TBK-1–TRAF3 interaction (Alff *et al.*, 2008). Because TRAF3 is required for IFNβ gene induction and associates with TBK-1 and IKKε (Oganesyan *et al.*, 2006), its sequestration by Gn cytoplasmic tail probably explains inhibition of IFNα/β responses by pathogenic hantaviruses.

Another mechanism by which hantaviruses evade innate immunity has also been described.

Hantaan virus infection prevents TNFα-induced NF-κB activation, and this phenomenon can be reproduced by expressing the N protein (Taylor *et al.*, 2009a). The defect in NF-κB activation was not at the level of IκB, as it was degraded normally as is typically seen following TNFα addition. Rather, tracking of the subcellular localization of the p65 subunit of NF-κB revealed a failure to translocate to the nucleus. Interestingly, the N protein was found to interact with karyopherin α needed to transport NF-κB to the nucleus (Taylor *et al.*, 2009a). This suggests a mechanism of NF-κB inhibition reminiscent of EBOV VP24 inhibition of STAT1. It should be noted, however, that different hantaviruses appear to have different capacities to block NF-κB activation (Taylor *et al.*, 2009b). Therefore, it will be of interest to determine whether N proteins from different hantaviruses exhibit differences in their interaction with karyopherin α and how strain-specific capabilities to evade or alter innate immune responses influence hantavirus pathogenesis.

CCHFV evasion of innate immunity

As with other bunyaviruses, replication of CCHFV is inhibited by IFNα treatment of human cells, and the IFN-induced antiviral MxA protein is a major effector of this antiviral effect (Andersson *et al.*, 2006). MxA mediates its antiviral effect most likely through its ability to interact with the viral nucleoprotein (Andersson *et al.*, 2004). However, once infection is established, the virus becomes largely resistant to the antiviral effects of IFN (Andersson *et al.*, 2008). CCHFV infection also impairs the activation of IFNα/β responses, most likely by impeding activation of IRF-3 (Andersson *et al.*, 2008).

CCHFV has recently been described to possess a unique means of innate immune antagonism. The CCHFV large (L) protein possesses motifs conserved among polymerases and is the catalytic subunit of the viral RNA-dependent RNA polymerase (Honig *et al.*, 2004). However, L also encodes a domain that is dispensable for viral polymerase function and which shows homology to ovarian tumour (OTU) domains (Bergeron *et al.*; Frias-Staheli *et al.*, 2007). These domains are found in a family of proteases among which are ubiquitin (Ub)-deconjugating enzymes. Exogenous expression of L in mammalian cells resulted in an overall decrease in levels of Ub-conjugated proteins, and when L was co-expressed with the enzymatic machinery required for ISG15ylation, levels of ISG15 conjugates were also decreased (Frias-Staheli *et al.*, 2007). An amino-terminal region of L containing the OTU domain was sufficient to mediate deubiquitylation and de-ISG15ylation, and this function was abrogated by mutations designed to disrupt the putative catalytic site of this presumptive protease domain (Frias-Staheli *et al.*, 2007). The deubiquitylation activity was found to be functional against K48 and K63-linked Ub conjugates and also removed ISG15 from conjugates. However, the viral OTU was not active against all Ub-like modifications; for example, it was not active towards poly-SUMO-2, poly-SUMO-3, pro-SUMO-1 or pro-Nedd8 (Frias-Staheli *et al.*, 2007).

ISG15 is antiviral against a variety of viruses, including influenza viruses, alphaviruses and herpes viruses (Lenschow *et al.*, 2005; Lenschow *et al.*, 2007; Zhang *et al.*, 2007; Giannakopoulos *et al.*, 2009). It has also been shown to inhibit HIV-1 replication and to inhibit the budding of HIV and EBOV (Kunzi and Pitha, 1996; Lu *et al.*, 2006; Okumura *et al.*, 2006, 2008; Harty *et al.*, 2009). To assess the potential impact of the L-encoded protease activity upon the antiviral effects of ISG15, transgenic mice expressing an amino-terminal domain of L that includes the OTU were generated and found to exhibit resistance of Sindbis virus infection (Frias-Staheli *et al.*, 2007). Further, although Sindbis viruses expressing ISG15 are attenuated *in vivo*, insertion of an enzymatically active CCHFV OTU reversed this attenuation whereas a catalytically inactive mutant did not, providing evidence that the CCHFV OTU can modulate virulence and can counteract the antiviral effects of the IFN-induced ISG15 protein (Frias-Staheli *et al.*, 2007). Exactly how OTU may modulate innate immune signalling remains to be fully elucidated; however, given the importance of ubiquitylation in numerous signalling pathways, it is likely to antagonize host cell signalling pathways. In one example, it was demonstrated that the CCHFV OTU can inhibit NF-κB activation (Frias-Staheli *et al.*, 2007). It is also clear that cellular deubiquitinating enzymes can act as negative regulators of the RIG-I pathway, suggesting another potential target of relevance to viral infection (e.g. Friedman *et al.*, 2008).

Rift Valley fever virus (RVFV) evasion of IFN responses

There is substantial evidence that the host IFN response is an important determinant of RVFV pathogenesis. Rhesus macaques administered RVFV intravenously developed disease that resembles human RVF with approximately 20% of animals developing a haemorrhagic fever (Morrill *et al.*, 1990). In this model, the early appearance of serum IFN correlated with recovery from infection and little sign of disease whereas later appearance of serum IFN correlated with severe disease and death (Morrill *et al.*, 1990). Further, the IFN-inducer polyriboinosinic-polyribocytidylic acid complexed with poly-L-lysine and carboxymethylcellulose (poly[ICLC]) could protect hamsters from RVFV disease (Peters *et al.*, 1986), and IFN-α could also prevent disease in infected non-human primates (Morrill *et al.*, 1989). The IFN-induced human MxA protein or its rat counterpart ratMx2 probably contribute to the anti-RVFV effects of IFNs in these species (Frese *et al.*, 1996; Sandrock *et al.*, 2001).

Substantial evidence now exists to support the view that the RVFV encoded non-structural protein NSs acts as an antagonist of IFN-α/β production. That NSs antagonizes IFN responses is evidenced by the behaviour of RVFV strains carrying NSs mutations. A virulent strain ZH548 is pathogenic in wild-type mice and in IFN-α/β receptor null mice. A mutant strain, clone 13, was attenuated in wild-type mice but virulent in IFN-α/β receptor null mice (Bouloy *et al.*, 2001). The clone 13 strain was found to be a potent inducer of IFN-α/β, relative to the virulent strain and, with reassortant viruses, the IFN-inducing phenotype was mapped to the S segment (Bouloy *et al.*, 2001). Because the NSs protein coding region was the portion of the S segment in which genetic differences were identified, this study directly implicated NSs as an IFN-antagonist protein that influences virulence. Genetically engineered MP12 viruses, in which NSs was deleted, supplied further evidence that NSs inhibits IFN-β production, as NSs mutant infected cells produced more IFNβ than cells infected with the parental virus (Ikegami *et al.*, 2006).

Two mechanisms have been described by which NSs inhibits IFN-α/β mRNA synthesis. Despite the cytoplasmic replication of bunyaviruses, NSs is a phosphoprotein that exhibits nuclear localization in mammalian cells (Kohl *et al.*, 1999). This localization appears to be related to the ability of wild-type NSs to act as an inhibitor of cellular transcription through interaction with the p44 subunit of TFIIH, a protein complex that plays a role in synthesis of mRNAs and rRNAs (Le May *et al.*, 2004). By generally inhibiting host cell gene expression, NSs thereby inhibits IFN-α/β gene expression (Billecocq *et al.*, 2004; Le May *et al.*, 2004).

Because the general inhibition of gene expression was found to occur 8 hours post-infection, while inhibition of IFN-β gene expression occurred as early as 3–6 hours post infection, additional mechanisms of inhibition of IFN-β gene expression were sought. It was thereby demonstrated that NSs can also interact with a host protein called SAP30 (Le May *et al.*, 2008). This was initially demonstrated by a yeast two-hybrid screen. SAP30 associates with transcriptional 'co-repressor complexes' and with the transcription factor YY1, which plays a role in IFN-β gene expression (Le May *et al.*, 2008). Immunofluorescence indicated a relocalization of SAP30 and YY1 with NSs in the nuclei of cells infected with a virulent virus but not with clone13-infected cells; infection with virulent virus resulted in recruitment of SAP30 to the IFN-β promoter, and this recruitment required the presence of YY1 (Le May *et al.*, 2008). This in turn appears to impair recruitment of CBP to the IFN-β promoter and acetylation of histones at the promoter, resulting in a loss of transcriptional activation (Le May *et al.*, 2008). Deletion from NSs of 20 amino acids that are required for SAP30 binding resulted in a RVFV that induced large amounts of IFN-β and that was avirulent in mice (Le May *et al.*, 2008).

In addition to its role as an inhibitor of cellular transcription, NSs also decreases levels of the IFN-induced, dsRNA-activated kinase PKR by promoting its degradation (Habjan *et al.*, 2009; Ikegami *et al.*, 2009). This function is required to maintain efficient viral protein synthesis (Ikegami *et al.*, 2009). Because PKR mediates antiviral effects and promotes maximal IFNβ production (McAllister and Samuel, 2009), inhibition of PKR may both help to suppress IFNα/β production and to evade the antiviral effects of IFNs.

Arenaviruses associated with VHF

Arenaviruses are associated with persistent infections in rodent species and often are transmitted to humans via rodent droppings. These viruses can be classified into two categories, the Old World and New World arenaviruses. Among the Old World viruses, Lassa virus causes Lassa fever, an often severe systemic infection that can lead to VHF characterized by shock, bleeding and death. Lymphocytic choriomeningitis virus (LCMV), the most extensively studied arenavirus, is less pathogenic to humans. Typically, LCMV infections are asymptomatic to mild although aseptic meningitis can occur. Transmission of LCMV to humans via transplanted organs has resulted in severe systemic infections that resulted in death of the organ recipient (Fischer *et al.*, 2006). The New World arenaviruses include a number of South American viruses that can cause haemorrhagic fever. These include Junin virus, the causative agent of Argentine haemorrhagic fever, Machupo virus, the causative agent of Bolivian haemorrhagic fever and Pichinde virus, which is avirulent in humans but which can cause lethal haemorrhagic fever in guinea pigs.

Arenaviruses are enveloped viruses which enter the cells by receptor mediated endocytosis and release their genomes, in the form of ribonucleoprotein complexes, into the cytoplasm following low pH-triggered fusion of the viral and cellular membranes. Assembly and release involves the budding of new particles from the host cell. Viral particles possess genomes consisting of two single-stranded RNA segments. The small (S) segment encodes the viral glycoprotein precursor GPC, which is cleaved into mature protein products Gn, Gc and the signal peptide, and the vial nucleoprotein. The large (L) segment encodes both the Z (or ring) protein and the L polymerase. The arenaviruses replicate within the cytoplasm of infected cells employing a strategy similar to other negative-sense RNA viruses in that the viruses package a polymerase and this polymerase is required for viral mRNA synthesis prior to production of viral proteins. The viral RNA synthesis reactions require both the NP and L proteins, with L providing the necessary enzymatic activities. The arenaviruses are unique, however, in that both genomic RNAs are ambisense. The two open reading frames on each RNA segment are non-overlapping, with one complementary to the viral RNA and one in the same sense as the viral RNA. The two segments encode four primary translation products but, although the viral genomic RNA contains the GPC and Z open reading frames in the positive sense orientation, it appears that the viral RNA does not serve directly as an mRNA. Rather, the antigenomic RNAs serve as template to synthesize mRNAs encoding these proteins.

Arenaviruses and IFN

A number of studies have assessed the levels of endogenous IFN during the course of infection *in vivo*. For example, in patients with Argentine haemorrhagic fever caused by Junin virus, subjects with the highest serum IFN levels during their second week of infection were more likely to die (Levis *et al.*, 1985). Passive administration to patients of convalescent sera was associated with decreased systemic IFN levels (Levis *et al.*, 1984). This suggests that high IFN levels in infected individuals are related to high levels of virus replication. Consistent with the progression of virus infection in the presence of very high systemic IFN levels, Lassa virus was found to be quite resistant to the antiviral effects of IFNs in cell culture (Peters *et al.*, 1989), although it is possible that this is a virus strain or host cell-dependent phenomenon (Asper *et al.*, 2004). Further supporting the general resistance of arenaviruses to IFNs was the observation that IFNα or an inducer of IFNα/β failed to exert a clinical benefit towards Pichinde virus infection in guinea pigs (Lucia *et al.*, 1989). Interpreting the role of endogenously produced IFNs upon viral pathogenesis may be complicated, however. While the high levels of systemic IFN appear ineffective at controlling infection, they may also exert pathogenic effects. For example, thrombocytopenia is a manifestation of VHF that may contribute to bleeding, and systemic IFN may contribute to thrombocytopenia by inhibiting platelet formation (Binder *et al.*, 1997; Borio *et al.*, 2002; Pozner *et al.*, 2010).

Evasion of RIG-I by the *Arenavirus* RNA replication strategy

Like the bunyaviruses, arenaviruses utilize a 'prime and realign' mechanism to initiate genome RNA replication (Marq *et al.*). For Tacaribe virus (TCRV), this results in addition of a single 5′pppG overhang which remains after genome replication is completed (Garcin and Kolakofsky, 1992). The additional nucleotide does not result in progressive increases in genome length as replication proceeds, because this extra nucleotide is not copied during subsequent rounds of replication (Garcin and Kolakofsky, 1992). The presence of a 5′ triphosphate raises the question as to how this genomic RNA interacts with the RIG-I pathway. There is evidence that a short double-stranded region is required, in addition to a 5′ triphosphate, for an RNA molecule to activate RIG-I (Schlee *et al.*, 2009; Schmidt *et al.*, 2009). This suggests that RIG-I activation by a genomic viral RNA would involve not only the 5′ triphosphate but also basepairing between the 5′ and 3′ ends of the arenaviral genomic RNA. This is possible, as arenavirus genomic RNA 5′ and 3′ ends show inverse complementarity and probably interact to serve as the 'promoter' for viral RNA synthesis. The hypothesis was tested that a single nucleotide 5′ overhang would prevent efficient recognition of a 5′ triphosphate-possessing, partially double-stranded hairpin RNA molecule. In addition, because the 'panhandle' formed by arenavirus genomic RNAs would contain a three-nucleotide 'bulge,' the impact of this imperfect complementarity was also assessed. Each of these features impaired the ability of RNAs to activate RIG-I, suggesting that this mode of genome RNA replication may serve as a means of evading detection by RIG-I (Marq *et al.*, 2010).

Inhibition of RIG-I signalling by *Arenavirus* NP and Z proteins

Despite the presence of a replication strategy that may dampen activation of RIG-I, there is also substantial evidence that arenaviruses encode proteins to antagonize the RIG-I pathway as well. This was first shown for the Old World arenavirus LCMV, which is not a cause of haemorrhagic fever, but which is a very well-studied model arenavirus. The NP of LCMV has been demonstrated to inhibit IFN-α/β production by antagonizing the activation of IRF-3 (Martinez-Sobrido *et al.*, 2006). Cells persistently infected with LCMV were found to be resistant to IFN-α/β production induced by transfection of DNA or by infection with Sendai virus. In this study, NP protein expression prevented IFN-α/β production following Sendai virus infection by preventing the nuclear accumulation of IRF-3 (Martinez-Sobrido *et al.*, 2006; Martinez-Sobrido *et al.*, 2009). A subsequent study profiled this ability in Lassa virus and several New World viruses (Connolly *et al.*, 1993). Each of these NP proteins impaired IRF-3 nuclear accumulation and IFN-α/β production (Martinez-Sobrido *et al.*, 2007). However, the NP of Tacaribe virus, which has only been detected in bats, did not (Martinez-Sobrido *et al.*, 2007). Thus, arenaviruses capable of causing disease in humans, including those that cause haemorrhagic fever in humans, possess NPs with IFN-antagonist function. However, there is not a strict relationship between relative virulence in humans and this function, as viruses such as LCMV, which typically exhibit low virulence, display a similar function. Whether there are differences in the potency of different NPs and the mechanisms by which arenavirus NPs inhibit IFN-α/β production remains to be determined.

Interestingly, the New World arenaviruses Guanarito virus, Junin virus, Machupo virus and Sabia viruses, associated with Venezuelan, Argentine, Bolivian and Sabia-associated haemorrhagic fevers respectively, were found to possess an additional mechanism to counter the RIG-I pathway. Their Z proteins, small ring finger proteins which can also suppress viral RNA synthesis and promote viral budding (Cornu and de la Torre, 2001a,b; Lopez *et al.*, 2001; Perez *et al.*, 2003, 2004; Hass *et al.*, 2004), were found to interact with and inhibit RIG-I, whereas the Z of Lassa virus did not (Fan *et al.*, 2010). The New World Z proteins were also able to inhibit IFN-β production induced by the RIG-I ligand 5′-triphosphate RNA and not only interacted with RIG-I but also disrupted RIG-I–MAVS interaction, resulting in impairment of activation of both IRF-3 and NF-κB (Fan *et al.,* 2010). The relative contribution of the different means by which arenaviruses are proposed to evade RIG-I, via the Z protein, NP or prime and align replication, remains to be worked out.

Flaviviruses causing VHF

Flaviviruses that cause haemorrhagic fever include dengue virus and yellow fever virus as well as the relatively under-studied Omsk and Kyasanur haemorrhagic fever viruses (Borio *et al.,* 2002; Bray, 2005). Flaviviruses are enveloped viruses with single-stranded, non-segmented positive-sense RNA genomes. Flavivirus genomes encode a single polyprotein which is processed by viral and cellular proteases into mature products. The amino-terminal end of the polyprotein gives rise to the viral structural proteins whereas the remaining portion, towards the carboxy-terminal end, encodes the viral non-structural proteins which carry out the enzymatic reactions needed for viral RNA synthesis. Flavivirus genomes possess a 5′ cap, which should prevent RIG-I activation, but lack a polyA tail, and their 5′ ends and 3′ ends consist of relatively long non-translated regions that can form significant secondary structure. Replication involves the copying of the positive-sense genomic RNA into a full-length reverse complementary negative-sense antigenome. The negative-sense RNA serves as a template for additional positive-sense RNAs which can serve as mRNAs or as genomes for newly synthesized viral particles. The replication is biased, such that approximately ten-fold more positive sense RNA is generated than negative sense RNA. Viral RNA synthesis occurs in association with viral 'replication complexes' in association with cellular membranes. The presence of significant RNA secondary structure as well as the possibility of generating dsRNA during replication provides opportunities for flaviviruses to activate antiviral sensors, such as RIG-I (Westaway *et al.,* 1999). However, association of viral RNA synthesis with these membrane-associated complexes may shield flaviviruses from the host antiviral IFN response (Hoenen *et al.,* 2007). Newly assembled virus particles bud intracellularly, probably into the endoplasmic reticulum, and passage through the cellular secretory pathway permits virus release.

Dengue virus

DENV is a mosquito-transmitted pathogen, primarily by *Aedes aegypti,* responsible for a very significant burden of human disease annually. DENV infections are associated with a range of outcomes including inapparent infections, a febrile illness known as dengue fever and a severe, life threatening syndrome, dengue haemorrhagic fever. There are four serotypes of virus, DENV 1–4. While infection with a particular serotype results in protective immunity towards that serotype, individuals remain susceptible to the remaining serotypes. In addition, prior infection with one serotype predisposes individuals to haemorrhagic disease upon infection with a second serotype, suggesting an immune-mediated component to dengue haemorrhagic fever.

The absence of animal models that mimic human infection have hampered the understanding of DENV pathogenesis. However, a number of studies have implicated the IFN-α/β response as important for virus replication and suggest that evasion of IFN responses may be important for DENV disease. For example, based on *in vitro* infections and *in vivo* measurements, IFN-γ and IFN-α/β appear to be relevant to human DENV infection (e.g. Kurane *et al.,* 1986, 1989; Kurane and Ennis, 1987, 1988). Also, microarray analysis of RNA harvested from cells of infected rhesus macaques displayed evidence of a systemic IFN-α/β response (Sariol *et al.,* 2007). Further, DENV is sensitive to the antiviral effects of IFN-α/β in cell culture (Diamond *et al.,* 2000; Diamond and Harris, 2001). In mice, deficiency of both the IFN-α/β and IFN-γ receptors results in susceptibility to lethal infection following intraperitoneal infection. In contrast, wild-type mice or individual IFN-α/β and IFNγ knockout mice did not show signs of illness (Johnson and Roehrig, 1999). These mouse models result in infection of the central nervous system, rather than an illness more reflective of human disease (Johnson and Roehrig, 1999). Therefore, these models do not permit the evaluation of viral functions on VHF. They have been used to evaluate the respective contributions of the IFN-α/β and IFNγ response to clearance of DENV infection. In this system, the IFN-α/β response is important for early control of virus replication and prevention of infection of the brain, while the IFN-γ response appears to be more important later during the course of infection, influencing the clearance of virus (Shresta *et al.*, 2004).

RIG-I has been demonstrated to be important for induction of IFN-α/β production by flaviviruses (Kato *et al.*, 2006; Fredericksen *et al.*, 2008; Loo *et al.*, 2008). In order to explore the importance of RLRs for production of IFN-α/β in response to DENV infection, the role of MAVS (referred to as CARDIF) in DENV infection was examined. MAVS knockout mice were infected with DENV, and the impact of the knockout on virus replication and IFN-α/β production was assessed in a variety of tissues. MAVS deficiency was associated with impaired IFN-α/β responses in the blood and in lymphoid tissues. Absence of MAVS was also associated with relatively higher viral titres compared with wt mice. Therefore, in mice, MAVS is important for IFN-α/β production but evidence for MAVS-independent IFN induction was also seen (Perry *et al.*, 2009).

DENV inhibits IFN-α/β responses in dendritic cells

DENV infection has been shown to inhibit dendritic cell (DC) IFN-α/β responses to infection and the virus is able to impair IFN-α/β signalling. For example, DENV-2 productively infects human monocyte-derived DCs, stimulating release of IL-6 and TNFα and upregulation of cell surface markers of DC activation. However, little to no IFN-α/β was produced by these cells, and while NF-κB activation could be documented in infected cultures, IRF-3 phosphorylation was not evident (Rodriguez-Madoz *et al.*, 2010). These data indicate that DENV-infection inhibits IFN-α/β production by targeting a step at or upstream of IRF-3 activation. As a consequence of this inhibition, DCs were not able to prime a Th1 T-cell response (Rodriguez-Madoz *et al.*, 2010). That DENV actively suppresses IFN-α/β production is suggested by the observation that DENV-infected DCs are impaired in their ability to produce IFN-α/β in response to infection with heterologous RNA viruses including Newcastle disease virus, Sendai virus or Semliki Forest virus, and in response to the TLR3 agonist poly(I:C; Rodriguez-Madoz *et al.*, 2010). The exact mechanism of inhibition remains to be elucidated; however, expression of DENV non-structural proteins identified the NS2B3 viral protease as sufficient to inhibit IFN-α/β production induced by NDV or Sendai virus infection, and an active protease was required for inhibition (Rodriguez-Madoz *et al.*, 2010).

DENV inhibits IFN signalling

Several studies now indicate that DENV can also inhibit IFN-α/β signalling. There may be variability among DENV strains in this capacity, and while this variation was not found to be associated with specific serotypes, it was strain specific (Umareddy *et al.*, 2008). A systematic evaluation of DENV-2 proteins identified NS2A, NS4A and NS4B as inhibitors of the IFN-α/β-induced antiviral response that could impair IFN-induced gene expression (Munoz-Jordan *et al.*, 2003). NS4B also prevented the appearance of phospho-STAT1 in cell nuclei following IFN-β or IFN-γ addition to cells. Further exploration also demonstrated an ability of DENV-2 NS4B to inhibit nuclear accumulation of STAT1 (Munoz-Jordan *et al.*, 2005). NS4B is expressed normally expressed with a signal peptide (2K) at its N-terminus. Although the DENV 2K could be replaced by a heterologous signal peptide, a signal peptide was required for NS4B IFN-antagonist function, suggesting cleavage and/or ER localization important for function. If NS4A was co-expressed with NS4B, enhanced inhibition of IFN signalling was seen. However, if NS4A and NS4B were expressed as a fused polyprotein, as would occur during DV infection, cleavage by the viral protease NS2B/NS3 was required for IFN signalling to be inhibited (Munoz-Jordan *et al.*, 2005).

At least part of the inhibition of IFN signalling may reflect the ability of DENV infection to down regulate STAT2 expression (Jones *et al.*, 2005). Degradation of STAT2 has been demonstrated to be mediated by the viral polymerase protein, NS5 (Ashour *et al.*, 2009). Expression of the non-structural genes of DENV was sufficient to greatly reduce cellular STAT2 levels, and this effect required NS5. NS5 was capable of binding to STAT2 while NS5 down-modulation of STAT2 expression requires the presence of a functioning proteasome, consistent with a model where NS5 promotes ubiquitin-dependent proteasome-mediated degradation of STAT2. Interestingly, NS5-induced STAT2 degradation requires that NS5 be expressed as a precursor protein that undergoes proteolytic processing, as occurs in infected cells (Ashour *et al.*, 2009).

Yellow fever virus

Yellow fever virus is another mosquito-borne virus. It is maintained in a cycle between non-human primates and mosquitoes, although an 'urban' transmission cycle can also occur in which viremic humans sustain the transmission cycle. YFV infections can give rise to a range of clinical presentations, including mild febrile illness to severe, life threatening infections. As suggested by its name, YFV infection can cause jaundice, a feature which distinguishes YFV infections from infections by other haemorrhagic fever viruses (Gubler *et al.*, 2007).

Studies have linked the host IFN response to the ability of YFV to cause disease in mice. Interestingly, whereas YFV infections of mice are typically associated with encephalitic disease, human infections are associated with systemic infection, symptoms of VHF and hepatitis. However, infection of mice lacking a functional IFN response results in manifestations of VHF and liver infection (Meier *et al.*, 2009). This response was seen when infection of wild-type YFV strains Asibi or Angola73 was compared with that of a live-attenuated vaccine strain called 17D-204 in mice with intact or disrupted IFN signalling (Meier *et al.*, 2009). Mice with intact IFN signalling pathways were resistant to disease from either virus. However, the wild-type viral strains caused lethality in mice lacking the IFN-α/β receptor, IFN-γ receptor or STAT1. The IFN-α/β receptor deficient animals had higher titres in several organs, indicating increased viral dissemination, and a more pronounced 'cytokine storm.' The IFN-α/β receptor deficient animals also showed enlarged livers and spleens, and their macrophages showed increased susceptibility to infection (Meier *et al.*, 2009). These data indicate a critical role for IFNs, particularly IFN-α/β, in control of YFV infection and regulation of cell tropism *in vivo*. Because YFV can cause disease in primates but not in wild-type mice, it is possible that the capacity of YFV to suppress IFN responses is superior in primates than mice. Therefore, understanding how YFV evades IFN responses may shed light on the ability of the virus to cause VHF in primate species. In this regard, yellow fever virus, like dengue virus, encodes mechanisms to suppress IFN-α/β signalling. Specifically, as is the case with DENV, expression in mammalian cells of the NS4B protein inhibits IFN-induced gene expression and impairs STAT1 nuclear translocation after IFN-β treatment of cells (Munoz-Jordan *et al.*, 2005). Interestingly, the NS4B of West Nile virus, which causes encephalitis, also suppresses IFN signalling, suggesting that NS4B may function as an IFN-signalling inhibitor for many flaviviruses (Munoz-Jordan *et al.*, 2005; Evans and Seeger, 2007). How these functions contribute to disease remains to be defined.

Conclusions

The recent studies outlined above have revealed novel molecular mechanisms by which haemorrhagic fever viruses induce and evade IFN responses (Table 7.2). Studies on the VP35 protein from EBOV and the NSs protein from RVFV have demonstrated the critical role the innate immune evasion mechanisms play in virulence, at least in rodent models. The connection of the filovirus IFN signalling inhibitors, EBOV VP24 and MARV VP40, with the ability of these viruses to cause disease in a host species-specific manner, suggests a role for these proteins in pathogenesis as well. Given their importance for virulence and the absence of drugs to fight haemorrhagic fever viruses, these IFN-antagonist proteins should be considered as targets for antiviral development. Ultimately, it will be important to extend studies on these factors to animal models that closely mimic human VHF. This goal may be most readily attainable for the filoviruses because of the availability of very well characterized and very reproducible non-human primate models of filovirus VHF as well as systems which permit the genetic manipulation of these viruses (Geisbert and Hensley, 2004; Martinez *et al.*, 2008; Prins *et al.*, 2010; Volchkov *et al.*, 2001). Such studies will define how specific innate immune evasion functions contribute to the common manifestations of this life threatening syndrome and suggest molecular mechanisms underlying viral pathogenesis.

References

Alff, P.J., Sen, N., Gorbunova, E., Gavrilovskaya, I.N., and Mackow, E.R. (2008). The NY-1 hantavirus Gn cytoplasmic tail coprecipitates TRAF3 and inhibits cellular interferon responses by disrupting TBK1–TRAF3 complex formation. J. Virol. *82*, 9115–9122.

Andersson, I., Bladh, L., Mousavi-Jazi, M., Magnusson, K.E., Lundkvist, A., Haller, O., and Mirazimi, A. (2004). Human MxA protein inhibits the replication of Crimean-Congo hemorrhagic fever virus. J. Virol. *78*, 4323–4329.

Andersson, I., Karlberg, H., Mousavi-Jazi, M., Martinez-Sobrido, L., Weber, F., and Mirazimi, A. (2008). Crimean-Congo hemorrhagic fever virus delays activation of the innate immune response. J. Med. Virol. *80*, 1397–1404.

Andersson, I., Lundkvist, A., Haller, O., and Mirazimi, A. (2006). Type I interferon inhibits Crimean-Congo hemorrhagic fever virus in human target cells. J. Med. Virol. *78*, 216–222.

Ashour, J., Laurent-Rolle, M., Shi, P.Y., and Garcia-Sastre, A. (2009). NS5 of dengue virus mediates STAT2 binding and degradation. J. Virol. *83*, 5408–5418.

Asper, M., Sternsdorf, T., Hass, M., Drosten, C., Rhode, A., Schmitz, H., and Gunther, S. (2004). Inhibition of different Lassa virus strains by alpha and gamma interferons and comparison with a less pathogenic arenavirus. J. Virol. *78*, 3162–3169.

Bamberg, S., Kolesnikova, L., Moller, P., Klenk, H.D., and Becker, S. (2005). VP24 of Marburg Virus Influences Formation of Infectious Particles. J. Virol. *79*, 13421–13433.

Basler, C.F., and Amarasinghe, G.K. (2009). Evasion of interferon responses by Ebola and Marburg viruses. J. Interferon Cytokine Res. *29*, 511–520.

Basler, C.F., Mikulasova, A., Martinez-Sobrido, L., Paragas, J., Muhlberger, E., Bray, M., Klenk, H.D., Palese, P., and Garcia-Sastre, A. (2003). The Ebola virus VP35 protein inhibits activation of interferon regulatory factor 3. J. Virol. *77*, 7945–7956.

Basler, C.F., Wang, X., Muhlberger, E., Volchkov, V., Paragas, J., Klenk, H.D., Garcia-Sastre, A., and Palese, P. (2000). The Ebola virus VP35 protein functions as a type I IFN antagonist. Proc. Natl. Acad. Sci. U.S.A. *97*, 12289–12294.

Bergeron, E., Albarino, C.G., Khristova, M.L., and Nichol, S.T. (2010). Crimean-Congo hemorrhagic fever virus-encoded ovarian tumor protease activity is dispensable for virus RNA polymerase function. J. Virol. *84*, 216–226.

Billecocq, A., Spiegel, M., Vialat, P., Kohl, A., Weber, F., Bouloy, M., and Haller, O. (2004). NSs protein of Rift Valley fever virus blocks interferon production by inhibiting host gene transcription. J. Virol. *78*, 9798–9806.

Binder, D., Fehr, J., Hengartner, H., and Zinkernagel, R.M. (1997). Virus-induced transient bone marrow aplasia: major role of interferon-alpha/beta during acute infection with the noncytopathic lymphocytic choriomeningitis virus. J. Exp. Med. *185*, 517–530.

Biron, C.A. (2001). Interferons alpha and beta as immune regulators – a new look. Immunity *14*, 661–664.

Borio, L., Inglesby, T., Peters, C.J., Schmaljohn, A.L., Hughes, J.M., Jahrling, P.B., Ksiazek, T., Johnson, K.M., Meyerhoff, A., O'Toole, T., *et al.* (2002). Hemorrhagic fever viruses as biological weapons: medical and public health management. JAMA *287*, 2391–2405.

Bouloy, M., Janzen, C., Vialat, P., Khun, H., Pavlovic, J., Huerre, M., and Haller, O. (2001). Genetic evidence for an interferon-antagonistic function of rift valley fever virus nonstructural protein NSs. J. Virol. *75*, 1371–1377.

Bray, M. (2001). The role of the Type I interferon response in the resistance of mice to filovirus infection. J. Gen. Virol. *82*, 1365–1373.

Bray, M. (2005). Pathogenesis of viral hemorrhagic fever. Curr. Opin. Immunol. *17*, 399–403.

Bray, M., Davis, K., Geisbert, T., Schmaljohn, C., and Huggins, J. (1998). A mouse model for evaluation of prophylaxis and therapy of Ebola hemorrhagic fever. J. Infect. Dis. *178*, 651–661.

Bray, M., and Geisbert, T.W. (2005). Ebola virus: the role of macrophages and dendritic cells in the pathogenesis of Ebola hemorrhagic fever. Int. J. Biochem. Cell Biol. *37*, 1560–1566.

Bray, M., Raymond, J.L., Geisbert, T., and Baker, R.O. (2002). 3-Deazaneplanocin A induces massively increased interferon-alpha production in Ebola virus-infected mice. Antiviral Res. *55*, 151–159.

Cardenas, W.B., Loo, Y.M., Gale, M., Jr., Hartman, A.L., Kimberlin, C.R., Martinez-Sobrido, L., Saphire, E.O., and Basler, C.F. (2006). Ebola virus VP35 protein binds double-stranded RNA and inhibits alpha/beta interferon production induced by RIG-I signaling. J. Virol. *80*, 5168–5178.

Chang, H.W., Watson, J.C., and Jacobs, B.L. (1992). The E3L gene of vaccinia virus encodes an inhibitor of the interferon-induced, double-stranded RNA-dependent protein kinase. Proc. Natl. Acad. Sci. U.S.A. *89*, 4825–4829.

Connolly, B.M., Jenson, A.B., Peters, C.J., Geyer, S.J., Barth, J.F., and McPherson, R.A. (1993). Pathogenesis of Pichinde virus infection in strain 13 guinea pigs: an immunocytochemical, virologic, and clinical chemistry study. Am. J. Trop. Med. Hyg. *49*, 10–24.

Connolly, B.M., Steele, K.E., Davis, K.J., Geisbert, T.W., Kell, W.M., Jaax, N.K., and Jahrling, P.B. (1999). Pathogenesis of experimental Ebola virus infection in guinea pigs. J. Infect. Dis. *179* Suppl *1*, S203–217.

Cornu, T.I., and de la Torre, J.C. (2001a). RING finger Z protein of lymphocytic choriomeningitis virus (LCMV) inhibits transcription and RNA replication of an LCMV S-segment minigenome. J. Virol. *75*, 9415–9426.

Cornu, T.I., and de la Torre, J.C. (2001b). RING finger Z protein of lymphocytic choriomeningitis virus (LCMV) inhibits transcription and RNA replication of an LCMV S-segment minigenome. J. Virol. *75*, 9415–9426.

Cortes, P., Ye, Z.S., and Baltimore, D. (1994). RAG-1 interacts with the repeated amino acid motif of the human homologue of the yeast protein SRP1. Proc. Natl. Acad. Sci. U.S.A. *91*, 7633–7637.

Cuomo, C.A., Kirch, S.A., Gyuris, J., Brent, R., and Oettinger, M.A. (1994). Rch1, a protein that specifically interacts with the RAG-1 recombination-activating protein. Proc. Natl. Acad. Sci. U.S.A. *91*, 6156–6160.

Diamond, M.S., and Harris, E. (2001). Interferon inhibits dengue virus infection by preventing translation of viral RNA through a PKR-independent mechanism. Virology *289*, 297–311.

Diamond, M.S., Roberts, T.G., Edgil, D., Lu, B., Ernst, J., and Harris, E. (2000). Modulation of Dengue virus infection in human cells by alpha, beta, and gamma interferons. J. Virol. *74*, 4957–4966.

Ebihara, H., Takada, A., Kobasa, D., Jones, S., Neumann, G., Theriault, S., Bray, M., Feldmann, H., and Kawaoka, Y. (2006). Molecular determinants of Ebola virus virulence in mice. PLoS Pathog. *2*, e73.

Elliott, L.H., Kiley, M.P., and McCormick, J.B. (1985). Descriptive analysis of Ebola virus proteins. Virology *147*, 169–176.

Evans, J.D., and Seeger, C. (2007). Differential effects of mutations in NS4B on West Nile virus replication and inhibition of interferon signaling. J. Virol. *81*, 11809–11816.

Fan, L., Briese, T., and Lipkin, W.I. (2010). Z proteins of New World arenaviruses bind RIG-I and interfere with type I interferon induction. J. Virol. *84*, 1785–1791.

Feldmann, H., Volchkov, V.E., Volchkova, V.A., Stroher, U., and Klenk, H.D. (2001). Biosynthesis and role of filoviral glycoproteins. J. Gen. Virol. *82*, 2839–2848.

Feng, Z., Cerveny, M., Yan, Z., and He, B. (2007). The VP35 protein of Ebola virus inhibits the antiviral effect mediated by double-stranded RNA-dependent protein kinase PKR. J. Virol. *81*, 182–192.

Fischer, S.A., Graham, M.B., Kuehnert, M.J., Kotton, C.N., Srinivasan, A., Marty, F.M., Comer, J.A., Guarner, J., Paddock, C.D., DeMeo, D.L., *et al.* (2006). Transmission of lymphocytic choriomeningitis virus by organ transplantation. N. Engl. J. Med. *354*, 2235–2249.

Fitzgerald, K.A., McWhirter, S.M., Faia, K.L., Rowe, D.C., Latz, E., Golenbock, D.T., Coyle, A.J., Liao, S.M., and Maniatis, T. (2003). IKKepsilon and TBK1 are essential components of the IRF3 signaling pathway. Nat. Immunol. *4*, 491–496.

Fredericksen, B.L., Keller, B.C., Fornek, J., Katze, M.G., and Gale, M., Jr. (2008). Establishment and maintenance of the innate antiviral response to West Nile Virus involves both RIG-I and MDA5 signaling through IPS-1. J. Virol. *82*, 609–616.

Frese, M., Kochs, G., Feldmann, H., Hertkorn, C., and Haller, O. (1996). Inhibition of bunyaviruses, phleboviruses, and hantaviruses by human MxA protein. J. Virol. *70*, 915–923.

Frias-Staheli, N., Giannakopoulos, N.V., Kikkert, M., Taylor, S.L., Bridgen, A., Paragas, J., Richt, J.A., Rowland, R.R., Schmaljohn, C.S., Lenschow, D.J., *et al.* (2007). Ovarian tumor domain-containing viral proteases evade ubiquitin- and ISG15-dependent innate immune responses. Cell Host Microbe *2*, 404–416.

Friedman, C.S., O'Donnell, M.A., Legarda-Addison, D., Ng, A., Cardenas, W.B., Yount, J.S., Moran, T.M., Basler, C.F., Komuro, A., Horvath, C.M., *et al.* (2008). The tumour suppressor CYLD is a negative regulator of RIG-I-mediated antiviral response. EMBO Rep. *9*, 930–936.

Gack, M.U., Kirchhofer, A., Shin, Y.C., Inn, K.S., Liang, C., Cui, S., Myong, S., Ha, T., Hopfner, K.P., and Jung, J.U. (2008). Roles of RIG-I N-terminal tandem CARD and splice variant in TRIM25-mediated antiviral signal transduction. Proc. Natl. Acad. Sci. U.S.A. *105*, 16743–16748.

Gack, M.U., Shin, Y.C., Joo, C.H., Urano, T., Liang, C., Sun, L., Takeuchi, O., Akira, S., Chen, Z., Inoue, S., *et al.* (2007). TRIM25 RING-finger E3 ubiquitin ligase is essential for RIG-I-mediated antiviral activity. Nature *446*, 916–920.

Gale, M., Jr. and Katze, M.G. (1998). Molecular mechanisms of interferon resistance mediated by viral-directed inhibition of PKR, the interferon-induced protein kinase. Pharmacol. Ther. *78*, 29–46.

Garcin, D., and Kolakofsky, D. (1992). Tacaribe arenavirus RNA synthesis *in vitro* is primer dependent and suggests an unusual model for the initiation of genome replication. J. Virol. *66*, 1370–1376.

Garcin, D., Lezzi, M., Dobbs, M., Elliott, R.M., Schmaljohn, C., Kang, C.Y., and Kolakofsky, D. (1995). The 5′ ends of Hantaan virus (Bunyaviridae) RNAs suggest a prime-and-realign mechanism for the initiation of RNA synthesis. J. Virol. *69*, 5754–5762.

Geimonen, E., Neff, S., Raymond, T., Kocer, S.S., Gavrilovskaya, I.N., and Mackow, E.R. (2002). Pathogenic and nonpathogenic hantaviruses differentially regulate endothelial cell responses. Proc. Natl. Acad. Sci. U.S.A. *99*, 13837–13842.

Geisbert, T.W., and Hensley, L.E. (2004). Ebola virus: new insights into disease aetiopathology and possible therapeutic interventions. Expert Rev. Mol. Med. *6*, 1–24.

Geisbert, T.W., Hensley, L.E., Larsen, T., Young, H.A., Reed, D.S., Geisbert, J.B., Scott, D.P., Kagan, E., Jahrling, P.B., and Davis, K. J. (2003). Pathogenesis of ebola hemorrhagic fever in cynomolgus macaques: evidence that dendritic cells are early and sustained targets of infection. Am. J. Pathol. *163*, 2347–2370.

Giannakopoulos, N.V., Arutyunova, E., Lai, C., Lenschow, D.J., Haas, A.L., and Virgin, H.W. (2009). ISG15 Arg151 and the ISG15-conjugating enzyme UbE1L are important for innate immune control of Sindbis virus. J. Virol. *83*, 1602–1610.

Gorlich, D., Henklein, P., Laskey, R.A., and Hartmann, E. (1996). A 41 amino acid motif in importin-alpha confers binding to importin-beta and hence transit into the nucleus. EMBO J. *15*, 1810–1817.

Gubler, D.J., Kuno, G., and Markoff, L. (2007). Flaviviruses. In Fields Virology, Knipe, D.M., Howley, P.M., *et al.*, eds. (Philadelphia, Lippincott Williams and Wilkins).

Gupta, M., Mahanty, S., Ahmed, R., and Rollin, P.E. (2001). Monocyte-derived human macrophages and peripheral blood mononuclear cells infected with ebola virus secrete MIP-1alpha and TNF-alpha and inhibit poly-IC-induced IFN-alpha *in vitro*. Virology *284*, 20–25.

Guschin, D., Rogers, N., Briscoe, J., Witthuhn, B., Watling, D., Horn, F., Pellegrini, S., Yasukawa, K., Heinrich, P., Stark, G.R., *et al.* (1995). A major role for the protein tyrosine kinase JAK1 in the JAK/STAT signal transduction pathway in response to interleukin-6. EMBO J. *14*, 1421–1429.

Habjan, M., Andersson, I., Klingstrom, J., Schumann, M., Martin, A., Zimmermann, P., Wagner, V., Pichlmair, A., Schneider, U., Muhlberger, E., *et al.* (2008). Processing of genome 5′ termini as a strategy of negative-strand RNA viruses to avoid RIG-I-dependent interferon induction. PLoS ONE 3, e2032.

Habjan, M., Pichlmair, A., Elliott, R.M., Overby, A.K., Glatter, T., Gstaiger, M., Superti-Furga, G., Unger, H., and Weber, F. (2009). NSs protein of rift valley fever virus induces the specific degradation of the double-stranded RNA-dependent protein kinase. J. Virol. *83*, 4365–4375.

Han, Z., Boshra, H., Sunyer, J.O., Zwiers, S.H., Paragas, J., and Harty, R.N. (2003). Biochemical and functional characterization of the Ebola virus VP24 protein: implications for a role in virus assembly and budding. J. Virol. *77*, 1793–1800.

Han, Z., Liu, Q.H., Paragas, J., Bray, M., Freedman, B.D., and Harty, R.N. (2002). The VP24 protein of Ebola (Zaire) virus possesses proton channel activity in mammalian cells (P26–1). Paper presented at: American Society for Virology, 21st Annual Meeting (Lexington, KY).

Harcourt, B.H., Sanchez, A., and Offermann, M.K. (1999). Ebola virus selectively inhibits responses to interferons, but not to interleukin-1beta, in endothelial cells. J. Virol. *73*, 3491–3496.

Hartman, A.L., Bird, B.H., Towner, J.S., Antoniadou, Z.A., Zaki, S.R., and Nichol, S.T. (2008a). Inhibition of IRF-3 activation by VP35 is critical for the high level of virulence of ebola virus. J. Virol. *82*, 2699–2704.

Hartman, A.L., Dover, J.E., Towner, J.S., and Nichol, S.T. (2006). Reverse genetic generation of recombinant Zaire Ebola viruses containing disrupted IRF-3 inhibitory domains results in attenuated virus growth *in vitro* and higher levels of IRF-3 activation without inhibiting viral transcription or replication. J. Virol. *80*, 6430–6440.

Hartman, A.L., Ling, L., Nichol, S.T., and Hibberd, M.L. (2008b). Whole genome expression profiling reveals that inhibition of host innate immune response pathways by Ebola virus can be reversed by a single amino acid change in the VP35 protein. J Virol. *82*, 5348–5358.

Hartman, A.L., Towner, J.S., and Nichol, S.T. (2004). A C-terminal basic amino acid motif of Zaire ebolavirus VP35 is essential for type I interferon antagonism and displays high identity with the RNA-binding domain of another interferon antagonist, the NS1 protein of influenza A virus. Virology *328*, 177–184.

Harty, R.N., Brown, M.E., Wang, G., Huibregtse, J., and Hayes, F.P. (2000). A PPxY motif within the VP40 protein of Ebola virus interacts physically and functionally with a ubiquitin ligase: implications for filovirus budding. Proc. Natl. Acad. Sci. U.S.A. *97*, 13871–13876.

Harty, R.N., Pitha, P.M., and Okumura, A. (2009). Antiviral activity of innate immune protein ISG15. J. Innate Immun. *1*, 397–404.

Hass, M., Golnitz, U., Muller, S., Becker-Ziaja, B., and Gunther, S. (2004). Replicon system for Lassa virus. J. Virol. *78*, 13793–13803.

He, B., Gross, M., and Roizman, B. (1997). The gamma(1)34.5 protein of herpes simplex virus 1 complexes with protein phosphatase 1alpha to dephosphorylate the alpha subunit of the eukaryotic translation initiation factor 2 and preclude the shutoff of protein synthesis by double-stranded RNA-activated protein kinase. Proc. Natl. Acad. Sci. U.S.A. *94*, 843–848.

Hensley, L.E., Young, H.A., Jahrling, P.B., and Geisbert, T.W. (2002). Proinflammatory response during Ebola virus infection of primate models: possible involvement of the tumor necrosis factor receptor superfamily. Immunol. Lett. *80*, 169–179.

Hiscott, J. (2007). Triggering the innate antiviral response through IRF-3 activation. J. Biol. Chem. *282*, 15325–15329.

Hiscott, J., Lacoste, J., and Lin, R. (2006). Recruitment of an interferon molecular signaling complex to the mitochondrial membrane: disruption by hepatitis C virus NS3-4A protease. Biochem. Pharmacol. *72*, 1477–1484.

Hoebe, K., and Beutler, B. (2004). LPS, dsRNA and the interferon bridge to adaptive immune responses: Trif, Tram, and other TIR adaptor proteins. J. Endotoxin Res. *10*, 130–136.

Hoenen, A., Liu, W., Kochs, G., Khromykh, A.A., and Mackenzie, J.M. (2007). West Nile virus-induced cytoplasmic membrane structures provide partial protection against the interferon-induced antiviral MxA protein. J. Gen. Virol. *88*, 3013–3017.

Honig, J.E., Osborne, J.C., and Nichol, S.T. (2004). Crimean-Congo hemorrhagic fever virus genome L RNA segment and encoded protein. Virology *321*, 29–35.

Hornung, V., Ellegast, J., Kim, S., Brzozka, K., Jung, A., Kato, H., Poeck, H., Akira, S., Conzelmann, K.K., Schlee, M., *et al.* (2006). 5′-Triphosphate RNA is the ligand for RIG-I. Science *314*, 994–997.

Huang, Y., Xu, L., Sun, Y., and Nabel, G. (2002). The assembly of ebola virus nucleocapsid requires virion-associated proteins 35 and 24 and posttranslational modification of nucleoprotein. Mol. Cell *10*, 307.

Hutchinson, K.L., and Rollin, P.E. (2007). Cytokine and chemokine expression in humans infected with Sudan Ebola virus. J. Infect. Dis. *196* Suppl. 2, S357–363.

Ikegami, T., Narayanan, K., Won, S., Kamitani, W., Peters, C.J., and Makino, S. (2009). Rift Valley fever virus NSs protein promotes post-transcriptional downregulation of protein kinase PKR and inhibits eIF2alpha phosphorylation. PLoS Pathog. *5*, e1000287.

Ikegami, T., Won, S., Peters, C.J., and Makino, S. (2006). Rescue of infectious rift valley fever virus entirely from cDNA, analysis of virus lacking the NSs gene, and expression of a foreign gene. J. Virol. *80*, 2933–2940.

Jahrling, P.B., Geisbert, T.W., Geisbert, J.B., Swearengen, J.R., Bray, M., Jaax, N.K., Huggins, J.W., LeDuc, J.W., and Peters, C.J. (1999). Evaluation of immune globulin and recombinant interferon-alpha2b for treatment of experimental Ebola virus infections. J. Infect. Dis. *179* Suppl *1*, S224–234.

Jiang, H., Wang, P.Z., Zhang, Y., Xu, Z., Sun, L., Wang, L.M., Huang, C.X., Lian, J.Q., Jia, Z.S., Li, Z.D., *et al.* (2008). Hantaan virus induces toll-like receptor 4 expression, leading to enhanced production of beta interferon, interleukin-6 and tumor necrosis factor-alpha. Virology *380*, 52–59.

Johnson, A.J., and Roehrig, J.T. (1999). New mouse model for dengue virus vaccine testing. J. Virol. 73, 783–786.

Jones, M., Davidson, A., Hibbert, L., Gruenwald, P., Schlaak, J., Ball, S., Foster, G.R., and Jacobs, M. (2005). Dengue virus inhibits alpha interferon signaling by reducing STAT2 expression. J. Virol. *79*, 5414–5420.

Kaletsky, R.L., Francica, J.R., Agrawal-Gamse, C., and Bates, P. (2009). Tetherin-mediated restriction of filovirus budding is antagonized by the Ebola glycoprotein. Proc. Natl. Acad. Sci. U.S.A. *106*, 2886–2891.

Kanerva, M., Melen, K., Vaheri, A., and Julkunen, I. (1996). Inhibition of puumala and tula hantaviruses in Vero cells by MxA protein. Virology *224*, 55–62.

Kash, J.C., Muhlberger, E., Carter, V., Grosch, M., Perwitasari, O., Proll, S.C., Thomas, M.J., Weber, F., Klenk, H.D., and Katze, M.G. (2006). Global suppression of the host antiviral response by Ebola- and Marburgviruses: increased antagonism of the type I interferon response is associated with enhanced virulence. J. Virol. *80*, 3009–3020.

Kato, H., Takeuchi, O., Sato, S., Yoneyama, M., Yamamoto, M., Matsui, K., Uematsu, S., Jung, A., Kawai, T., Ishii, K.J., *et al.* (2006). Differential roles of MDA5 and RIG-I helicases in the recognition of RNA viruses. Nature *441*, 101–105.

Kawai, T., Takahashi, K., Sato, S., Coban, C., Kumar, H., Kato, H., Ishii, K.J., Takeuchi, O., and Akira, S. (2005). IPS-1, an adaptor triggering RIG-I- and Mda5-mediated type I interferon induction. Nat. Immunol. *6*, 981–988.

Kimberlin, C.R., Bornholdt, Z.A., Li, S., Woods, V.L., Jr., MacRae, I.J., and Saphire, E.O. Ebolavirus VP35 uses a bimodal strategy to bind dsRNA for innate immune suppression. Proc. Natl. Acad. Sci. U.S.A. *107*, 314–319.

Kindzelskii, A.L., Yang, Z., Nabel, G.J., Todd, R.F., 3rd, and Petty, H.R. (2000). Ebola virus secretory glycoprotein (sGP) diminishes Fc gamma RIIIB-to-CR3 proximity on neutrophils. J. Immunol. *164*, 953–958.

Kitajewski, J., Schneider, R.J., Safer, B., Munemitsu, S.M., Samuel, C.E., Thimmappaya, B., and Shenk, T. (1986). Adenovirus VAI RNA antagonizes the antiviral action of interferon by preventing activation of the interferon-induced eIF-2 alpha kinase. Cell *45*, 195–200.

Kohl, A., di Bartolo, V., and Bouloy, M. (1999). The Rift Valley fever virus nonstructural protein NSs is phosphorylated at serine residues located in casein kinase II consensus motifs in the carboxy-terminus. Virology *263*, 517–525.

Kohler, M., Ansieau, S., Prehn, S., Leutz, A., Haller, H., and Hartmann, E. (1997). Cloning of two novel human importin-alpha subunits and analysis of the expression pattern of the importin-alpha protein family. FEBS Lett. *417*, 104–108.

Kohler, M., Speck, C., Christiansen, M., Bischoff, F.R., Prehn, S., Haller, H., Gorlich, D., and Hartmann, E. (1999). Evidence for distinct substrate specificities of importin alpha family members in nuclear protein import. Mol. Cell. Biol. *19*, 7782–7791.

Kraus, A.A., Raftery, M.J., Giese, T., Ulrich, R., Zawatzky, R., Hippenstiel, S., Suttorp, N., Kruger, D.H., and Schonrich, G. (2004). Differential antiviral response of endothelial cells after infection with pathogenic and nonpathogenic hantaviruses. J. Virol. *78*, 6143–6150.

Kunzi, M.S., and Pitha, P.M. (1996). Role of interferon-stimulated gene ISG-15 in the interferon-omega-mediated inhibition of human immunodeficiency virus replication. J. Interferon Cytokine Res. *16*, 919–927.

Kurane, I., and Ennis, F.A. (1987). Induction of interferon alpha from human lymphocytes by autologous, dengue virus-infected monocytes. J. Exp. Med. *166*, 999–1010.

Kurane, I., and Ennis, F.A. (1988). Production of interferon alpha by dengue virus-infected human monocytes. J. Gen. Virol. *69*, 445–449.

Kurane, I., Innis, B.L., Nisalak, A., Hoke, C., Nimmannitya, S., Meager, A., and Ennis, F.A. (1989). Human T cell responses to dengue virus antigens. Proliferative responses and interferon gamma production. J. Clin. Invest. *83*, 506–513.

Kurane, I., Meager, A., and Ennis, F.A. (1986). Induction of interferon alpha and gamma from human lymphocytes by dengue virus-infected cells. J. Gen. Virol. *67 (Pt 8)*, 1653–1661.

Le May, N., Dubaele, S., Proietti De Santis, L., Billecocq, A., Bouloy, M., and Egly, J.M. (2004). TFIIH transcription factor, a target for the Rift Valley hemorrhagic fever virus. Cell *116*, 541–550.

Le May, N., Mansuroglu, Z., Leger, P., Josse, T., Blot, G., Billecocq, A., Flick, R., Jacob, Y., Bonnefoy, E., and Bouloy, M. (2008). A SAP30 complex inhibits IFN-beta expression in Rift Valley fever virus infected cells. PLoS Pathog. *4*, e13.

Lenschow, D.J., Giannakopoulos, N.V., Gunn, L.J., Johnston, C., O'Guin, A.K., Schmidt, R.E., Levine,

B., and Virgin, H.W.t. (2005). Identification of interferon-stimulated gene 15 as an antiviral molecule during Sindbis virus infection *in vivo*. J. Virol. *79*, 13974–13983.

Lenschow, D.J., Lai, C., Frias-Staheli, N., Giannakopoulos, N.V., Lutz, A., Wolff, T., Osiak, A., Levine, B., Schmidt, R.E., Garcia-Sastre, A., *et al.* (2007). IFN-stimulated gene 15 functions as a critical antiviral molecule against influenza, herpes, and Sindbis viruses. Proc. Natl. Acad. Sci. U.S.A. *104*, 1371–1376.

Leroy, E.M., Kumulungui, B., Pourrut, X., Rouquet, P., Hassanin, A., Yaba, P., Delicat, A., Paweska, J.T., Gonzalez, J.P., and Swanepoel, R. (2005). Fruit bats as reservoirs of Ebola virus. Nature *438*, 575–576.

Leung, D.W., Ginder, N.D., Fulton, D.B., Nix, J., Basler, C.F., Honzatko, R.B., and Amarasinghe, G.K. (2009). Structure of the Ebola VP35 interferon inhibitory domain. Proc. Natl. Acad. Sci. U.S.A. *106*, 411–416.

Leung, D.W., Prins, K.C., Borek, D.M., Farahbakhsh, M., Tufariello, J.M., Ramanan, P., Nix, J.C., Helgeson, L.A., Otwinowski, Z., Honzatko, R.B., *et al.* (2010a). Structural basis for dsRNA recognition and interferon antagonism by Ebola VP35. Nat Struct Mol Biol *17*, 165–172.

Leung, D.W., Shabman, R.S., Farahbakhsh, M., Prins, K.C., Borek, D.M., Wang, T., Muhlberger, E., Basler, C.F., and Amarasinghe, G.K. (2010b). Structural and functional characterization of Reston Ebola VP35 interferon inhibitory domain. J Mol Biol. *399*, 347–357.

Levis, S.C., Saavedra, M.C., Ceccoli, C., Falcoff, E., Feuillade, M.R., Enria, D.A., Maiztegui, J.I., and Falcoff, R. (1984). Endogenous interferon in Argentine hemorrhagic fever. J. Infect. Dis. *149*, 428–433.

Levis, S.C., Saavedra, M.C., Ceccoli, C., Feuillade, M.R., Enria, D.A., Maiztegui, J.I., and Falcoff, R. (1985). Correlation between endogenous interferon and the clinical evolution of patients with Argentine hemorrhagic fever. J. Interferon Res. *5*, 383–389.

Licata, J.M., Johnson, R.F., Han, Z., and Harty, R.N. (2004). Contribution of ebola virus glycoprotein, nucleoprotein, and VP24 to budding of VP40 virus-like particles. J. Virol. *78*, 7344–7351.

Loo, Y.M., Fornek, J., Crochet, N., Bajwa, G., Perwitasari, O., Martinez-Sobrido, L., Akira, S., Gill, M.A., Garcia-Sastre, A., Katze, M.G., *et al.* (2008). Distinct RIG-I and MDA5 signaling by RNA viruses in innate immunity. J. Virol. *82*, 335–345.

Lopez, L.A., Yang, S.J., Hauser, H., Exline, C.M., Haworth, K.G., Oldenburg, J., and Cannon, P.M. Ebola virus glycoprotein counteracts BST-2/Tetherin restriction in a sequence-independent manner that does not require tetherin surface removal. J. Virol. *84*, 7243–7255.

Lopez, N., Jacamo, R., and Franze-Fernandez, M.T. (2001). Transcription and RNA replication of tacaribe virus genome and antigenome analogs require N and L proteins: Z protein is an inhibitor of these processes. J. Virol. *75*, 12241–12251.

Lu, C., Xu, H., Ranjith-Kumar, C.T., Brooks, M.T., Hou, T.Y., Hu, F., Herr, A.B., Strong, R.K., Kao, C.C., and Li, P. The Structural Basis of 5′ Triphosphate double-stranded RNA recognition by RIG-I C-Terminal Domain. Structure *18*, 1032–1043.

Lu, G., Reinert, J.T., Pitha-Rowe, I., Okumura, A., Kellum, M., Knobeloch, K.P., Hassel, B., and Pitha, P.M. (2006). ISG15 enhances the innate antiviral response by inhibition of IRF-3 degradation. Cell Mol. Biol. *52*, 29–41.

Lu, Y., Wambach, M., Katze, M.G., and Krug, R.M. (1995). Binding of the influenza virus NS1 protein to double-stranded RNA inhibits the activation of the protein kinase that phosphorylates the elF-2 translation initiation factor. Virology *214*, 222–228.

Lucia, H.L., Coppenhaver, D.H., and Baron, S. (1989). Arenavirus infection in the guinea pig model: antiviral therapy with recombinant interferon-alpha, the immunomodulator CL246,738 and ribavirin. Antiviral Res. *12*, 279–292.

Mahanty, S., and Bray, M. (2004). Pathogenesis of filoviral haemorrhagic fevers. Lancet Infect. Dis. *4*, 487–498.

Mahanty, S., Gupta, M., Paragas, J., Bray, M., Ahmed, R., and Rollin, P.E. (2003). Protection from lethal infection is determined by innate immune responses in a mouse model of Ebola virus infection. Virology *312*, 415–424.

Marq, J.B., Kolakofsky, D., and Garcin, D. (2010). Unpaired 5′ ppp-nucleotides, as found in arenavirus double-stranded RNA panhandles, are not recognized by RIG-I. J. Biol. Chem. *285*, 18208–18216.

Martinez-Sobrido, L., Emonet, S., Giannakas, P., Cubitt, B., Garcia-Sastre, A., and de la Torre, J.C. (2009). Identification of amino acid residues critical for the anti-interferon activity of the nucleoprotein of the prototypic arenavirus lymphocytic choriomeningitis virus. J. Virol. *83*, 11330–11340.

Martinez-Sobrido, L., Giannakas, P., Cubitt, B., Garcia-Sastre, A., and de la Torre, J.C. (2007). Differential inhibition of type I interferon induction by arenavirus nucleoproteins. J. Virol. *81*, 12696–12703.

Martinez-Sobrido, L., Zuniga, E.I., Rosario, D., Garcia-Sastre, A., and de la Torre, J.C. (2006). Inhibition of the type I interferon response by the nucleoprotein of the prototypic arenavirus lymphocytic choriomeningitis virus. J. Virol. *80*, 9192–9199.

Martinez, M.J., Biedenkopf, N., Volchkova, V., Hartlieb, B., Alazard-Dany, N., Reynard, O., Becker, S., and Volchkov, V. (2008). Role of Ebola virus VP30 in transcription reinitiation. J. Virol. *82*, 12569–12573.

Mateo, M., Reid, S.P., Leung, L.W., Basler, C.F., and Volchkov, V.E. Ebolavirus VP24 binding to karyopherins is required for inhibition of interferon signaling. J. Virol. *84*, 1169–1175.

McAllister, C.S., and Samuel, C.E. (2009). The RNA-activated protein kinase enhances the induction of interferon-beta and apoptosis mediated by cytoplasmic RNA sensors. J. Biol. Chem. *284*, 1644–1651.

McBride, K.M., Banninger, G., McDonald, C., and Reich, N.C. (2002). Regulated nuclear import of the STAT1 transcription factor by direct binding of importin-alpha. EMBO J. *21*, 1754–1763.

Meier, K.C., Gardner, C.L., Khoretonenko, M.V., Klimstra, W.B., and Ryman, K.D. (2009). A mouse model for studying viscerotropic disease caused by yellow fever virus infection. PLoS Pathog. *5*, e1000614.

Melen, K., Fagerlund, R., Franke, J., Kohler, M., Kinnunen, L., and Julkunen, I. (2003). Importin alpha nuclear localization signal binding sites for STAT1, STAT2, and influenza A virus nucleoprotein. J. Biol. Chem. *278*, 28193–28200.

Meylan, E., Curran, J., Hofmann, K., Moradpour, D., Binder, M., Bartenschlager, R., and Tschopp, J. (2005). Cardif is an adaptor protein in the RIG-I antiviral pathway and is targeted by hepatitis C virus. Nature *437*, 1167–1172.

Miyamoto, Y., Imamoto, N., Sekimoto, T., Tachibana, T., Seki, T., Tada, S., Enomoto, T., and Yoneda, Y. (1997). Differential modes of nuclear localization signal (NLS) recognition by three distinct classes of NLS receptors. J. Biol. Chem. *272*, 26375–26381.

Moller, P., Pariente, N., Klenk, H.D., and Becker, S. (2005). Homo-oligomerization of Marburgvirus VP35 is essential for its function in replication and transcription. J. Virol. *79*, 14876–14886.

Moroianu, J., Blobel, G., and Radu, A. (1995). Previously identified protein of uncertain function is karyopherin alpha and together with karyopherin beta docks import substrate at nuclear pore complexes. Proc. Natl. Acad. Sci. U.S.A. *92*, 2008–2011.

Morrill, J.C., Jennings, G.B., Cosgriff, T.M., Gibbs, P.H., and Peters, C.J. (1989). Prevention of Rift Valley fever in rhesus monkeys with interferon-alpha. Rev Infect Dis *11* Suppl. *4*, S815–825.

Morrill, J.C., Jennings, G.B., Johnson, A.J., Cosgriff, T.M., Gibbs, P.H., and Peters, C.J. (1990). Pathogenesis of Rift Valley fever in rhesus monkeys: role of interferon response. Arch. Virol. *110*, 195–212.

Muhlberger, E., Lotfering, B., Klenk, H.D., and Becker, S. (1998). Three of the four nucleocapsid proteins of Marburg virus, NP, VP35, and L, are sufficient to mediate replication and transcription of Marburg virus-specific monocistronic minigenomes. J. Virol. *72*, 8756–8764.

Muhlberger, E., Weik, M., Volchkov, V.E., Klenk, H.D., and Becker, S. (1999). Comparison of the transcription and replication strategies of marburg virus and Ebola virus by using artificial replication systems. J. Virol. 73, 2333–2342.

Muller, M., Briscoe, J., Laxton, C., Guschin, D., Ziemiecki, A., Silvennoinen, O., Harpur, A.G., Barbieri, G., Witthuhn, B.A., Schindler, C., *et al.* (1993). The protein tyrosine kinase JAK1 complements defects in interferon-alpha/beta and -gamma signal transduction. Nature *366*, 129–135.

Munoz-Jordan, J.L., Laurent-Rolle, M., Ashour, J., Martinez-Sobrido, L., Ashok, M., Lipkin, W.I., and Garcia-Sastre, A. (2005). Inhibition of alpha/beta interferon signaling by the NS4B protein of flaviviruses. J. Virol. *79*, 8004–8013.

Munoz-Jordan, J.L., Sanchez-Burgos, G.G., Laurent-Rolle, M., and Garcia-Sastre, A. (2003). Inhibition of interferon signaling by dengue virus. Proc. Natl. Acad. Sci. U.S.A. *100*, 14333–14338.

Nachury, M.V., Ryder, U.W., Lamond, A.I., and Weis, K. (1998). Cloning and characterization of hSRP1 gamma, a tissue-specific nuclear transport factor. Proc. Natl. Acad. Sci. U.S.A. *95*, 582–587.

Nam, J.H., Hwang, K.A., Yu, C.H., Kang, T.H., Shin, J.Y., Choi, W.Y., Kim, I.B., Joo, Y.R., Cho, H.W., and Park, K.Y. (2003). Expression of interferon inducible genes following Hantaan virus infection as a mechanism of resistance in A549 cells. Virus Genes *26*, 31–38.

Noda, T., Halfmann, P., Sagara, H., and Kawaoka, Y. (2007). Regions in Ebola virus VP24 that are important for nucleocapsid formation. J. Infect. Dis. *196* Suppl. *2*, S247–250.

Noda, T., Sagara, H., Suzuki, E., Takada, A., Kida, H., and Kawaoka, Y. (2002). Ebola virus VP40 drives the formation of virus-like filamentous particles along with GP. J. Virol. *76*, 4855–4865.

O'Neill, R.E., and Palese, P. (1995). NPI-1, the human homolog of SRP-1, interacts with influenza virus nucleoprotein. Virology *206*, 116–125.

Oelschlegel, R., Kruger, D.H., and Rang, A. (2007). MxA-independent inhibition of Hantaan virus replication induced by type I and type II interferon *in vitro*. Virus Res *127*, 100–105.

Oganesyan, G., Saha, S.K., Guo, B., He, J.Q., Shahangian, A., Zarnegar, B., Perry, A., and Cheng, G. (2006). Critical role of TRAF3 in the Toll-like receptor-dependent and -independent antiviral response. Nature *439*, 208–211.

Okumura, A., Lu, G., Pitha-Rowe, I., and Pitha, P.M. (2006). Innate antiviral response targets HIV-1 release by the induction of ubiquitin-like protein ISG15. Proc. Natl. Acad. Sci. U.S.A. *103*, 1440–1445.

Okumura, A., Pitha, P.M., and Harty, R.N. (2008). ISG15 inhibits Ebola VP40 VLP budding in an L-domain-dependent manner by blocking Nedd4 ligase activity. Proc. Natl. Acad. Sci. U.S.A. *105*, 3974–3979.

Perez, M., Craven, R.C., and de la Torre, J.C. (2003). The small RING finger protein Z drives arenavirus budding: implications for antiviral strategies. Proc. Natl. Acad. Sci. U.S.A. *100*, 12978–12983.

Perez, M., Greenwald, D.L., and de la Torre, J.C. (2004). Myristoylation of the RING finger Z protein is essential for arenavirus budding. J. Virol. *78*, 11443–11448.

Perry, S.T., Prestwood, T.R., Lada, S.M., Benedict, C.A., and Shresta, S. (2009). Cardif-mediated signaling controls the initial innate response to dengue virus *in vivo*. J. Virol. *83*, 8276–8281.

Peters, C.J., Liu, C.T., Anderson, G.W., Jr., Morrill, J.C., and Jahrling, P.B. (1989). Pathogenesis of viral hemorrhagic fevers: Rift Valley fever and Lassa fever contrasted. Rev Infect Dis *11* Suppl 4, S743–749.

Peters, C.J., Reynolds, J.A., Slone, T.W., Jones, D.E., and Stephen, E.L. (1986). Prophylaxis of Rift Valley fever with antiviral drugs, immune serum, an interferon inducer, and a macrophage activator. Antiviral Res. *6*, 285–297.

Pichlmair, A., Schulz, O., Tan, C.P., Naslund, T.I., Liljestrom, P., Weber, F., and Reis e Sousa, C. (2006). RIG-I-mediated antiviral responses to single-stranded RNA bearing 5′-phosphates. Science *314*, 997–1001.

Platanias, L.C. (2005). Mechanisms of type-I- and type-II-interferon-mediated signalling. Nat. Rev. Immunol. *5*, 375–386.

Pourrut, X., Delicat, A., Rollin, P.E., Ksiazek, T.G., Gonzalez, J.P., and Leroy, E.M. (2007). Spatial and temporal patterns of Zaire ebolavirus antibody prevalence in the possible reservoir bat species. J. Infect. Dis. *196* Suppl. 2, S176–183.

Pourrut, X., Souris, M., Towner, J.S., Rollin, P.E., Nichol, S.T., Gonzalez, J.P., and Leroy, E. (2009). Large serological survey showing cocirculation of Ebola and Marburg viruses in Gabonese bat populations, and a high seroprevalence of both viruses in Rousettus aegyptiacus. BMC Infect Dis *9*, 159.

Pozner, R.G., Ure, A.E., Jaquenod de Giusti, C., D'Atri, L.P., Italiano, J.E., Torres, O., Romanowski, V., Schattner, M., and Gomez, R.M. (2010). Junin virus infection of human hematopoietic progenitors impairs *in vitro* proplatelet formation and platelet release via a bystander effect involving type I IFN signaling. PLoS Pathog. *6*, e1000847.

Prins, K.C., Cardenas, W.B., and Basler, C.F. (2009a). Ebola virus protein VP35 impairs the function of interferon regulatory factor-activating kinases IKKepsilon and TBK-1. J. Virol. *83*, 3069–3077.

Prins, K.C., Cardenas, W.B., and Basler, C.F. (2009b). The Ebola virus protein VP35, impairs the function of the interferon regulatory factor-activating kinases, IKKε and TBK-1, 2009. J. Virol. *83*, 3069–3077.

Prins, K.C., Delpeut, S., Leung, D.W., Reynard, O., Volchkova, V.A., Reid, S.P., Ramanan, P., Cardenas, W.B., Amarasinghe, G.K., Volchkov, V.E., *et al.* (2010). Mutations abrogating VP35 interaction with double-stranded RNA render Ebola virus avirulent in guinea pigs. J. Virol. *84*, 3004–3015.

Ray, R.B., Basu, A., Steele, R., Beyene, A., McHowat, J., Meyer, K., Ghosh, A.K., and Ray, R. (2004). Ebola virus glycoprotein-mediated anoikis of primary human cardiac microvascular endothelial cells. Virology *321*, 181–188.

Reid, S.P., Cardenas, W.B., and Basler, C.F. (2005). Homo-oligomerization facilitates the interferon-antagonist activity of the ebolavirus VP35 protein. Virology *341*, 179–189.

Reid, S.P., Leung, L.W., Hartman, A.L., Martinez, O., Shaw, M.L., Carbonnelle, C., Volchkov, V.E., Nichol, S.T., and Basler, C.F. (2006). Ebola virus VP24 binds karyopherin alpha1 and blocks STAT1 nuclear accumulation. J. Virol. *80*, 5156–5167.

Reid, S.P., Valmas, C., Martinez, O., Sanchez, F.M., and Basler, C.F. (2007). Ebola virus VP24 proteins inhibit the interaction of NPI-1 subfamily karyopherin alpha proteins with activated STAT1. J. Virol. *81*, 13469–13477.

Rodig, S.J., Meraz, M.A., White, J.M., Lampe, P.A., Riley, J.K., Arthur, C.D., King, K.L., Sheehan, K.C., Yin, L., Pennica, D., *et al.* (1998). Disruption of the Jak1 gene demonstrates obligatory and nonredundant roles of the Jaks in cytokine-induced biologic responses. Cell *93*, 373–383.

Rodriguez-Madoz, J.R., Belicha-Villanueva, A., Bernal-Rubio, D., Ashour, J., Ayllon, J., and Fernandez-Sesma, A. (2010). Inhibition of type I ifn response in human dendritic cells by Dengue virus infection requires a catalytically active NS2B3 complex. J. Virol. *84*, 9760–9774.

Rodriguez-Madoz, J.R., Bernal-Rubio, D., Kaminski, D., Boyd, K., and Fernandez-Sesma, A. (2010). Dengue virus inhibits the production of type I interferon in primary human dendritic cells. J. Virol. *84*, 4845–4850.

Saito, T., Owen, D.M., Jiang, F., Marcotrigiano, J., and Gale, M., Jr. (2008). Innate immunity induced by composition-dependent RIG-I recognition of hepatitis C virus RNA. Nature *454*, 523–527.

Sanchez, A., Geisbert, T.W., and Feldmann, H. (2007). Filoviridae: Marburg and Ebola viruses. In Fields Virology, Knipe DM, Howley PM eds. (Philadelphia, Lippincott Williams and Wilkins), pp. 1410–1448.

Sandrock, M., Frese, M., Haller, O., and Kochs, G. (2001). Interferon-induced rat Mx proteins confer resistance to Rift Valley fever virus and other arthropod-borne viruses. J. Interferon Cytokine Res. *21*, 663–668.

Sariol, C.A., Munoz-Jordan, J.L., Abel, K., Rosado, L.C., Pantoja, P., Giavedoni, L., Rodriguez, I.V., White, L.J., Martinez, M., Arana, T., *et al.* (2007). Transcriptional activation of interferon-stimulated genes but not of cytokine genes after primary infection of rhesus macaques with dengue virus type 1. Clin. Vaccine Immunol. *14*, 756–766.

Schlee, M., Roth, A., Hornung, V., Hagmann, C.A., Wimmenauer, V., Barchet, W., Coch, C., Janke, M., Mihailovic, A., Wardle, G., *et al.* (2009). Recognition of 5′ triphosphate by RIG-I helicase requires short blunt double-stranded RNA as contained in panhandle of negative-strand virus. Immunity *31*, 25–34.

Schmaljohn, C., and Hjelle, B. (1997). Hantaviruses: a global disease problem. Emerg. Infect. Dis. *3*, 95–104.

Schmidt, A., Schwerd, T., Hamm, W., Hellmuth, J.C., Cui, S., Wenzel, M., Hoffmann, F.S., Michallet, M.C., Besch, R., Hopfner, K.P., *et al.* (2009). 5′-triphosphate RNA requires basepaired structures to activate antiviral signaling via RIG-I. Proc. Natl. Acad. Sci. U.S.A. *106*, 12067–12072.

Schumann, M., Gantke, T., and Muhlberger, E. (2009). Ebola virus VP35 antagonizes PKR activity through its C-terminal interferon inhibitory domain. J Virol. *83*, 8993–8997.

Seki, T., Tada, S., Katada, T., and Enomoto, T. (1997). Cloning of a cDNA encoding a novel importin-alpha homologue, Qip1: discrimination of Qip1 and Rch1 from hSrp1 by their ability to interact with DNA helicase Q1/RecQL. Biochem. Biophys. Res. Commun. *234*, 48–53.

Sekimoto, T., Imamoto, N., Nakajima, K., Hirano, T., and Yoneda, Y. (1997). Extracellular signal-dependent nuclear import of Stat1 is mediated by nuclear pore-targeting complex formation with NPI-1, but not Rch1. EMBO J. *16*, 7067–7077.

Seth, R.B., Sun, L., Ea, C.K., and Chen, Z.J. (2005). Identification and characterization of MAVS, a mitochondrial antiviral signaling protein that activates NF-kappaB and IRF 3. Cell *122*, 669–682.

Sharma, S., tenOever, B.R., Grandvaux, N., Zhou, G.P., Lin, R., and Hiscott, J. (2003). Triggering the interferon antiviral response through an IKK-related pathway. Science *300*, 1148–1151.

Shresta, S., Kyle, J.L., Snider, H.M., Basavapatna, M., Beatty, P.R., and Harris, E. (2004). Interferon-dependent immunity is essential for resistance to primary dengue virus infection in mice, whereas T- and B-cell-dependent immunity are less critical. J. Virol. *78*, 2701–2710.

Simmons, G., Wool-Lewis, R.J., Baribaud, F., Netter, R.C., and Bates, P. (2002). Ebola virus glycoproteins induce global surface protein down-modulation and loss of cell adherence. J. Virol. *76*, 2518–2528.

Stark, G.R., Kerr, I.M., Williams, B.R., Silverman, R.H., and Schreiber, R.D. (1998). How cells respond to interferons. Annu. Rev. Biochem. *67*, 227–264.

Stoltz, M., Ahlm, C., Lundkvist, A., and Klingstrom, J. (2007). Lambda interferon (IFN-lambda) in serum is decreased in hantavirus-infected patients, and *in vitro*-established infection is insensitive to treatment with all IFNs and inhibits IFN-gamma-induced nitric oxide production. J. Virol. *81*, 8685–8691.

Sullivan, N.J., Peterson, M., Yang, Z.Y., Kong, W.P., Duckers, H., Nabel, E., and Nabel, G.J. (2005). Ebola virus glycoprotein toxicity is mediated by a dynamin-dependent protein-trafficking pathway. J. Virol. *79*, 547–553.

Taylor, S.L., Frias-Staheli, N., Garcia-Sastre, A., and Schmaljohn, C.S. (2009a). Hantaan virus nucleocapsid protein binds to importin alpha proteins and inhibits tumor necrosis factor alpha-induced activation of nuclear factor kappa B. J. Virol. *83*, 1271–1279.

Taylor, S.L., Krempel, R.L., and Schmaljohn, C.S. (2009b). Inhibition of TNF-alpha-induced activation of NF-kappaB by hantavirus nucleocapsid proteins. Ann. N.Y. Acad. Sci. *1171* Suppl. 1, E86–93.

Timmins, J., Scianimanico, S., Schoehn, G., and Weissenhorn, W. (2001). Vesicular Release of Ebola Virus Matrix Protein VP40. Virology *283*, 1–6.

Towner, J.S., Pourrut, X., Albarino, C.G., Nkogue, C.N., Bird, B.H., Grard, G., Ksiazek, T.G., Gonzalez, J.P., Nichol, S.T., and Leroy, E.M. (2007). Marburg virus infection detected in a common African bat. PLoS One *2*, e764.

Umareddy, I., Tang, K.F., Vasudevan, S.G., Devi, S., Hibberd, M.L., and Gu, F. (2008). Dengue virus regulates type I interferon signalling in a strain-dependent manner in human cell lines. J. Gen. Virol. *89*, 3052–3062.

Valmas, C., Grosch, M.N., Schumann, M., Olejnik, J., Martinez, O., Best, S.M., Krahling, V., Basler, C.F., and Muhlberger, E. (2010). Marburg virus evades interferon responses by a mechanism distinct from ebola virus. PLoS Pathog. *6*, e1000721.

Villinger, F., Rollin, P.E., Brar, S.S., Chikkala, N.F., Winter, J., Sundstrom, J.B., Zaki, S.R., Swanepoel, R., Ansari, A.A., and Peters, C.J. (1999). Markedly elevated levels of interferon (IFN)-gamma, IFN-alpha, interleukin (IL)-2, IL-10, and tumor necrosis factor-alpha associated with fatal Ebola virus infection. J. Infect. Dis. *179* Suppl. 1, S188–191.

Volchkov, V.E., Becker, S., Volchkova, V.A., Ternovoj, V.A., Kotov, A.N., Netesov, S.V., and Klenk, H.D. (1995). GP mRNA of Ebola virus is edited by the Ebola virus polymerase and by T7 and vaccinia virus polymerases. Virology *214*, 421–430.

Volchkov, V.E., Chepurnov, A.A., Volchkova, V.A., Ternovoj, V.A., and Klenk, H.D. (2000). Molecular characterization of guinea pig-adapted variants of Ebola virus. Virology *277*, 147–155.

Volchkov, V.E., Volchkova, V.A., Muhlberger, E., Kolesnikova, L.V., Weik, M., Dolnik, O., and Klenk, H.D. (2001). Recovery of infectious Ebola virus from complementary DNA: RNA editing of the GP gene and viral cytotoxicity. Science *291*, 1965–1969.

Wahl-Jensen, V.M., Afanasieva, T.A., Seebach, J., Stroher, U., Feldmann, H., and Schnittler, H.J. (2005). Effects of Ebola virus glycoproteins on endothelial cell activation and barrier function. J. Virol. *79*, 10442–10450.

Wang, X., Hinson, E.R., and Cresswell, P. (2007). The interferon-inducible protein viperin inhibits influenza virus release by perturbing lipid rafts. Cell Host Microbe *2*, 96–105.

Wang, Y., Ludwig, J., Schuberth, C., Goldeck, M., Schlee, M., Li, H., Juranek, S., Sheng, G., Micura, R., Tuschl, T., *et al.* (2010). Structural and functional insights into 5′-ppp RNA pattern recognition by the innate immune receptor RIG-I. Nat Struct Mol Biol *17*, 781–787.

Warfield, K.L., Bradfute, S.B., Wells, J., Lofts, L., Cooper, M.T., Alves, D.A., Reed, D.K., VanTongeren, S.A., Mech, C.A., and Bavari, S. (2009). Development and characterization of a mouse model for Marburg hemorrhagic fever. J. Virol. *83*, 6404–6415.

Watanabe, S., Noda, T., Halfmann, P., Jasenosky, L., and Kawaoka, Y. (2007). Ebola virus (EBOV) VP24 inhibits transcription and replication of the EBOV genome. J. Infect. Dis. *196* Suppl. 2, S284–290.

Weis, K., Mattaj, I.W., and Lamond, A.I. (1995). Identification of hSRP1 alpha as a functional receptor for nuclear localization sequences. Science *268*, 1049–1053.

Weis, K., Ryder, U., and Lamond, A.I. (1996). The conserved amino-terminal domain of hSRP1 alpha is essential for nuclear protein import. EMBO J. *15*, 1818–1825.

Westaway, E.G., Khromykh, A.A., and Mackenzie, J.M. (1999). Nascent flavivirus RNA colocalized in situ with double-stranded RNA in stable replication complexes. Virology *258*, 108–117.

Xu, L.G., Wang, Y.Y., Han, K.J., Li, L.Y., Zhai, Z., and Shu, H.B. (2005). VISA is an adapter protein required for virus-triggered IFN-beta signaling. Mol. Cell *19*, 727–740.

Yang, Z., Delgado, R., Xu, L., Todd, R.F., Nabel, E.G., Sanchez, A., and Nabel, G.J. (1998). Distinct cellular interactions of secreted and transmembrane Ebola virus glycoproteins. Science *279*, 1034–1037.

Yoneyama, M., Kikuchi, M., Natsukawa, T., Shinobu, N., Imaizumi, T., Miyagishi, M., Taira, K., Akira, S., and

Fujita, T. (2004). The RNA helicase RIG-I has an essential function in double-stranded RNA-induced innate antiviral responses. Nat. Immunol. *5*, 730–737.

Zampieri, C.A., Fortin, J.F., Nolan, G.P., and Nabel, G.J. (2007). The ERK mitogen-activated protein kinase pathway contributes to Ebola virus glycoprotein-induced cytotoxicity. J. Virol. *81*, 1230–1240.

Zeng, W., Sun, L., Jiang, X., Chen, X., Hou, F., Adhikari, A., Xu, M., and Chen, Z.J. (2010). Reconstitution of the RIG-I pathway reveals a signaling role of unanchored polyubiquitin chains in innate immunity. Cell *141*, 315–330.

Zhang, Y., Burke, C.W., Ryman, K.D., and Klimstra, W.B. (2007). Identification and characterization of interferon-induced proteins that inhibit alphavirus replication. J. Virol. *81*, 11246–11255.

Influenza Virus and Interferons

8

Gijs A. Versteeg and Adolfo García-Sastre

Abstract

Influenza viruses are the aetiological agents of seasonal influenza outbreaks as well as three devastating influenza pandemics in the 20th century and the 2009 swine-origin subtype H1N1 pandemic. Like most viruses that cause significant disease, influenza viruses have developed means to circumvent the induction and effects of the innate immune system.

Unlike most other RNA viruses, influenza viruses replicate in the nucleus, rather than in the cytoplasm. This distinguishing feature makes the interactions of influenza viruses with their hosts both complex and unique, and requires a well-orchestrated manipulation of many cellular processes. This includes the interferon (IFN) response, a key innate immune pathway, critical for limiting virus replication. To cope with the IFN burden, influenza viruses express non-structural protein 1 (NS1), which is largely dedicated to antagonism of the host IFN response.

This chapter describes how influenza viruses induce the IFN response and the ample means they have developed to intersect with it at all three stages of the pathway. The molecular details of NS1-mediated IFN antagonism are discussed, as well as new vaccination and antiviral drug strategies that target NS1 to attenuate virus replication.

The influenza virus–host interface

Three distinct influenza virus types are currently recognized. Influenza A virus (IAV) infects humans as well as a wide variety of other mammals and birds, whereas influenza B (IBV) and C viruses mainly infect humans (reviewed in Palese and Shaw, 2007). Influenza A and B viruses are the causative agents of the annual influenza epidemics and intermittent pandemics, such as the 1918 'Spanish flu', 1957 'Asian flu', the 1968 'Hong Kong flu' and the recent 2009 swine-origin subtype H1N1 pandemic (reviewed in Subbarao *et al.*, 2006).

Patients with seasonal influenza most often present with acute respiratory illness of the upper respiratory tract (nose, throat, bronchi), which occasionally spreads to the lower respiratory tract, causing a primary viral pneumonia (reviewed in Bouvier and Palese, 2008). In people with compromised immune systems, such as the very young, the elderly and the immunosuppressed, secondary bacterial infection occasionally accompanies viral pneumonia, exacerbating the course of infection.

Unlike most other RNA viruses, influenza viruses replicate in the nucleus, rather than in the cytoplasm (reviewed in Bouvier and Palese, 2008). Nuclear replication necessitates transport across the nuclear membrane of incoming viral ribonucleoproteins (RNPs) containing the viral RNA (vRNA) templates, and export of viral mRNA and mature viral protein and vRNA (reviewed in Palese and Shaw, 2007). This distinguishing feature makes the interactions of influenza viruses with their hosts both complex and unique. They require well-orchestrated manipulation of many cellular processes in infected cells, such as transcription, transport and protein expression to facilitate RNA replication, packaging and budding of newly assembled virions. Additionally, they must do so in an environment where the host cell tries to limit virus replication by mounting potent innate immune responses (reviewed in Wolff and Ludwig, 2009).

The innate immune system forms the first line of defence against invading micro-organisms such as viruses. It dampens initial virus replication and spread. Moreover, it protects uninfected surrounding cells and stimulates leucocytes residing at the interface of innate and adaptive immunity, such as macrophages and dendritic cells (DCs; reviewed in Banchereau and Steinman, 1998; McGill *et al.*, 2009; Wolff and Ludwig, 2009; Wilkins and Gale Jr, 2010). The cells prod the adaptive immune system to mount a full, specialized response against the invading microbe.

Both at the cellular and the systemic level, the virus tries to outrun the innate immune response before the adaptive response is initiated. Influenza virus, like other viruses that cause significant disease, has a set of molecular tricks up its sleeve to tip the intricate balance between its own replication and the innate immune system in its favour. In addition to some advantages that come with its replication in the nucleus, such as the ability to cleave cap-structures from cellular mRNAs, influenza virus expresses non-structural protein 1 (NS1) with nearly the exclusive purpose of outrunning the 'star player' of the innate immune response: type I interferon (IFN).

This chapter describes the current knowledge on how influenza viruses modulate the innate immune system. It covers the latest insights into how the virus-encoded NS1 protein sidetracks the interferon machinery and how this knowledge is currently used to generate new generation live vaccines and antiviral compounds.

Influenza virus replication

Influenza viruses have segmented, negative-stranded RNA genomes (Fig. 8.1A), which are packaged in enveloped particles that are roughly 100 nm in size (reviewed in Palese and Shaw, 2007). The viral haemagglutinin (HA) and neuraminidase (NA) glycoproteins project from the outside of the virions and are involved in entry into target cells and release of progeny virus from them, respectively. HA and NA are the main targets for protective neutralizing antibodies. Based on their antigenic and genetic diversity, IAVs are classified in distinct sub-types, each of which contains different strains (reviewed in Palese and Shaw, 2007). Current and past human sub-types were all newly introduced during pandemics and include H1N1, H2N2 and H3N2 subtypes. Pathogenic avian H5N1 subtype strains are of significant concern because of their pandemic potential and association with lethal human infections (Beigel *et al.*, 2005; Abdel-Ghafar *et al.*, 2008).

Both influenza A and B viruses each express eight segments, which reside in the particles as tightly packed RNPs. The viral RNA is bound both by RNA-dependent RNA polymerase (RdRp) components as well as nucleoprotein (NP). A layer of M1 matrix protein forms an interface between the RNPs and the glycoprotein-containing viral membrane. The M2 membrane protein interacts with the M1 matrix and protrudes through the viral envelope, functioning as an ion channel (Khurana *et al.*, 2009).

Respiratory epithelial cells are the primary targets for human influenza viruses. Upon binding to sialic acids on the cell surface by the HA protein, the viral envelope fuses with the endosomal membrane (reviewed in Palese and Shaw, 2007). Influx of protons through the M2 ion channel releases the RNP from the M1 matrix into the cytoplasm, after which it enters the nucleus through active nucleocytoplasmic transport. Upon entry into the nucleus, the vRNA is transcribed into mRNA that can be translated into viral proteins. In addition, vRNA is replicated into complementary RNA (cRNA) that functions as a template to produce more genomic vRNA. Owing to the nature of their genome being complementary to mRNA orientation, the viral heterotrimeric RdRp is packaged together with the vRNA into virus particles to allow initiation of replication.

Eventually, after production of new viral proteins and genomic RNA, RNPs are exported out of the nucleus by binding to M1 and to the viral nuclear export protein (NEP). Progeny virus buds at the plasma membrane at the apical side of the polarized lung cells. NA protein on the released virions prevents them from sticking to the cell and to each other by cleaving sialic acids from the infected cell and from the viral glycoproteins.

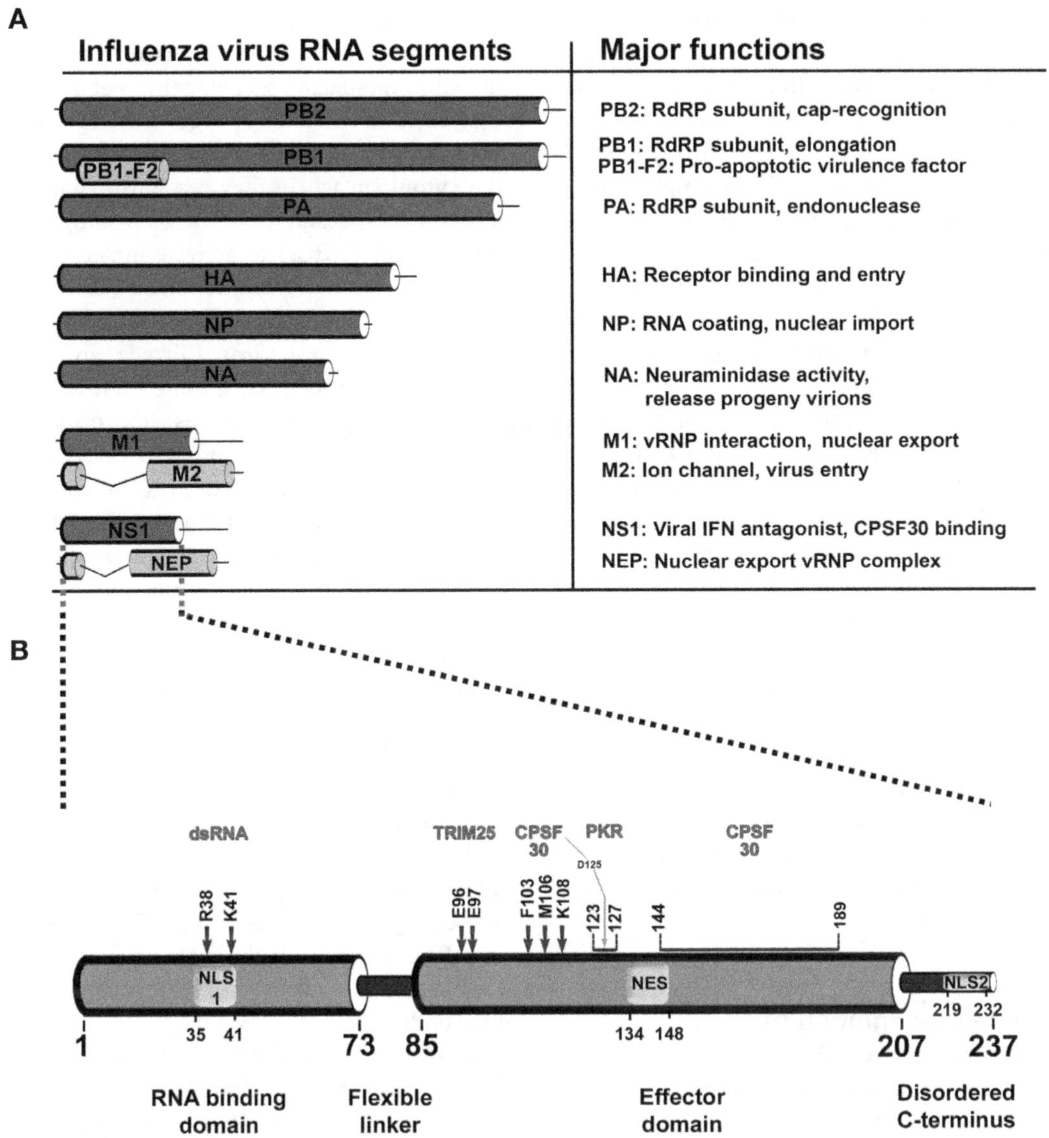

Figure 8.1 (A) Genome structure of prototypical influenza virus A/Puerto Rico/8/1934, displayed in positive orientation. Segment 2 encodes polymerase subunit PB1, as well as the PB1-F2 protein in a +1 ORF. Segments 7 and 8 mRNA are spliced into mRNAs that share a short 5′-end region with their unspliced counterparts. Unspliced mRNAs 8 encodes the NS1 protein. (B) Biochemical structure of the NS1 protein. Bold numbers indicate amino acid positions of the RNA-binding domain (RBD), effector domain (ED) and a C-terminal tail that is intrinsically disordered in the absence of binding to specific ligands. Nuclear localization (NLS) and export (NES) signals are indicated. Arrows and brackets above the structure indicate specific amino acids or amino acid regions involved in interaction with the cellular innate immune proteins indicated.

Influenza virus genome organization and protein expression

The first three influenza RNA segments encode the replicative proteins polymerase subunit basic 2 (PB2), PB1 and polymerase subunit acidic 1 (PA), respectively (Fig. 8.1A). Together, these proteins form the heterotrimeric RdRp complex, replicating the viral RNA in the nucleus. In addition, segment 2 encodes the virulence factor PB1-F2 in an alternative open-reading frame (Chen *et al.*, 2001). To assure efficient translation by the host cell protein synthesis machinery, influenza virus mRNAs contain both a 7-methylguanosine cap and a poly(A) tail. Whereas the viral poly(A) tail is synthesized by reiterative copying of the U track near the 5′ end of the vRNA template (Poon *et al.*, 1999), the cap structures are hijacked from cellular pre-mRNAs by the heterotrimeric RdRp complex in a process known as cap-snatching (Plotch *et al.*, 1981). The PA subunit possesses the actual endonuclease activity that cleaves the caps from cellular pre-mRNA (Dias *et al.*, 2009; Yuan *et al.*, 2009), whereas the basic subunits mediate binding of the polymerase complex to the cap-structures (Guilligay *et al.*, 2008; Sugiyama *et al.*, 2009). Once relieved, the cap-structure functions as a primer to initiate viral RNA synthesis.

The four subsequent RNA segments encode structural proteins (Fig. 8.1A). HA, NP and NA are expressed from segment 4, 5, and 6, respectively, whereas segment 7 expresses both M1, as well as M2 through splicing of the corresponding viral mRNA. The last segment yields a spliced mRNA too, from which the NEP is expressed. The unspliced mRNA encodes the NS1 protein, whose main function is circumventing the host innate immune response by antagonizing the type I IFN response.

Host cell recognition of influenza virus RNA

To prevent autoimmunity and at the same time fight pathogens and malignant processes, cells have stringent and complex innate immunity 'recognition and regulation' mechanisms in place. Innate immune responses are triggered when specialized pattern recognition receptors (PRRs) recognize pathogen-associated molecular patterns (PAMPs) that are unique to invading pathogens, and thus usually absent in the host under homeostatic conditions. Several classes of PRRs have so far been recognized: the Toll-like receptors (TLRs), C-type lectin receptors, RIG-I like receptors (RLRs) and NOD-like receptors (NLRs; reviewed in Wilkins and Gale Jr, 2010).

TLRs and RLRs are the main routes through which PAMPs induce a type I IFN response. Different TLRs recognize different PAMPs and are differentially expressed on various cell types. TLR localization on the plasma membrane and in endosomes facilitates recognition of molecular signatures on entering or phagocytosed pathogens (McGill *et al.*, 2009). RLRs exclusively localize in the cytoplasm where they recognize viral RNA. TLRs are especially important for antigen presenting cells (APCs) to activate innate immune responses. Plasmacytoid dendritic cells (pDCs) produce vast amounts of IFN-α in response to wild-type influenza virus (Takeda and Akira, 2005). The importance of most TLRs during influenza virus infection has not been completely established, yet studies with DCs from TLR7$^{-/-}$ mice indicate that TLR7 is critical for IFN production by pDCs during influenza virus infection (Diebold *et al.*, 2004; Lund *et al.*, 2004). Moreover, TLR4 signalling during infection with highly pathogenic subtype H5N1 viruses has been shown to exacerbate lung pathology, although it remains unclear if this is dependent on recognition of any of the viral surface glycoproteins (Imai *et al.*, 2008).

Retinoic acid-inducible gene I (RIG-I), melanoma differentiation-associated gene 5 (MDA-5) and laboratory of genetics and physiology 2 (LGP-2) comprise the RLR family (reviewed in Wilkins and Gale Jr, 2010). RIG-I and MDA-5 recognize viral RNA and activate the IFN response, whereas LGP-2 plays a regulatory role that is currently not fully understood.

Many studies on the substrates for RIG-I and MDA-5 have been performed, yet controversy about the exact nature of the activating molecules in infected cells remains (Kang *et al.*, 2002, 2003; Gitlin *et al.*, 2006; Marques *et al.*, 2006; Pichlmair *et al.*, 2006; Samanta *et al.*, 2006; Hornung *et al.*, 2006; Malathi *et al.*, 2007; Plumet *et al.*, 2007; Habjan *et al.*, 2008; Hausmann *et al.*, 2008; Kato *et al.*, 2008; Schmidt *et al.*, 2009; Uzri and Gehrke, 2009). Biochemical and functional studies suggest that RIG-I activation requires a free 5′-triphosphate moiety, to distinguish it from cellular RNAs that are capped or otherwise modified to remove the 5′-triphosphate (Hornung *et al.*, 2006; Pichlmair *et al.*, 2006; Plumet *et al.*, 2007; Schmidt *et al.*, 2009). Moreover, a certain degree of double-stranded RNA (dsRNA) seems to be required for full activation of RIG-I (Hausmann *et al.*, 2008; Kato *et al.*, 2008; Uzri and Gehrke, 2009). Substrates for MDA-5 have been less well characterized and seem to be long dsRNA molecules, unlike RIG-I substrates, which seem to consist of shorter dsRNA regions and adenosine-uridine-rich regions (Kang *et al.*, 2002, 2003; Gitlin *et al.*, 2006). Interestingly, cleavage products of the IFN-induced RNase L are also substrates for RIG-I (Malathi *et al.*, 2007).

Experiments using mouse embryonic fibroblasts from RIG-I and MDA-5 knock-out mice demonstrated that some viruses are preferentially recognized by either RIG-I or MDA-5 (Kato *et al.*, 2006). A lack of IFN induction in RIG-I$^{-/-}$ cells demonstrated that influenza viruses are exclusively recognized by RIG-I and not by MDA-5 (Kato *et al.*, 2006). Since influenza virus replicates in the nucleus and RIG-I is localized in the cytoplasm, it is to be expected that either viral mRNAs exported from the nucleus for translation or vRNA in RNPs exported from the nucleus for packaging into new virions are triggers for RIG-I. These latter RNAs are, however, covered by the RdRp complex and NP protein, theoretically masking the vRNA from RIG-I (reviewed by Palese and Shaw, 2007). Notably, RIG-I possesses helicase activity that could facilitate displacement of RNA-binding molecules (Takahasi *et al.*, 2008). Interestingly, Duesberg and co-workers have shown that vRNP complexes are sensitive to RNase treatment, suggesting that the RdRp-NP coat around vRNA is not impenetrable (Duesberg, 1969). On the other hand the cell may employ hitherto unknown mechanisms to feed samples of uncoated viral nuclear RNA to the cytoplasmic sensory machinery.

Many theories about natural inducers of RLRs, such as uncapped and/or structured non-translated regions (NTRs), replication-intermediates, viral mRNA and various forms of processed cellular RNA (e.g. by RNase L), have been proposed over the years (reviewed in Takeuchi and Akira, 2008). A recent study of influenza virus RNA synthesis using PA mutants that reduced either mRNA synthesis or replication, combined with size-fractionation of viral RNA from infected cells, suggested that genome-size, 5′-triphosphate bearing vRNA and cRNA are the major endogenous RIG-I agonists (Rehwinkel *et al.*, 2010). Also, RNA bound by the IFN inhibitor NS1 was mainly 5′-triphosphate containing genome-size RNA, rather than viral mRNA or short cellular RNA species such as found during RNase L activation. Similar results were reported for Sendai virus, also a negative stranded RNA virus of the family *Paramyxoviridae* (Rehwinkel *et al.*, 2010). Although viral RNA is tightly packed in RNP structures, and thus theoretically inaccessible to cellular dsRNA sensors, it is not unlikely that some of the billions of viral transcripts escape from coating with nucleoproteins and activate RIG-I.

Studies to find dsRNA species in infected cells that could contribute to RIG-I activation have been hindered by the fact that dsRNA structures can arise in isolated RNA as artefacts. *In situ* studies of cells infected with various positive- and negative-strand RNA viruses using dsRNA-specific monoclonal antibodies were the first to be performed under non-disruptive conditions (Weber *et al.*, 2006). The results indicated abundant accumulation of dsRNA during (+) RNA virus infection, yet no significant presence of dsRNA in cells infected with (–) RNA viruses such as IAV (Weber *et al.*, 2006). The question remains whether potentially undetectable amounts of dsRNA are present in (–) RNA virus infected cells that could be sufficient to trigger activation of RIG-I. Moreover, it indicates that major differences exist between the dsRNA species produced by viruses with different genome polarities and raises the question whether the dsRNA trigger in (–) RNA virus infected cells is different from that in (+) RNA virus-infected cells.

IFN induction and signalling

RIG-I bears three distinct domains: two N-terminal caspase activation and recruitment domains (CARDs), an ATP-dependent helicase domain in the central part of the protein and a C-terminal regulatory domain (RD; reviewed in Baum and Garcia-Sastre, 2009). Under non-induced conditions, RIG-I resides in the cytoplasm as a monomer in a state where the RD is thought to keep the CARD in an unexposed, inactive state. Upon binding of activating PAMPs to the RD, the protein conformation changes, resulting in dimerization and exposure of the CARDs (Cui *et al.*, 2008; Takahasi *et al.*, 2008). Subsequently, the E3 ubiquitin ligase tripartite motif protein 25 (TRIM25) binds the most N-terminal CARD and ubiquitinates lysine residue 172 in the second CARD (Gack *et al.*, 2008; Fig. 8.2). This modification and/or generation of polyubiquitin chains is critical for the interaction of RIG-I with its downstream signalling partner mitochondrial antiviral signalling adapter (MAVS; Gack *et al.*, 2007; Zeng *et al.*, 2010).

MAVS functions as a critical scaffolding protein in the recruitment of two distinct signalling complexes that ultimately activate the transcription factors nuclear factor κB (NFκB) and IFN-regulatory factor 3 (IRF-3; Seth *et al.*, 2005). Upon activation, NF-κB and IRF-3 translocate to the nucleus where they mediate the transcription of IFN-β (Fig. 8.2). Phosphorylation of IRF-3 is mediated by a complex consisting of TNF receptor-associated factor 3 (TRAF-3), TRAF family member-associated NF-κB activator (TANK), TANK-binding kinase 1 (TBK-1) and inhibitor of κB kinase (IκB) ε (IKKε; Seth *et al.*, 2005). The second complex phosphorylating IκB and thereby activating NF-κB consists of TRAF-6, receptor interacting protein 1 (RIP-1), NF-κB essential modulator (NEMO), TGF-β activated kinase 1 (TAK-1), IKKα and IKKβ (Seth *et al.*, 2005). The produced IFN-β is secreted from the cells and binds to the IFN-α receptor (IFNAR)

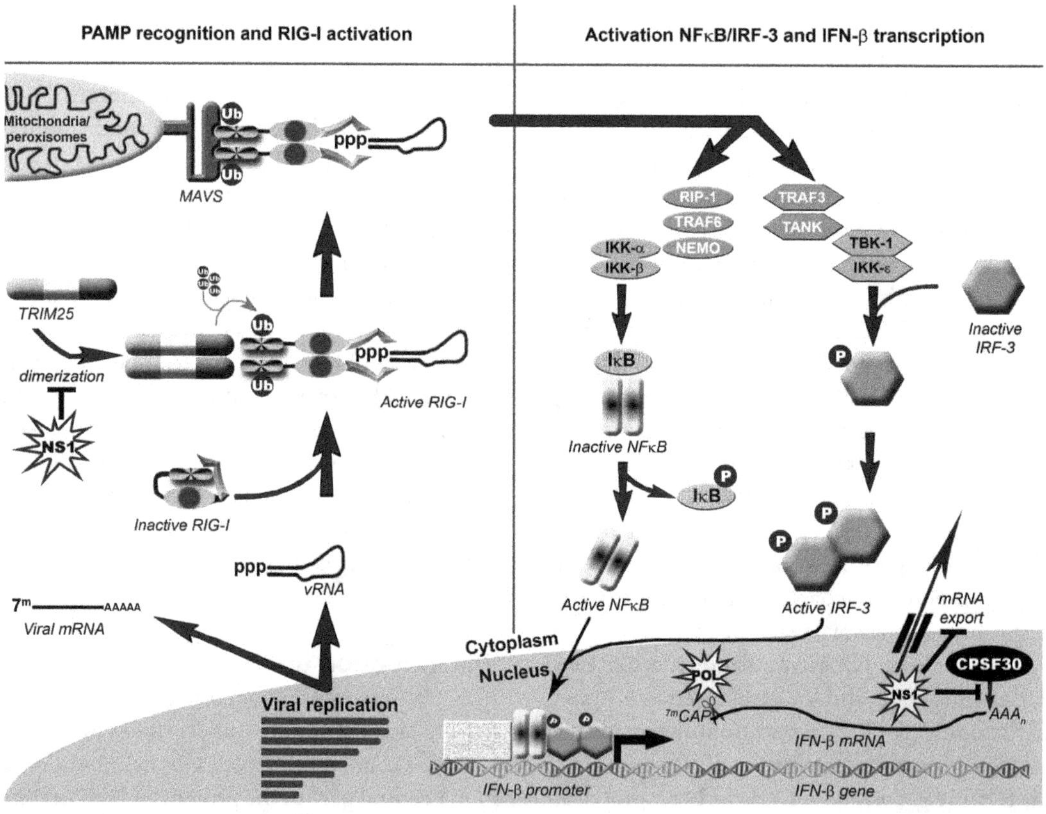

Figure 8.2 IFN-β induction and steps antagonized by influenza virus. Genome-size vRNA synthesized during replication in the nucleus, is transported to the cytoplasm. Unprotected 5′-triphosphates and double-stranded regions present in vRNA are recognized by RIG-I. RIG-I is ubiquitinated by E3 ligase TRIM25, subsequently interacts with MAVS on the mitochondrial membrane and thereby initiates the activation of transcription factors NFkB and IRF-3. Both activated transcription factors translocate to the nucleus and drive transcription of IFN-β mRNA. Mature mRNA is exported to the cytoplasm and translated. Secreted IFN-β confers an antiviral state by signalling the upregulation of IFN-stimulated genes through the IFN-α receptor (IFNAR) on the plasma membrane. Influenza virus NS1 prevents TRIM25-mediated activation of RIG-I and interferes with CPSF30-mediated pre-mRNA processing. The heterotrimeric polymerase complex cleaves cap-structures from cellular mRNA (cap-snatching).

present on the same or surrounding cells. The resulting IFN signalling eventually drives the expression of hundreds of IFN stimulated genes (ISGs) that confer antiviral protective effects (Randall and Goodbourn, 2008).

The IFNAR is a heteromeric receptor localized on the plasma membrane. Upon binding of extracellular IFN-α/β, it recruits Janus kinase 1 (JAK-1) and tyrosine kinase 2 (TYK-2) to its cytoplasmic domain. These kinases phosphorylate the key transcription factors signal transducers and activators of transcription (STAT) 1 and 2, which together with IRF-9 form the IFN-stimulated gene factor 3 (ISGF3) complex (Seth *et al.*, 2005). This active transcription complex translocates to the nucleus, where it drives the expression of ISGs by binding to their promoter regions named IFN-stimulated response elements. IFN signalling upregulates the transcription of 300–400 ISGs (Der *et al.*, 1998; Sadler and Williams, 2008). It is recognized that some viruses are more vulnerable to the action of certain ISGs than others and that the role of most ISGs remains so far unclear. The ISGs that have been shown to affect influenza virus replication include myxovirus resistance protein A (MxA), protein kinase R (PKR), 2′–5′ oligoadenylate synthetase, IFN-stimulated gene 15 (ISG15), members of the IFN-inducible trans-membrane protein (IFITM) family and viperin (Garcia *et al.*, 2006a; Haller *et al.*, 2006; Gack *et al.*, 2007; Silverman, 2007; Wang *et al.*, 2007; Brass *et al.*, 2009; Table 8.1B).

Contribution of type I, II and III IFNs to the control of influenza virus infection

Isaacs and Lindenmann were the first to discover the broadly antiviral cytokine now known as type I interferon (Isaacs and Lindenmann, 1957). Their experiments showed that treatment of cells with replication-deficient influenza virus resulted in the secretion of a substance in the supernatant of those cells that could subsequently protect new cells from infection with replication competent influenza virus. In the decades that followed, additional classes of IFNs were found that differed from the IFNs initially found by Isaacs and Lindenmann. Consequently, these different types of IFN were separated into several classes. The most important classes in the context of immune response during viral infection include the originally discovered type I IFNs (IFN-α and IFN-β), type II IFN (IFN-γ) and type III IFNs (IFN-λ).

Challenge studies with both seasonal and highly pathogenic strains of influenza virus indicate that prophylactic treatment with type I IFN protects mice, guinea pigs and ferrets from morbidity and mortality (Kugel *et al.*, 2009; Van Hoeven *et al.*, 2009; Steel *et al.*, 2010b). These studies show that influenza viruses are at least to some extent sensitive to the effect of IFN-induced antiviral effectors.

Accumulating evidence indicates that type I and type III interferon responses are triggered by the same PAMPs (Ank *et al.*, 2006; Uze and Monneron, 2006; Onoguchi *et al.*, 2007) and depend on the same signalling pathways for their function (Osterlund *et al.*, 2005; Yang *et al.*, 2005). The main difference lies in the use of a different cellular receptor for the two different cytokine classes. IFN-α and IFN-β bind and signal through the IFNAR. In contrast, IFN-λ signals through its own specific receptor, resulting in the activation of STAT1/STAT2 identical to type I signalling.

IFN-λ can be expressed by many different cell types, as is the case for type I IFNs. In the context of influenza virus infection, differentiated human alveolar type II cells are the predominant IFN-λ producers (Wang *et al.*, 2009). Studies using microinjection of plasmids encoding different types of IFN into mice suggested that the IFN-λ receptor is mainly confined to epithelial cells (Sommereyns *et al.*, 2008). Thus, since the IFNAR is ubiquitously expressed, distinct cells or tissues might be targeted by IFN-α/β and IFN-λ. This was confirmed by influenza virus infections in mice lacking either the IFNAR, the type III IFN receptor, or both (Mordstein *et al.*, 2008). The data demonstrated that although IFN-λ signalling confers substantial protection against influenza virus, it did not against hepatotropic virus infections (Mordstein *et al.*, 2010). Moreover, signalling through the IFNAR conferred the majority of the protection during influenza virus infection, although IFN-λ significantly contributes. Taken together, these observations suggest that the physiological role of IFN-λ is to protect

Table 8.1a Influenza virus-mediated antagonism of the type I IFN system

	Viral protein	By A	By B	Affected function/structure	Mechanism
IFN induction	NS1	Yes	Yes	Recognition of viral PAMPO	dsRNA sequestration
	NS1	Yes	ND	TRIM25 Ub E3 ligase activity	Binding TRIM25, preventing RIG-I ubiquitination
	PB2-PB1-PA	Yes	Yes	mRNA cap structure	'Cap snatching' from cellular mRNAs
	NS1	Yes	No	CPSF30	Processing Pre-mRNAs by binding CPSF30
	NS1	Yes	ND	mRNA export machinery	Formation of inhibitory complex with cellular export machinery
IFN signalling	PB2-PB1-PA	Yes	Yes	mRNA cap structure	'Cap snatching' from cellular mRNAs
	NS1	Yes	ND	mRNA export machinery	Formation of inhibitory complex with cellular export machinery
	Unknown	Yes	ND	STAT1 activation	NFkB-dependent SOCS3 activation thru 5′-triphos/RIG-I
Antiviral effectors	NS1	No	Yes	ISG15	Binding to N-terminal Ub-like domain of human ISG15
	NS1	Yes	Yes	PKR	Interaction with PKR
	Unknown	Yes	ND	PKR	Induction of cellular PKR inhibitor p58IPK
	NS1	Yes	Yes	PKR	sequestering dsRNA
	NS1	Yes	ND	OAS	sequestering dsRNA
	PB2-PB1-PA	Yes	ND	Mx1	Enhanced replication speed can overcome resistance in $Mx1^{+/+}$ mice

All currently known mechanisms of influenza A and B viruses to interfere with the three different steps of the IFN system are summarized. Differences in the ability of influenza A and B viruses to antagonize a particular function/structure ('by A' and 'by B') are indicated.

Table 8.1b IFN-induced antiviral effectors that limit influenza virus replication

	Effective against		
	A	B	Mechanism
MxA	Yes	ND	Possible binding of viral vRNA/NP complexes and sequestering in cytoplasm
PKR	Yes	Yes	Translation inhibition
2′,5′ OAS	Yes	ND	Activation of RNaseL, cleavage of viral RNA
ISG15	Yes	Yes	Unknown: possibly by modification of antiviral molecules or as cytokine
IFITM	Yes	ND	Inhibits viral entry
Viperin	Yes	ND	Inhibits lipid raft formation and thereby virus release from plasma membrane

Six IFN-stimulated gene (ISG) products particularly effective in limiting influenza viruses, and their modes of action, are listed. ND, no data.

the host from viral infections at mucosal membranes at many different body sites, including the lung.

IFN-γ is mainly produced by natural killer (NK) cells, NK T cells and after antigen stimulation by CD4+ T-helper cells and CD8+ cytotoxic T cells (CTL; reviewed in Schoenborn and Wilson, 2007). Although IFN-γ can under certain circumstances directly inhibit viral replication, its main antiviral capability is derived from its immunostimulatory and immunomodulatory effects (reviewed in Schoenborn and Wilson, 2007). It plays an important role in shaping the adaptive immune response by promoting antigen-driven T cell proliferation (Whitmire *et al.*, 2005) and homeostatic control of antigen-specific CTL numbers (Badovinac *et al.*, 2000; Dalton *et al.*, 2000; Whitmire *et al.*, 2005).

An effective CTL response was shown to be crucial for clearance of influenza virus infection from the lung of infected chimeric knock-out mice lacking combinations of genes crucial for development and function of a correct adaptive immune response (Bender *et al.*, 1992; Topham *et al.*, 1997). Since IFN-γ can regulate CTL maturation, it may play a role of importance during infection. However, contradicting reports exist on the significance of IFN-γ during influenza virus infection. *In vivo* administration of IFN-γ neutralizing antibodies delayed clearance of infection in mice (Sarawar *et al.*, 1994; Karupiah *et al.*, 1998), whereas studies in IFN-γ knock-out mice showed that IFN-γ did not significantly inhibit influenza virus replication or affect cytotoxic T-cell responses (Graham *et al.*, 1993). In contrast, recent detailed studies using influenza virus specific tetramer staining demonstrated aberrant CTL distribution during influenza virus infection. Infection in IFN-γ knock-out mice resulted in increased numbers of virus-specific CTLs in the spleen, yet reduced numbers of specific CTLs in the primary site of replication (lung; Turner *et al.*, 2007). These data suggest that IFN-γ directly acts on specific CD8+ CTLs to control their migration to the site of influenza virus infection.

Moreover, several studies have demonstrated that IFN-γ is critical to mount a proper memory T-cell response against homologous influenza virus challenge (Baumgarth and Kelso, 1996; Bot *et al.*, 1998) and that prophylactic treatment or treatment early during influenza virus infection may facilitate virus clearance by stimulating NK cell proliferation and function in infected lungs (Weiss *et al.*, 2010). Thus, although IFN-γ may inhibit influenza virus by recruitment of NK cells or even direct effects on replication, its main function seems to encompass the proper distribution of specific CTLs and the control of a proper memory T-cell response.

Lymphocytes at the innate immune interface during influenza virus infection

Natural killer cells

NK cells are innate immune lymphocytes that possess cytotoxic potential and produce several key inflammatory cytokines such as tumour necrosis factor α (TNF-α), macrophage inflammatory protein 1α (MIP-1α) and IFN-γ (reviewed in Culley, 2009). They play important roles in direct killing of various pathogens and secretion of inflammatory cytokines to recruit and activate other lymphocytes. Whereas pulmonary NK cells maintain a cytolytic phenotype, NK cells in draining lymph nodes display a cytokine-producing phenotype that may function in activating other immune cells at that site (Hayakawa and Smyth, 2006).

Several studies have indicated the importance of NK cells in controlling influenza virus infection. NK cell receptor 1 (NCR-1) knock-out mice, as well as mice and hamsters in whom NK cells were depleted by specific antibodies, showed increased morbidity and mortality during influenza virus infection (Stein-Streilein and Guffee, 1986; Gazit *et al.*, 2006). Moreover, NK cell receptor NKp46 was shown to recognize influenza HA proteins on infected cells and mediate their lysis (Mandelboim *et al.*, 2001; Draghi *et al.*, 2007).

Despite these indications that NK cells are important for containing influenza virus infection *in vivo*, it remains unclear if this is a result of regulation of cytolytic activity, or cytokine cross-talk to DCs and T cells in draining lymph nodes of infected lungs (He *et al.*, 2004; Gerosa *et al.*, 2005; Draghi *et al.*, 2007). Recent *in vitro* data suggest that IFN-α is required for cytotoxicity, whereas interleukin 12 (IL-12) in combination with NK cell receptor activation drives IFN-γ production (Draghi *et al.*, 2007).

Dendritic cells (DCs)

DCs are important antigen presenting cells (APCs) that form a crucial bridge between the innate and adaptive branches of the immune system. They are thought to be the main APC involved in the induction of influenza virus specific CTLs during infection. Upon direct influenza virus infection, or phagocytosis of infected cells, DCs produce a plethora of pro-inflammatory cytokines and chemokines, which include type I IFNs (reviewed in Banchereau and Steinman, 1998). These IFNs not only confer an antiviral state in surrounding stromal cells by upregulating ISGs, but also enhance respiratory DC (rDC) maturation, antigen cross-priming and stimulate CTL function and proliferation (Yamada *et al.*, 1986; Curtsinger *et al.*, 1999; Marrack *et al.*, 1999; Durbin *et al.*, 2000; Price *et al.*, 2000; Le Bon *et al.*, 2003; Le Bon *et al.*, 2006).

The influenza virus NS1 protein not only inhibits IFN induction in primary infected alveolar cells, but also in certain classes of DCs (Garcia-Sastre *et al.*, 1998; Wang *et al.*, 2000; Lopez *et al.*, 2003; Fernandez-Sesma *et al.*, 2006; Kochs *et al.*, 2007). NS1 expression in infected DCs prevents maturation *in vitro* and their ability to stimulate CTL responses (Fernandez-Sesma *et al.*, 2006). The inability of DCs to mature results from a lack of pro-inflammatory cytokine production (including type I IFNs), rather than an inability to respond to external cytokines (Lopez *et al.*, 2003; Fernandez-Sesma *et al.*, 2006). In line with these observations, pre-treatment of DCs with type I IFN before influenza virus infection overcomes NS1 antagonism and allows their activation during virus infection (Phipps-Yonas *et al.*, 2008). This implies that during infection *in vivo*, influenza virus NS1 only blocks maturation and cytokine secretion in rDCs that it actively infects (Fig. 8.3). rDCs that phagocytose non-infective virus antigen should be less susceptible to the disadvantageous effects of NS1. pDCs produce vast amounts of type I IFN during influenza virus infection *in vitro* and are not affected by infection with influenza (Phipps-Yonas *et al.*, 2008). Thus, potentially at later stages during influenza virus infection, type I IFN produced by pDCs could prime uninfected rDCs, such that when they are infected later on, they can still mature and contribute to an effective immune response.

Ex vivo exposure of DCs to influenza virus is sufficient to stimulate a virus-specific CTL response (Macatonia *et al.*, 1989; Bhardwaj *et al.*, 1994). Moreover, DCs are more effective than other APCs in stimulating specific CTL responses, even in the absence of exogenous cytokines (Nonacs *et al.*, 1992). Interestingly, whereas pDCs get activated during infection, acquire influenza virus antigen and are able to prime CTLs *in vitro*, they do not present influenza virus-specific antigen *in vivo* or *ex vivo* (reviewed in McGill *et al.*, 2009).

DCs must present their obtained influenza virus antigens to naive T-cells in draining lymph nodes of the infected lung. It is thought that DCs derive the antigens through direct infection of the DC itself (Macatonia *et al.*, 1989; Nonacs *et al.*, 1992; Bhardwaj *et al.*, 1994; Hamilton-Easton and Eichelberger, 1995) or via phagocytosis of apoptotic influenza virus-infected cells (Mori *et al.*, 1995; Albert *et al.*, 1998; Gallucci *et al.*, 1999; Sauter *et al.*, 2000; Brydon *et al.*, 2003). Neither model is mutually exclusive and evidence exists to support a role for both of them in influenza antigen acquirement by DCs (reviewed in McGill *et al.*, 2009).

Taken together, current knowledge indicates that rDCs phagocytose apoptotic infected cells in the lung and subsequently migrate to draining lymph nodes to mature (Fig. 8.3). Direct infection of rDCs prevents maturation by means of NS1. Upon maturation, non-infected rDCs prime CTLs and Th1 cells in lymph nodes by presenting antigen and secreting pro-inflammatory cytokines/chemokines. IFN-γ produced by NK cells at the site of infection attracts activated CTLs that contribute to virus clearance. In contrast, Th1 cells (together with IFN-α/β and IL-6 from mature rDCs and pDCs) stimulate naïve B-cells to develop into plasma cells. Antibody produced by the plasma cells neutralizes virus at the site of infection and prevents further dissemination (Fig. 8.3).

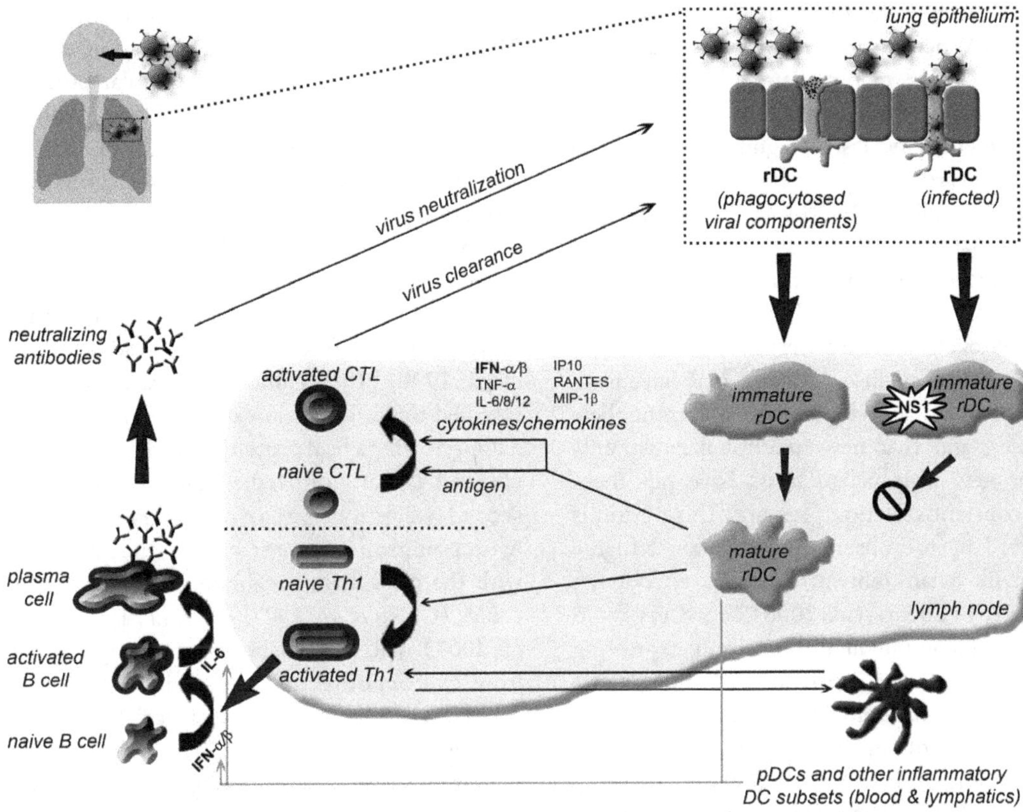

Figure 8.3 Influenza virus NS1 dysregulates induction of adaptive immunity by preventing DC maturation. Lung-resident rDCs migrate to draining lymph nodes upon encounter of viral antigen by phagocytosis or infection. Subsequently, rDCs with phagocytosed influenza virus antigens mature, whereas maturation is blocked in infected cells by the NS1 protein. Mature rDCs activate CD8+ cytotoxic lymphocytes and CD4+ T-helper cells by presenting influenza virus antigen and production of inflammatory cytokines and chemokines. Other DC subsets may contribute to T-helper cell activation, which subsequently results in the maturation of naïve B-cells to neutralizing antibody-producing plasma cells. CTL and antibody responses are critical for efficient virus clearance and neutralization, respectively. IFN-γ secreted by lymphocytes at the site of infection attracts CTLs to the site of infection and regulates memory responses.

Structure and biochemistry of the influenza virus NS1 protein

Most viruses that cause significant disease have developed means to circumvent the IFN response. The large genomes of dsDNA viruses allow them to encode various different proteins to evade the IFN hurdle at multiple points in the pathway. Despite the fact that RNA viruses have much more limited coding capacity, they require evasion of the effects of IFN to a similar extent. Hence, RNA viruses encode highly pleiotropic proteins that usually fulfil an active role in viral replication as well as evasion of the IFN response (reviewed in Randall and Goodbourn, 2008). Although several studies indicate a direct role of NS1 in splicing of viral transcripts, temporal regulation of vRNA synthesis and preferential translation of viral mRNA (reviewed in Hale *et al.*, 2008b), it seems that influenza viruses are one of the few RNA viruses that encode a separate protein that is largely dedicated to antagonism of the host IFN response. The ability to antagonize the IFN response is conserved between influenza A and B viruses. However, notable differences exist in the nature of antagonism and interaction with the host (Table 8.1A).

The IAV NS1 can be divided into two major groups on phylogenetic grounds, named 'allele

A' and 'allele B' (Treanor *et al.*, 1989; Ludwig *et al.*, 1991). Whereas allele A includes avian influenza species and mammalian influenza species (human, swine, horse), allele B only contains avian influenza species. A significant degree of sequence divergence exists between the NS1 proteins of both alleles. Moreover, genetic analysis indicates that the NS1 protein of allele A viruses is under constant selection pressure, whereas this is less so for allele B NS1 (Ludwig *et al.*, 1991).

Taken together these observations suggest that NS1 sequences in allele A viruses may have been derived from an ancestor similar to current allele B viruses, and that new functional constraints in their new mammalian hosts have put them under constant selection pressure. This notion is supported by the observation that most highly pathogenic avian influenza viruses encode an allele A NS1 (Zohari *et al.*, 2008). Moreover, replication of a recombinant allele A strain, expressing an allele B NS1 was attenuated in the lungs of a non-human primate model (Treanor *et al.*, 1989).

IAV NS1 proteins are between 230 and 237 amino acids long. Two distinct domains are recognized in the NS1 protein (Fig. 8.1B). The first 73 amino acids constitute a dsRNA-binding domain (RBD), whereas residues 85–207 constitute the effector domain (ED). The last 30 residues form a C-terminal disordered tail (Fig. 8.1B). NS1 forms homodimers, through dimerization of both the RBD and ED with the same domain of a complementary NS1 monomer (Nemeroff *et al.*, 1995). The domains are relatively well conserved within strains, yet the flexible linker between them and the disordered C-terminus display significant sequence variability.

The flexible linker region influences the relative orientations of the RBD and ED and may thus play a role in pathogenesis (Bornholdt and Prasad, 2008). More than 90% of the recent subtype H5N1 viruses have a five amino acid (residues 80–84) deletion in their linker region that has been associated with high virulence (Lipatov *et al.*, 2005; Long *et al.*, 2008). Also, the D92E mutation in that region has been associated with enhanced subtype H5N1 virulence (Heui Seo *et al.*, 2002). Subtype H1N1 viruses prior to 1940 contained 230 amino acid long NS1 proteins. Around 1940, a point mutation extended the tail by 7 amino acids (Melen *et al.*, 2007). The longer NS1 contained additional localization signals and potential binding sites for cellular proteins. Although the full significance of those features remains ill-defined, the extension was transmitted to subtypes H2N2 and H3N2 viruses, until it was again lost in the 1980s (Melen *et al.*, 2007).

The RBD binds dsRNA *in vitro* in a sequence-independent manner (Hatada and Fukuda, 1992; Qian *et al.*, 1995; Chien *et al.*, 2004). Dimerization of the RDB is required for dsRNA binding (Wang *et al.*, 1999). The crystal structure of the RDB revealed three helices in each monomer, to form a dimeric, six-helical protein fold when dimerized (Fig. 8.4A). An anti-parallel conformation of the second helices forms a face of conserved basic and hydrophilic residues that facilitates the interaction with the phosphate backbone of the RNA molecules (Chien *et al.*, 1997; Liu *et al.*, 1997; Yin *et al.*, 2007). dsRNA is anchored by hydrogen bonds from two opposite arginine residues (R38-R38'; Fig. 8.4A). Residue R38 is critical for dsRNA binding, whereas K41 is required, but not essential for effective dsRNA binding. Underneath the R38 pair lays a conserved deep pocket that forms a promising target for antiviral drug design. RBD crystal structures of several virus strains have been solved, and all of them demonstrate a strong conservation of their dimeric structure.

A recently solved full-length NS1 structure of the highly pathogenic A/Vietnam/1203/2004 (H5N1) strain indicated that alternating dimers of RBDs and EDs result in the formation of a chain of NS1 molecules instead of distinct dimeric units (Bornholdt and Prasad, 2008). Three such chains in the crystal interact with one another extensively to form a tunnel that fits dsRNA and could provide a model explaining the ability of NS1 to sequester dsRNA of different lengths (Bornholdt and Prasad, 2008). The conserved residues R38 and K41, implicated in dsRNA binding, project into the tubular structure to interact with dsRNA.

Whereas the RDB is involved in binding of dsRNA, the ED predominantly interacts with cellular proteins to regulate their functions (reviewed in Hale *et al.*, 2008b). The ED is crucial for RBD positioning, NS1 nuclear export and inactivation of the 30-kDa RNA-processing protein cleavage and polyadenylation specificity factor (CPSF30;

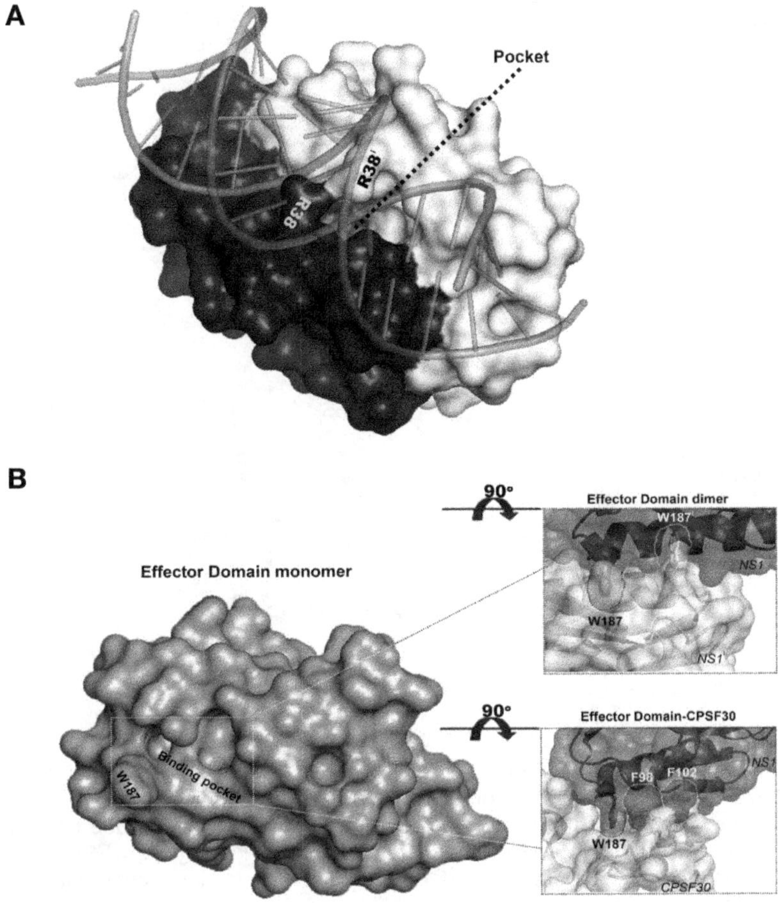

Figure 8.4 (A) Conserved binding pocket in the NS1 RBD as a promising target for small-molecule inhibitors. Two anti-parallel α-helices in the NS1 RBD homodimer form a concave surface facilitating dsRNA binding. dsRNA is anchored by hydrogen bonds from two opposite arginine residues (R38-R38′). Underneath the R38 pair lays a conserved deep pocket that forms a promising target for antiviral drug design. PyMol rendering of A/Udorn/307/1972 (H3N2) NS1 RBD (1–73) using PDB file 2ZKO. (B) Conserved binding pocket in the NS1 ED as a promising target for small-molecule inhibitors. Hydrophobic residues line the conserved binding pocket indicated. Residue W187 is critical for NS1 ED homodimer formation and CPSF30 binding. Binding of W187′ on the opposite ED monomer and residues F98 and F102 in the CPSF30 F2F3 zinc-finger region are shown (insets). The conserved pocket is an attractive target for small-molecule compounds since it would disrupt both ED dimer formation and CPSF30 interaction, decreasing the chance of escape mutants. Insets: ED residues surrounding W187, 90 rotated over X-axis, interacting with opposite ED monomer (top) or F2F3 zinc-finger region of CPSF30 (bottom). PyMol rendering of A/Udorn/307/1972 (H3N2) NS1 ED (84–205) using PDB file 3EE9.

reviewed in Hale *et al.*, 2008b). Several ED crystal structures have been solved that show similar crescent shape-like folds for the monomers (Fig. 8.4B), but distinct orientation of the ED dimers (Bornholdt and Prasad, 2006; Das *et al.*, 2008; Hale *et al.*, 2008a). This indicates that virus strain or ligand-specific conformations exist, which are discussed below.

IAV NS1 inhibits processing of cellular pre-mRNAs by sequestering CPSF30. This effectively results in inhibition of protein expression (discussed below in more detail). The A/Udorn/307/1972 (H3N2; allele A) NS1 forms a hetero-tetramer with CPSF30 zinc-finger domains F2F3 (Fig. 8.5). Each of the two CPSF30 zinc finger pairs wraps around two NS1 EDs that

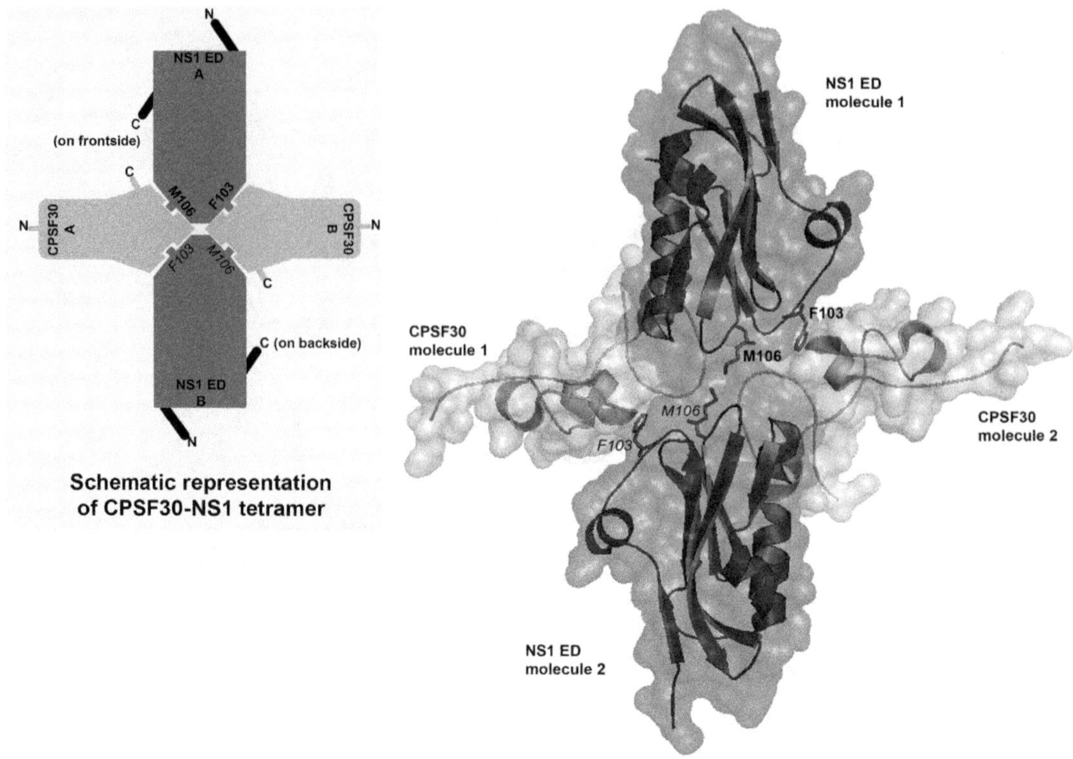

Figure 8.5 Crystal structure of CPSF30-NS1 ED tetramer. Two CPSF30 zinc-finger domains (F2F3) embrace two A/Udorn/307/1972 (H3N2) NS1 effector domains (ED). The EDs are positioned in a head-to-head orientation that facilitates docking of the CPSF30 F2F3 zinc-finger regions. Residue F103 in the NS ED mediates interaction with hydrophobic amino acids L72, Y88 and P111 on the CPSF30 molecule. NS1 residue M106 is positioned at the tetrameric interface and is critical for interaction with the other three molecules in the complex.

interact with each other head-to-head (Fig. 8.5; Das *et al.*, 2008). Residue W187 is critical for interaction with CPSF30 (Xia *et al.*, 2009). It resides adjacent to a conserved hydrophobic pocket in NS1 in which critical residues F98 and F102 from CPSF30 bind (Fig. 8.4B, bottom inset).

In contrast, the NS1 protein of the avian influenza virus strain A/Duck/Albany/1976 (H12N5; allele B) forms a homodimer, in which the two monomers are roughly perpendicularly juxtaposed near a central α-helix containing W187 (Hale *et al.*, 2008a). Biochemical experiments showed that reciprocal packing of this W187 residue into a hydrophobic pocket found in the adjacent monomer is essential for dimer formation (Hale *et al.*, 2008a). Although the monomeric structure of this avian NS1 ED resembles the structure of EDs of human isolates, its dimeric form differs from that of the following allele A human viruses. The NS1 ED of A/Puerto Rico/8/1934 (H1N1; allele A) and A/Udorn/1972 (H3N2; allele A) both form antiparallel homodimers through interactions in β-strands located near their N-termini (Fig. 8.4B, top inset; Bornholdt and Prasad, 2006; Xia *et al.*, 2009). Similarly, the ED in the full-length NS1 structure of the highly pathogenic A/Vietnam/1203/2004 (H5N1; allele A) strain adopts a fold and dimerization interface resembling the A/Puerto Rico/8/1934 ED structure. Nevertheless, the H5N1 subtype ED dimer subunits are more twisted with respect to each other, resulting in electrostatic interactions on the H5N1 subtype ED dimer surface, rather than the primarily hydrophobic interactions that constitute the A/Puerto Rico/8/1934 dimer interface (Bornholdt and Prasad, 2008).

Taken together, these structural studies indicate that the functional form of NS1 is a dimer that can adopt different conformations (depending on viral strain or bound ligand) that are mutually exclusive. It remains unclear what other biologically relevant oligomeric forms of NS1 exist, and if some dimer confirmations are artefacts from single domain crystallization, or crystallization artefacts such as crystal packing. Nevertheless, it emphasizes the multifunctional and dynamic nature of the NS1 protein. The study of more full-length NS1 structures co-crystallized with their cellular interactors (e.g. PKR and TRIM25, see below) may provide further insight into this conundrum.

NS1 localization

Many studies have described the localization of NS1. Although some inconsistencies exist that could be attributed to differences in experimental procedures and timing, they indicate that NS1 resides predominantly in the nucleus of infected cells. Nevertheless, a substantial cytoplasmic population is present, especially at later times during infection.

Current data suggest that the localization of NS1 proteins is determined by nuclear localization signals (NLS) and a nuclear export signal (NES) that reside in the NS1 protein (Fig. 8.1B). IAVs from before 1989 encode NS1 proteins that contain two NLSs, whereas the A/WSN/1933 isolate and all isolates from after 1989 have only the first NLS (Privalsky and Penhoet, 1981; Greenspan *et al.*, 1988; Melen *et al.*, 2007). NLS1 is located in the RBD and spans residues 35–41 that bind importin-α and facilitate transport through the nuclear membrane. Both Arg-38 and Lys-41 are critical for importin-α binding and NLS activity since their mutation to alanines disrupts importin-α interaction (Melen *et al.*, 2007). The same residues are important for dsRNA binding. Although Arg-38 is critical for dsRNA binding, Lys-41 is important but not essential (Hatada and Fukuda, 1992; Wang *et al.*, 1999; Zhao *et al.*, 2010). Nevertheless, NS1 can interact with both importin-α and dsRNA simultaneously, indicating that the NS1 binding sites are not fully overlapping (Melen *et al.*, 2007). Interestingly, the tubular dsRNA sequestering structures proposed by Bornholdt and Prasad (2008) would allow for a 'free' RBD at the end each tubule, that could interact with importin-α. It is currently unclear if bound dsRNA is imported into the nucleus.

The second NLS is located in the NS1 C-terminal disordered tail of subtype H3N2 and subtype H2N2 viruses and spans key residues K219/R220 and R231/R232 in a bipartite manner (Melen *et al.*, 2007). In addition, this region also functions as a nucleolar localization signal (NoLS), thereby localizing these NS1 proteins to the nucleolus (Melen *et al.*, 2007). Basic arginine or lysine residues at positions 219, 220, 231, and 232 in the A/Udorn/1972 NS1 protein are required for a functional NLS2/NoLS (Melen *et al.*, 2007). The function of nucleolar localization remains unclear. However, viruses with inactivating mutations in their second NLS were not attenuated in their ability to replicate in tissue culture (Melen *et al.*, 2007), although early studies suggest that nucleolar NS1 may inhibit rRNA synthesis (Krug and Soeiro, 1975). Thus, additional *in vivo* studies are required to determine the influence of NLS2/NoLS in replication fitness of different influenza virus strains.

IAV NS1s also contain a latent NES, which is located in the amino end of the ED (residues 138–147). The NS1 protein of NES-deletion viruses localizes almost exclusively to the nucleus, whereas a substantial amount of cytoplasmic NS1 is present in wild-type virus infections (Li *et al.*, 1998). Interestingly, overexpressed NS1 primarily has a nuclear localization (Li *et al.*, 1998). Mutational studies on residues surrounding the NES indicated that a balance between the effects of the NLSs and NES may exist, determining the localization of NS1 proteins in infected cells. The data suggest that adjacent residues can mask the NES, and that other viral proteins or virus-induced factors may unmask the NES and shift the balance of NS1 localization towards nuclear export during infection (Li *et al.*, 1998). Similarly, certain viral factors could mask the NLS sequences and shift the balance to nuclear export even further.

Although the precise mechanism of NS1 localization remains to be determined, the current models all suggest an important role of either yet unknown factors or modifications in the regulation of NS1 localization. Interestingly, a temperature sensitive virus bearing a mutation in the potentially phosphorylated NS1 S195 residue, showed mainly nuclear NS1 localization (Garaigorta *et al.*, 2005; Bornholdt and Prasad, 2006). The true effect of this mutation and its resulting predominant nuclear NS1 localization on virus replication remains obscure. However, it indicates that both protein interaction and post-translational modifications may be involved in NS1 localization, which in turn may be essential for the ability of NS1 to perform its many different functions during infection.

Pre-transcriptional antagonism of the influenza IFN antagonist NS1

Studies using recombinant IAVs with full or partial deletions in their NS1 gene provided the first evidence that the NS1 protein is involved in regulation of the IFN response (Garcia-Sastre *et al.*, 1998; Talon *et al.*, 2000a). Whereas wild-type viruses only induce minute amounts of type I IFNs, viruses lacking a functional NS1 protein induce high levels of IFN and are severely attenuated in tissue culture and various animal models (Garcia-Sastre *et al.*, 1998; Talon *et al.*, 2000a; Donelan *et al.*, 2003; Ferko *et al.*, 2004; Falcon *et al.*, 2005; Quinlivan *et al.*, 2005; Solorzano *et al.*, 2005; Richt *et al.*, 2006; Baskin *et al.*, 2007; Vincent *et al.*, 2007; Chambers *et al.*, 2009; Steel *et al.*, 2009). Restoration of replication and virulence of these viruses in STAT1$^{-/-}$ mice provided the closing evidence that attenuation of NS1 mutant viruses resulted from an inability to antagonize the type I IFN response (Garcia-Sastre *et al.*, 1998).

All natural influenza A and B virus isolates express NS1 proteins and the IFN antagonistic function seems to be conserved amongst both virus types (Garcia-Sastre *et al.*, 1998; Talon *et al.*, 2000a; Donelan *et al.*, 2004; Dauber *et al.*, 2004; Table 8.1A). Although A/NS1 and B/NS1 share less than 25% sequence identity, they both contain a conserved N-terminal RBD. Mutation of residues R38 and K41 in the NS1 RBD abolishes dsRNA binding and partially abolishes virulence and the ability to prevent IFN induction (Donelan *et al.*, 2003). Moreover, overexpression of NS1 prevented activation and nuclear translocation of IRF-3, NF-κB and ATF-2/c-Jun and hence the transcription of IFN-β (Talon *et al.*, 2000b; Wang *et al.*, 2000; Ludwig *et al.*, 2002). These observations led to the conclusion that NS1 blocks recognition of certain replicative dsRNA species by binding to them. The nature of these RNA species remains uncertain as very little or no dsRNA seems to be present in influenza virus infected cells (Talon *et al.*, 2000b; Wang *et al.*, 2000; Ludwig *et al.*, 2002; Weber *et al.*, 2006). In addition, NS1 expression was able to block activation of the IFN-β promoter by a constitutively active form of RIG-I, suggesting that additional means existed for NS1 to interfere with IFN induction that were independent of potential shielding of viral RNA from recognition by RIG-I (Guo *et al.*, 2007; Mibayashi *et al.*, 2007; Opitz *et al.*, 2007; Table 8.1A).

Recent studies have shown that TRIM25, an E3 ubiquitin ligase critical for RIG-I activation, is a direct target for IAV NS1 (Gack *et al.*, 2009). All TRIM proteins share a conserved N-terminal tripartite domain consisting of a 'really interesting new gene' (RING) domain, B-boxes and a coiled-coil domain (Ozato *et al.*, 2008). The C-terminal domains are unique to each individual class of TRIM proteins. TRIM25 possesses additional splA and ryanodine receptor (SPRY) domain in its C-terminus that mediates the interaction with the most N-terminal CARD of RIG-I.

The coiled-coil domain mediates dimerization of TRIM25, essential for its activity as an E3 ligase for RIG-I. IAV NS1 binds to the TRIM25 coiled-coil domain and thereby prevents TRIM25 dimerization and subsequent RIG-I ubiquitination and activation (Gack *et al.*, 2009). Mutational studies indicated that the E96/E97 residues in the A/NS1 protein interact with TRIM25. Interestingly, NS1 protein from human, porcine and avian IAVs interact with TRIM25, suggesting that antagonism with IFN induction through TRIM25 binding may well be a conserved trait (Gack *et al.*, 2009). So far, the potential of B/NS1 to block TRIM25-mediated RIG-I activation has not been tested, although studies show that B/NS1 can also block IFN-β induction at a pre-transcriptional level (Dauber *et al.*, 2006). Also, it remains to be determined if NS1 interacts with other members of the TRIM family, which will be of significant interest as other members of the TRIM family have also been involved in immunity against viruses (Ozato *et al.*, 2008). Moreover, TRIM25 conjugates ubiquitin and the IFN-induced ubiquitin-like modifier ISG15 to other proteins than RIG-I. The effect of NS1 interaction with TRIM25 on those targets is currently also unclear.

Inhibition of cellular mRNA maturation and export

In addition to specifically targeting the IFN induction pathway, most IAV NS1 proteins also block general gene expression. This trait is not shared by the IBV NS1 protein, or the NS1 protein of several IAVs, including the new pandemic subtype H1N1 virus (Hayman *et al.*, 2006; Hale *et al.*, 2010b). Binding of the A/NS1 effector domain to CPSF30 prevents its normal function in splicing and processing of 3′ end of pre-mRNA. As a consequence, a wide range of cellular pre-mRNAs are retained in the nucleus and prevented from being translated into potentially antiviral proteins.

An X-ray structure of the A/Udorn/307/1972 (H3N2) NS1 co-crystallized with a domain of CPSF30 comprising its second and third zinc finger motifs revealed a tetrameric complex in which the two NS1 effector domains are arranged in a head-to-head confirmation. CPSF30 is wrapped around the NS1 dimer (Das *et al.*, 2008). Aromatic residues in the third CPSF30 zinc finger fit in a hydrophobic pocket of the NS1 effector domain comprised of amino acid residues K110, I117, I119, Q121, V180, G183, G184, and W187. Amino acids F103 and M106 in NS1 fall outside of the hydrophobic binding pocket, yet were shown to be essential for the formation of the tetrameric complex. These latter residues are highly conserved amongst seasonal strains. In fact, only five strains have ever been isolated with other amino acids at these residues. Interestingly, these isolates include the A/Puerto Rico/8/1934 (H1N1) strain. These structural data explain some of the conflicting data showing that PR8 NS1 is unable to bind and inhibit CPSF30, whereas all tested natural isolates could (Das *et al.*, 2008). The PR8 NS1 protein has a hydrophilic serine residue at position 103 that eliminates binding to CPSF30 (Kochs *et al.*, 2007). Interestingly, the new pandemic subtype H1N1 virus also possesses an NS1 unable to bind to CPSF because of amino acid differences at positions 108, 125 and 189 (Hale *et al.*, 2010b). Mutations that restore CPSF binding in the pandemic subtype H1N1 NS1 do not increase virulence of this virus, indicating that CPSF binding is not required by all strains of influenza virus for pathogenicity.

In addition to CPSF30 binding, NS1 negatively affects gene expression by forming an inhibitory complex with components of the cellular RNA export machinery and thus preventing export of mature mRNA (Satterly *et al.*, 2007). Moreover, NS1 has reported poly(A) binding capability and could prevent mature mRNA from being exported from the nucleus (Qian *et al.*, 1994; Qiu and Krug, 1994).

A general block of gene expression is a feature seen in various RNA viruses other than IAV and seems to provide an efficient means to prevent synthesis of not only IFN, but also additional antiviral molecules detrimental to virus replication (reviewed in Lyles, 2000). Whereas host shut-off in some large dsDNA viruses (such as herpes simplex virus) is in part related to its necessity to temporally regulate expression of its own genes (reviewed in Smith *et al.*, 2005), RNA viruses seem to mainly block viral gene expression to circumvent the expression of antiviral proteins.

Both (+) RNA and (–) RNA viruses have been described to block general gene expression both pre- and post-transcriptionally. Unlike influenza viruses, all of them replicate in the cytoplasm, which requires viral proteins that interfere at a pre-transcriptional level, to translocate to the nucleus. Firstly, vesicular stomatitis virus (VSV), poliovirus (PV) and the two bunya viruses Rift Valley fever virus (RVFV) and Bunyamwera virus (BUN) interfere with general transcription (Smith *et al.*, 2005). The VSV M protein inactivates transcription factor subunit IID without affecting its total concentration and thereby inhibits transcription *in vitro* by >90% (Yuan *et al.*, 1998). The poliovirus 3C protease catalytically cleaves a TFIID subunit and thereby renders it inactive (Das and Dasgupta, 1993; Yalamanchili *et al.*, 1996). Despite the fact that RVFV and BUN belong to the same virus family, they have evolved slightly different ways to inhibit transcription by their NSs protein. RVFV NSs binds and sequesters subunits of TFIIH in filamentous structures (Le *et al.*, 2004), whereas BUN NSs prevents phosphorylation of a serine residue in the C-terminal domain of RNA polymerase II, thereby blocking the transition from transcription initiation to elongation (Thomas *et al.*, 2004).

Secondly, nuclear export studies have shown that the VSV M protein also interferes with nuclear export of mRNA by binding to the mRNA export factor Rae1/mrnp41 (Her *et al.*, 1997; Faria *et al.*, 2005). Additionally, the Leader protein of Theiler's murine encephalitis virus (TMEV) blocks transport of proteins from the cytoplasm to the nucleus, which mainly seems to affect nuclear transport of transcription factors during antiviral response (such as IRF-3), rather than general gene transcription (Delhaye *et al.*, 2004).

Thirdly, most picornaviruses block translation through catalytic cleavage of translation factor eIF4G by their 2A protease (Lpro in case of foot and mouth disease virus), which is required for efficient cap-dependent translation. Their own mRNAs contain an internal ribosome entry site (IRES) and are thus efficiently translated in the absence of functional eIF4G (reviewed in Porter, 1993). Moreover, non structural protein 1 (nsp1) of SARS-coronavirus has been shown to interact with the 40 ribosomal subunit in mRNAs and stall translation and potentially marking mRNA for degradation by cellular RNases (Kamitani *et al.*, 2006, 2009; Narayanan *et al.*, 2008).

In conclusion, influenza viruses are not the only viruses inhibiting global gene expression. However, since their RNA is synthesized in the nucleus and they do not possess an IRES element, they do so by their own unique mechanisms of cap-snatching and by preventing pre-mRNA processing by CPSF30 binding and nuclear export of mature mRNA.

IFN-induced antiviral effectors: influenza virus sensitivity and antagonism

In addition to preventing type I IFN induction at the transcriptional level, influenza viruses also antagonize the function of several prominent IFN-induced antiviral effectors (Table 8.1A). Since infected cells do not produce significant amounts of type I IFNs, this line of defence is especially important for viral spread *in vivo* at the stage where lymphocytes (such as pDCs) excrete large amounts of type I IFNs, which spread systemically to confer an antiviral state in uninfected cells (Table 8.1B).

The NS1 protein has been reported to interact with many other viral and cellular proteins than described above and in Table 8.1A, thereby potentially influencing viral protein expression, apoptosis and cell cycle regulation (reviewed in Hale *et al.*, 2008b). Since these proteins do not directly influence the IFN system and thus fall outside the scope of this chapter, they will not be discussed further in detail. Nevertheless, their multitude once more underpins the multifunctional nature of the influenza virus NS1 protein.

Mx proteins

Mx proteins are IFN-induced members of the dynamin superfamily of large GTPases. They inhibit a wide range of viruses by blocking early stages in their replication (Haller *et al.*, 2006). The human MxA protein confers an antiviral effect against most influenza virus strains (Haller *et al.*, 2006; Dittmann *et al.*, 2008). Interestingly, the effectiveness of Mx-mediated antiviral activity is different for each viral strain and is thus regarded as an important virulence factor (Dittmann *et al.*, 2008). Avian influenza strains are very sensitive to MxA action, whereas most contemporary influenza virus strains are mildly inhibited by MxA. In combination with the observation that the highly pathogenic A/Brevig Mission/1/1918 'Spanish influenza' strain is completely insensitive to MxA, this underpins the importance of MxA as a restriction factor (Dittmann *et al.*, 2008).

So far, no direct means in influenza virus to counteract the effects of Mx proteins have been described. Nevertheless, recent studies indicate that enhanced replication and dissemination speed resulting from mutations in the viral polymerase and HA protein, can tip the balance between viral replication and antiviral responses in favour of the virus, effectively enabling it to outrun the innate immune system (Grimm *et al.*, 2007; Rolling *et al.*, 2009). This observation underlines the importance of initial control of virus replication by the IFN system.

Double-stranded RNA induced 2′,5′-oligoadenylate synthetase (OAS) and protein kinase R (PKR)

NS1 can directly block the function of the two IFN-induced antiviral proteins 2′,5′-oligoadenylate synthetase (OAS), which activates the latent antiviral RNaseL, and the dsRNA-dependent protein kinase R (PKR; Min and Krug, 2006; Min *et al.*, 2007). OAS/RNase L and PKR can restrict viral transcription and translation and participate in IFN-β induction (reviewed in Garcia *et al.*, 2006a,b; Sen and Peters, 2007; Silverman, 2007).

OAS synthesizes 2′,5′ linked oligoadenylates in the presence of dsRNA. In turn, these chains activate RNase L, which inhibits virus replication by degradation of RNA (Sen and Peters, 2007; Silverman, 2007). Infection studies with recombinant influenza viruses with NS1 proteins deficient for dsRNA binding, showed that the replication of these viruses was 1000-fold inhibited by IFN treatment compared with wild-type viruses. RNAi of OAS and OAS knock-out cells demonstrated that the sensitivity was primarily the result of the activation of RNase L and that the dsRNA binding capacity of NS1 is predominantly important to prevent OAS/RNaseL activation (Min and Krug, 2006). Although several studies suggested that NS1-mediated OAS inactivation also contributes to suppression of IFN-β synthesis (Donelan *et al.*, 2003; Talon *et al.*, 2000), recent experiments to identify RIG-I activating RNAs during IAV infection indicate that 2′,5′ oligoadenylate chains do not play a significant role in type I IFN induction (Rehwinkel *et al.*, 2010).

Similar to RIG-I, PKR is also activated by 5′-triphosphate bearing RNA and dsRNA (Garcia *et al.*, 2006a,b; Nallagatla *et al.*, 2007; Dauber *et al.*, 2009). PKR is also a direct target of the PACT protein, which can activate in the absence of RNA. Upon binding of activating RNA, auto-inhibition is released and PKR activated. Activated PKR predominantly represses translation by phosphorylating eukaryotic translation initiation factor 2α (eIF2α; reviewed in Garcia *et al.*, 2006a,b). Mutant IAVs expressing RNA binding-defective NS1 proteins fail to efficiently limited PKR activation in response to dsRNA or PACT (Min and Krug, 2006). Although *in vitro* experiments initially suggested that this resulted from lack of dsRNA sequestering by NS1 and hence limiting the pool of PKR activators (Lu *et al.*, 1995; Hatada *et al.*, 1999), these studies showed that direct interaction of A/NS1 RBD residues 123–127 with PKR antagonizes PKR function in a dsRNA-independent manner (Tan and Katze, 1998; Li *et al.*, 2006; Min *et al.*, 2007). Mapping studies indicate A/NS1 interaction with a linker region in PKR, thereby preventing a conformational change required for release of PKR auto-inhibition (Li *et al.*, 2006). Such a mechanism would allow A/NS1 to circumvent both dsRNA- and PACT-mediated inhibition of translation by PKR. Also IBV mutants expressing only the NS1 RDB were able to prevent PKR activation (Dauber *et al.*, 2006). However, these mutants were unable to block IFN induction. The data suggested that B/NS1 antagonizes PKR by interacting with it in a dsRNA dependent manner (Dauber *et al.*, 2006).

In addition to PKR inhibition by NS1 interaction, several studies have shown that the cellular PKR inhibitor p58(IPK) is activated during IAV infection and contributes to negative regulation of PKR by direct protein–protein interaction (Lee *et al.*, 1990, 1992, 1994; Melville *et al.*, 1997, 1999). It remains unclear if this mode of PKR inhibition by cellular components is a general cellular stress-induced negative feedback loop also activated during infection with other viruses or an influenza virus-specific means to counter the detrimental effects of PKR. Although the full molecular mechanism by which p58(IPK) inhibits PKR is not fully understood, recent studies suggest that influenza virus infection promotes dissociation of the p58(IPK) inhibitor heat shock protein 70 (HSP70) from it (Melville *et al.*, 1999).

Ubiquitin-like modifier IFN-stimulated gene 15

ISG15 is an IFN-induced ubiquitin-like molecule that is conjugated to target proteins by specific E1-activating, E2-conjugating, and E3-ligase enzymes (that are also IFN-inducible; reviewed in Harty *et al.*, 2009). ISG15 exhibits potent antiviral activity against several viruses, including influenza viruses (Lenschow *et al.*, 2007). Although the precise functions of ISG15 during IFN induction remain poorly understood, a recent report suggests that it may be conjugated to NLS1 in the NS1 protein of IAVs, thereby inhibiting its nuclear import and nuclear functions (Zhao *et al.*, 2010).

Conjugation of ISG15 to target proteins is essential for its anti-viral function (Lenschow *et al.*, 2007; Harty *et al.*, 2009), and several viruses counteract host ISGylation by encoding ISG15-deconjugating enzymes (Lindner *et al.*, 2005; Yang *et al.*, 2005; Frias-Staheli *et al.*, 2007; Chang *et al.*, 2008; Sommereyns *et al.*, 2008; Harty *et al.*, 2009). The NS1 protein of IBV NS1 protein prevents ISGylation by binding unconjugated ISG15 (Yuan and Krug, 2001). The precise mechanism by which B/NS1 binding to ISG15 prevents its conjugation is unclear, yet specific inhibition of ISG15–E1 adduct formation (Yuan and Krug, 2001) and/or a steric block of the ISG15:E3 interaction have been proposed (Randall and Goodbourn, 2008). Recent studies indicate that the B/NS1 protein only interacts with human ISG15, but not mouse ISG15 (Sridharan *et al.*, 2010; Versteeg *et al.*, 2010). These data could provide an explanation for the observation that ISG15 inhibits IBV replication in mice (Lenschow *et al.*, 2007).

Novel live-attenuated vaccine candidates with disruptions in the IFN antagonist NS1

The host immune response to influenza virus infection is largely determined by the immune system's recognition of HA and NA. Currently, two types of influenza vaccines are FDA approved, both of which act by eliciting neutralizing HA and NA specific antibodies. The first one is the trivalent inactivated vaccine that consists of partially purified HA and NA proteins from two circulating IAVs and one IBV. This vaccine is administered through injection. The second vaccine is the live attenuated influenza vaccine. This vaccine is also trivalent and contains cold-adapted viruses that are genetically modified to express the HA and NA molecules of circulating strains. The cold adaptations promote replication in the colder nasal mucosa and attenuate the virus, minimizing symptoms of illness.

The influenza virus RdRp is very error-prone and can successfully generate HA and NA escape mutants (antigenic drift) that are inefficiently neutralized by existing antibodies. This occurs with such a speed that influenza vaccines with updated HA and NA variants need to be formulated every two to five years (Boni, 2008). Although current influenza vaccines are effective if correctly matched to circulating strains, various new approaches are currently developed to generate second-generation vaccines that are less affected by antigenic drift. In addition to attempts to generate neutralizing antibodies to highly conserved epitopes in the influenza virus surface (Jegerlehner *et al.*, 2004), HA epitopes that will elicit cross-neutralizing antibodies (Steel *et al.*, 2010a; Wang *et al.*, 2010) and NP/M1/M2 DNA vaccination (Rao *et al.*, 2010), also influenza viruses with adaptations in their NS1 genes are developed as live attenuated vaccines (reviewed in Richt and Garcia-Sastre, 2009).

Recombinant influenza viruses expressing truncated or mutated NS1 proteins express high levels of IFN and are attenuated in tissue culture and various animal models (Donelan *et al.*, 2003; Richt *et al.*, 2006; Vincent *et al.*, 2007; Steel *et al.*, 2009). Moreover, these NS1 mutants do not repress IFN induction, dendritic cell maturation and subsequent adaptive immune responses required for efficient generation of protective and memory antibody responses (Lopez *et al.*, 2003; Fernandez-Sesma *et al.*, 2006; Phipps-Yonas *et al.*, 2008; Longhi *et al.*, 2009). Indeed, NS1 truncation mutants have been shown not to block DC maturation, allowing T cell priming (Haye *et al.*, 2009) and enhanced elicitation of immunogenicity (Talon *et al.*, 2000a; Ferko *et al.*, 2004; Steel *et al.*, 2009).

The efficacy of recombinant influenza viruses with truncated NS1 genes has been demonstrated in various infection models, including macaques (Donelan *et al.*, 2003; Ferko *et al.*, 2004; Quinlivan *et al.*, 2005; Solorzano *et al.*, 2005; Richt *et al.*, 2006; Vincent *et al.*, 2007; Baskin *et al.*, 2007; Chambers *et al.*, 2009; Steel *et al.*, 2009) and seems an efficient strategy to generate live-adapted vaccines for both influenza A and B viruses (Hai *et al.*, 2008; Wressnigg *et al.*, 2009a,b).

Current optimization of the NS1 truncation or mutation viruses for use as vaccines focuses on achieving an optimal balance between minimal clinical symptoms and maximum replication in the nasal mucosa. The latter would require decreased doses of vaccine and is thought to stimulate mucosal immunity that is especially effective against respiratory viruses such as influenza virus (Yuki and Kiyono, 2009). In addition, current immunological studies are aimed to better understand how immune responses can be skewed towards desired neutralizing antibody production in combination with a potent cytolytic and memory response (Katz *et al.*, 2004). Using combined immunization strategies with novel adjuvants, immunomodulators and immunogens may prove to be critical for developing broadly neutralizing influenza vaccines (Katz *et al.*, 2004; Wang *et al.*, 2010).

NS1 attenuation provides promising means to generate an attenuated virus that can be grown in large quantities in IFN-deficient hosts. Nevertheless, there are some drawbacks that indicate that we are still not close to the ideal influenza vaccine. For instance, the strategy only generates an attenuated viral backbone, which implies that new vaccines with 'updated' HA and NA genes will still need to be generated. The current live attenuated vaccine can be used by individuals up to 49 years, but is contraindicated in older populations for whom annual influenza vaccination is recommended (McCarthy and Kockler, 2004). Future studies are needed to determine if NS1-attenuation-based live influenza vaccines can be safely and efficiently used in older individuals.

Novel antivirals targeting the influenza virus type I IFN antagonist NS1

Influenza vaccination is the most effective method for preventing influenza virus infection and disease. However, during pandemics with novel influenza viruses against which no prior immunity or vaccine exists, in immune compromised patients and during years of severe influenza epidemics owing to vaccine mismatch with the circulating strains, antiviral drugs are critical to treat acute life-threatening cases while bridging the time it takes to produce matching vaccines.

Two classes of influenza specific antivirals are currently licensed in the US. However, resistance of currently circulating seasonal viruses exists. The first group consists of the NA inhibitors oseltamivir and zanamivir. They interact with the NA proteins, thereby interfering with release of progeny virus particles from infected cells and subsequent spread. All circulating seasonal subtype H1N1 strains previous to the 2009 pandemic were resistant to oseltamivir, yet remained sensitive to zanamivir (CDC weekly U.S. influenza surveillance report week 12, ending 27 March 2010: http://www.cdc.gov/flu/weekly/). On the other hand, the novel swine-origin 2009 subtype H1N1 virus remains largely sensitive to both oseltamivir (~1.2% resistant) and zanamivir (no resistance). Currently circulating subtype H3N2 and IBVs remain sensitive to both classes of NA inhibitors. The highly pathogenic avian subtype H5N1 strains are still predominantly sensitive to the current NA inhibitors, although some oseltamivir-resistant mutants have been reported (Le *et al.*, 2005).

The second class of antivirals consists of the adamantanes (amantadine and rimantadine). They bind the M2 protein in virions and thereby prevent proper proton channel function critical for uncoating of entering viruses. Adamantanes are only effective against IAVs. Also, all currently circulating subtype H3N2 viruses and the swine-origin 2009 subtype H1N1 virus (98%) are resistant to adamantanes. In contrast, seasonal subtype H1N1 viruses remain sensitive.

Thus, although most strains remain sensitive to at least one of the classes of antivirals, treatment requires time-consuming strain sub-typing or treatment with a combination of different antivirals. With rising resistance against current antivirals of seasonal and potentially pandemic strains, it will be important to develop novel drugs. Several classes of new influenza virus antivirals are currently being developed and include (i) inhibitors of HA cleavage, (ii) inhibitors of fusion and entry, (iii) inhibitors of viral RNPs, polymerase function or endonuclease activity, (iv) inhibitors of host kinases required for efficient viral replication, (v) inhibitors of NA and (vi) NS1 inhibitors (Hsieh and Hsu, 2007; Beigel and Bray, 2008). Here we discuss inhibition of the IFN antagonist NS1 as a potential antiviral approach.

In theory, NS1 could be targeted by small-molecule compounds in all the parts of the protein that have been shown to contribute to subversion of critical host-cell responses. However, not all cellular interaction partners or the NS1 residues they interact with are known. Moreover, some of the NS1 functions are attributed to the disordered C-terminal tail. The lack of a stable structure in this region makes it virtually impossible to rationally design small compounds. Perhaps, more stable structures of the C-terminal region in complex with specific ligands will be solved in the future. However, so far the RBD and the ED are the best targets for new antiviral approaches since both their structures have been solved and influenza viruses expressing NS1 proteins lacking specific functions of these domains are attenuated *in vitro* and *in vivo* (Donelan *et al.*, 2003; Noah *et al.*, 2003; Dauber *et al.*, 2004, 2006; Hale *et al.*, 2006, 2010a; Min and Krug, 2006; Min *et al.*, 2007; Shin *et al.*, 2007; Das *et al.*, 2008; Ehrhardt and Ludwig, 2009; Gack *et al.*, 2009).

The NS1 RNA-binding domain

The NS1 RBD is mainly involved in dsRNA sequestration to prevent PKR, OAS/RNasL and IFN-β promoter activation by sequestering dsRNA (Dauber *et al.*, 2004; Dauber *et al.*, 2006; Li *et al.*, 2006; Min and Krug, 2006; Kochs *et al.*, 2007). Since viruses in which these functions are blocked by point mutations are severely attenuated, the RBD is a promising target for antiviral compounds. Moreover, NMR and crystal structures of several influenza A and B virus RBDs are available allowing rational design of antiviral compounds (Chien *et al.*, 1997, 2004; Liu *et al.*, 1997; Yin *et al.*, 2007; Bornholdt and Prasad, 2008). The structural data indicate that the six-helical symmetric head-to-tail homodimer structure of the RDB is found in both influenza A and B viruses (Yin *et al.*, 2007). Moreover, this fold is absent in all known cellular proteins (Liu *et al.*, 1997), hence making potential RDB-inhibitors effective against both virus subtypes and reducing the chance of toxic effects by perturbing function of cellular proteins.

The two anti-parallel tracks formed by the second alpha-helices of both NS1 molecules in the dimer form a hydrophilic, charged binding surface that facilitates interaction with dsRNA (Fig. 8.4A). Particularly the R38 residue penetrating into the helix from both molecules to form hydrogen bonds with the RNA phosphate backbone is critical for dsRNA binding (Yin *et al.*, 2007). This pocket is an inviting target for antiviral drug design (Fig. 8.4A). Although only R38 is critical for dsRNA binding, several additional residues in this pocket positively contribute to this function, decreasing the chance of generating escape mutants.

The NS1 effector domain

The NS1 ED dimerization stabilizes RBD structure and is required for efficient formation of dsRNA binding tubules by the RBD (Bornholdt and Prasad, 2008). Moreover, it inhibits (i) PKR function by direct interaction, (ii) RIG-I activation by binding to TRIM25 and (iii) gene expression by interaction with CPSF30 (Nemeroff *et al.*, 1998; Noah *et al.*, 2003; Twu *et al.*, 2006; Min *et al.*, 2007; Das *et al.*, 2008; Gack *et al.*, 2009). Although the residues in the NS1 protein required for interaction with PKR and TRIM25 have been mapped (Fig. 8.1B), no crystal structures exist of NS1 bound to either of those two cellular proteins. Future co-crystal structures will have to provide further insight in potentially important binding surfaces and pockets. The structure of the NS1 protein in complex with CPSF30 revealed a tetramer with the second and third zinc-finger of CPSF30 (Das *et al.*, 2008). This previously unpredicted conformation underpins the necessity to solve the structure of PKR and TRIM25 in complex with NS1.

Recombinant IAVs expressing NS1 proteins unable to interact with CPSF30 are attenuated and do not inhibit pre-mRNA processing (Das *et al.*, 2008; Krug and Aramini, 2009). A largely hydrophobic binding pocket in NS1 containing several highly conserved residues forms the interaction surface with aromatic side chains from the F3 zinc finger of CPSF30 (Das *et al.*, 2008).

The same hydrophobic binding pocket is critical for homodimerization of the ED. Dimerization requires key residue W187 on each NS1 ED monomer to reciprocally dock into the hydrophobic pocket of the neighbouring ED monomer (Hale *et al.*, 2008a; Xia *et al.*, 2009). These two mutually exclusive conformations, both critical for efficient influenza virus replication, require binding in the in the same hydrophobic pocket, making it a prime target for antiviral small molecules (Fig. 8.4B). Moreover, the requirement of the same pocket for two different viral functions makes it less likely that viral escape mutants will occur. The first proof of principle that competitive binding with the hydrophobic pocket in the ED can perturb NS1 function and attenuate viral replication was generated by overexpressing a nucleus-targeted F2F3 region of CPSF30 (Twu *et al.*, 2006). Replication and expression studies showed that this fragment efficiently competes for NS1 binding with endogenous CPSF30, thereby lowering IAV replication and increasing IFN-β expression without affecting cellular mRNA processing. Nevertheless, the fact that specific influenza virus strains are able to cause disease despite the inability of their NS1 proteins to inhibit CPSF would suggest that influenza virus could overcome antiviral inhibition that targets NS1 binding to CPSF.

Another approach currently used to identify new antiviral lead compounds is high-throughput screening with small-compound libraries that interfere with particular functions of viral proteins such as NS1. Using this approach, a recent study identified four compounds that reversed NS1-mediated growth inhibition in yeast (Basu *et al.*, 2009). Follow-up experiments confirmed that the compounds blocked the ability of NS1 to inhibit IFN-β promoter activation by dsRNA, but not the activity of proteins from other RNA viruses (Basu *et al.*, 2009). Although the potential of these compounds to inhibit virus replication *in vivo* remains unclear, they prove that targeting NS1 function, at least in tissue culture, is a viable strategy for the development of antiviral drugs. Moreover, solving their mode of action may reveal molecular details on NS1 structure and function that could form the basis for rational design of a next generation of NS1 inhibitors.

Concluding remarks

Differences in pathogenicity in different animals indicate a large host component contributing to influenza virus pathogenicity (Tumpey *et al.*, 2005, 2007a,b; Kobasa *et al.*, 2007; Lipatov *et al.*, 2008, 2009; Weingartl *et al.*, 2009; Van Hoeven *et al.*, 2009). Although NS1 contributes significantly to pathogenesis, other factors such as compatibility of the viral polymerase complex with a host and HA-compatibility for efficient spread are crucial as well (Tumpey *et al.*, 2005, 2007a). Also recruitment of different lymphocytes and excretion of cytokines contribute to pathogenicity (Peper and Van Campen, 1995; Xu *et al.*, 2004; Jayasekera *et al.*, 2006; Snelgrove *et al.*, 2006; Conenello *et al.*, 2007; Salomon *et al.*, 2007; Deng *et al.*, 2008; Schmolke *et al.*, 2009; Garigliany *et al.*, 2010; Nakamura *et al.*, 2010). These factors and others highlighted in this chapter underpin the complexity of virus–host interactions and determinants of pathogenicity.

Like for all RNA viruses, the influenza virus RdRp has high mutation rates compared with DNA polymerases. The human genome is copied with less than 1 mistake per 10^7–10^8 synthesized basepairs (reviewed in Kunkel, 2004). In contrast, the influenza virus RdRp makes roughly 1 mistake per 10^4 synthesized bases (Drake, 1993), which translates to roughly one mutation per synthesized genome. As a result of this high mutation rate, virtually none of the genomes in the cell is identical and the pool of genomes exists as a so-called 'quasispecies' (Bull *et al.*, 2005; Domingo and Gomez, 2007). Studies with polioviruses expressing an RdRp with increased fidelity and decreased quasispecies complexity showed that quasispecies diversity determines pathogenesis through cooperative interactions in a viral population (Vignuzzi *et al.*, 2006).

Studies on how intracellular virus diversity contributes to replication fitness of influenza virus and the multifunctional nature of proteins such as NS1 have been long hampered by the inability to sequence all viral genomes in infected cells. However new techniques such as affordable high-throughput sequencing will aid the studies on the dynamics of virus populations and their impact on transmission and host manipulation. Part of the multifunctional nature of NS1 is determined by differential interaction to various cellular and viral ligands (Bornholdt and Prasad, 2008; Das *et al.*, 2008). However, an interesting hypothesis is that NS1 function, at the level of localization and binding surfaces, could be regulated by directed mutations in specific areas of the genome. The RdRp could preferentially accumulate mutations in the NLS or NES regions, thereby potentially affecting the localization of a subset of NS1 molecules. Similarly, mutations could specifically accumulate in one of the two functional domains to fine-tune the interaction of a subset of NS1 molecules with particular cellular partners.

High through-put siRNA and small molecule screens have provided an increasing wealth of data on cellular restriction factors and cellular factors required for viral replication (Hao *et al.*, 2008; Brass *et al.*, 2009; Karlas *et al.*, 2010; Konig *et al.*, 2010). The latter could be potential drug targets, whereas cellular restriction factors could be suppressed to facilitate enhanced virus production for vaccine purposes. These screens were performed in immortalized cells and even cells from non-mammalian species. Although some of the host-factors required for influenza virus replication were confirmed by siRNA experiments, the function of these genes *in vivo* remains to be tested. There is an increasing realization that many cell-type and tissue-specific interactions exist during influenza virus infection *in vivo*. To answer questions on pathogenicity determinants, increasing use of primary cell culture and *in vivo* systems will be crucial to better understand the delicate balance between host and viral factors. Moreover, a better understanding of how influenza virus regulates different cell populations and secretion of cytokines will (in combination with a better understanding of adjuvants and immunogens) contribute to developing improved vaccines and antivirals that limit the action of the influenza virus IFN antagonist NS1.

Acknowledgements

Research in the A.G.-S. laboratory is partly supported by NIH funding: R01 AI46954, U54 AI57158 (North East Biodefense Center), U19AI83025, U01AI070469, U01AI074539, P01AI058113, P01AI082325 and by CRIP (Center for Research on Influenza Pathogenesis, NIAID contract HHSN266200700010C).

Disclosure

Mount Sinai School of Medicine owns intellectual property in influenza virus reverse genetics and vaccines. A.G.-S. owns equity and receives financial compensation for providing consulting services and for serving on the advisory board of Vivaldi BioScience, a biotechnology company that develops influenza vaccines based on NS-1-modification technology.

The authors have no other relevant affiliations or financial involvement with any organization or entity with a financial interest in, or financial conflict with, the subjects discussed in the manuscript apart from those disclosed. No writing assistance was utilized in the production of this manuscript.

References

Abdel-Ghafar, A.N., Chotpitayasunondh, T., Gao, Z., Hayden, F.G., Nguyen, D.H., de Jong, M.D., Naghdaliyev, A., Peiris, J.S., Shindo, N., Soeroso, S., and Uyeki, T.M. (2008). Update on AVIAN INFLUENZA A (H5N1) virus infection in humans. N. Engl. J. Med. *358*, 261–273.

Albert, M.L., Sauter, B., and Bhardwaj, N. (1998). Dendritic cells acquire antigen from apoptotic cells and induce class I-restricted CTLs. Nature *392*, 86–89.

Ank, N., West, H., Bartholdy, C., Eriksson, K., Thomsen, A.R., and Paludan, S.R. (2006). Lambda interferon (IFN-lambda), a type III IFN, is induced by viruses and IFNs and displays potent antiviral activity against select virus infections *in vivo*. J. Virol. *80*, 4501–4509.

Badovinac, V.P., Tvinnereim, A.R., and Harty, J.T. (2000). Regulation of antigen-specific CD8+ T cell homeostasis by perforin and interferon-gamma. Science *290*, 1354–1357.

Banchereau, J., and Steinman, R.M. (1998). Dendritic cells and the control of immunity. Nature *392*, 245–252.

Baskin, C.R., Bielefeldt-Ohmann, H., Garcia-Sastre, A., Tumpey, T.M., Van Hoeven, N., Carter, V.S., Thomas, M.J., Proll, S., Solorzano, A., Billharz, R., Fornek, J.L., Thomas, S., Chen, C.H., Clark, E.A., Murali-Krishna, K., and Katze, M.G. (2007). Functional genomic and serological analysis of the protective immune response resulting from vaccination of macaques with an NS1-truncated influenza virus. J. Virol. *81*, 11817–11827.

Basu, D., Walkiewicz, M.P., Frieman, M., Baric, R.S., Auble, D.T., and Engel, D.A. (2009). Novel influenza virus NS1 antagonists block replication and restore innate immune function. J. Virol. *83*, 1881–1891.

Baum, A., and Garcia-Sastre, A. (2010). Induction of type I interferon by RNA viruses: cellular receptors and their substrates. Amino Acids *38*, 1283–1299.

Baumgarth, N., and Kelso, A. (1996). *In vivo* blockade of gamma interferon affects the influenza virus-induced humoral and the local cellular immune response in lung tissue. J. Virol. *70*, 4411–4418.

Beigel, J.H., Farrar, J., Han, A.M., Hayden, F.G., Hyer, R., de Jong, M.D., Lochindarat, S., Nguyen, T.K., Nguyen, T.H., Tran, T.H., Nicoll, A., Touch, S., and Yuen, K.Y. (2005). Avian Influenza A (H5N1) Infection in Humans. N. Engl. J. Med. *353*, 1374–1385.

Beigel, J., and Bray, M. (2008). Current and future antiviral therapy of severe seasonal and avian influenza. Antiviral. Res. *78*, 91–102.

Bender, B.S., Croghan, T., Zhang, L., and Small, P.A., Jr. (1992). Transgenic mice lacking class I major histocompatibility complex-restricted T cells have delayed viral clearance and increased mortality after influenza virus challenge. J. Exp. Med. *175*, 1143–1145.

Bhardwaj, N., Bender, A., Gonzalez, N., Bui, L.K., Garrett, M.C., and Steinman, R.M. (1994). Influenza virus-infected dendritic cells stimulate strong proliferative and cytolytic responses from human CD8+ T cells. J. Clin. Invest. *94*, 797–807.

Boni, M.F. (2008). Vaccination and antigenic drift in influenza. Vaccine *26* Suppl 3, C8–14.

Bornholdt, Z.A., and Prasad, B.V.V. (2006). X-ray structure of influenza virus NS1 effector domain. Nat. Struct. Mol. Biol. *13*, 559–560.

Bornholdt, Z.A., and Prasad, B.V.V. (2008). X-ray structure of NS1 from a highly pathogenic H5N1 influenza virus. Nature *456*, 985–988.

Bot, A., Bot, S., and Bona, C.A. (1998). Protective role of gamma interferon during the recall response to influenza virus. J. Virol. *72*, 6637–6645.

Bouvier, N.M., and Palese, P. (2008). The biology of influenza viruses. Vaccine *26* Suppl 4, D49-D53.

Brass, A.L., Huang, I.C., Benita, Y., John, S.P., Krishnan, M.N., Feeley, E.M., Ryan, B.J., Weyer, J.L., van der Weyden, L., Fikrig, E., Adams, D.J., Xavier, R.J., Farzan, M., and Elledge, S.J. (2009b). The IFITM proteins mediate cellular resistance to influenza A H1N1 virus, West Nile virus, and dengue virus. Cell *139*, 1243–1254.

Brydon, E.W.A., Smith, H., and Sweet, C. (2003). Influenza A virus-induced apoptosis in bronchiolar epithelial (NCI-H292) cells limits pro-inflammatory cytokine release. J. Gen. Virol. *84*, 2389–2400.

Bull, J.J., Meyers, L.A., and Lachmann, M. (2005). Quasispecies Made Simple. PLoS Comput. Biol. *1*, e61.

Chambers, T.M., Quinlivan, M., Sturgill, T., Cullinane, A., Horohov, D.W., Zamarin, D., Arkins, S., Garcia-Sastre, A., and Palese, P. (2009). Influenza A viruses with truncated NS1 as modified live virus vaccines: pilot studies of safety and efficacy in horses. Equine Vet. J. *41*, 87–92.

Chang, Y.G., Yan, X.Z., Xie, Y.Y., Gao, X.C., Song, A.X., Zhang, D.E., and Hu, H.Y. (2008). Different roles for two ubiquitin-like domains of ISG15 in protein modification. J. Biol. Chem. *283*, 13370–13377.

Chen, W., Calvo, P.A., Malide, D., Gibbs, J., Schubert, U., Bacik, I., Basta, S., O'Neill, R., Schickli, J., Palese, P., Henklein, P., Bennink, J.R., and Yewdell, J.W. (2001). A novel influenza A virus mitochondrial protein that induces cell death. Nat. Med. *7*, 1306–1312.

Chien, C.Y., Tejero, R., Huang, Y., Zimmerman, D.E., Rios, C.B., Krug, R.M., and Montelione, G.T. (1997). A novel RNA-binding motif in influenza A virus non-structural protein Nat. Struct. Mol. Biol. *4*, 891–895.

Chien, C.Y., Xu, Y., Xiao, R., Aramini, J.M., Sahasrabudhe, P.V., Krug, R.M., and Montelione, G.T. (2004). Biophysical characterization of the complex between double-stranded RNA and the N-terminal domain of the NS1 protein from influenza A virus: evidence for a novel RNA-binding mode. Biochemistry *43*, 1950–1962.

Conenello, G.M., Zamarin, D., Perrone, L.A., Tumpey, T., and Palese, P. (2007). A single mutation in the PB1-F2 of H5N1 (HK/97) and 1918 influenza A viruses contributes to increased virulence. PLoS Pathog. 3, 1414–1421.

Cui, S., Eisenacher, K., Kirchhofer, A., Brzozka, K., Lammens, A., Lammens, K., Fujita, T., Conzelmann, K.K., Krug, A., and Hopfner, K.P. (2008). The C-terminal regulatory domain is the RNA 5'-triphosphate sensor of RIG-I. Mol. Cell *29*, 169–179.

Culley, F.J. (2009). Natural killer cells in infection and inflammation of the lung. Immunology *128*, 151–163.

Curtsinger, J.M., Schmidt, C.S., Mondino, A., Lins, D.C., Kedl, R.M., Jenkins, M.K., and Mescher, M.F. (1999). Inflammatory Cytokines Provide a Third Signal for Activation of Naive CD4+ and CD8+ T Cells. J. Immunol. *162*, 3256–3262.

Dalton, D.K., Haynes, L., Chu, C.Q., Swain, S.L., and Wittmer, S. (2000). Interferon gamma Eliminates Responding CD4 T Cells during Mycobacterial Infection by Inducing Apoptosis of Activated CD4 T Cells. J. Exp. Med. *192*, 117–122.

Das, K., Ma, L.C., Xiao, R., Radvansky, B., Aramini, J., Zhao, L., Marklund, J., Kuo, R.L., Twu, K.Y., Arnold, E., Krug, R.M., and Montelione, G.T. (2008). Structural basis for suppression of a host antiviral response by influenza A virus. Proc. Natl. Acad. Sci. U.S.A. *105*, 13093–13098.

Das, S., and Dasgupta, A. (1993). Identification of the cleavage site and determinants required for poliovirus 3CPro-catalyzed cleavage of human TATA-binding transcription factor TBP. J. Virol. *67*, 3326–3331.

Dauber, B., Martinez-Sobrido, L., Schneider, J., Hai, R., Waibler, Z., Kalinke, U., Garcia-Sastre, A., and Wolff, T. (2009). Influenza B virus ribonucleoprotein is a potent activator of the antiviral kinase PKR. PLoS Pathog. *5*, e1000473.

Dauber, B., Heins, G., and Wolff, T. (2004). The Influenza B Virus Nonstructural NS1 Protein Is Essential for Efficient Viral Growth and Antagonizes Beta Interferon Induction. J. Virol. *78*, 1865–1872.

Dauber, B., Schneider, J., and Wolff, T. (2006). Double-stranded rna binding of influenza b virus nonstructural NS1 protein inhibits protein kinase R but is not essential to antagonize production of alpha/beta interferon. J. Virol. *80*, 11667–11677.

Delhaye, S., van Pesch, V., and Michiels, T. (2004). The leader protein of Theiler's virus interferes with nucleocytoplasmic trafficking of cellular proteins. J. Virol. 78, 4357–4362.

Deng, R., Lu, M., Korteweg, C., Gao, Z., McNutt, M.A., Ye, J., Zhang, T., and Gu, J. (2008). Distinctly different expression of cytokines and chemokines in the lungs of two H5N1 avian influenza patients. J. Pathol. *216*, 328–336.

Der, S.D., Zhou, A., Williams, B.R., and Silverman, R.H. (1998). Identification of genes differentially regulated by interferon alpha, beta, or gamma using oligonucleotide arrays. Proc. Natl. Acad. Sci. U.S.A. *95*, 15623–15628.

Dias, A., Bouvier, D., Crepin, T., McCarthy, A.A., Hart, D.J., Baudin, F., Cusack, S., and Ruigrok, R.W.H. (2009). The cap-snatching endonuclease of influenza virus polymerase resides in the PA subunit. Nature *458*, 914–918.

Diebold, S.S., Kaisho, T., Hemmi, H., Akira, S., and Reis e Sousa, C. (2004). Innate antiviral responses by means of TLR7-mediated recognition of single-stranded RNA. Science *303*, 1529–1531.

Dittmann, J., Stertz, S., Grimm, D., Steel, J., Garcia-Sastre, A., Haller, O., and Kochs, G. (2008). Influenza A virus strains differ in sensitivity to the antiviral action of Mx-GTPase. J. Virol. *82*, 3624–3631.

Domingo, E., and Gomez, J. (2007). Quasispecies and its impact on viral hepatitis. Virus Res .*127*, 131–150.

Donelan, N.R., Basler, C.F., and Garcia-Sastre, A. (2003). A recombinant influenza A virus expressing an RNA-binding-defective NS1 protein induces high levels of beta interferon and is attenuated in mice. J. Virol. *77*, 13257–13266.

Donelan, N.R., Dauber, B., Wang, X., Basler, C.F., Wolff, T., and Garcia-Sastre, A. (2004). The N- and C-terminal domains of the NS1 protein of influenza B virus can independently inhibit IRF-3 and beta interferon promoter activation. J. Virol. *78*, 11574–11582.

Draghi, M., Pashine, A., Sanjanwala, B., Gendzekhadze, K., Cantoni, C., Cosman, D., Moretta, A., Valiante, N.M., and Parham, P. (2007). NKp46 and NKG2D recognition of infected dendritic cells is necessary for NK cell activation in the human response to influenza infection. J. Immunol. *178*, 2688–2698.

Drake, J.W. (1993). Rates of spontaneous mutation among RNA viruses. Proc. Natl. Acad. Sci. U.S.A. *90*, 4171–4175.

Duesberg, P.H. (1969). Distinct subunits of the ribonucleoprotein of influenza virus. J. Mol. Biol. *42*, 485–499.

Durbin, J.E., Fernandez-Sesma, A., Lee, C.K., Rao, T.D., Frey, A.B., Moran, T.M., Vukmanovic, S., Garcia-Sastre, A., and Levy, D.E. (2000). Type I IFN modulates innate and specific antiviral immunity. J. Immunol. *164*, 4220–4228.

Ehrhardt, C., and Ludwig, S. (2009). A new player in a deadly game: influenza viruses and the PI3K/Akt signalling pathway. Cell Microbiol. *11*, 863–871.

Falcon, A.M., Fernandez-Sesma, A., Nakaya, Y., Moran, T.M., Ortin, J., and Garcia-Sastre, A. (2005). Attenuation and immunogenicity in mice of temperature-sensitive influenza viruses expressing truncated NS1 proteins. J. Gen. Virol. *86*, 2817–2821.

Faria, P.A., Chakraborty, P., Levay, A., Barber, G.N., Ezelle, H.J., Enninga, J., Arana, C., van, D.J., and Fontoura, B.M. (2005). VSV disrupts the Rae1/mrnp41 mRNA nuclear export pathway. Mol Cell *17*, 93–102.

Ferko, B., Stasakova, J., Romanova, J., Kittel, C., Sereinig, S., Katinger, H., and Egorov, A. (2004). Immunogenicity and protection efficacy of replication-deficient influenza A viruses with altered NS1 genes. J. Virol. *78*, 13037–13045.

Fernandez-Sesma, A., Marukian, S., Ebersole, B.J., Kaminski, D., Park, M.S., Yuen, T., Sealfon, S.C., Garcia-Sastre, A., and Moran, T.M. (2006). influenza virus evades innate and adaptive immunity via the NS1 protein. J. Virol. *80*, 6295–6304.

Frias-Staheli, N., Giannakopoulos, N.V., Kikkert, M., Taylor, S.L., Bridgen, A., Paragas, J., Richt, J.A., Rowland, R.R., Schmaljohn, C.S., Lenschow, D.J., Snijder, E.J., Garcia-Sastre, A., and Virgin, H.W. (2007). Ovarian tumor domain-containing viral proteases evade ubiquitin- and ISG15-dependent innate immune responses. Cell Host. Microbe 2, 404–416.

Gack, M.U., Albrecht, R.A., Urano, T., Inn, K.S., Huang, I.C., Carnero, E., Farzan, M., Inoue, S., Jung, J.U., and Garcia-Sastre, A. (2009). Influenza A virus NS1 targets the ubiquitin ligase TRIM25 to evade recognition by the host viral RNA sensor RIG-I. Cell Host Microbe *5*, 439–449.

Gack, M.U., Kirchhofer, A., Shin, Y.C., Inn, K.S., Liang, C., Cui, S., Myong, S., Ha, T., Hopfner, K.P., and Jung, J.U. (2008). Roles of RIG-I N-terminal tandem CARD and splice variant in TRIM25-mediated antiviral signal transduction. Proc. Natl. Acad. Sci. U.S.A. *105*, 16743–16748.

Gack, M.U., Shin, Y.C., Joo, C.H., Urano, T., Liang, C., Sun, L., Takeuchi, O., Akira, S., Chen, Z., Inoue, S., and Jung, J.U. (2007). TRIM25 RING-finger E3 ubiquitin ligase is essential for RIG-I-mediated antiviral activity. Nature *446*, 916–920.

Gallucci, S., Lolkema, M., and Matzinger, P. (1999). Natural adjuvants: endogenous activators of dendritic cells. Nat. Med. *5*, 1249–1255.

Garaigorta, U., Falcon, A.M., and Ortin, J. (2005). Genetic analysis of influenza virus NS1 gene: a temperature-sensitive mutant shows defective formation of virus particles. J. Virol. *79*, 15246–15257.

Garcia, M.A., Gil, J., Ventoso, I., Guerra, S., Domingo, E., Rivas, C., and Esteban, M. (2006a). Impact of protein kinase PKR in cell biology: from antiviral to antiproliferative action. Microbiol. Mol. Biol. Rev. *70*, 1032–1060.

Garcia, M.A., Meurs, E.F., and Esteban, M. (2006b). The dsRNA protein kinase PKR: virus and cell control. Biochimie *89*, 799–811.

Garcia-Sastre, A., Egorov, A., Matassov, D., Brandt, S., Levy, D.E., Durbin, J.E., Palese, P., and Muster, T. (1998). Influenza A virus lacking the NS1 gene replicates in interferon-deficient systems. Virology *252*, 324–330.

Garigliany, M.M., Habyarimana, A., Lambrecht, B., Van de Paar, E., Cornet, A., van den Berg, T., and Desmecht, D. (2010). Influenza A strain-dependent pathogenesis in fatal H1N1 and H5N1 subtype infections of mice. Emerg. Infect. Dis. *16*, 595–603.

Gazit, R., Gruda, R., Elboim, M., Arnon, T.I., Katz, G., Achdout, H., Hanna, J., Qimron, U., Landau, G., Greenbaum, E., Zakay-Rones, Z., Porgador, A., and Mandelboim, O. (2006). Lethal influenza infection in the absence of the natural killer cell receptor gene Ncr1. Nat. Immunol. *7*, 517–523.

Gerosa, F., Gobbi, A., Zorzi, P., Burg, S., Briere, F., Carra, G., and Trinchieri, G. (2005). The reciprocal interaction of nk cells with plasmacytoid or myeloid dendritic cells profoundly affects innate resistance functions. J. Immunol. *174*, 727–734.

Gitlin, L., Barchet, W., Gilfillan, S., Cella, M., Beutler, B., Flavell, R.A., Diamond, M.S., and Colonna, M. (2006). Essential role of mda-5 in type I IFN responses to polyriboinosinic:polyribocytidylic acid and encephalomyocarditis picornavirus. Proc. Natl. Acad. Sci. U.S.A. *103*, 8459–8464.

Graham, M.B., Dalton, D.K., Giltinan, D., Braciale, V.L., Stewart, T.A., and Braciale, T.J. (1993). Response to influenza infection in mice with a targeted disruption in the interferon gamma gene. J. Exp. Med. *178*, 1725–1732.

Greenspan, D., Palese, P., and Krystal, M. (1988). Two nuclear location signals in the influenza virus NS1 nonstructural protein. J. Virol. *62*, 3020–3026.

Grimm, D., Staeheli, P., Hufbauer, M., Koerner, I., Martinez-Sobrido, L., Solirzano, A., Garcia-Sastre, A., Haller, O., and Kochs, G. (2007). Replication fitness determines high virulence of influenza A virus in mice carrying functional Mx1 resistance gene. Proc. Natl. Acad. Sci. U.S.A. *104*, 6806–6811.

Guilligay, D., Tarendeau, F., Resa-Infante, P., Coloma, R., Crepin, T., Sehr, P., Lewis, J., Ruigrok, R.W.H., Ortin, J., Hart, D.J., and Cusack, S. (2008). The structural basis for cap binding by influenza virus polymerase subunit PB2. Nat. Struct. Mol. Biol. *15*, 500–506.

Guo, Z., Chen, L.M., Zeng, H., Gomez, J.A., Plowden, J., Fujita, T., Katz, J.M., Donis, R.O., and Sambhara, S. (2007). NS1 protein of influenza A virus inhibits the function of intracytoplasmic pathogen sensor, RIG-I. Am. J. Respir. Cell Mol. Biol. *36*, 263–269.

Habjan, M., Andersson, I., Klingstrom, J., Schumann, M., Martin, A., Zimmermann, P., Wagner, V., Pichlmair, A., Schneider, U., Muhlberger, E., Mirazimi, A., and Weber, F. (2008). Processing of genome 5′ termini as a strategy of negative-strand RNA viruses to avoid RIG-I-dependent interferon induction. PLoS One. *3*, e2032.

Hai, R., Martinez-Sobrido, L., Fraser, K.A., Ayllon, J., Garcia-Sastre, A., and Palese, P. (2008). Influenza B virus NS1-truncated mutants: live-attenuated vaccine approach. J. Virol. *82*, 10580–10590.

Hale, B.G., Jackson, D., Chen, Y.H., Lamb, R.A., and Randall,R.E. (2006). Influenza A virus NS1 protein binds p85 and activates phosphatidylinositol-3-kinase signaling. Proc. Natl. Acad. Sci. U.S.A. *103*, 14194–14199.

Hale, B.G., Barclay, W.S., Randall, R.E., and Russell, R.J. (2008a). Structure of an avian influenza A virus NS1 protein effector domain. Virology *378*, 1–5.

Hale, B.G., Randall, R.E., Ortin, J., and Jackson, D. (2008b). The multifunctional NS1 protein of influenza A viruses. J. Gen. Virol. *89*, 2359–2376.

Hale, B.G., Kerry, P.S., Jackson, D., Precious, B.L., Gray, A., Killip, M.J., Randall, R.E., and Russell, R.J. (2010a). Structural insights into phosphoinositide 3-kinase activation by the influenza A virus NS1 protein. Proc. Natl. Acad. Sci. U.S.A. *107*, 1954–1959.

Hale, B.G., Steel, J., Medina, R.A., Manicassamy, B., Ye, J., Hickman, D., Hai, R., Schmolke, M., Lowen, A.C., Perez, D.R., and Garcia-Sastre, A. (2010b). Inefficient control of host gene expression by the 2009 pandemic H1N1 influenza A virus NS1 protein. J Virol. *84*, 6909–6922.

Haller, O., Staeheli, P., and Kochs, G. (2006). Interferon-induced Mx proteins in antiviral host defense. Biochimie *89*, 812–818.

Hamilton-Easton, A., and Eichelberger, M. (1995). Virus-specific antigen presentation by different subsets of cells from lung and mediastinal lymph node tissues of influenza virus-infected mice. J. Virol. *69*, 6359–6366.

Hao, L., Sakurai, A., Watanabe, T., Sorensen, E., Nidom, C.A., Newton, M.A., Ahlquist, P., and Kawaoka, Y. (2008). Drosophila RNAi screen identifies host genes important for influenza virus replication. Nature *454*, 890–893.

Harty, R.N., Pitha, P.M., and Okumura, A. (2009). Antiviral Activity of Innate Immune Protein ISG15. J. Innate Immun. *1*, 397–404.

Hatada, E., and Fukuda, R. (1992). Binding of influenza A virus NS1 protein to dsRNA *in vitro*. J. Gen. Virol. 73, 3325–3329.

Hatada, E., Saito, S., and Fukuda, R. (1999). Mutant Influenza Viruses with a Defective NS1 Protein Cannot Block the Activation of PKR in Infected Cells. J. Virol. 73, 2425–2433.

Hausmann, S., Marq, J.B., Tapparel, C., Kolakofsky, D., and Garcin, D. (2008). RIG-I and dsRNA-induced IFNbeta activation. PLoS One. *3*, e3965.

Hayakawa, Y., and Smyth, M.J. (2006). CD27 dissects mature NK cells into two subsets with distinct responsiveness and migratory capacity. J. Immunol. *176*, 1517–1524.

Haye, K., Burmakina, S., Moran, T., Garcia-Sastre, A., and Fernandez-Sesma, A. (2009). The NS1 protein of a human influenza virus inhibits type I interferon production and the induction of antiviral responses in primary human dendritic and respiratory epithelial cells. J. Virol. *83*, 6849–6862.

Hayman, A., Comely, S., Lackenby, A., Murphy, S., McCauley, J., Goodbourn, S., and Barclay, W. (2006). Variation in the ability of human influenza A viruses to induce and inhibit the IFN-beta pathway. Virology *347*, 52–64.

He, X.S., Draghi, M., Mahmood, K., Holmes, T.H., Kemble, G.W., Dekker, C.L., Arvin, A.M., Parham, P., and Greenberg, H.B. (2004). T cell-dependent production of IFN-gamma by NK cells in response to influenza A virus. J. Clin. Invest. *114*, 1812–1819.

Her, L.S., Lund, E., and Dahlberg, J.E. (1997). Inhibition of ran guanosine triphosphatase-dependent nuclear transport by the matrix protein of vesicular stomatitis virus. Science *276*, 1845–1848.

Heui Seo, S., Hoffmann, E., and Webster, R.G. (2002). Lethal H5N1 influenza viruses escape host anti-viral cytokine responses. Nat. Med. *8*, 950–954.

Hornung, V., Ellegast, J., Kim, S., Brzozka, K., Jung, A., Kato, H., Poeck, H., Akira, S., Conzelmann, K.K., Schlee, M., Endres, S., and Hartmann, G. (2006). 5′-Triphosphate RNA is the ligand for RIG-I. Science *314*, 994–997.

Hsieh, H.P., and Hsu, J.T. (2007). Strategies of development of antiviral agents directed against influenza virus replication. Curr. Pharm. Des *13*, 3531–3542.

Imai, Y., Kuba, K., Neely, G.G., Yaghubian-Malhami, R., Perkmann, T., van, L.G., Ermolaeva, M., Veldhuizen, R., Leung, Y.H., Wang, H, *et al.* (2008). Identification of oxidative stress and Toll-like receptor 4 signaling as a key pathway of acute lung injury. Cell *133*, 235–249.

Isaacs, A., and Lindenmann, J. (1957). Virus interference. I. The interferon. Proc. R. Soc. Lond. B Biol. Sci. *147*, 258–267.

Jayasekera, J.P., Vinuesa, C.G., Karupiah, G., and King, N.J.C. (2006). Enhanced antiviral antibody secretion and attenuated immunopathology during influenza virus infection in nitric oxide synthase-2-deficient mice. J. Gen. Virol. *87*, 3361–3371.

Jegerlehner, A., Schmitz, N., Storni, T., and Bachmann, M.F. (2004). Influenza A vaccine based on the extracellular domain of M2: weak protection mediated via antibody-dependent NK cell activity. J. Immunol. *172*, 5598–5605.

Kamitani, W., Huang, C., Narayanan, K., Lokugamage, K.G., and Makino, S. (2009). A two-pronged strategy to suppress host protein synthesis by SARS coronavirus Nsp1 protein. Nat Struct Mol Biol *16*, 1134–1140.

Kamitani, W., Narayanan, K., Huang, C., Lokugamage, K., Ikegami, T., Ito, N., Kubo, H., and Makino, S. (2006). Severe acute respiratory syndrome coronavirus nsp1 protein suppresses host gene expression by promoting host mRNA degradation. Proc. Natl. Acad. Sci. U.S.A. *103*, 12885–12890.

Kang, D.C., Gopalkrishnan, R.V., Lin, L., Randolph, A., Valerie, K., Pestka, S., and Fisher, P.B. (2003). Expression analysis and genomic characterization of human melanoma differentiation associated gene-5, mda-5: a novel type I interferon-responsive apoptosis-inducing gene. Oncogene 23, 1789–1800.

Kang, D.C., Gopalkrishnan, R.V., Wu, Q., Jankowsky, E., Pyle, A.M., and Fisher, P.B. (2002). mda-5: an interferon-inducible putative RNA helicase with double-stranded RNA-dependent ATPase activity and melanoma growth-suppressive properties. Proc. Natl. Acad. Sci. U.S.A. *99*, 637–642.

Karlas, A., Machuy, N., Shin, Y., Pleissner, K.P., Artarini, A., Heuer, D., Becker, D., Khalil, H., Ogilvie, L.A., Hess, S., Maurer, A.P., Muller, E., Wolff, T., Rudel, T., and Meyer, T.F. (2010). Genome-wide RNAi screen identifies human host factors crucial for influenza virus replication. Nature *463*, 818–822.

Karupiah, G., Chen, J.H., Mahalingam, S., Nathan, C.F., and MacMicking, J.D. (1998). Rapid Interferon-gamma-dependent clearance of influenza A virus and protection from consolidating pneumonitis in nitric oxide synthase 2-deficient mice. J. Exp. Med. *188*, 1541–1546.

Kato, H., Takeuchi, O., Mikamo-Satoh, E., Hirai, R., Kawai, T., Matsushita, K., Hiiragi, A., Dermody, T.S., Fujita, T., and Akira, S. (2008). Length-dependent recognition of double-stranded ribonucleic acids by retinoic acid-inducible gene-I and melanoma differentiation-associated gene 5. J. Exp. Med. *205*, 1601–1610.

Kato, H., Takeuchi, O., Sato, S., Yoneyama, M., Yamamoto, M., Matsui, K., Uematsu, S., Jung, A., Kawai, T., Ishii, K.J., Yamaguchi, O., Otsu, K., Tsujimura, T., Koh, C.S., Reis e Sousa, C., Matsuura, Y., Fujita, T., and Akira, S. (2006). Differential roles of MDA5 and RIG-I helicases in the recognition of RNA viruses. Nature *441*, 101–105.

Katz, J.M., Plowden, J., Renshaw-Hoelscher, M., Lu, X., Tumpey, T.M., and Sambhara, S. (2004). Immunity to influenza: the challenges of protecting an aging population. Immunol Res *29*, 113–124.

Khurana, E., Peraro, M.D., DeVane, R., Vemparala, S., DeGrado, W.F., and Klein, M.L. (2009). Molecular dynamics calculations suggest a conduction mechanism for the M2 proton channel from influenza A virus. Proc. Natl. Acad. Sci. U.S.A. *106*, 1069–1074.

Kobasa, D., Jones, S.M., Shinya, K., Kash, J.C., Copps, J., Ebihara, H., Hatta, Y., Hyun Kim, J., Halfmann, P., Hatta, M., Feldmann, F., Alimonti, J.B., Fernando, L., Li, Y., Katze, M.G., Feldmann, H., and Kawaoka, Y. (2007). Aberrant innate immune response in lethal infection of macaques with the 1918 influenza virus. Nature *445*, 319–323.

Kochs, G., Garcia-Sastre, A., and Martinez-Sobrido, L. (2007). Multiple anti-interferon actions of the influenza A virus NS1 protein. J. Virol. *81*, 7011–7021.

Konig, R., Stertz, S., Zhou, Y., Inoue, A., Hoffmann, H.-H., Bhattacharyya, S., Alamares, J.G., Tscherne, D.M., Ortigoza, M.B., Liang, Y., Gao, Q., Andrews, S.E., Bandyopadhyay, S., De Jesus, P., Tu, B.P., Pache, L., Shih, C., Orth, A., Bonamy, G., Miraglia, L., Ideker, T., Garcia-Sastre, A., Young, J.A.T., Palese, P., Shaw, M.L., and Chanda, S.K. (2010). Human host factors required for influenza virus replication. Nature *463*, 813–817.

Krug, R.M., and Aramini, J.M. (2009). Emerging antiviral targets for influenza A virus. Trends Pharm. Sci. *30*, 269–277.

Krug, R.M., and Soeiro, R. (1975). Studies on the intranuclear localization of influenza virus-specific proteins. Virology *64*, 378–387.

Kugel, D., Kochs, G., Obojes, K., Roth, J., Kobinger, G.P., Kobasa, D., Haller, O., Staeheli, P., and von Messling, V. (2009). Intranasal Administration of Alpha Interferon Reduces Seasonal Influenza A Virus Morbidity in Ferrets. J. Virol. *83*, 3843–3851.

Kunkel, T.A. (2004). DNA Replication Fidelity. J. Biol. Chem. *279*, 16895–16898.

Le Bon, A., Durand, V., Kamphuis, E., Thompson, C., Bulfone-Paus, S., Rossmann, C., Kalinke, U., and Tough, D.F. (2006). Direct stimulation of T cells by type I IFN enhances the CD8+ T cell response during cross-priming. J. Immunol. *176*, 4682–4689.

Le Bon, A., Etchart, N., Rossmann, C., Ashton, M., Hou, S., Gewert, D., Borrow, P., and Tough, D.F. (2003). Cross-priming of CD8+ T cells stimulated by virus-induced type I interferon. Nat. Immunol. *4*, 1009–1015.

Le, M.N., Dubaele, S., Proietti De, S.L., Billecocq, A., Bouloy, M., and Egly, J.M. (2004). TFIIH transcription factor, a target for the Rift Valley hemorrhagic fever virus. Cell *116*, 541–550.

Le, Q.M., Kiso, M., Someya, K., Sakai, Y.T., Nguyen, T.H., Nguyen, K.H.L., Pham, N.D., Ngyen, H.H., Yamada, S., Muramoto, Y., Horimoto, T., Takada, A., Goto, H., Suzuki, T., Suzuki, Y., and Kawaoka, Y. (2005). Avian flu: isolation of drug-resistant H5N1 virus. Nature *437*, 1108.

Lee, T.G., Tang, N., Thompson, S., Miller, J., and Katze, M.G. (1994). The 58,000-dalton cellular inhibitor of the interferon-induced double-stranded RNA-activated protein kinase (PKR) is a member of the tetratricopeptide repeat family of proteins. Mol. Cell. Biol. *14*, 2331–2342.

Lee, T.G., Tomita, J., Hovanessian, A.G., and Katze, M.G. (1990). Purification and partial characterization of a cellular inhibitor of the interferon-induced protein kinase of Mr 68,000 from influenza virus-infected cells. Proc. Natl. Acad. Sci. U.S.A. *87*, 6208–6212.

Lee, T.G., Tomita, J., Hovanessian, A.G., and Katze, M.G. (1992). Characterization and regulation of the 58,000-dalton cellular inhibitor of the interferon-induced, dsRNA-activated protein kinase. J. Biol. Chem. *267*, 14238–14243.

Lenschow, D.J., Lai, C., Frias-Staheli, N., Giannakopoulos, N.V., Lutz, A., Wolff, T., Osiak, A., Levine, B., Schmidt, R.E., Garcia-Sastre, A., Leib, D.A., Pekosz, A., Knobeloch, K.P., Horak, I., and Virgin, H.W. (2007). IFN-stimulated gene 15 functions as a critical antiviral molecule against influenza, herpes, and Sindbis viruses. Proc. Natl. Acad. Sci. U.S.A. *104*, 1371–1376.

Li, S., Min, J.Y., Krug, R.M., and Sen, G.C. (2006). Binding of the influenza A virus NS1 protein to PKR mediates the inhibition of its activation by either PACT or double-stranded RNA. Virology *349*, 13–21.

Li, Y., Yamakita, Y., and Krug, R.M. (1998). Regulation of a nuclear export signal by an adjacent inhibitory sequence: the effector domain of the influenza virus NS1 protein. Proc. Natl. Acad. Sci. U.S.A. *95*, 4864–4869.

Lindner, H.A., Fotouhi-Ardakani, N., Lytvyn, V., Lachance, P., Sulea, T., and Menard, R. (2005). The papain-like protease from the severe acute respiratory syndrome coronavirus is a deubiquitinating enzyme. J. Virol. *79*, 15199–15208.

Lipatov, A.S., Kwon, Y.K., Pantin-Jackwood, M.J., and Swayne, D.E. (2009). Pathogenesis of H5N1 influenza virus infections in mice and ferret models differs according to respiratory tract or digestive system exposure. J Infect. Dis. *199*, 717–725.

Lipatov, A.S., Kwon, Y.K., Sarmento, L.V., Lager, K.M., Spackman, E., Suarez, D.L., and Swayne, D.E. (2008). Domestic pigs have low susceptibility to H5N1 highly pathogenic avian influenza viruses. PLoS Pathog. *4*, e1000102.

Lipatov, A.S., Andreansky, S., Webby, R.J., Hulse, D.J., Rehg, J.E., Krauss, S., Perez, D.R., Doherty, P.C., Webster, R.G., and Sangster, M.Y. (2005). Pathogenesis of Hong Kong H5N1 influenza virus NS gene reassortants in mice: the role of cytokines and B- and T-cell responses. J. Gen. Virol. *86*, 1121–1130.

Liu, J., Lynch, P.A., Chien, C.Y., Montelione, G.T., Krug, R.M., and Berman, H.M. (1997). Crystal structure of the unique RNA-binding domain of the influenza virus NS1 protein. Nat Struct Mol Biol *4*, 896–899.

Long, J.X., Peng, D.X., Liu, Y.L., Wu, Y.T., and Liu, X.F. (2008). Virulence of H5N1 avian influenza virus enhanced by a 15-nucleotide deletion in the viral nonstructural gene. Virus Genes *36*, 471–478.

Longhi, M.P., Trumpfheller, C., Idoyaga, J., Caskey, M., Matos, I., Kluger, C., Salazar, A.M., Colonna, M., and Steinman, R.M. (2009). Dendritic cells require a systemic type I interferon response to mature and induce CD4+ Th1 immunity with poly-IC as adjuvant. J. Exp. Med. *206*, 1589–1602.

Lopez, C.B., Garcia-Sastre, A., Williams, B.R., and Moran, T.M. (2003). Type I Interferon Induction Pathway, but Not Released Interferon, Participates in the Maturation of Dendritic Cells Induced by Negative Strand RNA Viruses. J. Infect. Dis. *187*, 1126–1136.

Lu, Y.U.A.N., Wamback, M.A.R.L., Katze, M.G., and Krug, R.M. (1995). Binding of the Influenza Virus NS1 Protein to Double-Stranded RNA Inhibits the Activation of the Protein Kinase That Phosphorylates the eIF-2 Translation Initiation Factor. Virology *214*, 222–228.

Ludwig, S., Schultz, U., Mandler, J., Fitch, W.M., and Scholtissek, C. (1991). Phylogenetic relationship of the nonstructural (NS) genes of influenza A viruses. Virology *183*, 566–577.

Ludwig, S., Wang, X., Ehrhardt, C., Zheng, H., Donelan, N., Planz, O., Pleschka, S., Garcia-Sastre, A., Heins, G., and Wolff, T. (2002). The influenza A virus NS1 protein inhibits activation of Jun N-terminal kinase and AP-1 transcription factors. J. Virol. *76*, 11166–11171.

Lund, J.M., Alexopoulou, L., Sato, A., Karow, M., Adams, N.C., Gale, N.W., Iwasaki, A., and Flavell, R.A. (2004). Recognition of single-stranded RNA viruses by Toll-like receptor 7. Proc. Natl. Acad. Sci. U.S.A. *101*, 5598–5603.

Lyles, D.S. (2000). Cytopathogenesis and inhibition of host gene expression by RNA viruses. Microbiol. Mol. Biol. Rev. *64*, 709–724.

Macatonia, S.E., Taylor, P.M., Knight, S.C., and Askonas, B.A. (1989). Primary stimulation by dendritic cells induces antiviral proliferative and cytotoxic T cell responses *in vitro*. J. Exp. Med. *169*, 1255–1264.

Malathi, K., Dong, B., Gale, M., and Silverman, R.H. (2007). Small self-RNA generated by RNase L amplifies antiviral innate immunity. Nature *448*, 816–819.

Mandelboim, O., Lieberman, N., Lev, M., Paul, L., Arnon, T.I., Bushkin, Y., Davis, D.M., Strominger, J.L., Yewdell, J.W., and Porgador, A. (2001). Recognition of haemagglutinins on virus-infected cells by NKp46 activates lysis by human NK cells. Nature *409*, 1055–1060.

Marques, J.T., Devosse, T., Wang, D., Zamanian-Daryoush, M., Serbinowski, P., Hartmann, R., Fujita, T., Behlke, M.A., and Williams, B.R. (2006). A structural basis for discriminating between self and nonself double-stranded RNAs in mammalian cells. Nat. Biotech. *24*, 559–565.

Marrack, P., Kappler, J., and Mitchell, T. (1999). Type I Interferons Keep Activated T Cells Alive. J. Exp. Med. *189*, 521–530.

McCarthy, M.W., and Kockler, D.R. (2004). Trivalent intranasal influenza vaccine, live. Ann. Pharmacother. *38*, 2086–2093.

McGill, J., Heusel, J.W., and Legge, K.L. (2009). Innate immune control and regulation of influenza virus infections. J. Leukoc. Biol. *86*, 803–812.

Melen, K., Kinnunen, L., Fagerlund, R., Ikonen, N., Twu, K.Y., Krug, R.M., and Julkunen, I. (2007). Nuclear and nucleolar targeting of influenza A virus NS1 protein: striking differences between different virus subtypes. J. Virol. *81*, 5995–6006.

Melville, M., Hansen, W., Freeman, B., Welch, W., and Katze, M. (1997). The molecular chaperone hsp40 regulates the activity of P58IPK, the cellular inhibitor of PKR. Proc. Natl. Acad. Sci. U.S.A. *94*, 97–102.

Melville, M.W., Tan, S.L., Wambach, M., Song, J., Morimoto, R.I., and Katze, M.G. (1999). The cellular inhibitor of the PKR protein kinase, P58IPK, is an influenza virus-activated co-chaperone that modulates heat shock protein 70 activity. J. Biol. Chem. *274*, 3797–3803.

Mibayashi, M., Martinez-Sobrido, L., Loo, Y.M., Cardenas, W.B., Gale, M., Jr., and Garcia-Sastre, A. (2007). Inhibition of retinoic acid-inducible gene I-mediated induction of beta interferon by the NS1 protein of influenza a virus. J. Virol. *81*, 514–524.

Min, J.Y., and Krug, R.M. (2006). The primary function of RNA binding by the influenza A virus NS1 protein in infected cells: inhibiting the 2–5 oligo (A) synthetase/RNase L pathway. Proc. Natl. Acad. Sci. U.S.A. *103*, 7100–7105.

Min, J.Y., Li, S., Sen, G.C., and Krug, R.M. (2007). A site on the influenza A virus NS1 protein mediates both inhibition of PKR activation and temporal regulation of viral RNA synthesis. Virology *363*, 236–243.

Mordstein, M., Kochs, G., Dumoutier, L., Renauld, J.C., Paludan, S.R., Klucher, K., and Staeheli, P. (2008). Interferon-lambda contributes to innate immunity of mice against influenza A virus but not against hepatotropic viruses. PLoS Pathog. *4*, e1000151.

Mordstein, M., Neugebauer, E., Ditt, V., Jessen, B., Rieger, T., Falcone, V., Sorgeloos, F., Ehl, S., Mayer, D., Kochs, G., *et al.* (2010). Interferon-lambda renders epithelial cells of respiratory and gastrointestinal tract resistant to viral infections. J. Virol. *84*, 5670–5677.

Mori, I., Komatsu, T., Takeuchi, K., Nakakuki, K., Sudo, M., and Kimura, Y. (1995). *In vivo* induction of apoptosis by influenza virus. J. Gen. Virol. *76*, 2869–2873.

Nakamura, R., Maeda, N., Shibata, K., Yamada, H., Kase, T., and Yoshikai, Y. (2010). Interleukin-15 is critical in the pathogenesis of influenza A virus-induced acute lung injury. J. Virol. *84*, 5574–5582.

Nallagatla, S.R., Hwang, J., Toroney, R., Zheng, X., Cameron, C.E., and Bevilacqua, P.C. (2007). 5′-Triphosphate-Dependent Activation of PKR by RNAs with Short Stem–loops. Science *318*, 1455–1458.

Narayanan, K., Huang, C., Lokugamage, K., Kamitani, W., Ikegami, T., Tseng, C.T., and Makino, S. (2008). Severe acute respiratory syndrome Coronavirus nsp1 suppresses host gene expression, including that of type I interferon, in infected cells. J. Virol. *82*, 4471–4479.

Nemeroff, M.E., Barabino, S.M.L., Li, Y., Keller, W., and Krug, R.M. (1998). Influenza virus NS1 protein interacts with the cellular 30 kDa subunit of CPSF and inhibits 3′ end formation of cellular pre-mRNAs. Mol. Cell *1*, 991–1000.

Nemeroff, M.E., Qian, X.Y., and Krug, R.M. (1995). The influenza virus NS1 Protein forms multimers *in vitro* and *in vivo*. Virology *212*, 422–428.

Noah, D.L., Twu, K.Y., and Krug, R.M. (2003). Cellular antiviral responses against influenza A virus are countered at the posttranscriptional level by the viral NS1A protein via its binding to a cellular protein required for the 3′ end processing of cellular pre-mRNAS. Virology *307*, 386–395.

Nonacs, R., Humborg, C., Tam, J.P., and Steinman, R.M. (1992). Mechanisms of mouse spleen dendritic cell function in the generation of influenza-specific, cytolytic T lymphocytes. J. Exp. Med. *176*, 519–529.

Onoguchi, K., Yoneyama, M., Takemura, A., Akira, S., Taniguchi, T., Namiki, H., and Fujita, T. (2007). Viral infections activate types I and III Interferon genes through a common mechanism. J. Biol. Chem. *282*, 7576–7581.

Opitz, B., Rejaibi, A., Dauber, B., Eckhard, J., Vinzing, M., Schmeck, B., Hippenstiel, S., Suttorp, N., and Wolff, T. (2007). IFNbeta induction by influenza A virus is mediated by RIG-I which is regulated by the viral NS1 protein. Cell Microbiol. *9*, 930–938.

Osterlund, P., Veckman, V., Siren, J., Klucher, K.M., Hiscott, J., Matikainen, S., and Julkunen, I. (2005). Gene Expression and Antiviral Activity of Alpha/Beta Interferons and Interleukin-29 in Virus-Infected Human Myeloid Dendritic Cells. J. Virol. *79*, 9608–9617.

Ozato, K., Shin, D.M., Chang, T.H., and Morse, H.C. (2008). TRIM family proteins and their emerging roles in innate immunity. Nat. Rev. Immunol. *8*, 849–860.

Palese, P., and Shaw, M.L. (2007). Orthomyxoviridae: the viruses and their replication. In Fields Virology, Knipe, D.M. and Howley, P.M., eds. Philadelphia, Lippincott Williams &Wilkins, pp. 1647–1689.

Peper, R.L., and Van Campen, H. (1995). Tumor necrosis factor as a mediator of inflammation in influenza A viral pneumonia. Microbial Pathog. *19*, 175–183.

Phipps-Yonas, H., Seto, J., Sealfon, S.C., Moran, T.M., and Fernandez-Sesma, A. (2008). Interferon pretreatment of conventional and plasmacytoid human dendritic cells enhances their activation by influenza virus. PLoS Pathog. *4*, e1000193.

Pichlmair, A., Schulz, O., Tan, C.P., Naslund, T.I., Liljestrom, P., Weber, F., and Reis e Sousa, C. (2006). RIG-I-mediated antiviral responses to single-stranded RNA bearing 5′-phosphates. Science *314*, 997–1001.

Plotch, S.J., Bouloy, M., Ulmanen, I., and Krug, R.M. (1981). A unique cap(m7GpppXm)-dependent influenza virion endonuclease cleaves capped RNAs to generate the primers that initiate viral RNA transcription. Cell *23*, 847–858.

Plumet, S., Herschke, F., Bourhis, J.M., Valentin, H., Longhi, S., and Gerlier, D. (2007). Cytosolic 5′-triphosphate ended viral leader transcript of measles virus as activator of the RIG I-mediated interferon response. PLoS One 2, e279.

Poon, L.L., Pritlove, D.C., Fodor, E., and Brownlee, G.G. (1999). Direct evidence that the poly(A) tail of influenza A virus mRNA is synthesized by reiterative copying of a U track in the virion RNA template. J. Virol. *73*, 3473–3476.

Porter, A.G. (1993). Picornavirus nonstructural proteins: emerging roles in virus replication and inhibition of host cell functions. J. Virol. *67*, 6917–6921.

Price, G.E., Gaszewska-Mastarlarz, A., and Moskophidis, D. (2000). The role of alpha/beta and gamma interferons in development of immunity to influenza A virus in mice. J. Virol. *74*, 3996–4003.

Privalsky, M.L., and Penhoet, E.E. (1981). The structure and synthesis of influenza virus phosphoproteins. J. Biol. Chem. *256*, 5368–5376.

Qian, X.Y., Alonso-Caplen, F., and Krug, R.M. (1994). Two functional domains of the influenza virus NS1 protein are required for regulation of nuclear export of mRNA. J. Virol. *68*, 2433–2441.

Qian, X.Y., Chien, C.Y., Lu, Y., Montelione, G.T., and Krug, R.M. (1995). An amino-terminal polypeptide fragment of the influenza virus NS1 protein possesses specific RNA-binding activity and largely helical backbone structure. RNA. *1*, 948–956.

Qiu, Y., and Krug, R.M. (1994). The influenza virus NS1 protein is a poly(A)-binding protein that inhibits nuclear export of mRNAs containing poly(A). J. Virol. *68*, 2425–2432.

Quinlivan, M., Zamarin, D., Garcia-Sastre, A., Cullinane, A., Chambers, T., and Palese, P. (2005). Attenuation of equine influenza viruses through truncations of the NS1 protein. J. Virol. *79*, 8431–8439.

Randall, R.E., and Goodbourn, S. (2008). Interferons and viruses: an interplay between induction, signalling, antiviral responses and virus countermeasures. J. Gen. Virol. *89*, 1–47.

Rao, S.S., Kong, W.P., Wei, C.J., Van, H.N., Gorres, J.P., Nason, M., Andersen, H., Tumpey, T.M., and Nabel, G.J. (2010). Comparative efficacy of hemagglutinin, nucleoprotein, and matrix 2 protein gene-based vaccination against H5N1 influenza in mouse and ferret. PLoS One *5*, e9812.

Rehwinkel, J., Tan, C.P., Goubau, D., Schulz, O., Pichlmair, A., Bier, K., Robb, N., Vreede, F., Barclay, W., Fodor, E., and Reis e Sousa, C. (2010). RIG-I detects viral genomic RNA during negative-strand RNA virus infection. Cell *140*, 397–408.

Richt, J.A., and Garcia-Sastre, A. (2009). Attenuated influenza virus vaccines with modified NS1 proteins. Curr. Top. Microbiol. Immunol *333*, 177–195.

Richt, J.A., Lekcharoensuk, P., Lager, K.M., Vincent, A.L., Loiacono, C.M., Janke, B.H., Wu, W.H., Yoon, K.J., Webby, R.J., Solorzano, A., *et al.* (2006). Vaccination of pigs against swine influenza viruses by using an NS1-truncated modified live-virus vaccine. J. Virol. *80*, 11009–11018.

Rolling, T., Koerner, I., Zimmermann, P., Holz, K., Haller, O., Staeheli, P., and Kochs, G. (2009). Adaptive mutations resulting in enhanced polymerase activity contribute to high virulence of influenza a virus in mice. J. Virol. *83*, 6673–6680.

Sadler, A.J., and Williams, B.R.G. (2008). Interferon-inducible antiviral effectors. Nat. Rev. Immunol. *8*, 559–568.

Salomon, R., Hoffmann, E., and Webster, R.G. (2007). Inhibition of the cytokine response does not protect against lethal H5N1 influenza infection. Proc. Natl. Acad. Sci. U.S.A. *104*, 12479–12481.

Samanta, M., Iwakiri, D., Kanda, T., Imaizumi, T., and Takada, K. (2006). EB virus-encoded RNAs are recognized by RIG-I and activate signaling to induce type I IFN. EMBO J. *25*, 4207–4214.

Sarawar, S.R., Sangster, M., Coffman, R.L., and Doherty, P.C. (1994). Administration of anti-IFN-gamma antibody to beta 2-microglobulin-deficient mice delays influenza virus clearance but does not switch the response to a T helper cell 2 phenotype. J. Immunol. *153*, 1246–1253.

Satterly, N., Tsai, P.L., van Deursen, J., Nussenzveig, D.R., Wang, Y., Faria, P.A., Levay, A., Levy, D.E., and Fontoura, B.M.A. (2007). Influenza virus targets the mRNA export machinery and the nuclear pore complex. Proc. Natl. Acad. Sci. U.S.A. *104*, 1853–1858.

Sauter, B., Albert, M.L., Francisco, L., Larsson, M., Somersan, S., and Bhardwaj, N. (2000). Consequences of Cell Death. J. Exp. Med. *191*, 423–434.

Schmidt, A., Schwerd, T., Hamm, W., Hellmuth, J.C., Cui, S., Wenzel, M., Hoffmann, F.S., Michallet, M.C., Besch, R., Hopfner, K.P., Endres, S., and Rothenfusser, S. (2009). 5-triphosphate RNA requires basepaired structures to activate antiviral signaling via RIG-I. Proc. Natl. Acad. Sci. U.S.A. *106*, 12067–12072.

Schmolke, M., Viemann, D., Roth, J., and Ludwig, S. (2009). Essential impact of NF-kB signaling on the H5N1 influenza A virus-induced transcriptome. J. Immunol. *183*, 5180–5189.

Schoenborn, J.R., and Wilson, C.B. (2007). Regulation of interferon-gamma during innate and adaptive immune responses. Adv. Immunol. *96*, 41–101.

Sen, G.C., and Peters, G.A. (2007). Viral stress-inducible genes. Adv. Virus Res. *70*, 233–263.

Seth, R.B., Sun, L., Ea, C.K., and Chen, Z.J. (2005). Identification and characterization of MAVS, a mitochondrial antiviral signaling protein that activates NF-kappaB and IRF 3. Cell *122*, 669–682.

Shin, Y.K., Li, Y., Liu, Q., Anderson, D.H., Babiuk, L.A., and Zhou, Y. (2007). SH3 binding motif 1 in influenza A virus NS1 Protein is essential for PI3K/Akt signaling pathway activation. J. Virol. *81*, 12730–12739.

Silverman, R.H. (2007). Viral encounters with 2′,5′-oligoadenylate synthetase and RNase L during the interferon antiviral response. J. Virol. *81*, 12720–12729.

Smith, R.W.P., Malik, P., and Clements, J.B. (2005). The herpes simplex virus ICP27 protein: a multifunctional post-transcriptional regulator of gene expression. Biochem. Soc. Trans. *33*, 499–501.

Snelgrove, R.J., Edwards, L., Rae, A.J., and Hussell, T. (2006). An absence of reactive oxygen species improves the resolution of lung influenza infection. Eur. J. Immunol. *36*, 1364–1373.

Solorzano, A., Webby, R.J., Lager, K.M., Janke, B.H., Garcia-Sastre, A., and Richt, J.A. (2005). Mutations in the NS1 protein of swine influenza virus impair anti-interferon activity and confer attenuation in pigs. J. Virol. *79*, 7535–7543.

Sommereyns,C., Paul,S., Staeheli,P., and Michiels,T. (2008). IFN-lambda (IFN-lambda) is expressed in a tissue-dependent fashion and primarily acts on epithelial cells *in vivo*. PLoS. Pathog. *4*, e1000017.

Sridharan, H., Zhao, C., and Krug, R.M. (2010). Species specificity of the NS1 protein of influenza B virus: it binds only human and non-human primate ubiquitin-like ISG15 proteins. J. Biol. Chem. *285*, 7852–7856.

Steel, J., Lowen, A.C., Pena, L., Angel, M., Solorzano, A., Albrecht, R., Perez, D.R., Garcia-Sastre, A., and Palese, P. (2009). Live attenuated influenza viruses containing NS1 truncations as vaccine candidates against H5N1 highly pathogenic avian influenza. J. Virol. *83*, 1742–1753.

Steel, J., Lowen, A.C., Wang, T.T., Yondola, M., Gao, Q., Haye, K., Garcia-Sastre, A., and Palese, P. (2010a). influenza virus vaccine based on the conserved hemagglutinin stalk domain. mBio *1*, e00018–10.

Steel, J., Staeheli, P., Mubareka, S., Garcia-Sastre, A., Palese, P., and Lowen, A.C. (2010b). Transmission of pandemic H1N1 influenza virus and impact of prior exposure to seasonal strains or interferon treatment. J. Virol. *84*, 21–26.

Stein-Streilein, J., and Guffee, J. (1986). *In vivo* treatment of mice and hamsters with antibodies to asialo GM1 increases morbidity and mortality to pulmonary influenza infection. J. Immunol. *136*, 1435–1441.

Subbarao, K., Swayne, D.E., and Olsen, C.W. (2006). Influenza Virology: Current Topics. Kawaoka Y, ed., pp. 229–280. (Caister Academic Press, Wymondham).

Sugiyama, K., Obayashi, E., Kawaguchi, A., Suzuki, Y., Tame, J.R.H., Nagata, K., and Park, S.Y. (2009). Structural insight into the essential PB1-PB2 subunit contact of the influenza virus RNA polymerase. EMBO J. *28*, 1803–1811.

Takahasi, K., Yoneyama, M., Nishihori, T., Hirai, R., Kumeta, H., Narita, R., Gale, M., Jr., Inagaki, F., and Fujita, T. (2008). Nonself RNA-sensing mechanism of RIG-I helicase and activation of antiviral immune responses. Mol. Cell *29*, 428–440.

Takeda, K., and Akira, S. (2005). Toll-like receptors in innate immunity. Int. Immunol. *17*, 1–14.

Takeuchi, O., and Akira, S. (2008). MDA5/RIG-I and virus recognition. Curr. Opin. Immunol. *20*, 17–22.

Talon, J., Salvatore, M., O'Neill, R.E., Nakaya, Y., Zheng, H., Muster, T., Garcia-Sastre, A., and Palese, P. (2000a). Influenza A and B viruses expressing altered NS1 proteins: a vaccine approach. Proc. Natl. Acad. Sci. U.S.A. *97*, 4309–4314.

Talon, J., Horvath, C.M., Polley, R., Basler, C.F., Muster, T., Palese, P., and Garcia-Sastre, A. (2000b). Activation of interferon regulatory factor 3 is inhibited by the influenza A virus NS1 protein. J. Virol. *74*, 7989–7996.

Tan, S.L., and Katze, M.G. (1998). Biochemical and genetic evidence for complex formation between the influenza A virus NS1 protein and the interferon-induced PKR protein kinase. J. Interferon Cytokine Res. *18*, 757–766.

Thomas, D., Blakqori, G., Wagner, V., Banholzer, M., Kessler, N., Elliott, R.M., Haller, O., and Weber, F. (2004). Inhibition of RNA Polymerase II Phosphorylation by a Viral Interferon Antagonist. J. Biol. Chem. *279*, 31471–31477.

Topham, D.J., Tripp, R.A., and Doherty, P.C. (1997). CD8+ T cells clear influenza virus by perforin or Fas-dependent processes. J. Immunol. *159*, 5197–5200.

Treanor, J.J., Snyder, M.H., London, W.T., and Murphy, B.R. (1989). The B allele of the NS gene of avian influenza viruses, but not the A allele, attenuates a human influenza a virus for squirrel monkeys. Virology *171*, 1–9.

Tumpey, T.M., Basler, C.F., Aguilar, P.V., Zeng, H., Solorzano, A., Swayne, D.E., Cox, N.J., Katz, J.M., Taubenberger, J.K., Palese, P., and Garcia-Sastre, A. (2005). Characterization of the reconstructed 1918 Spanish influenza pandemic virus. Science *310*, 77–80.

Tumpey, T.M., Maines, T.R., Van Hoeven, N., Glaser, L., Solorzano, A., Pappas, C., Cox, N.J., Swayne, D.E., Palese, P., Katz, J.M., and Garcia-Sastre, A. (2007a). A two-amino acid change in the hemagglutinin of the 1918 influenza virus abolishes transmission. Science *315*, 655–659.

Tumpey, T.M., Szretter, K.J., Van Hoeven, N., Katz, J.M., Kochs, G., Haller, O., Garcia-Sastre, A., and Staeheli, P. (2007b). The Mx1 gene protects mice against the pandemic 1918 and highly lethal human h5n1 influenza viruses. J. Virol. *81*, 10818–10821.

Turner, S.J., Olivas, E., Gutierrez, A., Diaz, G., and Doherty, P.C. (2007). Disregulated influenza a virus-specific CD8+ T cell homeostasis in the absence of IFN-gamma signaling. J. Immunol. *178*, 7616–7622.

Twu, K.Y., Noah, D.L., Rao, P., Kuo, R.L., and Krug, R.M. (2006). The CPSF30 binding site on the NS1A protein of influenza A virus is a potential antiviral target. J. Virol. *80*, 3957–3965.

Uze, G., and Monneron, D. (2006). IL-28 and IL-29: newcomers to the interferon family. Biochimie *89*, 729–734.

Uzri, D., and Gehrke, L. (2009). Nucleotide Sequences and Modifications That Determine RIG-I/RNA Binding and Signaling Activities. J. Virol. *83*, 4174–4184.

Van Hoeven, N., Belser, J.A., Szretter, K.J., Zeng, H., Staeheli, P., Swayne, D.E., Katz, J.M., and Tumpey, T.M. (2009). Pathogenesis of 1918 pandemic and H5N1 influenza virus infections in a guinea pig model: antiviral potential of exogenous alpha interferon to reduce virus shedding. J. Virol. *83*, 2851–2861.

Versteeg, G.A., Hale, B.G., van Boheemen, S., Wolff, T., Lenschow, D.J., and Garcia-Sastre, A. (2010). Species-specific antagonism of host ISGylation by the influenza B virus NS1 protein. J. Virol. *84*, 5423–5430.

Vignuzzi, M., Stone, J.K., Arnold, J.J., Cameron, C.E., and Andino, R. (2006). Quasispecies diversity determines pathogenesis through cooperative interactions in a viral population. Nature *439*, 344–348.

Vincent, A.L., Ma, W., Lager, K.M., Janke, B.H., Webby, R.J., Garcia-Sastre, A., and Richt, J.A. (2007). Efficacy of intranasal administration of a truncated NS1 modified live influenza virus vaccine in swine. Vaccine *25*, 7999–8009.

Wang, J., Oberley-Deegan, R., Wang, S., Nikrad, M., Funk, C.J., Hartshorn, K.L., and Mason, R.J. (2009). Differentiated human alveolar type II cells secrete antiviral IL-29 (IFN-lambda-1) in response to influenza A infection. J. Immunol. *182*, 1296–1304.

Wang, T.T., Tan, G.S., Hai, R., Pica, N., Petersen, E., Moran, T.M., and Palese, P. (2010). Broadly protective monoclonal antibodies against H3 influenza viruses following sequential immunization with different hemagglutinins. PLoS Pathog. *6*, e1000796.

Wang, W., Riedel, K., Lynch, P., Chien, C.Y., Montelione, G.T., and Krug, R.M. (1999). RNA binding by the novel helical domain of the influenza virus NS1 protein requires its dimer structure and a small number of specific basic amino acids. RNA. *5*, 195–205.

Wang, X., Hinson, E.R., and Cresswell, P. (2007). The interferon-inducible protein viperin inhibits influenza virus release by perturbing lipid rafts. Cell Host Microbe *2*, 96–105.

Wang, X., Li, M., Zheng, H., Muster, T., Palese, P., Beg, A.A., and Garcia-Sastre, A. (2000). Influenza A virus NS1 protein prevents activation of NF-kappa B and induction of alpha/beta interferon. J. Virol. *74*, 11566–11573.

Weber, F., Wagner, V., Rasmussen, S.B., Hartmann, R., and Paludan, S.R. (2006). Double-stranded RNA is produced by positive-strand RNA viruses and DNA viruses but not in detectable amounts by negative-strand RNA viruses. J. Virol. *80*, 5059–5064.

Weingartl, H.M., Albrecht, R.A., Lager, K.M., Babiuk, S., Marszal, P., Neufeld, J., Embury-Hyatt, C., Lekcharoensuk, P., Tumpey, T.M., Garcia-Sastre, A., and Richt, J.A. (2009). Experimental infection of pigs with the human 1918 pandemic influenza virus. J. Virol. *83*, 4287–4296.

Weiss, I.D., Wald, O., Wald, H., Beider, K., Abraham, M., Galun, E., Nagler, A., and Peled, A. (2010). IFN-gamma treatment at early stages of influenza virus infection protects mice from death in a NK cell-dependent manner. J. Interferon Cytokine Res. *30*, 439–449.

Whitmire, J.K., Tan, J.T., and Whitton, J.L. (2005). Interferon-gamma acts directly on CD8+ T cells to increase their abundance during virus infection. J. Exp. Med. *201*, 1053–1059.

Wilkins, C., and Gale Jr, M. (2010). Recognition of viruses by cytoplasmic sensors. Curr. Opin. Immunol. *22*, 41–47.

Wolff, T., and Ludwig, S. (2009). Influenza viruses control the vertebrate type I interferon system: factors, mechanisms, and consequences. J. Interferon Cytokine Res. *29*, 549–557.

Wressnigg, N., Shurygina, A.P., Wolff, T., Redlberger-Fritz, M., Popow-Kraupp, T., Muster, T., Egorov, A., and Kittel, C. (2009a). Influenza B mutant viruses with truncated NS1 proteins grow efficiently in Vero cells and are immunogenic in mice. J. Gen. Virol. *90*, 366–374.

Wressnigg, N., Voss, D., Wolff, T., Romanova, J., Ruthsatz, T., Mayerhofer, I., Reiter, M., Nakowitsch, S., Humer, J., Morokutti, A., Muster, T., Egorov, A., and Kittel, C. (2009b). Development of a live-attenuated influenza B Delta NS1 intranasal vaccine candidate. Vaccine *27*, 2851–2857.

Xia, S., Monzingo, A.F., and Robertus, J.D. (2009). Structure of NS1A effector domain from the influenza A/Udorn/72 virus. Acta Crystallogr. D. Biol. Crystallogr. *65*, 11–17.

Xu, L., Yoon, H., Zhao, M.Q., Liu, J., Ramana, C.V., and Enelow, R.I. (2004). Cutting edge: pulmonary immunopathology mediated by antigen-specific expression of TNF-alpha by antiviral CD8+ T cells. J. Immunol. *173*, 721–725.

Yalamanchili, P., Harris, K., Wimmer, E., and Dasgupta, A. (1996). Inhibition of basal transcription by poliovirus: a virus-encoded protease (3Cpro) inhibits formation of TBP-TATA box complex *in vitro*. J. Virol. *70*, 2922–2929.

Yamada, Y.K., Meager, A., Yamada, A., and Ennis, F.A. (1986). Human Interferon Alpha and gamma production by lymphocytes during the generation of influenza virus-specific cytotoxic T lymphocytes. J. Gen. Virol. *67*, 2325–2334.

Yang, K., Puel, A., Zhang, S., Eidenschenk, C., Ku, C.L., Casrouge, A., Picard, C., von, B.H., Senechal, B., Plancoulaine, S., *et al.* (2005). Human TLR-7-, –8-, and -9-mediated induction of IFN-alpha/beta and -lambda

Is IRAK-4 dependent and redundant for protective immunity to viruses. Immunity *23*, 465–478.

Yin, C., Khan, J.A., Swapna, G.V.T., Ertekin, A., Krug, R.M., Tong, L., and Montelione, G.T. (2007). Conserved Surface Features Form the Double-stranded RNA Binding Site of Non-structural Protein 1 (NS1) from Influenza A and B Viruses. J. Biol. Chem. *282*, 20584–20592.

Yuan, H., Yoza, B.K., and Lyles, D.S. (1998). Inhibition of Host RNA Polymerase II-Dependent Transcription by Vesicular Stomatitis Virus Results from Inactivation of TFIID. Virology *251*, 383–392.

Yuan, P., Bartlam, M., Lou, Z., Chen, S., Zhou, J., He, X., Lv, Z., Ge, R., Li, X., Deng, T., *et al.* (2009). Crystal structure of an avian influenza polymerase PAN reveals an endonuclease active site. Nature *458*, 909–913.

Yuan, W., and Krug, R.M. (2001). Influenza B virus NS1 protein inhibits conjugation of the interferon (IFN)-induced ubiquitin-like ISG15 protein. EMBO J. *20*, 362–371.

Yuki, Y., and Kiyono, H. (2009). Mucosal vaccines: novel advances in technology and delivery. Exp. Rev. Vaccines *8*, 1083–1097.

Zeng, W., Sun, L., Jiang, X., Chen, X., Hou, F., Adhikari, A., Xu, M., and Chen, Z.J. (2010). Reconstitution of the RIG-I pathway reveals a signaling role of unanchored polyubiquitin chains in innate immunity. Cell *141*, 315–330.

Zhao, C., Hsiang, T.Y., Kuo, R.L., and Krug, R.M. (2010). ISG15 conjugation system targets the viral NS1 protein in influenza A virus infected cells. Proc. Natl. Acad. Sci. U.S.A. *107*, 2253–2258.

Zohari, S., Gyarmati, P., Thoren, P., Czifra, G.R., Brojer, C., Belak, S., and Berg, M. (2008). Genetic characterization of the NS gene indicates co-circulation of two sub-lineages of highly pathogenic avian influenza virus of H5N1 subtype in Northern Europe in 2006. Virus Genes *36*, 117–125.

Hepatitis C Virus Regulation of Interferon Antiviral Defences

9

Helene M. Liu and Michael Gale Jr.

Abstract

Mammalian cells respond to virus challenge by initiating an intracellular innate immune response that is designed to limit virus replication and to inform and modulate the ensuing adaptive immune response. Innate immune defences are characterized by pathogen recognition receptor signalling that mediates the expression of antiviral gene products, the production of interferon (IFN)-α/β and interferon-stimulated genes, and the secretion of other proinflammatory cytokines from the site of infection. Hepatitis C virus (HCV) confers a chronic infection of the liver in nearly 200 million people around the world. Hepatic innate immune defences impose the front line of protection against HCV replication and pathogenesis. HCV infection is treated with IFN-based therapy. However, HCV most often evades hepatic innate immunity and responds overall poorly to therapy to thereby persist and run a chronic disease course. HCV persistence has been linked to a complex combination of virus–host interactions that disrupt intracellular innate immune signalling pathways and attenuate the antiviral actions of IFN. Viral regulation of these processes breaks a critical cross-talk between innate and adaptive immunity to attenuate antiviral immune defences and provide a foundation for chronic HCV infection.

Introduction

Hepatitis C virus (HCV) was first discovered through molecular biology approaches approximately twenty years ago (Choo *et al.*, 1989). Besides hepatitis B virus, HCV is the most important aetiological agent of blood-borne viral hepatitis in the world. It has been estimated that 3% of the worldwide population are infected with HCV (Lauer and Walker, 2001). HCV transient infection induces acute hepatitis, which may be without symptoms or may result in hepatitis accompanied by jaundice, but fulminant liver failure is rare. However, 70–80% of acute infected patients develop persistent infection wherein HCV causes chronic liver disease that can associate with constitutive hepatitis, liver cirrhosis, and hepatocellular carcinoma. The high frequency of persistent HCV infection in immunocompetent adults indicates that HCV interferes with many aspects of immunity. There is no existing valid vaccine for HCV, and the current interferon (IFN)-based treatments for HCV cause unpleasant side-effects (McHutchison *et al.*, 1998; McHutchison and Patel, 2002). Thus, HCV is a major contemporary public health problem. Understanding the virus–host interactions that regulate immunity and support HCV infection will allow for defining new therapeutic approaches aimed at suppressing HCV-associated disease.

HCV is spread by parenteral transmission of contaminated blood products. Chronic HCV is problematically associated with a gradual loss of liver function through progressive fibrosis that can proceed to cirrhosis and end point disease requiring liver transplantation. Moreover, HCV infection is considered a leading aetiology of liver cancer. The risk factors for acquiring HCV infection are similar to the risk factors for acquisition of the human immunodeficiency virus (HIV), and in many population centres the two viruses now track together as a dual infection such that liver dysfunction due to HCV has become a leading cause of death among people infected with HIV (Shepard *et al.*, 2005; Iorio *et al.*, 2010; Mallet *et al.*, 2010). Thus, HCV infection continues to be a major public health problem that presents new challenges for treatment and disease management. The ability of HCV to mediate persistent, life-long infection in its human host is linked to the evasive nature of the virus to overcome the host immune system and to resist the antiviral actions of IFN-based therapy. Molecular studies of HCV–host interactions have revealed several levels of immune regulation and evasion directed by HCV interactions with host immune programmes. Among these interactions, HCV directs the regulation of intracellular innate immune programmes that would otherwise suppress virus replication and cell–cell spread within the liver. This chapter serves as an update to previous literature and will provide an overview of the virus–host interactions that govern hepatic innate immunity and the actions of IFN therapy against HCV.

Innate intracellular immune defences

In response to virus infection, signalling pathways within mammalian cells direct a variety of intracellular events that generate an antiviral state directly within the infected cell. This antiviral response, termed the 'innate immune response' to virus infection, represents our first line of immune defence against viruses. If this response is successful, exposure to the virus will render a self-limiting, abortive infection. It is the hepatic innate immune response that imposes initial antiviral immune defences against HCV infection (Gale, Jr., 2003). The innate immune response is triggered when the infected cell recognizes a molecular signature within the invading virus, called a pathogen-associated molecular pattern (PAMP). A PAMP is typically a physical characteristic of the virus that is recognized and engaged by specific pathogen recognition receptor proteins (PRRs) within the host cell (O'Neill, 2004). The PAMP/PRR interaction initiates signalling cascades that induce the expression of antiviral effector genes (Sen, 2001). For RNA viruses, protein and nucleic acid products of infection comprise an array of PAMP signatures that can engage specific PAMP receptors, including the RIG-I-like receptors (RLRs), Toll-like receptors (TLRs) and other nucleic acid binding proteins such as the Nod-like receptors (NLRs; Cook *et al.*, 2004; Iwasaki and Medzhitov, 2004; Fig. 9.1). The HCV RNA contains specific PAMP signatures that include a free 5′ triphosphate (5′-ppp) and a poly-uridine/cytidine (poly-U/UC) motif as well as stem–loop double-stranded RNA (dsRNA) structures within its single-stranded RNA genome (Tuplin *et al.*, 2002; Simmonds *et al.*, 2004). HCV RNA is sufficient to induce the innate immune response in cultured hepatocyte-derived cell lines and *in vivo* (McCormick *et al.*, 2004; Sumpter *et al.*,

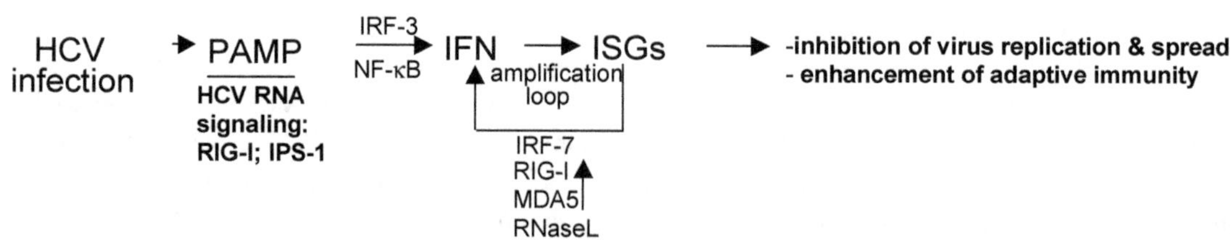

Figure 9.1 Triggering the hepatic innate immune response to HCV infection. HCV triggers the innate immune response through the process pathogen-associated molecular pattern (PAMP)/PAMP receptor engagement. The poly-U/UC RNA motif within the 3′ NTR of the HCV genome presents PAMP attributes that are recognized by RIG-I. Signalling through RIG-I leads to the activation of IRF-3, IFN production and ISG expression. Autocrine and paracrine signalling by IFN can potentiate viral PAMP signalling and serves to amplify the host response through a feedback loop involving IRF-7 and further PAMP signalling from products of RNaseL, an endoribonuclease that is activated during the innate immune response. IFNs also enhance immune cell maturation and effector function, resulting in an immune response aimed at controlling virus infection.

2005; Saito *et al.*, 2008). HCV RNA is recognized as non-self through engagement of the viral 5′-ppp and poly-U/UC motifs by the host PRR, retinoic acid inducible gene I (RIG-I). RIG-I is the charter member of the RLR family of cytosolic RNA helicases, which includes melanoma differentiation antigen (MDA5) and the laboratory of gene and protein expression 2 protein (LGP2). RIG-I and MDA5 encode tandem caspase activation and recruitment domains (CARDs) that confer their signalling activities. RIG-I is the essential and primary PRR of HCV and several other RNA viruses (Yoneyama *et al.*, 2004; Sumpter *et al.*, 2005), whereas MDA5 serves as a PRR for a distinct set of RNA viruses while providing a semi-redundant role as a PRR for certain other RIG-I-specific viruses (Loo *et al.*, 2008). The role of LGP2 as a PRR is not fully understood. Unlike the other RLRs, LGP2 lacks CARDs and may function to regulate the CARD signalling actions of RIG-I and MDA5 (Saito *et al.*, 2006). After engagement of a non-self (viral) RNA ligand, the RLRs undergo modification and specific confirmation changes that confer interaction with a common adaptor protein termed IPS-1 (Kawai *et al.*, 2005). IPS-1 is a single transmembrane protein that encodes a CARD and other modulator domains serving to recruit signalling proteins that drive downstream transcription activation. In hepatocytes, both RIG-I and IPS-1 are essential for triggering innate immune defences against HCV infection (Loo *et al.*, 2006). TLR signalling programmes have also been implicated in HCV infection, wherein TLR3 may function as a PRR for HCV. However, a role for TLR3 in triggering hepatic innate immunity against HCV *in vivo* has yet to be formally proven (Li *et al.*, 2005a).

In the infected hepatocyte, PAMP RNA/RIG-I engagement signals the activation of a variety of transcription factors, resulting in the immediate expression of innate immune response genes (Malmgaard, 2004; Fig. 9.2). IFN regulatory factor (IRF)-3 and nuclear factor-kappa B (NF-κB) are triggered in response to virus infection, and each are activated upon cellular recognition of the HCV PAMP by RIG-I and signalling through the IPS-1 protein (Au *et al.*, 1995; Lin *et al.*, 1999; Fredericksen *et al.*, 2001; Richmond, 2002; Prabhu *et al.*, 2004). IPS-1 signalling bifurcates to drive IRF-3 and NF-κB activation and nuclear translocation. Other IRF family members, including IRF-5 and IRF-7, are essential components of the innate immune response to virus infection in specific cell types, including certain dendritic cell subsets and macrophages (Barnes *et al.*, 2001; Kawai *et al.*, 2004), though the specific role of each in HCV infection and hepatic immune defences has not been characterized. Virus-induced signalling events that activate activating transcription factor 2 (ATF-2) and further direct chromatin remodelling contribute to the building of a virus-triggered enhanceosome on the IFN-β promoter. The IFN-β enhanceosome includes IRF-3 and NF-κB, which produce a transcriptional response resulting in IFN-β expression and secretion from the infected cell (Sen, 2001). NF-κB activation and function is central to the chemokine and proinflammatory cytokine response arm to virus infection, which functions side by side with IFN to modulate the ensuing adaptive immune response (Tai *et al.*, 2000). Secreted levels of IFN-β drive autocrine and paracrine signalling processes by binding the IFN-α/β receptors of the infected cell and local surrounding tissue, respectively. This results in activation of the Jak-STAT pathway. Here, receptor-associated Jak and Tyk1 protein kinases phosphorylate signal transducer and activator of transcription (STAT) proteins on critical serine and tyrosine residues to confer STAT activation, STAT association with IRF-9, and nuclear localization of the resulting ISGF3 transcription factor complex. ISGF3 is the central transcription factor that promotes high-level expression of IFN stimulated genes (ISGs) by binding to the IFN-stimulated response element (ISRE) within the ISG promoter/enhancer region. IFN binding to the IFN-α/β receptor and signalling of the Jak-STAT pathway drives a second wave of transcriptional activity initiated by virus infection and denoted by the expression of ISGs.

ISG products have antiviral functions

The human genome encodes hundreds of ISGs, most of which remain uncharacterized (Der *et al.*, 1998). ISG products can directly regulate aspects of virus replication, immune defence, host

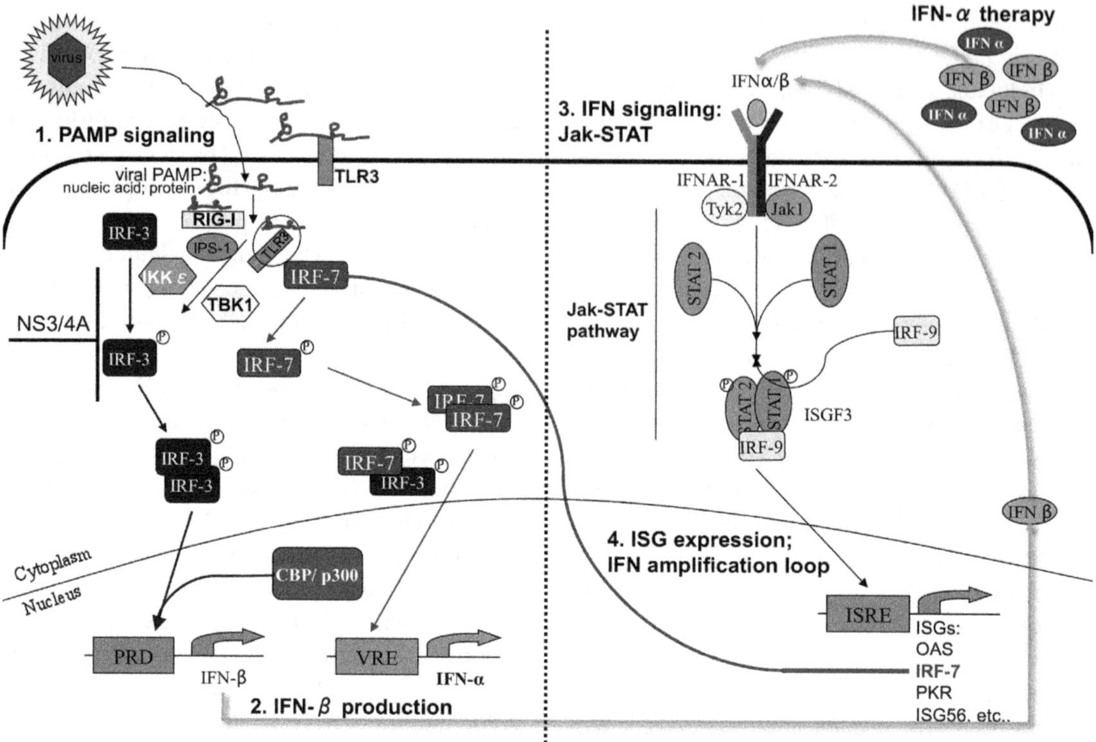

Figure 9.2 Signalling the hepatic innate immune response to HCV infection. 1. Viral PAMP (HCV RNA/poly-U/UC motif) binding to RIG-I results in signalling the downstream phosphorylation and activation of IRF-3 by the TBK1 or IKKε protein kinases. The dimer of phospho-IRF-3 translocates to the cell nucleus, interacts with its transcription partners (Yoneyama *et al.*, 1998), and binds to the cognate DNA positive regulatory domain (PRD) in the promoter region of IRF-3 target genes, including IFN-β. 2. IRF-3 activation drives IFN-β production. 3. IFN-β binding to the IFN-α/β receptor signals the activation of the associated Tyk2 and Jak1 protein kinases and the phosphorylation and assembly of a STAT1/STAT2 heterodimer and trimeric ISGF3 (Sen, 2001). 4. ISGs are induced by ISGF3. ISGs are the genetic effectors of the host response, and IRF-7 is itself and ISG and a critical transcription factor whose actions are turned on through viral PAMP signalling pathways that overlap with the pathways of IRF-3 activation. IRF-7 phosphorylation, dimerization and heterodimerization with IRF-3 directs its binding to the virus-responsive element (VRE) in the promoter region of IFN-α genes, resulting in IFN-α subtypes expression (Au *et al.*, 2001). This increases the abundance of RIG-I and viral PAMP signalling components to amplify host response. The therapeutic administration of IFN-α provides antiviral action against HCV by signalling ISG expression through the IFN-α/β receptor and the Jak-STAT pathway. The NS3/4A protease prevents IFN-β production by disrupting RIG-I signalling through the targeted proteolysis of IPS-1.

metabolism, gene expression, apoptosis, and other actions that can serve to support an antiviral state that limits infection (Loo and Gale, Jr., 2007). In terms of HCV, ISGs have been shown to interrupt virus replication through processes that include suppression of viral protein synthesis, inhibition of the viral negative strand replicative intermediate, RNA processing, and amplification of innate immune signalling (Guo *et al.*, 2001; Shimazaki *et al.*, 2002; Wang *et al.*, 2002; Prabhu *et al.*, 2004). The main feature resulting from the secretion of IFN from infected cells into the local tissue is the paracrine induction of a tissue-wide response that blocks cell–cell spread of the virus. Many components of the innate immune response pathways are themselves ISGs, and though expressed basally at low levels that facilitate surveillance and response to virus infection, their abundance will increase in response to IFN signalling. In the liver, paracrine signalling of IFN serves to enhance overall responsiveness of cellular signalling pathways to potentiate the innate immune response to infection in a tissue-wide manner. IFN signalling thus provides an amplification loop to further promote the host response against HCV (Foy *et al.*, 2005; Loo *et al.*, 2006; Lau *et al.*, 2008). Recent studies have shown that the amplification loop is dependent upon the processing of viral and/or self RNA by ribonuclease-L (RNase-L) to small RNA ligands of RIG-I signalling (Malathi *et al.*, 2007), and that the synthesis and activity of IRF-7 may further support this amplification (Honda *et al.*, 2005). IRF-7 is an ISG and its expression in the liver is IFN-dependent (Smith *et al.*, 2003; Honda *et al.*, 2005). IRF-7 promotes the expression of the various IFN-α subtypes, wherein IFN-α production and secretion can drive further signalling to mediate prolonged IFN production (Honda *et al.*, 2005). IFN-based therapy for HCV infection (described below) exploits these actions of IFN-α to limit HCV replication and spread (McHutchison and Patel, 2002; Fig. 9.2). IFN-α also signals the maturation of immune effector cells, antigen presenting cells and dendritic cells. And it potentiates the production of other proinflammatory cytokines by resident hepatic cells to indirectly modulate cell-mediated defences and adaptive immunity (Biron, 1999; Kolumam *et al.*, 2005; Havenar-Daughton *et al.*, 2006; Thompson *et al.*, 2006). Viral triggering and control of the innate immune response will therefore define cellular permissiveness for HCV RNA replication and influence the outcome of infection.

IFN therapy

The current approved therapy for HCV infection is parenteral administration of IFN-α in combination with oral ribavirin, a nucleoside analogue (McHutchison and Patel, 2002). IFN treatment of HCV infection was first described in 1986 by Hoofnagle *et al.* (1986). In this early study, recombinant human IFN-α2b was administered to HCV-infected patients for up to 12 months. It was found that serum levels of the liver enzyme alanine amino transferase (ATL), an indicator of liver damage, were decreased in HCV patients as a result of IFN therapy. However, in some patients, elevated serum ALT levels promptly returned after therapy was stopped. Patients who demonstrated a sustained virological response (SVR) after therapy ended (non-detectable serum HCV levels) exhibited sustained normal ALT levels, demonstrating that IFN therapy can offer both antiviral and disease-protective benefits. Problematically, today only 55% of patients achieve an SVR.

Contemporary IFN-based therapeutic regimens most often utilize IFN-α2a or -2b, though consensus IFN (CIFN), comprising the consensus sequence derived from the different IFN-α isoforms, is also approved for treatment of HCV infection (McHutchison and Patel, 2002; Table 9.1). Therapeutic IFN is injected by the patient either subcutaneously or by intramuscular injection once to three times weekly, depending on

Table 9.1 Approved IFN therapy regimen for HCV infection

Product	Name (manufacturer)	Dosing[1]
Interferon alfa-2a	Roferon-A (Roche)	3 MU TIW, SC
Interferon alfa-2b	Intron A (Schering-Plough)	3 MU TIW, SC
Interferon alfacon-1	Infergen (InterMune)	9 μg TIW, SC 15 μg TIW, SC†
Peginterferon alfa-2a	Pegasys (Roche)	180 μg QW, SC
Peginterferon alfa-2b	Peg-Intron (Schering-Plough)	1.0–1.5 μg/kg QW, SC
Ribavirin*	Copegus (Roche) Rebetrol (Schering-Plough)	0.8–1.4 g/day 0.8–1.4 g/day

[1]Ribavirin is not approved for monotherapy, but as part of combination with IFN-α.

MU, million units; TIW, three injections weekly; SC, subcutaneous; QW, weight-based dosing.

Ribavirin is taken orally.

the formulation used. The therapeutic effects are greatly enhanced with daily intake of Ribavirin capsules, wherein Ribavirin is incorporated into the viral RNA during genome replication and is thought to drive viral mutation to error catastrophe that results in suppressed HCV levels (Pawlotsky, 2009). CIFN and IFN-β have been shown to have better antiviral activities *in vitro* and in brief clinical trials; however, these treatments are not widely used owing to their requirement for increased injection schedules that are unpleasant to patients (Erickson *et al.*, 2008). It has also been shown that IFN-λ could inhibit HCV replication *in vitro* (Robek *et al.*, 2005). Since IFN-λ is a type III IFN and utilizes different receptors from IFN-α/β, the addition of IFN-λ to current standard IFN treatment may possibly activate another group of antiviral genes and therefore enhance or extend the effectiveness of antiviral therapy for HCV infection. Of importance is that *Ifn-λ* gene polymorphisms have been implicated in the outcome of HCV in infection, suggesting an important role for this gene product in HCV immunity (Ge *et al.*, 2009; Suppiah *et al.*, 2009; Tanaka *et al.*, 2009; Thomas *et al.*, 2009; Rallon *et al.*, 2010; Rauch *et al.*, 2010; Thompson *et al.*, 2010). Overall the current IFN-based therapy for HCV presents side-effects to the patient of skin reactions such as redness, rash, itching, a lump, swelling, or bruising at the site of injection. Moreover, the side-effects of therapy are at times too difficult for the patients to fully comply with their treatment schedule. Indeed, a major side-effect is presented as 'flu-like symptoms' (including fever, chills, muscle aches, joint pain, headaches), and – along with possible upset stomach (such as nausea, taste changes, diarrhoea), hair loss, and mental depression – IFN therapy is difficult to experience. The balance between the effectiveness of the treatment and the quality of life of the HCV patient remains a very challenging area of clinical translational efforts to effectively treat HCV infection. Understanding how IFN works to suppress HCV and to enhance antiviral immunity is a major focus of current research aimed at improving therapeutic outcome for HCV infection.

IFN and the cross-talk between innate and adaptive immune defences.

IFN-α is a potent immunomodulator that influences the onset of the cellular immune response and adaptive immunity to HCV infection. IFN-α modulates natural killer (NK) cell activity towards lysis of infected target cells by promoting NK cell activation and proliferation, and supporting their survival through the induction of IL-15 (Biron, 1999; Loza and Perussia, 2004). Activated NK cells produce IFN-γ (Shoukry *et al.*, 2004); in the HCV-infected liver, NK cell homing and local secretion of IFN-γ may serve to limit HCV replication directly. Indeed, IFN-γ has been shown to mediate direct antiviral effects against HCV RNA replication *in vitro* (Frese *et al.*, 2002), and its production by immune effector cells in the liver associates with viral clearance in the chimpanzee model of HCV infection (Thimme *et al.*, 2001; Su *et al.*, 2002). IFN-α also influences the maturation of dendritic cells and modulates their presentation of viral antigens (Colonna *et al.*, 2004; Barth *et al.*, 2005). Antigen presentation by dendritic cells under the influence of IFN-α contributes to the differentiation of CD4 T cells towards the Th1 phenotype and importantly, a Th1-predominate response is associated with clearance of HCV infection (Shoukry *et al.*, 2004). IFN-α production during HCV infection may therefore indirectly contribute towards directing the maturation of CD4 T cells to the Th1 phenotype. Dendritic cells are also involved in the cross priming and IFN-γ production of CD8 T cells. In this context the co-stimulatory signals presented at the time of antigen cross-presentation to CD8 T cells can determine whether the cells are cross-primed for a cytotoxic response or cross-tolerized for anergy (Cooper *et al.*, 1999; Shoukry *et al.*, 2004). TLR3 plays an essential role in cross priming by subset-specific dendritic cells, in which it signals the production of IFN-α triggered by viral PAMPs within products of phagocytosis (Schulz *et al.*, 2005). Moreover, the RLRs play a role in directing cytokine production from specific dendritic cell subsets. IFN-α produced by dendritic cells, including plasmacytoid dendritic cells, induces the expression of co-stimulatory molecules and cytokines to promote the cross-priming and activation of CD8-cells, thus linking innate immune response signalling processes to induction of the adaptive immune response. The link between innate, IFN-induced antiviral pathways and the adaptive immune response is not entirely understood and further studies are required to define the linkage of these processes with the outcome of HCV infection.

IFN-γ activation of the antiviral response

IFN-γ may serve to complete the signalling loop between infected hepatocytes and immune effector cells. The IFN-γ receptor is expressed by hepatocytes and cells of most other tissues. In the liver, IFN-γ is produced by infiltrating NK cells and activated T cells, where it can bind its receptor in a paracrine fashion to induce a hepatic IFN-γ response. IFN-γ receptor binding leads to phosphorylation of STAT1 through the receptor-bound Jak1 and Jak2 protein kinases (Stark *et al.*, 1998). The phosphorylated STAT then forms a homodimer that translocates to the cell nucleus and acts on the gamma-activated sequence (GAS) elements of target genes to promote their transcription (Der *et al.*, 1998). In hepatocytes the genes under the control of the GAS element have significant overlap with those expressed in response to IFN-α/β and under control of the ISRE (Cheney *et al.*, 2002). GAS elements are also found within genes whose products are involved in antigen processing and presentation, immune effector action and apoptosis, thus deriving a variety of antiviral actions. A role for GAS element genes in host defence against HCV is supported by studies of the chimpanzee model for HCV infection, where the high expression of IFN-γ responsive genes in the liver is associated with the resolution of acute infection (Su *et al.*, 2002). Together with IFN-α, the actions of IFN-γ provide additional levels of innate and adaptive immune response cross-talk that serves to limit HCV infection.

Hepatic innate immune defences are triggered by HCV

Functional genomic analyses from cohorts of patients infected with HCV have shown that infection associates with a hepatic gene expression profile marked by ISGs whose expression levels vary widely among patients (Smith *et al.*, 2003). These observations indicate that HCV can induce and regulate the hepatic innate immune response to infection. Gene profiling studies of HCV-infected chimpanzees have demonstrated that acute resolving HCV infection is associated with a hepatic innate immune response characterized by high level hepatic ISG expression (Bigger *et al.*, 2001). In similar studies, 'outcome predictor' gene sets were identified by the overall hepatic expression level of certain ISGs and virus-responsive genes. This gene set was accurately defined as those virus-responsive genes whose high expression associated with low viral load and effective viral clearance but whose low expression correlated with progression to chronic HCV infection (Su *et al.*, 2002). Like many virus-responsive genes and ISGs, the various products of the 'outcome predictor' gene set are known to interact with components of T cell immunity, again demonstrating the complex cross-talk that goes on during the immune response to infection and between parameters of innate and adaptive immunity. These observations demonstrate that the hepatic innate immune response is triggered during HCV infection but differentially regulated in association with disease course, and that HCV mediates persistence through strategies to regulate and/or evade the antiviral actions of this response.

Triggering the innate immune response to HCV infection

The processes by which HCV initiates and controls the innate immune response have been addressed in cell culture models of HCV RNA replication, HCV infection, and viral protein expression. Earlier studies demonstrated that genome-length or specific subgenomic fragments of HCV RNA are sufficient to trigger IFN-β expression and production when introduced into cultured human hepatoma cells (Fredericksen *et al.*, 2001; McCormick *et al.*, 2004), and this effect was attributed to the 3′ non-translated region (NTR) of the HCV genome. This region encodes a series of highly conserved stem–loop/dsRNA structures along with a variable length poly-uridine region containing interspersed cytosines (poly-U/UC region; Tuplin *et al.*, 2002). In cultured cells the HCV 3′ NTR presents PAMP structures that serve as potent agonists of the innate immune response. This suggests that during infection these RNA motifs are recognized and engaged by PAMP receptor(s) that trigger host immune defences (Sumpter *et al.*, 2005).

The nature of the major HCV RNA PAMP receptor was revealed through studies of the Huh7-derived (human hepatoma) cell line, termed Huh7.5. This cell line does not mount an intracellular innate immune response to virus infection or transfected HCV RNA and it is highly permissive for HCV infection and viral RNA replication (Blight *et al.*, 2002; Sumpter *et al.*, 2005; Saito *et al.*, 2008). Complementation studies identified RIG-I as an HCV RNA PAMP receptor that binds HCV dsRNA motifs and signals the activation of IRF-3 and NF-κB. This was shown to induce IFN-β expression and onset of the innate immune response (Sumpter *et al.*, 2005). Signalling by RIG-I directs a response that suppresses HCV RNA replication through the actions of expressed ISGs such as ISG56 (Foy *et al.*, 2005; Sumpter *et al.*, 2005). The permissiveness of Huh7.5 cells for HCV infection and RNA replication has been attributed to a point mutation in the RIG-I CARD motif that ablated downstream IRF-3 phosphorylation and NF-κB activation (Sumpter *et al.*, 2005). Virus-induced signalling by RIG-I and the resulting host response may therefore influence the outcome of HCV infection. TLR3 is a dsRNA PAMP receptor that signals a host response after it engages the dsRNA PAMP ligand (Alexopoulou *et al.*, 2001; Fig. 9.2). TLR3 signals the activation of IRF-3 and NF-κB through MyD88-independent processes that require the Toll-IL-1 receptor domain-containing adaptor inducing IFN-β (TRIF) protein (Yamamoto *et al.*, 2003). Genetic and biochemical studies have now defined prominent roles for RIG-I, and indicate a role for TLR3 pathways in signalling the innate immune response to virus infection in hepatocytes (described below), though the role of TLR3 in natural HCV infection remains to be evaluated.

Protein products of infection may also stimulate the innate immune response and antiviral defences against HCV. Expression of the HCV NS5A protein induces cellular stress signalling pathways to activate STAT3 (Gong *et al.*, 2001). Similar to the IFN-α/β receptor signalling, STAT3 promotes gene expression through processes that involve the Jak-STAT pathway (Sarcar *et al.*, 2004). This results in a gene expression profile that includes ISGs and proinflammatory cytokines that may influence the overall level of HCV RNA replication (Zhu *et al.*, 2003). Moreover, the HCV core protein has been shown to activate the catalytic activity of the dsRNA dependent protein kinase R (PKR), an ISG effector component of the host response to virus infection (Delhem *et al.*, 2001). The activation of PKR is induced by binding to dsRNA, resulting in eukaryotic initiation factor 2 phosphorylation and inhibition of local protein synthesis. Active PKR also signals the DNA binding activity of NF-κB and induces IRF-1 transcription-effector action (Williams, 1999). Activation of PKR by the HCV core protein is most likely attributed to the 5′ NTR viral RNA binding property of the core protein (Tanaka *et al.*, 2000), which would provide the PKR activator dsRNA PAMP required for kinase activation. Cell interaction with virus particles may also trigger signalling events that induce IFN production. HCV pseudo particle binding to dendritic cells has been shown to mediate particle uptake and dendritic cell activation (Barth *et al.*, 2005). Since dendritic cell subsets constitute a major source of IFN production during viral infection, HCV modulation of dendritic cell function may influence HCV infection by local modulation or systemic IFN levels (Colonna *et al.*, 2004).

Regulation and evasion of the innate immune response and IFN defences by HCV

HCV utilizes a variety of strategies to regulate and evade host innate immunity. Studies of the HCV/host interface have revealed PAMP-responsive signalling pathways that impart IRF-3 activation, IFN-α/β receptor signalling, and ISG effector action as major sites of control and evasion of the host response (Katze *et al.*, 2002). The HCV NS3/4A protease has been identified as an antagonist of virus-induced IRF-3 activation and IFN-β expression. NS3/4A mediates a block to RIG-I signalling of IRF-3 activation triggered either by

HCV infection, endogenous replicating HCV replicon RNA or by exogenous virus infection of cells harbouring a replicating HCV genomic RNA (Foy *et al.*, 2003). The IRF-3 blockade was attributed to the ability of the NS3/4A protein complex to target and cleave IPS-1 from intracellular membranes, thus ablating RIG-I signalling (Loo *et al.*, 2006; Johnson *et al.*, 2007). The end result of this regulation is that IPS-1 proteolysis ends up preventing virus-induced IRF-3 phosphorylation, thus retaining IRF-3 in an inactive, cytoplasmic-bound state in the HCV-infected cell. NS3 is a bifunctional enzyme that encodes a serine protease within its amino-terminal domain and a RNA helicase within its carboxyl-terminal domain (Reed and Rice, 1998). Importantly, the NS3/4A complex constitutes the essential viral protease, and it releases the non-structural proteins from the HCV polyprotein during virus replication (De Francesco and Steinkuhler, 2000). The helicase activity of NS3 is not required for the control of IRF-3 activation but the NS3/4A protease activity is required for IPS-1 proteolysis (Foy *et al.*, 2003; Johnson *et al.*, 2007). Indeed, NS3/4A regulation of RIG-I signalling through IPS-1 proteolysis has been identified as a causal link of the IRF-3 phosphorylation blockade directed by NS3/4A and a major attribute of persistent HCV infection (Loo *et al.*, 2006). The NS3/4A block to RIG-I signalling additionally ablates virus activation of NF-κB (Foy *et al.*, 2005). This dual regulation of IRF-3 and NF-κB indicates that NS3/4A may regulate multiple pathways within the cell that involve IRF-3 and/or NF-κB actions (Loo *et al.*, 2006).

TLR3 signalling has been reported to be targeted and regulated by NS3/4A *in vitro*. In this case NS3/4A protease activity has been shown to cleave the TRIF adaptor protein between amino acids C372 and S373 (Li *et al.*, 2005b). This cleavage site is homologous with the HCV NS5A/B polyprotein cleavage site. Structure/function studies show that NS3/4A recognizes TRIF through binding of a proline-rich region adjacent to the site of cleavage (Ferreon *et al.*, 2005). TRIF is essential for signalling by TLR3 but it does not play a role in the RIG-I pathway, thus TRIF cleavage by NS3/4A does not provide a mechanism of RIG-I pathway regulation (Li *et al.*, 2005a; Foy *et al.*, 2005). NS3/4A cleavage of TRIF prevents TLR3 signalling, thus blocking IRF-3 and NF-κB activation and preventing IFN production (Li *et al.*, 2005b; Fig. 9.2). The targeting of the RIG-I and TLR3 pathways by NS3/4A allows HCV to evade two major arms of IFN production and host defence. This observation has been further validated through pharmacological studies that used a peptidomimetic active site NS3 protease inhibitor to evaluate the requirement for protease activity in the regulation of host response signalling by NS3/4A. Treatment of cells that expressed wild type, functional NS3/4A showed that the protease inhibitor effectively removed the blockade to RIG-I and TLR3 signalling imposed by HCV, thereby restoring virus-induced IRF-3 phosphorylation and activation, and the activation NF-κB (Foy *et al.*, 2003, 2005). Protease inhibitor treatment of cells also protected TRIF from cleavage by NS3/4A (Li *et al.*, 2005b). However, the significance of TLR3 regulation by HCV and its occurrence within infected patients is not yet known.

What are the implications resulting from HCV disruption of RIG-I? First, viral control of the RIG-I pathway serves to attenuate major programmes of IFN production by infected cells and tissues. Second, many of the components of these pathways are IFN-responsive and though expressed at low levels, their expression is increased after exposure of cells and tissues to IFN-α/β. The IFN-responsiveness of these factors provides an amplification loop to enhance the strength and length of the host response. The signalling blockade imposed by NS3/4A cleavage of IPS-1 breaks this IFN amplification loop (Foy *et al.*, 2005) and may limit the level and diversity of ISG expression induced in response to IFN therapy. Third, HCV attenuation of the RIG-I pathway and the innate immune response to infection is expected to cause alterations in antigen presentation by the affected hepatic tissue, potentially leading to inefficient activation of cytolytic T cells and an inability of the adaptive immune response to clear HCV-infected hepatocytes by disrupting the cross-talk between the innate and the adaptive arms of immunity. This may provide a causal link between high level intrahepatic ISG expression and a vigorous T cell response to diverse viral epitopes, both of which are associated with viral clearance and inversely correlate with chronic infection (Su *et al.*, 2002; Shoukry *et al.*, 2004). Fourth, IRF-3 has been ascribed proapoptotic and tumour suppressor functions (Heylbroeck *et al.*, 2000; Duguay *et al.*, 2002). A prolonged block to IRF-3 activation might disrupt these actions and cause a tumorigenic phenotype within infected cells. This could provide a link between chronic HCV infection and hepatocellular carcinoma (Liang and Heller, 2004). Finally, the blockade of NF-κB function may interfere with a variety of chemokine and cytokine genes whose expression is dependent on NF-κB and which serve to drive the inflammatory response to virus infection (Zhu and Liu, 2003; Foy *et al.*, 2005). HCV regulation of NF-κB may therefore contribute to the systemic immune defects associated with HCV infection.

HCV regulation of IFN signalling

The overall low response rate of HCV to IFN therapy, particularly among patients with genotype 1 HCV infection (McHutchison and Patel, 2002), provides clinical evidence that HCV can effectively evade IFN actions *in vivo*. Many studies have focused on defining the molecular mechanisms by which HCV evades and resists IFN actions. These studies have shown that HCV protein expression associates with inhibition of STAT1 function, and can occur independently of STAT tyrosine phosphorylation (Heim *et al.*, 1999). Analysis of transgenic mice that express HCV proteins in their hepatocytes showed that the regulation of STAT1 occurs below the level of STAT tyrosine phosphorylation, resulting in a defective hepatic IFN response (Blindenbacher *et al.*, 2003). STAT dysfunction might be attributed to protein phosphatase 2A (Fig. 9.3), which confers STAT1 hypomethylation and complex formation with the protein inhibitor of activated STAT1 (PIAS). This modification was found to prevent STAT1 assembly into the ISGF3 complex and to attenuate ISG expression (Duong *et al.*, 2004). The mechanism by which protein phosphatase 2A triggers these events is not known. Others have shown that expression of the HCV core protein associates with increased levels of suppressor of cytokine signalling (SOCS)-3 (Bode *et al.*, 2003). SOCS3 belongs to a family of SOCS proteins that are negative regulators and inhibitors of Jak-STAT signalling. SOCS proteins mediate a classical negative feedback loop on IFN α/β receptor signalling events; SOCS-1 and SOCS-3 confer reduced levels of ISG expression (Alexander, 2002). While induction of SOCS-3 by the HCV core protein might impart evasion from IFN actions, it should be noted that the overall role of SOCS-3 in HCV infection is not known and it is unclear if regulation of SOCS-3 by the core is a result of the potential for the core protein to stimulate IFN signalling events itself (Miller *et al.*, 2004) or through induction of yet undefined signalling pathways that stimulate SOCS expression.

Regulation of ISG expression or function

Molecular studies have linked HCV evasion of the host response and IFN therapy with various strategies directed by viral proteins to control ISG expression or function (Table 9.2). The HCV NS5A protein has been identified as an IFN antagonist. Several studies have shown that expression of NS5A alone can attenuate IFN-α actions and rescue the replication of IFN-sensitive viruses (Macdonald and Harris, 2004). Microarray analyses have shown that NS5A expression can confer a general attenuation of ISG expression, though the mechanism of this regulation was not defined (Geiss *et al.*, 2003). NS5A induces interleukin (IL)–8 expression and secretion. This has implications for IFN therapy because IL-8 is a proinflammatory chemokine whose actions interfere with IFN (Fig. 9.3). The mechanism(s) of IL-8's anti-IFN action may include attenuation of IFN signalling and ISG expression, and/or direct inhibition of select ISGs (Khabar *et al.*, 1997). Serum IL-8 levels are found elevated in patients with chronic HCV, and NS5A has been shown to stimulate IL-8 production through transactivation of the IL-8 promoter, possibly involving NF-κB and AP-1 transcription factor activation by other cytokines (Polyak *et al.*, 2001a,b). IRFs may also drive IL-8 production when activated during virus infection (Casola *et al.*, 2000). This latter point is another example that serves to demonstrate the link between innate antiviral and

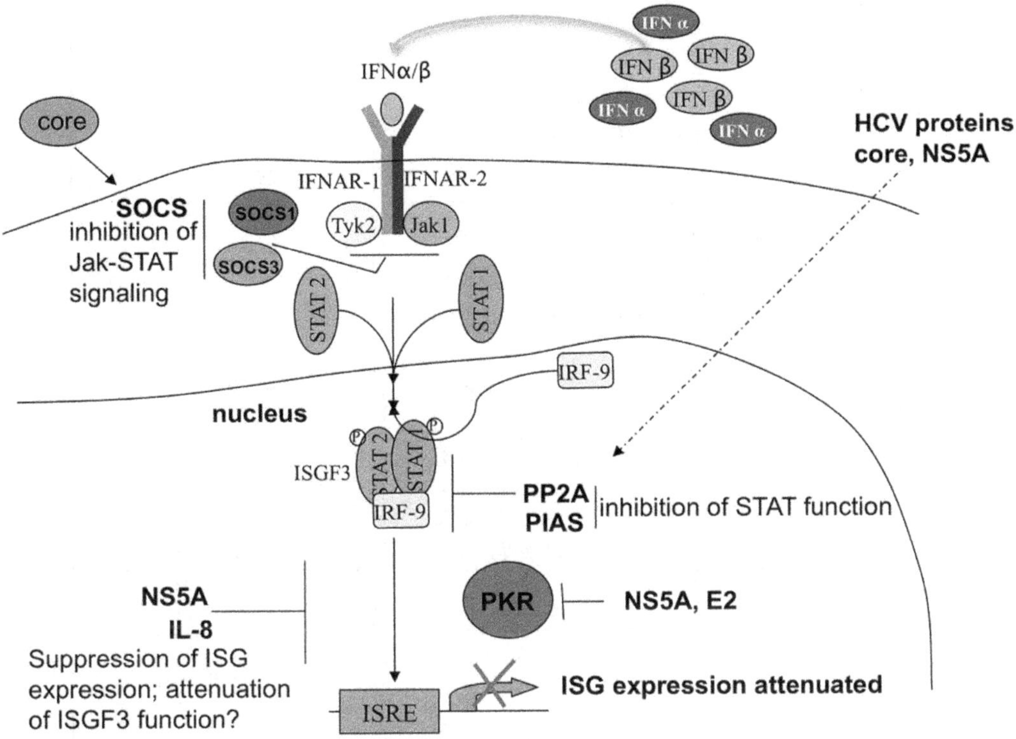

Figure 9.3 HCV attenuation of IFN signalling. IFN-α/β receptor signalling by IFN from autocrine/paracrine and therapeutic sources is subject to feedback inhibition by suppressor of cytokine signalling (SOCS) proteins. The HCV core protein can induce the expression of SOCS-3 within hepatocytes, which suppresses Jak-STAT signalling events (Alexander, 2002). Expression of the protein inhibitor of activated STAT (PIAS) is induced by HCV proteins, possibly mediated by protein phosphatase 2A (PP2A) signalling events and STAT demethylation (Duong *et al.*, 2004). This blocks STAT1 function. Some patients with chronic HCV infection exhibit aberrantly high levels of serum IL-8 (Polyak *et al.*, 2001b). The biological activity of IL-8 interferes with IFN signalling events (Khabar *et al.*, 1997). HCV modulation of IFN signalling allows the virus to evade the antiviral actions of the host response and IFN therapy. NS5A and E2 proteins of HCV have each been identified as inhibitors of PKR activity, thus suppressing an important ISG effector of IFN actions and further supporting HCV replication.

pro-inflammatory responses. In this case the link is exploited by NS5A as a means of evading the host response to HCV infection.

The NS5A and E2 proteins of HCV have both been identified as inhibitors of PKR (Gale, Jr. *et al.*, 1998; Taylor *et al.*, 1999; Noguchi *et al.*, 2001). Inhibition of PKR may allow HCV to evade, in part, the translational-suppressive and signalling actions of PKR (Katze *et al.*, 2002). This regulation is not universal and is subject to alteration through viral genetic variation, such that not all NS5A sequences have the ability to bind and inhibit PKR (Gimenez-Barcons *et al.*, 2005). Thus, HCV evasion of PKR-independent processes of the host response must also contribute to HCV resistance to IFN. The product of the IFN-induced ISG56 gene (also known as IFIT1 or the 561 gene), p56, can suppress HCV RNA translation (Wang *et al.*, 2002). Viral attenuation of ISG56 expression has been shown to associate with a level of resistance to HCV RNA replication from the antiviral actions of IFN in the HCV replicon cell culture model. ISG56 is both an ISG and an IRF-3 target gene (Grandvaux *et al.*, 2002), and the IFN-resistant phenotype in this case is associated with viral genetic adaptations that enhanced the NS3/4A blockade to IRF-3 signalling (Sumpter *et al.*, 2004). This raises the possibility

Table 9.2 Processes of ISG regulation or control by HCV

Viral strategy	Mechanism of action	Implications	Reference
IL-8 induction	NS5A induces IL-8 production through processes involving NFkB and AP-1 transcription factor activation	Attenuates ISG expression	Polyak *et al.* (2001a)
Induction of SOCS expression	The HCV core protein can induce expression of SOCS-1 and SOCS-3	Blocks Jak-STAT signalling action through the IFN-α receptor	Bode *et al.* (2003)
PKR inhibition	NS5A and E2 proteins bind PKR and inhibit its catalytic activity	Disruption of PKR-dependent translational control and signalling actions	Noguchi *et al.* (2001), Gimenez-Barcons *et al.* (2005), Taylor *et al.* (1999)
IRF-1 regulation	NS5A blocks ds RNA induced IRF-1 action through inhibition of PKR signalling	Relieves IRF-1 suppression of HCV RNA replication	Pflugheber *et al.* (2002), Kanazawa *et al.* (2004)
Evasion of 2′,5′ OAS/RNase L pathway	HCV genome sequence	The HCV genome encodes a paucity of RNase L recognition sites, which allow protection from nucleolytic processing	Han *et al.* (2004), Han and Barton (2002)
Disruption of STAT1 function.	HCV proteins	HCV proteins induce PP2A expression and STAT1 hypomethylation to attenuate ISG expression	Heim *et al.* (1999), Duong *et al.* (2004)
Suppression of ISG56 expression	HCV non-structural proteins	*In vitro*: NS3/4A and non-structural proteins disrupt virus signalling to the ISG56 promoter. Removes the ISG56 block to viral RNA translation	Sumpter *et al.* (2004), Wang *et al.* (2002)
Regulation of RIG-I signalling	NS3/4A protease blockade of signalling	Blockade of RIG-I signalling breaks an IFN amplification loop that otherwise enhances ISG expression	Foy *et al.* (2005)
Regulation of TLR3 signalling	NS3/4A protease cleavage of TRIF	Disruption of a TLR3-pathway IFN amplification loop	Li *et al.* (2005b)
Activation of STAT3	The NS5A protein induces a stress response that activates STAT3	STAT3-target genes, including some ISGs, are expressed	Gong *et al.* (2001)

that HCV control of IRF-3 activation pathways may attenuate ISG expression by preventing the cross-talk of cellular pathways that converge on ISG promoter elements. Studies that have evaluated HCV interactions with the IFN-induced 2′,5′ oligoadenylate synthetase (OAS)/RNase L antiviral pathway have shown that HCV proteins interact with this pathway (Taguchi *et al.*, 2004). When activated, this pathway directs RNase L, an endoribonuclease, to cleave the HCV genome RNA into non-functional products (Han and Barton, 2002). RNase L cleaves HCV RNA only at certain UU and UA dinucleotide sites (Han *et al.*, 2004), and genotype 1 HCV sequences have overall fewer RNase L cleavage sites than HCV genotypes 2 or 3 (Han *et al.*, 2004). This could provide a genetic mechanism for how, in part, HCV 1a and 1b infections resist IFN therapy (Table 9.2). However, as described above and owing to the pleiotropic nature of IFN effects mediated by the hundreds of ISGs, it is likely that HCV evades IFN action through multiple strategies to redirect ISG functions.

Viral genetics impact the host response to HCV infection

The viral polymerase of HCV lacks proofreading function (Reed and Rice, 1998) and during persistent infection, error-prone virus replication generates a repertoire of highly related but genetically distinct viral variants or 'quasispecies'. This provides a remarkable adaptive potential to HCV and has been implicated in evasion and control of the host response and IFN therapy (Farci, 2001). A hostile host environment may drive the outgrowth of HCV 'evasion variants' from a pre-existing quasispecies pool or through viral genetic adaptation. This idea is supported by *in vivo* studies that evaluated viral sequences from the E1 and E2 coding regions within patients with HCV infection. This work demonstrated viral genetic patterns that associated with infection outcome and in which the resolution of acute HCV infection consistently associated with an overall reduction in viral quasispecies complexity (Farci *et al.*, 2000; Farci *et al.*, 2002). In contrast, progression to chronic infection and resistance to IFN therapy is associated with increased viral genetic complexity, suggesting that host immune pressure drives the outgrowth or selection of viral evasion variants able to resist IFN action and persist. Analysis of the HCV NS5A coding region has also identified specific domains that exhibit sequence variation in patients with differential outcomes to IFN therapy. Meta-analyses and long-term follow-up of these studies now provide support for NS5A sequence variation within a 40 amino acid 'IFN sensitivity determining region' (ISDR) that associates with IFN therapy outcome (Enomoto *et al.*, 1996; Pascu *et al.*, 2004; Schinkel *et al.*, 2004). The ISDR is within a genetically flexible domain that is a major site of viral adaptations among HCV RNA replicons (Blight *et al.*, 2000; Appel *et al.*, 2005). Thus, ISDR variation may affect the HCV replication fitness and the host response to infection.

Exogenous induction of antiviral hepatic defences

Various studies have described an absence or only a low level expression of IFN-α/β within liver tissue from patients with chronic HCV infection (Mihm *et al.*, 2004). This lack of IFN α/β gene expression within the HCV-infected liver provides indirect evidence that HCV imposes a blockade to IRF-3 activation *in vivo* and may explain why some patients with chronic infection do not express significant levels of hepatic ISGs. However, it fails to explain why other patients exhibit broad and abundant ISG expression despite only low level IFN α/β expression in the infected liver (Smith *et al.*, 2003). The fact that hepatic ISG expression has associated with the level and extent of liver pathology (Smith *et al.*, 2003) suggests that ISGs might be induced indirectly as a result of cellular stress from fibrosis and/or cirrhosis. ISGs are also induced through STAT3 signalling; studies have shown that STAT3 is activated by NS5A, and that STAT3 activation occurs concomitantly with HCV RNA replication (Gong *et al.*, 2001; Waris *et al.*, 2005). Viral activation of STAT3, possibly mediated through stress-responsive signalling events, may contribute to ISG expression during HCV infection (Fig. 9.4). ISGs may also be induced through TLR engagement exogenously by extracellular products of damaged tissue or viral replication. It is

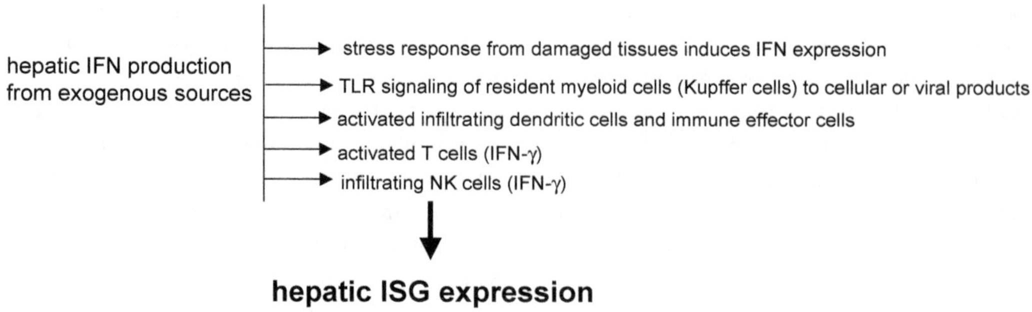

Figure 9.4 Processes of hepatic IFN production from exogenous sources. Various processes can result in hepatic ISG expression despite a block to RIG-I signalling imposed by NS3/4A and other HCV proteins. In most of the examples shown, IFN will be produced from non-infected (exogenous) cell sources and not from the chronic HCV-infected hepatocyte, which studies have shown do not express appreciable levels of IFN (Mihm *et al.*, 2004). NS5A triggering of STAT3 signalling (Waris *et al.*, 2005) would produce ISGs in the absence of IFN, and this effect would be limited to infected cells. Moreover, resident macrophages of the liver, the Kupffer cell, may respond to TLR ligands to produce hepatic-endogenous IFN during chronic HCV infection. This would support basal levels of ISG expression and may explain why chronic HCV associates with ISG expression.

also noted that stress-induced cytokines, including TNF-α and IL-1 can trigger IRF-1 expression and transactivation function, resulting in IFN-β production (Fujita *et al.*, 1986). This could contribute to hepatic ISG expression even when the RIG-I or TLR3 pathways are blocked by the HCV NS3/4A protease. Exogenous immune effector cells that infiltrate the liver may also contribute to hepatic ISG expression. In particular, IFN production by tissue macrophages and dendritic cells that have infiltrated the infected tissue could lead to a paracrine IFN response and ISG expression during the processes of antigen cross-presentation *in vivo* (Schulz *et al.*, 2005). By this model hepatic ISG levels would vary with the composition and extent of immune cell infiltration, which has been observed (Smith *et al.*, 2003). Secretion of IFN-γ by T cells and NK cells that have infiltrated the infected liver also contributes to ISG expression but the pattern of expressed ISGs only partially overlaps with those induced by IFN-α/β (Der *et al.*, 1998).

The HCV/host interface model of viral evasion

Fig. 9.5 depicts a model of the virus–host interface and host response regulation that forms a foundation for persistent HCV infection. The transmission event of HCV infection presents unique pressures for the virus to adapt to the new host environment. HCV adaptation to a new host will involve fine tuning of viral strategies to control and evade the host response to infection. The transmission event results in an acute infection that involves viral regulation of the host response though RIG-I, TLR3 and other host defence signalling pathways within the infected cell (Li *et al.*, 2005a; Sumpter *et al.*, 2005). Highly fit variants of HCV will mediate signalling interference at the virus–host interface. This involves the actions of the NS3/4A protease to block RIG-I and TLR3 signalling pathways. However, this regulation is influenced by viral genetic variation, and genetic distinctions among the many different quasispecies of HCV that are replicating in various cells at various times post-infection resulting in differential levels of control and activation of this response. During acute infection the differential activation and control of the host response by viral genetic variants will lead to the production of IFN and ISG expression to mediate an antiviral state in the local hepatic tissue (Bigger *et al.*, 2001). 15–25% of all exposures to HCV typically result in acute/resolving infection (McHutchison, 2004), indicating that a robust hepatic host response could provide protection against the replication and spread of HCV. The host response and the ensuing adaptive immune response present enormous

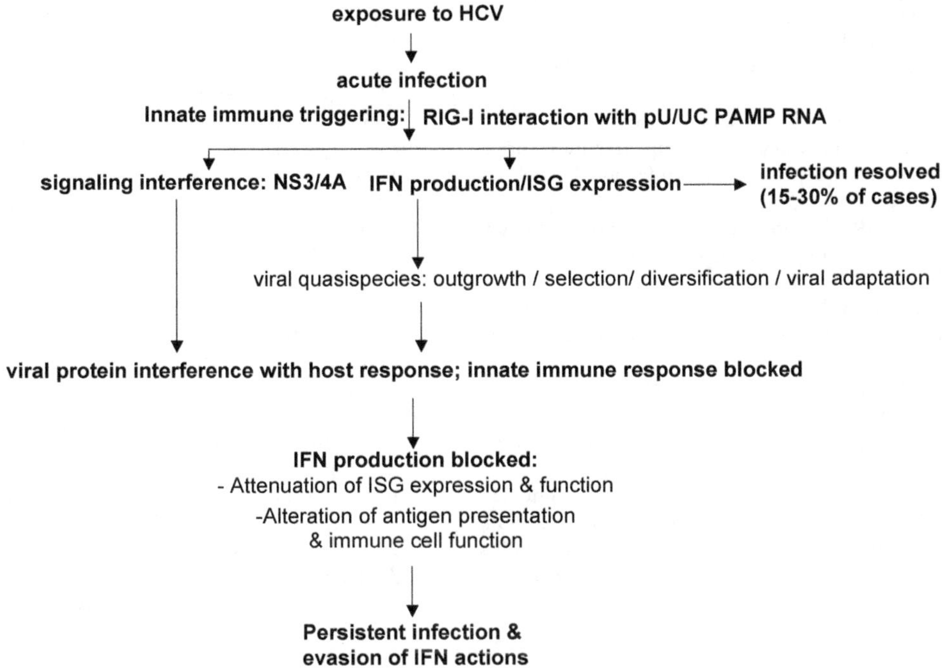

Figure 9.5 HCV/host interactions regulate infection outcome. The model of events in which the virus–host interface and interactions within the hepatic innate immune response decide the fate of HCV infection. By this model, HCV disruption of the innate immune response to infection provides a framework for viral persistence and chronic infection, and may confer resistance to IFN therapy. Details of the model are outlined in the text.

pressures that will select for the outgrowth of viral quasispecies that can evade and successfully control the host response and immune defences (Farci *et al.*, 2000; Sumpter *et al.*, 2004). HCV/host interactions at key sites of host defence signalling serve to suppress the host response, attenuate the actions of IFN therapy, and provide a solid foundation for persistent HCV infection. This model incorporates the important, variable and perhaps unpredictable aspect of viral adaptation and quasispecies selection within the evasion and control strategies by which HCV limits the host response to infection. Further studies to identify the viral genetic elements (including viral PAMPs and PAMP structure), signalling factors and ISG effectors that regulate the host response to infection will certainly increase our understanding of host defence and the HCV/host interface that controls infection outcome.

Acknowledgements

The Gale laboratory is supported by grants from the NIH and the Burroughs Wellcome Fund.

References

Alexander, W.S. (2002). Suppressors of cytokine signalling (SOCS) in the immune system. Nat. Rev. Immunol. 2, 410–416.

Alexopoulou, L., Holt, A.C., Medzhitov, R., and Flavell, R.A. (2001). Recognition of double-stranded RNA and activation of NF-kappaB by Toll-like receptor 3. Nature *413*, 732–738.

Appel, N., Pietschmann, T., and Bartenschlager, R. (2005). Mutational analysis of hepatitis C virus nonstructural protein 5A: potential role of differential phosphorylation in RNA replication and identification of a genetically flexible domain. J. Virol. *79*, 3187–3194.

Au, W.C., Moore, P.A., Lowther, W., Juang, Y.T., and Pitha, P.M. (1995). Identification of a member of the interferon regulatory factor family that binds to the interferon-stimulated response element and activates expression of interferon-induced genes. Proc. Natl. Acad. Sci. U. S.A. *92*, 11657–11661.

Au, W.C., Yeow, W.S., and Pitha, P.M. (2001). Analysis of functional domains of interferon regulatory factor 7 and its association with IRF-3. Virology *280*, 273–282.

Barnes, B.J., Moore, P.A., and Pitha, P.M. (2001). Virus-specific activation of a novel interferon regulatory factor, IRF-5, results in the induction of distinct interferon alpha genes. J. Biol. Chem. *276*, 23382–23390.

Barth, H., Ulsenheimer, A., Pape, G.R., Diepolder, H.M., Hoffmann, M., Neumann-Haefelin, C., Thimme, R.,

Henneke, P., Klein, R., Paranhos-Baccala, G., *et al.* (2005). Uptake and presentation of hepatitis C virus-like particles by human dendritic cells. Blood *105*, 3605–3614.

Bigger, C.B., Brasky, K.M., and Lanford, R.E. (2001). DNA microarray analysis of chimpanzee liver during acute resolving hepatitis C virus infection. J. Virol. *75*, 7059–7066.

Biron, C.A. (1999). Initial and innate responses to viral infections – pattern setting in immunity or disease. Curr. Opin. Micribiol. *2*, 374–381.

Blight, K.J., Kolykhalov, A.A., and Rice, C.M. (2000). Efficient initiation of HCV RNA replication in cell culture. Science *290*, 1972–1974.

Blight, K.J., McKeating, J.A., and Rice, C.M. (2002). Highly permissive cell lines for subgenomic and genomic hepatitis C virus RNA replication. J. Virol. *76*, 13001–13014.

Blindenbacher, A., Duong, F.H., Hunziker, L., Stutvoet, S.T., Wang, X., Terracciano, L., Moradpour, D., Blum, H.E., Alonzi, T., *et al.* (2003). Expression of hepatitis c virus proteins inhibits interferon alpha signaling in the liver of transgenic mice. Gastroenterology *124*, 1465–1475.

Bode, J.G., Ludwig, S., Ehrhardt, C., Albrecht, U., Erhardt, A., Schaper, F., Heinrich, P.C., and Haussinger, D. (2003). IFN-alpha antagonistic activity of HCV core protein involves induction of suppressor of cytokine signaling-3. FASEB J. *17*, 488–490.

Casola, A., Garofalo, R.P., Jamaluddin, M., Vlahopoulos, S., and Brasier, A.R. (2000). Requirement of a novel upstream response element in respiratory syncytial virus-induced IL-8 gene expression. J. Immunol. *164*, 5944–5951.

Cheney, I.W., Lai, V.C., Zhong, W., Brodhag, T., Dempsey, S., Lim, C., Hong, Z., Lau, J.Y., and Tam, R.C. (2002). Comparative analysis of anti-hepatitis C virus activity and gene expression mediated by alpha, beta, and gamma interferons. J. Virol. *76*, 11148–11154.

Colonna, M., Trinchieri, G., and Liu, Y.J. (2004). Plasmacytoid dendritic cells in immunity. Nat. Immunol. *5*, 1219–1226.

Cook, D.N., Pisetsky, D.S., and Schwartz, D.A. (2004). Toll-like receptors in the pathogenesis of human disease. Nat. Immunol. *5*, 975–979.

Cooper, S., Erickson, A.L., Adams, E.J., Kansopon, J., Weiner, A.J., Chien, D.Y., Houghton, M., Parham, P., and Walker, C.M. (1999). Analysis of a successful immune response against hepatitis C virus. Immunity *10*, 439–449.

De Francesco, R., and Steinkuhler, C. (2000). Structure and function of the hepatitis C virus NS3-NS4A serine proteinase. Curr. Top. Microbiol. Immunol. *242*, 149–169.

Delhem, N., Sabile, A., Gajardo, R., Podevin, P., Abadie, A., Blaton, M.A., Kremsdorf, D., Beretta, L., and Brechot, C. (2001). Activation of the interferon-inducible protein kinase PKR by hepatocellular carcinoma derived-hepatitis C virus core protein. Oncogene *20*, 5836–5845.

Der, S.D., Zhou, A., Williams, B.R.G., and Silverman, R.H. (1998). Identification of genes differentially regulated by interferon alpha, beta, or gamma using oligonucleotide arrays. Proc. Natl. Acad. Sci. U.S.A. *95*, 15623–15628.

Duguay, D., Mercier, F., Stagg, J., Martineau, D., Bramson, J., Servant, M., Lin, R., Galipeau, J., and Hiscott, J. (2002). *In vivo* interferon regulatory factor 3 tumor suppressor activity in B16 melanoma tumors. Cancer Res. *62*, 5148–5152.

Duong, F.H., Filipowicz, M., Tripodi, M., La Monica, N., and Heim,M.H. (2004). Hepatitis C virus inhibits interferon signaling through up-regulation of protein phosphatase 2A. Gastroenterology *126*, 263–277.

Enomoto, N., Sakuma, I., Asahina, Y., Kurosaki, M., Murakami, T., Yamamoto, C., Ogura, Y., Izumi, N., Maruno, F., and Sato, C. (1996). Mutations in the nonstructural protein 5A gene and response to interferon in patients with chronic hepatitis C virus 1b infection. N. Engl. J. Med. *334*, 77–81.

Erickson, A.K., Seiwert, S., and Gale, M., Jr. (2008). Antiviral potency analysis and functional comparison of consensus interferon, interferon-alpha2a and pegylated interferon-alpha2b against hepatitis C virus infection. Antivir. Ther. *13*, 851–862.

Farci, P. (2001). Hepatitis C virus. The importance of viral heterogeneity. Clin. Liver Dis. *5*, 895–916.

Farci, P., Shimoda, A., Coiana, A., Diaz, G., Peddis, G., Melpolder, J.C., Strazzera, A., Chien, D.Y., Munoz, S.J., Balestrieri, A., *et al.* (2000). The outcome of acute hepatitis C predicted by the evolution of the viral quasispecies. Science *288*, 339–344.

Farci, P., Strazzera, R., Alter, H.J., Farci, S., Degioannis, D., Coiana, A., Peddis, G., Usai, F., Serra, G., Chessa, L., *et al.* (2002). Early changes in hepatitis C viral quasispecies during interferon therapy predict the therapeutic outcome. Proc. Natl. Acad. Sci. U.S.A. *99*, 3081–3086.

Ferreon, J.C., Ferreon, A.C., Li, K., and Lemon, S.M. (2005). Molecular determinants of TRIF proteolysis mediated by the hepatitis C virus NS3/4A protease. J. Biol. Chem. *280*, 20483–20492.

Foy, E., Li, K., Sumpter, R., Jr., Loo, Y.M., Johnson,C.L., Wang, C., Fish, P.M., Yoneyama, M., Fujita, T., Lemon, S.M., and Gale, M., Jr. (2005). Control of antiviral defenses through hepatitis C virus disruption of retinoic acid-inducible gene-I signaling. Proc. Natl. Acad. Sci. U.S.A. *102*, 2986–2991.

Foy, E., Li, K., Wang, C., Sumpter,R., Ikeda, M., Lemon, S.M., and Gale, M., Jr. (2003). Regulation of interferon regulatory factor-3 by the hepatitis C virus serine protease. Science *300*, 1145–1148.

Fredericksen, B., Akkaraju, G., Foy, E., Wang, C., Pflugheber, J., Chen, Z.J., and Gale, M., Jr. (2001). Activation of the interferon-β promoter during hepatitis C virus RNA replication. Viral Immunol. *15*, 29–40.

Frese, M., Schwarzle, V., Barth, K., Krieger, N., Lohmann, V., Mihm, S., Haller, O., and Bartenschlager, R. (2002). Interferon-gamma inhibits replication of subgenomic and genomic hepatitis C virus RNAs. Hepatology *35*, 694–703.

Fujita, T., Reis, L.F., Watanabe, N., Kimura, Y., and Taniguchi, T. (1986). Induction of the transcription factor IRF-1 and interferon-beta mRNAs by cytokines and activators of second-messenger pathways. Proc. Natl. Acad. Sci. U.S.A. *86*, 9936–9940.

Gale, M., Jr. (2003). Effector genes of interferon action against hepatitis C virus. Hepatology *37*, 975–978.

Gale, M., Jr., Blakely, C.M., Kwieciszewski, B., Tan, S.-L., Dossett, M., Korth, M.J., Polyak, S.J., Gretch, D.R., and Katze, M.G. (1998). Control of PKR protein kinase by hepatitis C virus nonstructural 5A protein: molecular mechanisms of kinase regulation. Mol. Cell Biol. *18*, 5208–5218.

Ge, D., Fellay, J., Thompson, A.J., Simon, J.S., Shianna, K.V., Urban, T.J., Heinzen, E.L., Qiu, P., Bertelsen, A.H., Muir, A.J., *et al.* (2009). Genetic variation in IL28B predicts hepatitis C treatment-induced viral clearance. Nature *461*, 399–401.

Geiss, G.K., Carter, V.S., He, Y., Kwieciszewski, B.K., Holzman, T., Korth, M.J., Lazaro, C.A., Fausto, N., Bumgarner, R.E., and Katze, M.G. (2003). Gene expression profiling of the cellular transcriptional network regulated by alpha/beta interferon and its partial attenuation by the hepatitis C virus nonstructural 5A protein. J. Virol. *77*, 6367–6375.

Gimenez-Barcons, M., Wang, C., Chen, M., Sanchez-Tapias, J.M., Saiz, J.C., and Gale, M., Jr. (2005). The oncogenic potential of hepatitis C virus NS5A sequence variants is associated with PKR regulation. J. Interferon Cytokine Res. *25*, 152–164.

Gong, G., Waris, G., Tanveer, R., and Siddiqui, A. (2001). Human hepatitis C virus NS5A protein alters intracellular calcium levels, induces oxidative stress, and activates STAT-3 and NF-kappa B. Proc. Natl. Acad. Sci. U.S.A. *98*, 9599–9604.

Grandvaux, N., Servant, M.J., tenOever, B., Sen, G.C., Balachandran, S., Barber, G.N., Lin, R., and Hiscott, J. (2002). Transcriptional profiling of interferon regulatory factor 3 target genes: direct involvement in the regulation of interferon-stimulated genes. J. Virol. *76*, 5532–5539.

Guo, J., Bichko, V., and Seeger, C. (2001). Effect of alpha interferon on the hepatitis C virus replication. J. Virol. *75*, 8516–8523.

Han, J.Q., and Barton, D.J. (2002). Activation and evasion of the antiviral 2′-5′ oligoadenylate synthetase/ribonuclease L pathway by hepatitis C virus mRNA. RNA *8*, 512–525.

Han, J.Q., Wroblewski, G., Xu, Z., Silverman, R.H., and Barton, D.J. (2004). Sensitivity of hepatitis C virus RNA to the antiviral enzyme ribonuclease L is determined by a subset of efficient cleavage sites. J. Interferon Cytokine Res. *24*, 664–676.

Havenar-Daughton, C., Kolumam, G.A., and Murali-Krishna, K. (2006). Cutting edge: the direct action of type I IFN on CD4 T cells is critical for sustaining clonal expansion in response to a viral but not a bacterial infection. J. Immunol. *176*, 3315–3319.

Heim, M.H., Moradpour, D., and Blum, H.E. (1999). Expression of hepatitis C virus proteins inhibits signal transduction through the Jak-STAT pathway. J. Virol. *73*, 8469–8475.

Heylbroeck, C., Balachandran, S., Servant, M.J., DeLuca, C., Barber, G.N., Lin,R., and Hiscott, J. (2000). The IRF-3 transcription factor mediates Sendai virus-induced apoptosis. J. Virol. *74*, 3781–3792.

Honda, K., Yanai, H., Negishi, H., Asagiri, M., Sato, M., Mizutani, T., Shimada, N., Ohba, Y., Takaoka, A., Yoshida, N. *et al.* (2005). IRF-7 is the master regulator of type-I interferon-dependent immune responses. Nature. *434*, 772–777.

Hoofnagle, J.H., Mullen, K.D., Jones, D.B., Rustgi, V., Di Bisceglie, A., Peters, M., Waggoner, J.G., Park, Y., Jones, E.A. (1986). Treatment of chronic non-A, non-B hepatitis with recombinant human alpha interferon. A preliminary report. N. Engl. J. Med. *315*, 1575–1578.

Iorio, A., Marchesini, E., Awad, T., and Gluud, L.L. (2010). Antiviral treatment for chronic hepatitis C in patients with human immunodeficiency virus. Cochrane Database Syst. Rev. CD004888.

Iwasaki, A., and Medzhitov, R. (2004). Toll-like receptor control of the adaptive immune responses. Nat. Immunol. *5*, 987–995.

Johnson, C.L., Owen, D.M., and Gale, M., Jr. (2007). Functional and therapeutic analysis of hepatitis C virus NS3/4A protease control of antiviral immune defense. J. Biol Chem. *282*, 10792–10804.

Kanazawa, N., Kurosaki, M., Sakamoto, N., Enomoto, N., Itsui, Y., Yamashiro, T., Tanabe, Y., Maekawa, S., Nakagawa, M., Chen, C.H., *et al.* (2004). Regulation of hepatitis C virus replication by interferon regulatory factor 1. J. Virol *78*, 9713–9720.

Katze, M.G., He, Y., and Gale, M.Jr. (2002). Viruses and interferon: a fight for supremacy. Nature Rev. Immunol. *2*, 675–667.

Kawai, T., Sato, S., Ishii, K.J., Coban, C., Hemmi, H., Yamamoto, M., Terai, K., Matsuda, M., Inoue, J., Uematsu, S., *et al.* (2004). Interferon-alpha induction through Toll-like receptors involves a direct interaction of IRF7 with MyD88 and TRAF6. Nat. Immunol. *5*, 1061–1068.

Kawai, T., Takahashi, K., Sato, S., Coban, C., Kumar, H., Kato, H., Ishii, K.J., Takeuchi, O., and Akira, S. (2005). IPS-1, an adaptor triggering RIG-I- and Mda5-mediated type I interferon induction. Nat. Immunol. *6*, 981–988.

Khabar, K.S., Al Zoghaibi, F., Al Ahdal, M.N., Murayama, T., Dhalla, M., Mukaida, N., Taha, M., Al Sedairy, S.T., Siddiqui, Y., Kessie, G., *et al.* (1997). The alpha chemokine, interleukin 8, inhibits the antiviral action of interferon alpha. J. Exp. Med. *186*, 1077–1085.

Kolumam, G.A., Thomas, S., Thompson, L.J., Sprent, J., and Murali-Krishna, K. (2005). Type I interferons act directly on CD8 T cells to allow clonal expansion and memory formation in response to viral infection. J. Exp. Med. *202*, 637–650.

Lau, D.T., Fish, P.M., Sinha, M., Owen, D.M., Lemon, S.M., and Gale, M., Jr. (2008). Interferon regulatory factor-3 activation, hepatic interferon-stimulated gene expression, and immune cell infiltration in hepatitis C virus patients. Hepatology *47*, 799–809.

Li, K., Chen, Z., Kato, N., Gale, M., Jr., and Lemon, S.M. (2005a). Distinct poly-I: C and virus-activated signaling pathways leading to interferon-beta production in hepatocytes. J. Biol. Chem. *280*, 16739–16747.

Li, K., Foy, E., Ferreon, J.C., Nakamura, M., Ferreon, A.C., Ikeda, M., Ray, S.C., Gale, M., Jr., and Lemon, S.M. (2005b). Immune evasion by hepatitis C virus NS3/4A protease-mediated cleavage of the Toll-like receptor 3 adaptor protein TRIF. Proc. Natl. Acad. Sci. U.S.A. *102*, 2992–2997.

Liang, T.J., and Heller, T. (2004). Pathogenesis of hepatitis C-associated hepatocellular carcinoma. Gastroenterology *127*, S62-S71.

Lin, R., Heylbroeck, C., Genin, P., Pitha, P.M., and Hiscott, J. (1999). Essential role of interferon regulatory factor 3 in direct activation of RANTES chemokine transcription. Mol. Cell Biol. *19*, 959–966.

Loo, Y.M., Fornek, J., Crochet, N., Bajwa, G., Perwitasari, O., Martinez-Sobrido, L., Akira, S., Gill, M.A., Garcia-Sastre, A., Katze, M.G., *et al.* (2008). Distinct RIG-I and MDA5 signaling by RNA viruses in innate immunity. J. Virol. *82*, 335–345.

Loo, Y.M., and Gale, M., Jr. (2007). Viral regulation and evasion of the host response. Curr. Top. Microbiol. Immunol. *316*, 295–313.

Loo, Y.M., Owen, D.M., Li, K., Erickson, A.K., Johnson, C.L., Fish, P.M., Carney, D.S., Wang, T., Ishida, H., Yoneyama, M., *et al.* (2006). Viral and therapeutic control of IFN-beta promoter stimulator 1 during hepatitis C virus infection. Proc. Natl. Acad. Sci. U.S.A.*103*, 6001–6006.

Loza, M.J., and Perussia, B. (2004). Differential regulation of NK cell proliferation by type I and type II IFN. Int. Immunol. *16*, 23–32.

Macdonald, A., and Harris, M. (2004). Hepatitis C virus NS5A: tales of a promiscuous protein. J. Gen. Virol *85*, 2485–2502.

Malathi, K., Dong, B.S.S., Gale, M., Jr., and Silverman, R.H. (2007). Role of small RNAs produced by RNase L in antiviral innate immunity. Nature *448*, 816–819.

Mallet, V., Vallet-Pichard, A., and Pol, S. (2010). New trends in hepatitis C management. Presse Med. *39*, 446–451.

Malmgaard, L. (2004). Induction and regulation of IFNs during viral infections. J. Interferon Cytokine Res. *24*, 439–454.

McCormick, C.J., Challinor, L., Macdonald, A., Rowlands, D.J., and Harris, M. (2004). Introduction of replication-competent hepatitis C virus transcripts using a tetracycline-regulable baculovirus delivery system. J. Gen. Virol. *85*, 429–439.

McHutchison, J.G. (2004). Understanding hepatitis C. Am. J. Manag. Care *10*, S21-S29.

McHutchison, J.G., Gordon, S.C., Schiff, E.R., Shiffman, M.L., Lee, W.M., Goodman, Z.D., Ling, M.H., Cort, S., Albrecht, J.K., for the Interferon Interventional Therapy Group (1998). Interferon alfa-2b alone or in combination with ribavirin as initial treatment for chronic hepatitis C. N. Engl. J. Med. *339*, 1485–1492.

McHutchison, J.G., and Patel, K. (2002). Future therapy of hepatitis C. Hepatology *36*, S245–S252.

Mihm, S., Frese, M., Meier, V., Wietzke-Braun, P., Scharf, J.G., Bartenschlager, R., and Ramadori, G. (2004). Interferon type I gene expression in chronic hepatitis C. Lab. Invest. *84*, 1148–1159.

Miller, K., Mcardle, S., Gale, M.J., Jr., Geller, D.A., tenOever, B., Hiscott, J., Gretch, D.R., and Polyak,S.J. (2004). Effects of the hepatitis C virus core protein on innate cellular defense pathways. J. Interferon Cytokine Res. *24*, 391–402.

Noguchi, T., Satoh, S., Noshi, T., Hatada, E., Fukuda, R., Kawai, A., Ikeda, S., Hijikata, M., and Shimotohno, K. (2001). Effects of mutation in hepatitis C virus nonstructural protein 5A on interferon resistance mediated by inhibition of PKR kinase activity in mammalian cells. Microbiol. Immunol. *45*, 829–840.

O'Neill, L.A. (2004). Immunology. After the toll rush. Science *303*, 1481–1482.

Pascu, M., Martus, P., Hohne, M., Wiedenmann, B., Hopf, U., Schreier, E., and Berg, T. (2004). Sustained virological response in hepatitis C virus type 1b infected patients is predicted by the number of mutations within the NS5A-ISDR: a meta-analysis focused on geographical differences. Gut *53*, 1345–1351.

Pawlotsky, J.M. (2009). Hepatitis: HCV variability, the immune system and resistance to antiviral drugs. Nat. Rev. Gastroenterol. Hepatol. *6*, 383–385.

Pflugheber, J., Fredericksen, B., Sumpter, R., Wang, C., Ware, F., Sodora, D., and Gale, M. Jr. (2002). Regulation of PKR and IRF-1 during hepatitis C virus RNA replication. Proc. Natl. Acad. Sci. U.S.A. *99*, 4650–4655.

Polyak, S.J., Khabar, K.S., Paschal, D.M., Ezelle, H.J., Duverlie, G., Barber, G.N., Levy, D.E., Mukaida, N., and Gretch, D.R. (2001a). Hepatitis C virus nonstructural 5A protein induces interleukin-8, leading to partial inhibition of the interferon-induced antiviral response. J. Virol. *75*, 6095–6106.

Polyak, S.J., Khabar, K.S., Rezeiq, M., and Gretch, D.R. (2001b). Elevated levels of interleukin-8 in serum are associated with hepatitis C virus infection and resistance to interferon therapy. J. Virol. *75*, 6209–6211.

Prabhu, R., Joshi, V., Garry, R.F., Bastian, F., Haque, S., Regenstein, F., Thung, S., and Dash, S. (2004). Interferon alpha-2b inhibits negative-strand RNA and protein expression from full-length HCV1a infectious clone. Exp. Mol. Pathol. *76*, 242–252.

Rallon, N.I., Naggie, S., Benito, J.M., Medrano, J., Restrepo, C., Goldstein, D., Shianna, K.V., Vispo, E., Thompson, A., McHutchison, J., *et al.* (2010). Association of a single nucleotide polymorphism near the interleukin-28B gene with response to hepatitis C therapy in HIV/hepatitis C virus-coinfected patients. AIDS *24*, F23-F29.

Rauch, A., Kutalik, Z., Descombes, P., Cai, T., Di Iulio, J., Mueller, T., Bochud, M., Battegay, M., Bernasconi,

E., Borovicka, J., *et al.* (2010). Genetic variation in IL28B is associated with chronic hepatitis C and treatment failure: a genome-wide association study. Gastroenterology *138*, 1338–45, 1345.

Reed, K.E., and Rice, C.M. (1998). Molecular characterization of hepatitis C virus. In Hepatitis C Virus, Reesink, H.W., ed. (Karger, Basel), pp. 1–37.

Richmond, A. (2002). Nf-kappa B, chemokine gene transcription and tumour growth. Nat. Rev. Immunol. *2*, 664–674.

Robek, M.D., Boyd, B.S., and Chisari, F.V. (2005). Lambda interferon inhibits hepatitis B and C virus replication. J. Virol. *79*, 3851–3854.

Saito, T., Hirai, R., Loo, Y.M., Owen, D., Johnson, C.L., Sinha, S.C., Akira, S., Fujita, T., and Gale, M., Jr. (2006). Regulation of innate antiviral defenses through a shared repressor domain in RIG-I and LGP2. Proc. Natl. Acad. Sci. U.S.A. *104*, 582–587.

Saito, T., Owen, D.M., Jiang, F., Marcotrigiano, J., and Gale, M., Jr. (2008). Innate immunity induced by composition-dependent RIG-I recognition of hepatitis C virus RNA. Nature *454*, 523–527.

Sarcar, B., Ghosh, A.K., Steele, R., Ray, R., and Ray, R.B. (2004). Hepatitis C virus NS5A mediated STAT3 activation requires co-operation of Jak1 kinase. Virology *322*, 51–60.

Schinkel, J., Spoon, W.J., and Kroes, A.C. (2004). Meta-analysis of mutations in the NS5A gene and hepatitis C virus resistance to interferon therapy: uniting discordant conclusions. Antivir. Ther. *9*, 275–286.

Schulz, O., Diebold, S.S., Chen, M., Naslund, T.I., Nolte, M.A., Alexopoulou, L., Azuma, Y.T., Flavell, R.A., Liljestrom, P., and Reis e Sousa, C. (2005). Toll-like receptor 3 promotes cross-priming to virus-infected cells. Nature *433*, 887–892.

Sen, G.C. (2001). Viruses and interferons. Annu. Rev. Microbiol. *55*, 255–281.

Shepard, C.W., Finelli, L., and Alter, M.J. (2005). Global epidemiology of hepatitis C virus infection. Lancet Infect. Dis. *5*, 558–567.

Shimazaki, T., Honda, M., Kaneko, S., and Kobayashi, K. (2002). Inhibition of internal ribosomal entry site-directed translation of HCV by recombinant IFN-alpha correlates with a reduced La protein. Hepatology *35*, 199–208.

Shoukry, N.H., Cawthon, A.G., and Walker, C.M. (2004). Cell-mediated immunity and the outcome of hepatitis C virus infection. Annu. Rev. Microbiol. *58*, 391–424.

Simmonds, P., Tuplin, A., and Evans, D.J. (2004). Detection of genome-scale ordered RNA structure (GORS) in genomes of positive-stranded RNA viruses: implications for virus evolution and host persistence. RNA *10*, 1337–1351.

Smith, M.W., Yue, Z.N., Korth, M.J., Do, H.A., Boix, L., Fausto, N., Bruix, J., Carithers, R.L., Jr., and Katze, M.G. (2003). Hepatitis C virus and liver disease: global transcriptional profiling and identification of potential markers. Hepatology *38*, 1458–1467.

Stark, G.R., Kerr, I.M., Williams, B.R.G., Silverman, R.H., and Schreiber, R.D. (1998). How cells respond to interferons. Annu. Rev. Biochem. *67*, 227–264.

Su, A.I., Pezacki, J.P., Wodicka, L., Brideau, A.D., Supekova, L., Thimme, R., Wieland, S., Bukh, J., Purcell, R.H., Schultz, P.G., *et al.* (2002). Genomic analysis of the host response to hepatitis C virus infection. Proc. Natl. Acad. Sci. U.S.A. *99*, 15669–15674.

Sumpter, R., Loo, Y.-M., Foy, E., Li, K., Yoneyama, M., Fujita, T., Lemon, S.M., and Gale, M.Jr. (2005). Regulating intracellular anti-viral defense and permissiveness to hepatitis C virus RNA replication through a cellular RNA helicase, RIG-I. J. Virol. *79*, 2689–2699.

Sumpter, R., Wang, C., Foy, E., Loo, Y.-M., and Gale, M. Jr. (2004). Viral evolution and interferon resistance of hepatitis C virus RNA replication in a cell culture model. J. Virol. *78*, 11591–11604.

Suppiah, V., Moldovan, M., Ahlenstiel, G., Berg, T., Weltman, M., Abate, M.L., Bassendine, M., Spengler, U., Dore, G.J., Powell, E., *et al.* (2009). IL28B is associated with response to chronic hepatitis C interferon-alpha and ribavirin therapy. Nat. Genet. *41*, 1100–1104.

Taguchi, T., Nagano-Fujii, M., Akutsu, M., Kadoya, H., Ohgimoto, S., Ishido, S., and Hotta, H. (2004). Hepatitis C virus NS5A protein interacts with 2′,5′-oligoadenylate synthetase and inhibits antiviral activity of IFN in an IFN sensitivity-determining region-independent manner. J. Gen. Virol *85*, 959–969.

Tai, D.I., Tsai, S.L., Chen, Y.M., Chuang, Y.L., Peng, C.Y., Sheen, I.S., Yeh, C.T., Chang, K.S., Huang, S.N., Kuo, G.C., *et al.* (2000). Activation of nuclear factor kappaB in hepatitis C virus infection: implications for pathogenesis and hepatocarcinogenesis. Hepatology *31*, 656–664.

Tanaka, Y., Nishida, N., Sugiyama, M., Kurosaki, M., Matsuura, K., Sakamoto, N., Nakagawa, M., Korenaga, M., Hino, K., Hige, S., *et al.* (2009). Genome-wide association of IL28B with response to pegylated interferon-alpha and ribavirin therapy for chronic hepatitis C. Nat. Genet. *41*, 1105–1109.

Tanaka, Y., Shimoike, T., Ishii, K., Suzuki, R., Suzuki, T., Ushijima, H., Matsuura, Y., and Miyamura, T. (2000). Selective binding of hepatitis C virus core protein to synthetic oligonucleotides corresponding to the 5′ untranslated region of the viral genome. Virology *270*, 229–236.

Taylor, D.R., Shi, S.T., Romano, P.R., Barber, G.N., and Lai, M.M.C. (1999). Inhibition of the interferon-inducible protein kinase PKR by HCV E2 protein. Science *285*, 107–110.

Thimme, R., Oldach, D., Chang, K.M., Steiger, C., Ray, S.C., and Chisari, F.V. (2001). Determinants of viral clearance and persistence during acute hepatitis C virus infection. J. Exp. Med. *194*, 1395–1406.

Thomas, D.L., Thio, C.L., Martin, M.P., Qi, Y., Ge, D., O'Huigin, C., Kidd, J., Kidd, K., Khakoo, S.I., Alexander, G., *et al.* (2009). Genetic variation in IL28B and spontaneous clearance of hepatitis C virus. Nature *461*, 798–801.

Thompson, A.J., Muir, A.J., Sulkowski, M.S., Ge, D., Fellay, J., Shianna, K.V., Urban, T., Afdhal, N.H., Jacobson, I.M., Esteban, R., *et al.* (2010). IL28B polymorphism improves viral kinetics and is the strongest pre-treatment predictor of SVR in HCV-1 patients. Gastroenterology *139*, 120–129. e18.

Thompson, L.J., Kolumam, G.A., Thomas, S., and Murali-Krishna, K. (2006). Innate inflammatory signals induced by various pathogens differentially dictate the IFN-I dependence of CD8 T cells for clonal expansion and memory formation. J. Immunol. *177*, 1746–1754.

Tuplin, A., Wood, J., Evans, D.J., Patel, A.H., and Simmonds, P. (2002). Thermodynamic and phylogenetic prediction of RNA secondary structures in the coding region of hepatitis C virus. RNA *8*, 824–841.

Wang, C., Pflugheber, J., Sumpter, R., Sodora, D., Hui, D., Sen, G.C., and Gale, M., Jr. (2002). Alpha interferon induces distinct translational control programs to suppress hepatitis C virus RNA replication. J. Virol. *77*, 3898–3912.

Waris, G., Turkson, J., Hassanein, T., and Siddiqui, A. (2005). Hepatitis C virus (HCV) constitutively activates STAT-3 via oxidative stress: role of STAT-3 in HCV replication. J. Virol *79*, 1569–1580.

Williams, B.R. (1999). PKR; a sentinel kinase for cellular stress. Oncogene *18*, 6112–6120.

Yamamoto, M., Sato, S., Hemmi, H., Hoshino, K., Kaisho, T., Sanjo, H., Takeuchi, O., Sugiyama, M., Okabe, M., Takeda, K., *et al.* (2003). Role of adaptor TRIF in the MyD88-independent toll-like receptor signaling pathway. Science *301*, 640–643.

Yoneyama, M., Kikuchi, M., Natsukawa, T., Shinobu, N., Imaizumi, T., Miyagishi, M., Taira, K., Akira, S., and Fujita, T. (2004). The RNA helicase RIG-I has an essential function in double-stranded RNA-induced innate antiviral responses. Nat. Immunol. *5*, 730–737.

Yoneyama, M., Suhara, W., Fukuhara, Y., Fukuda, M., Nishida, E., and Fujita, T. (1998). Direct triggering of the type I interferon system by virus infection: activation of a transcription factor complex containing IRF-3 and CBP/p300. EMBO J. *17*, 1087–1095.

Zhu, H., and Liu, C. (2003). Interleukin-1 inhibits hepatitis C virus subgenomic RNA replication by activation of extracellular regulated kinase pathway. J. Virol. *77*, 5493–5498.

Zhu, H., Zhao, H., Collins,C.D., Eckenrode, S.E., Run, Q., McIndoe, R.A., Crawford, J.M., Nelson, D.R., She, J.X., and Liu, C. (2003). Gene expression associated with interferon alfa antiviral activity in an HCV replicon cell line. Hepatology *37*, 1180–1188.

Clinical Application of Interferons

10

Ben X. Wang, Ramtin Rahbar and Eleanor N. Fish

Abstract

In this chapter, the clinical uses of interferons (IFNs), predominantly the IFN-α(s), will be reviewed in the context of virus infections. The last 30 years have seen an accumulation of clinical studies evaluating the potential safety and efficacy of IFN treatment for acute and chronic virus infections, most notably hepatitis C virus. Moreover, given the pleiotropic effects of type I IFNs in terms of their antiproliferative and apoptotic effects and their ability to modulate an immune response specifically activating dendritic cells, cytolytic T cells and NK cells, their therapeutic potential for the treatment of a wide variety of virus infections continues to merit considerable investigation.

IFNs as antiviral agents in clinical use

Despite considerable evidence that the type 1 IFNs are the primary innate immune response to virus infections, exerting both direct antiviral effects to inhibit virus infection at multiple stages in the replicative cycles of different viruses and influencing the recruitment and activation of immune cells at the site of an infection, their clinical use as antivirals has been limited. With the frequency of global outbreaks of newly emerging and re-emerging viruses and with limited availability of targeted vaccines, the therapeutic potential of IFNs as broad-spectrum antivirals merits consideration. Their use is of interest since pathogen-specific antiviral drugs invariably lead to the emergence of drug-resistant virus strains.

IFN for the treatment of hepatitis C virus (HCV) infection

HCV infection causes a chronic liver disease that can lead to liver cirrhosis and hepatocellular carcinoma and affects over 200 million worldwide, an overall incidence of 3.3% of the world's population (Hoofnagle, 2002). HCV is a single-stranded RNA virus belonging to the family *Flaviviridae* and genus *Flavivirus*. To date there is no vaccine to prevent HCV infection and presently the most common reason for liver transplantation is prior HCV infection. There are 6 genotypes of HCV and 15 subtypes, which vary in prevalence. Each of these major genotypes differs in terms of replication and mutation rates, the type and severity of liver damage they invoke and treatment outcomes.

Therapy for HCV has improved considerably since IFNs were first introduced as approved therapy in the 1990s. 15–20% of patients treated with IFN alone were able to eradicate virus (Lin *et al.*, 1995; Poynard *et al.*, 1996; Hoofnagle and di Bisceglie, 1997; McHutchison *et al.*, 1998). The current standard of care (SOC) is pegylated IFN (Peg-IFN) in combination with weight-based ribavirin. Ribavirin interferes with viral RNA metabolism and enhances the early response to Peg-IFN while reducing the risk of viral relapse (Feld *et al.*, 2010; Morello *et al.*, 2010). Peg-IFNs are composed of IFN covalently linked to a water-soluble polymer of polyethylene glycol (Peg). Presently there are two Food and Drug Administration (FDA)-approved pegylated formulations of Peg-IFN, the linear Peg-IFN-α2a (Pegasys®, Hoffmann-La Roche Ltd.) and the branched Peg-IFN-α2b (PEG-intron®, Merck/Schering-Plough Corp.; Caliceti, 2004). The major advantage of using these modified IFNs for the treatment of HCV infections is their favourable pharmacokinetics. Peg-IFNs remain in the circulation for extended periods of time, allowing doses to be administered at longer intervals (Formann *et al.*, 2003). However, the covalently linked Peg reduces the specific activity of the IFN molecule by compromising receptor interaction sites, thereby affecting receptor affinity and subsequent biopotency. Fig. 10.1 summarizes the affects of current treatments on HCV replication.

The typical Peg-IFN and ribavirin treatment regimen for HCV patients is weekly subcutaneous dosing over the course of 24–72 weeks, depending on the HCV genotype (McHutchison *et al.*, 1998, 2009b; Manns *et al.*, 2001; Fried *et al.*, 2002; Hadziyannis *et al.*, 2004; Mangia *et al.*, 2005; Dalgard *et al.*, 2008; Shiffman *et al.*, 2007). Patients with HCV genotypes 2, 3 and 4 are treated for 24–48 weeks, whereas patients with HCV genotype 1 may benefit from extended treatment up to 72 weeks (Pearlman *et al.*, 2007; Ikeda *et al.*, 2009; Farnik *et al.*, 2010). Serum HCV RNA levels are used to evaluate the clinical efficacy of IFN therapies (Sarrazin *et al.*, 2010), with the objective of complete viral elimination leading to a sustained virological response (SVR), thereby preventing progression to end-stage liver disease, hepatocellular carcinoma and mortality (Butt *et al.*, 2009; Kurokawa *et al.*, 2009; Singal *et al.*, 2010a,b). Notably, SVR rates of 76–82% are observed in individuals infected with genotypes 2 or 3 (Mangia *et al.*, 2005; Shiffman *et al.*, 2007; Yu *et al.*, 2007; Dalgard *et al.*, 2008), whereas lower SVR rates of approximately 67% and 40–50% are obtained in individuals infected with genotype 4 (Varghese *et al.*, 2009) and 1 (Manns *et al.*, 2001; Fried *et al.*, 2002; Hadziyannis *et al.*, 2004;

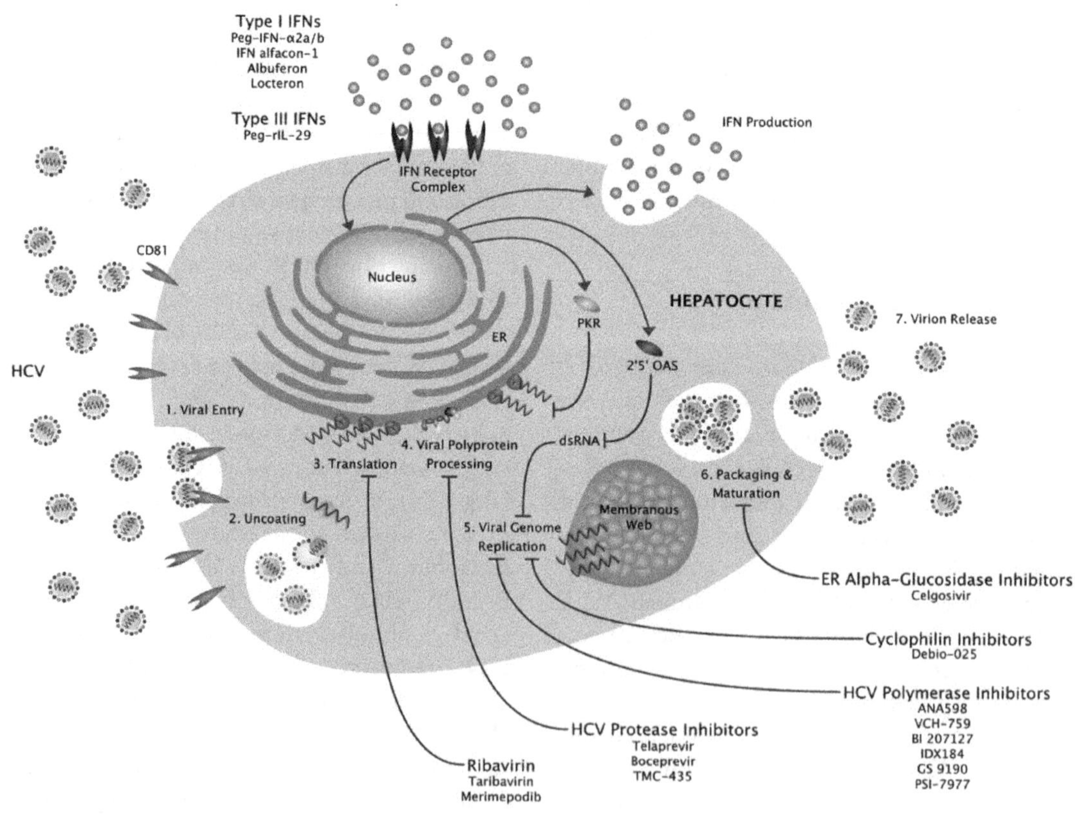

Figure 10.1 Schematic of effects of IFN, ribavirin and DAAs on HCV replication.

Legrand-Abravanel *et al.*, 2009), respectively (Table 10.1). It is also notable that IFN-α therapy is more effective if it is started early during the acute phase rather than the chronic phase of disease (Kamal *et al.*, 2006a,b; Wiegand *et al.*, 2006), suggesting that there is progressive damage to the host defence system with prolonged infection. Furthermore, HCV infection can be effectively treated with IFN-α in patients with haemophilia, an inherited bleeding disorder that can lead to cirrhosis (Takahashi *et al.*, 1993; Ko *et al.*, 2001; Posthouwer *et al.*, 2007a,b). Although Peg-IFN and ribavirin therapy greatly reduces the incidence of end-stage liver disease, the rate of SVR is significantly lower in patients who have late stage fibrosis and cirrhosis at the time of start of therapy (Bruno *et al.*, 2010). IFN therapy is also effective in children and adolescents. Independent clinical studies in children and adolescents have shown that over 93% of paediatric patients with HCV genotype 2 or 3 and 53–57% with HCV genotype 1 achieve an SVR following Peg-IFN and ribavirin therapy; a modestly enhanced response rate compared with adult patients (Sokal *et al.*, 2010; Wirth *et al.*, 2010). In elderly patients, aged 65 years and older, with HCV genotypes 2 or 3, the rate of SVR is comparable to that of patients aged 40–64 years (Hiramatsu *et al.*, 2006; Antonucci *et al.*, 2007; Huang *et al.*, 2010a). However, there is a significant decrease in the rate of SVR in elderly patients with HCV genotype 1 (Hiramatsu *et al.*, 2006; Huang *et al.*, 2010a). HCV can also be treated effectively in patients co-infected with hepatitis B virus (HBV) by Peg-IFN and ribavirin therapy (Liu *et al.*, 2003; Potthoff *et al.*, 2008, 2009, 2010). However, HBV infection may relapse following HCV SOC in these HCV/HBV co-infected patients.

Predictors of sensitivity to Peg-IFN plus ribavirin

Considerable efforts have been made to better predict the outcome of Peg-IFN/ribavirin therapy in order to optimize clinical efficacy (Gao *et al.*, 2004; Kurosaki *et al.*, 2010). In a retrospective study of several clinical trials, the relationship between treatment-related changes in serum HCV RNA and the effect of IFN therapy was analysed. The serum HCV RNA in treatment-naïve patients (Tong *et al.*, 1997), non-responders and relapsing patients (Heathcote *et al.*, 1998) revealed that 48% of patients who achieved an SVR had serum HCV RNA levels below 100 copies/ml by week 2 of therapy (Lee *et al.*, 1998). This percentage doubled to 78% by week 4, and ultimately to 96% by week 12. In an independent study, 92% of patients with HCV genotype 2 who achieved an SVR experienced a rapid virological response (RVR; Inoue *et al.*, 2010). The importance of viral kinetics is such that even Peg-IFN monotherapy is sufficient to induce an SVR in patients who have low baseline serum HCV RNA and who achieve serum negativity by week 1 (Yada *et al.*, 2010). An RVR whereby serum HCV RNA negativity is achieved by week 4 of treatment is now considered to be the best predictor of an SVR in genotype 1 patients (Pattullo *et al.*, 2010; Puoti *et al.*, 2010). Age, baseline serum alanine aminotransferase

Table 10.1 Rate of SVR associated with approved SOC

SOC	Company	Rate of SVR (%)		
		HCV 1	HCV 2 and 3	HCV 4
Peg-IFN-α2a (*Pegasys*) + Ribavirin (*Copegus*)	Hoffman-LaRoche	40–50	76–82	60–67
Peg-IFN-α2b (*PEG-Intron*) + Ribavirin (*Rebetol*)	Merck/Schering-Plough Corp.			
IFN alfacon-1* (*Infergen*) + Ribavirin	Three Rivers Pharmaceuticals	Up to 17		

*For retreatment of SOC non-responders.

(ALT), and the degree of cirrhosis are predictors of an RVR (Rodriguez-Torres *et al.*, 2010). HCV genotype 1 patients who do not experience a reduction of serum HCV RNA by weeks 2, 4, and 8 of treatment are most likely to be non-responders (Neumann *et al.*, 2009b; Lukasiewicz *et al.*, 2010). Likewise, in HCV genotype 4 patients, failure to achieve a 2 log decrease in serum HCV RNA by week 12 can be an indicator of non-response (Elefsiniotis *et al.*, 2009). Based on these studies of HCV kinetics during IFN therapy, the duration of combined therapy can be adjusted (Shiffman, 2010). For example, in HCV genotype 2 and 3 patients with an RVR, 14 weeks of combined therapy is as effective as 24 weeks (Dalgard *et al.*, 2010) and 48 weeks is optimal for HCV genotype 1 patients with an RVR, in comparison to 72 weeks for intermediate or slow responders (Moreno *et al.*, 2010).

Inherited host factors also influence the outcome of IFN therapy. Host hepatic expression of the suppressor of cytokine signalling 3 (SOCS3) protein, an important suppressor of IFN signalling (Wormald and Hilton, 2004), has been identified as an important predictor of an SVR (Kim *et al.*, 2009a). Analysis of the baseline levels of SOCS3 expression in two independent cohorts of non-responders showed that SOCS3 mRNA and protein levels were significantly higher in both groups, especially in individuals with SOCS3–4878 AA genotype (Persico *et al.*, 2008). Earlier clinical studies have also shown that there is a significant correlation between SOCS3 expression and serum ALT levels in patients undergoing IFN therapy (Miyaaki *et al.*, 2009). In addition, polymorphisms in *MxA* (Hijikata *et al.*, 2000) and *IFNAR1* (Matsuyama *et al.*, 2003) gene promoter regions also correlate with patient responsiveness to IFN therapy. More recently, polymorphisms in the upstream non-coding region of the *IL-28B* gene, which encodes IFN-λ3, have been linked to an SVR in response to Peg-IFN and ribavirin combined therapy and can be as strong a predictor as RVR (Thompson *et al.*, 2010). Specifically, a C/C genotype (single nucleotide rs12979860) is associated with a greater than two-fold increase in SVR in European-Americans and Hispanics, and a three-fold increase in SVR in African-Americans, compared with a T/T genotype (Ge *et al.*, 2009). Expression of the major T/T genotype has been associated with reduced ISG expression and more severe liver fibrosis versus either T/C or C/C genotypes (Abe *et al.*, 2010; Honda *et al.*, 2010). In addition to rs12979860, several other important single nucleotide polymorphisms have been identified on chromosome 19 in independent clinical studies (Suppiah *et al.*, 2009; Tanaka *et al.*, 2009).

Of the 6 HCV genotypes, patients infected with genotypes 2 or 3 are almost two times more likely to respond to Peg-IFN and ribavirin therapy (Mangia *et al.*, 2005; Shiffman *et al.*, 2007; Yu *et al.*, 2007; Dalgard *et al.*, 2008) in comparison to patients infected with genotype 1 (Manns *et al.*, 2001; Fried *et al.*, 2002; Hadziyannis *et al.*, 2004; Legrand-Abravanel *et al.*, 2009). In order to characterize the link between HCV genotype and susceptibility to IFN therapy, an extensive phylogenetic analysis of 345 available HCV genomes was completed (Pang *et al.*, 2009). A rooted consensus cladogram mapping the relative evolutionary age of each HCV genotype revealed a remarkable correlation with the rate of SVR (Pang *et al.*, 2009). According to the analysis, HCV genotypes branched off in the order of 2, then 6, 3 and 5, and then 4 and 1. Using this new evolutionary model, one can predict that individuals with HCV genotypes 5 or 6 will have a rate of SVR similar to individuals with HCV genotype 3 following IFN therapy. In a retrospective study, 67% of individuals infected with HCV genotype 5 had an SVR (Antaki *et al.*, 2008), which is in agreement with this evolutionary model (Pang *et al.*, 2009). Moreover, the increased resistance of HCV genotypes 1 and 4 to SOC may be a result of a lack of variation in the gene sequence encoding the protein kinase R (PKR) binding domain on the HCV E2 and NS5A proteins (Pang *et al.*, 2009). In a separate clinical study, more HCV NS5A variants were isolated from patients who developed an RVR than those with a slow virological response (positive serum HCV RNA after 4 weeks of therapy) during IFN therapy (Jain *et al.*, 2009). There was an increase in the number of amino acid substitutions in the carboxy terminus of HCV NS5A proteins, which encodes the PKR-binding domain responsible for attenuating host IFN-mediated immune responses (Gale *et al.*, 1997, 1998a,b), isolated after Peg-IFN and ribavirin treatment in early responders (Jain *et al.*, 2009). HCV NS5A proteins isolated prior to IFN treatment in the same early responders were initially less diverse at their amino and carboxy termini in comparison to slow responders. More recent studies have also identified the number of mutations in the IFN sensitivity determining region (ISDR) of HCV genotype 1b as potentially contributing to patient responsiveness to SOC (Nakagawa *et al.*, 2010).

Adverse events (Table 10.2) and the rate of viral relapse present major hurdles for current and future anti-HCV therapies and encourage personalization of treatment regimens that also consider RVR and host genetics (Bailly *et al.*, 2010; Teoh *et al.*, 2010; Zeuzem and Poordad, 2010). Adverse events are a significant cause of Peg-IFN and ribavirin therapy modification. The most common side-effects associated with Peg-IFN-α2a/b are fatigue, headache, pyrexia, rigors, myalgia, nausea, abdominal pain, depression, anxiety, and insomnia (Negro, 2010). Serious adverse events including neutropenia and thrombocytopenia can inevitably force patients to discontinue treatment. IFN therapy has also been linked with cases of hypo- and hyperthyroidism, irreversible pulmonary hypertension (Dhillon *et al.*, 2010), pancreatitis (Kim *et al.*, 2009b) and type 1 diabetes mellitus (Yamazaki *et al.*, 2010). Elderly patients are more at risk for adverse events including cytopenia during Peg-IFN and ribavirin therapy and discontinue treatment more frequently as a consequence (Hiramatsu *et al.*, 2006; Nudo *et al.*, 2006).

The most severe adverse event associated with ribavirin is haemolytic anaemia. The use of growth factors, such as erythropoietin, counteracts this anaemia. Taribavirin (Valeant Pharmaceuticals International) is an oral prodrug of ribavirin that is being investigated in phase III clinical trials as a safer alternative (Kearney *et al.*, 2008). In a randomized phase II study of 180 patients, the rates of SVR were lower, but not significantly, with Peg-IFN-α2a and taribavirin in comparison to Peg-IFN-α2a and ribavirin (Gish *et al.*, 2007).

Table 10.2 Adverse events associated with IFN therapy

Peg-IFN-α2a/b	Ribavirin
Fatigue	Fatigue
Headache	Headache
Pyrexia	Nausea
Rigours	Diarrhoea
Myalgia	Cough
Weight loss	Chest pain
Nausea	Pruritus
Abdominal pain	Haemolytic anaemia
Hair loss	Teratogenicity
Dermatitis	
Pruritus	
Neutropenia	
Thrombocytopenia	
Hyper/hypothyroidism	
Pulmonary hypertension	
Pancreatitis	
Type I diabetes	
Depression	
Psychosis	
Anxiety	
Irritability	
Insomnia	

Data acquired from Negro (2010).

Nevertheless, there were far fewer incidences of severe anaemia in patients who received taribavirin (4%) in contrast to those who received ribavirin (27%). Phase III studies (*ViSER1/2*) also provide evidence of significantly reduced rates of haemoglobin events and anaemia with taribavirin (Benhamou *et al.*, 2009). Similar to previous phase II studies, however, there appears to be no improvement with respect to producing an SVR. Merimepodib (Vertex Pharmaceuticals Inc.) is another ribavirin analogue that is effective with Peg-IFN in ribavirin non-responders (Marcellin *et al.*, 2007). However phase II studies revealed that there is a higher risk of anaemia associated with merimepodib (Rustgi *et al.*, 2009). Unfortunately, reducing the duration of combined therapy or the dosage of ribavirin in patients suffering from adverse events increases the risk of viral relapse and treatment failure (Hiramatsu *et al.*, 2009; Singal and Anand, 2010). Notably, In HCV genotype 1 patients, the rate of relapse can be as high as 30% (Hadziyannis *et al.*, 2004). The RVR is an important indicator of relapse. Patients with HCV genotype 1 with high baseline serum HCV RNA and who fail to achieve a significant decrease in serum HCV RNA by week 12 are more likely to relapse after 48 weeks of combined therapy (Poordad and Flamm, 2009; Deschenes *et al.*, 2010). Moreover, transient reappearance of serum HCV RNA can be used as a predictor of relapse in patients who have achieved an SVR at week 48 (Toyoda *et al.*, 2010).

The options available for non-responders and relapse patients have been limited to retreatment using SOC and considering alternate IFNs (Pearlman and Sjogren, 2010). Retreatment using SOC is effective in approximately 17% of patients who failed non-pegylated IFN-α and ribavirin therapy, probably a consequence of non-compliance during the initial therapy (Poynard *et al.*, 2009). However, retreatment using Peg-IFN-α2a and ribavirin is only effective in approximately 7–9% of patients who failed Peg-IFN-α2b and ribavirin therapy (*REPEAT*; Bacon *et al.*, 2009). The novel synthetic consensus IFN (cIFN), IFN alfacon-1 (INFERGEN®, Three Rivers Pharmaceuticals), is an IFN-α with a superior biopotency in comparison to IFN-α2a/b, owing to a greater binding affinity to the type I IFN receptor (Blatt *et al.*, 1996; Klein *et al.*, 1996). cIFN is more effective than IFN-α2b and earlier clinical data suggested that cIFN was at least as effective as Peg-IFN for the retreatment of HCV genotype 1 patients when used in combination with ribavirin (Sjogren *et al.*, 2005; Sjogren *et al.*, 2007). As with treatment-naïve patients, the same factors also influence sensitivity to cIFN therapy, including HCV genotype and the severity of liver disease (Gonzalez and Keeffe, 2009). Based on substantial clinical evidence from phase III trials (*DIRECT*), cIFN is now approved for retreatment of patients who have previously failed Peg-IFN and ribavirin therapy (Bacon *et al.*, 2009).

Evaluation of direct antiviral agents (DAAs) in combination with IFN-α

Owing to the relatively low cure rate (SVR) in patients with HCV genotype 1, the rate of viral relapse (20–30%) and the prevalence of non-responders (20%) associated with the FDA approved SOC, novel antiviral agents in addition to Peg-IFN-α2a/b and ribavirin have moved to the forefront of anti-HCV clinical research (Kronenberger and Zeuzem, 2008; Thompson *et al.*, 2009; Lange *et al.*, 2010; Rustgi, 2010). DAAs are currently in development to block almost every stage of the HCV life cycle. Table 10.3 outlines the success of DAAs in combination with the current standard of care.

Phase II and phase III trials of HCV NS3/4A protease inhibitors in combination with SOC have been very promising (Chary and Holodniy, 2010). Telaprevir (Vertex Pharmaceuticals Inc. and Johnson and Johnson) when combined with Peg-IFN and ribavirin for the treatment of HCV genotype 1 patients increased the SVR over SOC therapy (*PROVE1/2/3, ADVANCE, ILLUMINATE, REALIZE*): all patients showed early reduction in viral load during the first two weeks of treatment (Forestier *et al.*, 2007; Kieffer *et al.*, 2007; Lawitz *et al.*, 2008; Suzuki *et al.*, 2009; Thompson *et al.*, 2009; Lange *et al.*, 2010). In another clinical trial, 41% of HCV genotype 1 patients receiving Peg-IFN and ribavirin responded to treatment, but the response was even higher when telaprevir was added to the treatment regimen and 67% of patients achieved an SVR (Hezode *et al.*, 2009). Telaprevir triple therapy is

Table 10.3 Success of DAAs in combination with SOC

DAA in combination with SOC	Target	Company	Phase	HCV genotype	Rate of SVR
Telaprevir (VX-950)	NS3/4A Protease	Vertex Pharmaceuticals Inc./Johnson and Johnson	III	1/4	67–75%[1]
				2	> 80%[2]
				3	67%[2]
Boceprevir (SCH503034)	NS3/4A Protease	Merck/Schering-Plough Corp.	III	1	75%
TMC-435	NS3/4A Protease	Tibotec Pharmaceuticals/Medavir	II	Reduces HCV RNA titre in HCV genotype 1 patients[3]	
Celgosivir	Host ER α-glucosidase	Migenix Inc.	II	Reduces HCV RNA titre in SOC non-responders	
Alisporivir (Debio-025)	Host cyclophilin	Debiopharm Group/Novartis International AG	II	Reduces HCV 1/4 RNA in treatment naïve patients	
Silibinin (Legalon SIL)	HCV replication machinery?	Madaus Inc.	II	Reduces HCV RNA titre in HCV 1/4 SOC non-responders	
Nitazoxanide (Alinia)	PKR	Romark Laboratories L.C.	II	4	Up to 79%

[1]75% obtained from most recent phase III *ADVANCE* trial data, (http://www.natap.org/2010/HCV/052510_01.htm).

[2]http://www.hivandhepatitis.com/2010_conference/easl/docs/0514_2010_b.html (Foster *et al.*, 2010).

[3]Tsantrizos (2009) and Reesink *et al.* (2010).

also a more effective method of retreatment than SOC for HCV genotype 1 relapse patients, or patients who were unresponsive to previous SOC therapy (McHutchison *et al.*, 2010). However, a number of patients discontinued triple therapy with telaprevir because of adverse events, such as an increased risk of ribavirin-related haemolytic anaemia (McHutchison *et al.*, 2009a; Gentile *et al.*, 2010). Boceprevir (Merck/Schering-Plough Corp.), another HCV NS3 protease inhibitor, in phase III trials (*SPRINT-9, RESPOND-2*) demonstrated a synergism with Peg-IFN and ribavirin (Berman and Kwo, 2009; Mederacke *et al.*, 2009; Susser *et al.*, 2009). One of the biggest challenges facing the use of DAAs such as telaprevir and boceprevir in the treatment of HCV infection is the risk of emerging DAA-resistant viruses that can compromise treatment efficacy (Susser *et al.*, 2009; Gentile *et al.*, 2010; Kieffer *et al.*, 2010).

In addition to inhibitors specific for viral proteins, inhibitors of host cell factors that are essential for HCV replication have also been examined in phase II clinical trials. Host endoplasmic reticulum α-glucosidase inhibitor celgosivir (Migenix Inc.) prevents virion assembly (Whitby *et al.*, 2004; Durantel *et al.*, 2007, 2009). Celgosivir monotherapy was not effective, but in triple therapy with Peg-IFN and ribavirin, therapeutic benefits were shown (Durantel, 2009). As with other DAAs, celgosivir monotherapy may lead to the emergence of resistant HCV strains (Puyang *et al.*, 2010). Cyclosporin A inhibits HCV replication by blocking the interactions between host cyclophilin and HCV-NS5A protein, by directly binding with the host protein (Chatterji *et al.*, 2010a,b; Flisiak *et al.*, 2007). Cyclosporin A is an immunosuppressive and anti-inflammatory agent, and when combined with IFN-α increases the SVR (Inoue *et al.*, 2003). 55% of patients treated with IFN-α and cyclosporin A had an SVR in comparison to 32% of patients treated with IFN-α alone (Inoue *et al.*, 2003). Notably, 42% of patients with HCV genotype 1 had an SVR with combined therapy, compared with only 5% on IFN-α monotherapy. Alisporivir (Debio-025, Debiopharm Group and Novartis International AG) is a non-immunosuppressive cyclosporin A analogue that is more effective at inhibiting HCV replication (Kaul *et al.*, 2009; Gaither *et al.*, 2010; Landrieu *et al.*, 2010; Watashi, 2010). A combination of Debio-025 and Peg-IFN synergizes against HCV genotype 1 and 4 infections (Crabbe *et al.*, 2009; Flisiak *et al.*, 2009). SCY-635 (Scynexis Inc.) and NIM-811 (Novartis International AG) are also cyclosporin A analogues currently under study in phase I trials. *In vitro*, both SCY-635 (Hopkins *et al.*, 2010) and NIM-811 (Ma *et al.*, 2006) inhibit HCV replication.

The mechanism of action of other antiviral drugs effective against HCV infection, most notably silymarin and nitazoxanide (Alinia®, Romark Laboratories, L.C.), remain unclear. Silymarin is composed of a mixture of flavonolignans, most notably silibinin, extracted from the *Silybum marianum* milk thistle (Lee and Liu, 2003). *In vitro* studies have shown that silibinin is a direct inhibitor of HCV RNA-dependent RNA polymerase (RdRp), HCV genotype 1b replicon replication, and HCV genotype 2a JFH1 replication (Ahmed-Belkacem *et al.*, 2010). However there have been conflicting studies showing that silymarin, but not silibinin, inhibits HCV genotype 2a RdRp (Wagoner *et al.*, 2010). Moreover, silymarin inhibits virion release *in vitro*, and could therefore play a role in limiting HCV infection (Wagoner *et al.*, 2010). In earlier clinical studies, 5–20 mg/kg/day of silibinin (Legalon® SIL, Madaus Inc.) administered intravenously for 7 days followed by Peg-IFN and ribavirin therapy starting day 8, reduced serum HCV RNA levels in a dose-dependent manner in non-responders with primarily HCV genotype 1 and 4 infections (Ferenci *et al.*, 2008). At day 14, the combination of Peg-IFN-α2a, ribavirin and silibinin had further decreased HCV RNA, at which point only silibinin was discontinued in this particular study (Ferenci *et al.*, 2008). A case study has also been presented in which silibinin monotherapy successfully prevented HCV re-infection after orthotopic liver transplant (Neumann *et al.*, 2010). A phase I clinical study to test the tolerance and dosing of orally administered Silymarin in chronic HCV patients has been completed (Hawke *et al.*, 2010). Doses of up to 2.1 g per day were well tolerated with no adverse events, but also no indication of reduced serum HCV RNA or ALT levels. Nitazoxanide is a potent anti-infection agent (Keeffe and Rossignol, 2009; Mederacke and Wedemeyer, 2009) that inhibits HCV and HBV replication *in vitro*, with enhanced activity when used in combination with IFN (Korba *et al.*, 2008b). Its mechanism of action is unknown; however, recent studies have revealed that nitazoxanide may function to activate protein kinase R (PKR; Elazar *et al.*, 2009). In a clinical study, 30.4% of patients with HCV genotype 4 experienced a virological response and had undetectable levels of serum HCV RNA after 24 weeks of nitazoxanide, Peg-IFN, and ribavirin therapy (Rossignol *et al.*, 2008). In a separate open-label study, 4 weeks of nitazoxanide monotherapy was followed by nitazoxanide and Peg-IFN combined therapy in patients with primarily HCV genotype 4 (Rossignol *et al.*, 2009b). Here 79% of patients had an SVR. These results were similar to findings in another study using a 12-week lead-in with nitazoxanide followed by Peg-IFN with or without ribavirin (Rossignol *et al.*, 2009a). The ability for HCV to develop resistance against nitazoxanide is under investigation (Korba *et al.*, 2008a).

Evaluation of immunomodulating drugs in combination with IFN-α

Chronic HCV infection is characterized by a weakened and functionally compromised immune response (Rehermann and Nascimbeni, 2005). In order to control HCV infection, strong HCV-specific CD4+ and CD8+ T cell responses are necessary (Shoukry *et al.*, 2004). In an independent study, host T cell populations were monitored at weeks 2, 4, 12, 24, 36, and 48 during SOC, in an attempt to elucidate the effects of therapy on the immune response (Barnes *et al.*, 2002). Enhanced CD4+ T cell proliferation was observed during treatment, but it was not maintained after therapy terminated. In contrast, no significant changes in the number of CD8+ T cells were observed and IFN-γ production was low for the duration of combined therapy. Thymalfasin (thymosin alpha 1, TA1) is an immunomodulating agent that increases the NK cell population early on during IFN therapy (Grandini *et al.*, 2010). While TA1, Peg-IFN-α2a and ribavirin triple therapy was well tolerated (Camerini *et al.*, 2007), a phase III clinical trial showed that inclusion of TA1 did not improve the SVR (Ciancio and Rizzetto, 2010; Sherman, 2010). However, HCV patients receiving triple therapy exhibited a significantly reduced rate of viral relapse (Ciancio and Rizzetto, 2010). GI-5005 (GlobeImmune Inc.) is a novel therapeutic vaccine incorporating a recombinant yeast strain that expresses an HCV NS3–Core fusion protein. Yeast (*Saccharomyces cerevisiae*) proteins (i) promote pro-inflammatory gene expression and (ii) are presented in the context of MHC class I and II resulting in an antigen-specific T cell response (Stubbs *et al.*, 2001; Tada *et al.*, 2002; Ikeda *et al.*, 2005). *In vivo* mouse studies have shown that GI-5005 is a positive inducer of Th1 pro-inflammatory cytokine production from splenocytes. In addition, immunization with GI-5005 protected the mice from subsequent administration of HCV NS3-expressing tumour cells (Haller *et al.*, 2007). GI-5005 monotherapy is well tolerated in patients with chronic HCV and preliminary data from a completed phase II clinical trial also indicate that triple therapy with GI-5005, Peg-IFN and ribavirin improves the early virological response (EVR) in treatment naïve patients (Habersetzer *et al.*, 2009).

Beyond Peg-IFN-α2a/b

Beyond pegylation of the recombinant IFNs IFN-α2a/b, efforts have been directed towards the development of alternative long-acting IFN formulations (summarized in Table 10.4). Recently, a phase II clinical trial was conducted to examine the activity and tolerance of controlled-release IFN-α2b Locteron® (BLX-883, Biolex Therapeutics) in combination with ribavirin (Dzyublyk *et al.*, 2010). Locteron/BLX-833 is composed of unmodified recombinant human IFN-α2b in *Lemna*-derived poly(ether-ester) microspheres and allows IFN to be released over the course of 2 weeks (van Dijkhuizen-Radersma *et al.*, 2005; De Leede *et al.*, 2008; Dzyublyk *et al.*, 2010). Initial phase I clinical studies on the pharmacokinetics and tolerability of Locteron in healthy subjects revealed that a subcutaneous injection of 320 μg resulted in an elimination half-life of almost twice as long as 80 μg of Peg-IFN-α2b (De Leede *et al.*, 2008). Specifically, for both formulations IFN-α reached a peak serum concentration at 24 hours post-injection; however, a high level of serum IFN-α2b was maintained for up to 14 days in patients receiving 320 μg Locteron. In comparison, Peg-IFN-α2b concentrations decreased

Table 10.4 Beyond Peg-IFN-α2a/b: new IFNs for HCV therapy

IFN in combination with ribavirin	Company	Phase	HCV genotype	Rate of SVR	Advantage over Peg-IFN
Albuferon	Human Genome Sciences Inc./ Novartis International AG	Phase III	1-Apr	Comparable to Peg-IFN-α2a/ ribavirin	Prolonged circulating half-life
Locteron (BLX-883)	Biolex Therapeutics	Phase II	Reduces HCV RNA in HCV genotype 1 treatment naïve patients		Prolonged circulating half-life, fewer adverse events
IFN-β-1a (Rebif)	EMD Serono Inc.	Phase III[1]	1 (2)	46.3%[3] (87.5%)[4]	
Peg-rIL-29 (IFN-λ)	ZymoGenetics Inc.	Phase II[2]	Reduces HCV RNA in HCV genotype 1 treatment naïve and SOC relapse patients		Fewer adverse events, reduced risk of neutropenia

[1]http://clinicaltrialsfeeds.org/clinical-trials/show/NCT00249860

[2]http://clinicaltrials.gov/ct2/show/NCT01001754

[3]Chan *et al.* (2007).

[4]Using natural human IFN-β (Feron®, Toray Industries Inc.).

significantly by day 2 and were undetectable by day 7. Locteron was well tolerated in comparison to Peg-IFN-α2b. All of the patients who were treated with 80 μg of Peg-IFN-α2b experienced mild to severe influenza-like symptoms, that did not occur in patients treated with either 20 or 80 μg of Locteron (De Leede *et al.*, 2008). In subsequent phase II clinical studies of 32 HCV genotype 1 patients, 62% of patients who received 320, 480 or 640 μg of Locteron biweekly, and ribavirin daily, had serum HCV RNA negativity after week 12 of treatment (Dzyublyk *et al.*, 2010). At week 4, 75% of patients receiving 320, 480, or 640 μg of Locteron experienced ≥ than a 2 log drop in serum HCV RNA. Importantly, antibodies against *Lemna*-derived aquatic plant proteins were not detected.

In vitro and *in vivo* pre-clinical studies with Albuferon® (Human Genome Sciences Inc. and Novartis International AG), composed of recombinant IFN-α2b fused to human serum albumin (Subramanian *et al.*, 2007), revealed that Albuferon has a prolonged half-life compared with Peg-IFN (Osborn *et al.*, 2002; Liu *et al.*, 2007; Subramanian *et al.*, 2007). Phase I and phase II clinical studies revealed that Albuferon was well tolerated in HCV patients up to doses of 900 μg, with a mean elimination half-life of 159 hours (Balan *et al.*, 2006b), potentially allowing for biweekly or monthly dosing. IFN-induced expression of OAS1 and IRF7 was measurable for 28 days after a single 40 μg dose of Albuferon (Balan *et al.*, 2006a). 47% (37/78) of patients receiving a single subcutaneous injection (120–900 μg) had a ≥ 1 log reduction in HCV RNA, compared with 59% (16/27) of patients who received two injections (400–900 μg) in a 2 week interval (Balan *et al.*, 2006b). In a second phase II study, 56 HCV patients received two injections of Albuferon, 14 days apart, in escalating doses ranging from 200 to 1200 μg (Bain *et al.*, 2006). A biphasic response was observed, where 1.7 log and 0.75 log reductions (mean values) in HCV RNA occurred following the initial and second injections, respectively at doses of 900–1200 μg. At week 4, patients who received 900–1200 μg of Albuferon exhibited, on average, a 3.8 log reduction in HCV RNA. The first-phase reduction in serum HCV RNA and the RVR associated with Albuferon is a good predictor of SVR, as for Peg-IFN (Neumann *et al.*, 2009a,b). Some of the most common adverse events reported in both clinical trials with Albuferon were headache (56–73%), fatigue (52–61%), joint pain (32–55%), and fever (27–43%). Preliminary analysis of results from a phase III multicenter clinical trial, dosing every 14 days with 900 μg of Albuferon or once weekly with 180 μg Peg-IFN-α2a, (*ACHIEVE 1*, efficacy endpoint at 72 weeks) and HCV genotypes 2 or 3 (*ACHIEVE 2/3*, efficacy endpoint at 48 weeks; Rustgi, 2009) suggest no improved antiviral activity over SOC, with adverse pulmonary toxicity, which have affected FDA and EU approval.

Based on preliminary data, IFN-λ1 (IL-29), in the form of pegylated recombinant IL-29 (Peg-rIL-29, ZymoGenetics Inc.) may have therapeutic potential for the treatment of HCV (Ank *et al.*, 2006a,b; Kotenko *et al.*, 2003; Miller *et al.*, 2009). *In vitro*, Peg-rIL-29 does not inhibit granulocyte-macrophage (GM) colony formation, in contrast to Peg-IFN-α2a and Peg-IFN-α2b. Receptors for IL-29 exhibit restricted expression on peripheral blood leucocytes (Doyle *et al.*, 2006) suggesting that there will be fewer incidences of neutropenia. This idea was supported in phase I clinical trials, where consistent absolute neutrophil counts were maintained for 4 weeks in patients treated with PEG-rIL-29 (Muir *et al.*, 2010). PEG-rIL-29 was well tolerated in healthy volunteers, up to 5 μg per kg of body weight, without any reports of influenza-like symptoms. At 1.5 μg/kg, all HCV patients experienced a 2-log reduction in HCV RNA, and 4 of 6 patients had less than 1000 IU/ml of HCV RNA after 4 weeks of PEG-rIL-29 therapy (Miller *et al.*, 2009). Notably, the absence of IL-29 receptors on leucocytes precludes the involvement of an immune response in the clearance of HCV, in contrast to IFN-α therapy, since all leucocytes express IFN-α receptors and therefore IFN-α treatment will invoke an immune response that contributes to viral clearance. In another recent clinical study evaluating the therapeutic potential of human IFN-β and ribavirin for the treatment of HCV genotype 2 infection, 87.5% of patients had an SVR (Arase *et al.*, 2010).

IFN for the treatment of HBV infection

Approximately 350 million people worldwide are chronically infected with hepatitis B virus, resulting in 600,000 deaths each year, attributed in the main to cirrhosis and liver cancer. Unlike HCV, an HBV vaccine is available. In addition to IFN-α and Peg-IFN-α, lamivudine (Epivir®-HBV, GlaxoSmithKline Inc.), adefovir (Hepsera®, Gilead Sciences Inc.), and entecavir (Baraclude®, Bristol-Myers Squibb Company) are available for the treatment of chronic HBV (Straley and Terrault, 2004; Takkenberg *et al.*, 2010). IFN-α treatment effectively delays and prevents cirrhosis and hepatocellular carcinoma in patients chronically infected with HBV (Yang *et al.*, 2009). 39% of patients treated for 6 months with IFN-α2b achieved HBeAg seroconversion, but 84% relapsed, identified as HBeAg reversion or an increase in serum HBV RNA levels (Senturk *et al.*, 2010). Peg-IFN is a more effective treatment for chronic HBV infection (Brunetto *et al.*, 2009; Caruntu *et al.*, 2009). In a long-term study, 42% of patients treated with Peg-IFN seroconverted, 12% cleared HBV one year post-treatment, while inactive HBV was detected in another 17% of treated patients (Keating, 2009). In a separate clinical study to evaluate the durability of Peg-IFN treatment, 85 HBeAg-positive patients who were treated with Peg-IFN-α2b and lamivudine were followed for 5 years post-treatment termination (Wong *et al.*, 2010). The rate of HBeAg seroconversion rose from 37% at the end of treatment to 60% at 5 years. Notably, Peg-IFN-α is less active in chronic HBV patients who are HBeAg-positive and are lamivudine resistant (Leemans *et al.*, 2006), yet treatment with Peg-IFN-α2a until serum HCV RNA is less than 1000 copies/ml can prevent the emergence of lamivudine-resistant HBV mutants (Villa *et al.*, 2009). In clinical studies of lamivudine-resistant patients, serum HBV RNA and ALT levels initially decreased during Peg-IFN therapy, but patients relapsed on termination of therapy (Leemans *et al.*, 2006). Similar to lamivudine, a combination therapy of Peg-IFN-α2a and adefovir is safe, capable of reducing HBV-related liver disease and viral load (Jones *et al.*, 2009; Piccolo *et al.*, 2009). In 2009, a multicentre trial was conducted to compare the efficacy of treatment with 9 μg of cIFN to treatment with 50 μg of IFN-α1b, three-times weekly for 24 weeks. No significant differences between the two IFN-αs in terms of HBV RNA reduction and ALT normalization were identified, yet dosing was considerably less for cIFN (Zheng *et al.*, 2009). Pilot studies in patients with chronic HBV treated with thymalfasin and IFN resulted in a 70% SVR (Ciancio and Rizzetto, 2010). As for HCV, nitazoxanide exhibits anti-HBV activity *in vitro* (Korba *et al.*, 2008b).

IFN treatment for HCV and HBV with HIV co-infection

Over 33 million people live with HIV/AIDS worldwide and 2.5 million are newly infected each year. Approximately 25% of all HIV infected individuals in North America and Europe are co-infected with HCV. Treatment with recombinant IFN-γ was evaluated in a phase III clinical trial as a strategy to reduce opportunistic infections in HIV positive individuals and the data revealed a trend towards decreased opportunistic infections and increased survival (Riddell *et al.*, 2001). Currently, highly active anti-retroviral therapy (HAART) has had a significant impact on reducing the number of opportunistic infections associated with HIV/AIDs, yet can be hepatotoxic, accelerating liver fibrosis, making treatment of HCV infection all the more critical in HIV co-infected patients (Maier and Wu, 2002). Notably, the response to treatment with Peg-IFN and ribavirin in HIV co-infected individuals is inferior to the response in patients with HCV mono-infection. SVRs for HIV co-infected patients are 38% for HCV genotype 1 and 75% for HCV genotype 2 or 3 infections, with an almost halved 5 year survival rate in comparison to HCV mono-infected patients (Singal and Anand, 2009). Similar combined therapy for HBV in HIV co-infected patients results in a reduced response compared with HBV mono-infection (Di Martino *et al.*, 1996, 2002a,b). Although one study reported a 0.5 log increase in HIV viral load in 14% of co-infected patients treated with Peg-IFN and ribavirin (Landau *et al.*, 2000; Zylberberg *et al.*, 2000), the mechanism of interactions between the viruses is not understood. Peg-IFN plus ribavirin is effective and well tolerated in patients co-infected with HCV and HIV with severe immunosuppression or haemophilia (Sauleda *et al.*, 2000, 2001; Mira *et al.*, 2009). Moreover, there is no significant difference in the response rate whether HIV/HCV co-infected patients are treated with Peg-IFN-α2a plus ribavirin or Peg-IFN-α2b and ribavirin (Berenguer *et al.*, 2009). As with SOC for HCV mono-infection, the RVR and the EVR are predictors of outcome of Peg-IFN plus ribavirin therapy in HCV/HIV co-infected patients (Van den Eynde *et al.*, 2010). As for HCV mono-infection, HIV/HCV co-infected responders to Peg-IFN and ribavirin therapy experience more frequent IFN-related adverse events than non-responders (Osinusi *et al.*, 2010).

IFN treatment for herpes simplex virus (HSV) infection

HSV-1 and -2 establish lifelong latent infections capable of causing recurrent infections. HSV is transmitted during viral shedding in saliva, genital secretions and ulcerating lesions on the skin or mucus membranes. Approximately 37% of individuals seropositive for HSV are clinically symptomatic with lesions (Langenberg *et al.*, 1999). HSV causes genital herpes and although not a life-threatening disease, with the exception of mother to newborn transmission, it is prevalent as a sexually transmitted infection (STI). Most importantly, recent evidence suggests that HSV genital infection may be a critical factor for risk for HIV infection, the data revealing that HSV/HIV co-infection greatly increases HIV viral load (Johnson and Lewis, 2008; Ward and Ronn, 2010). Indeed, suppression of HSV in HIV co-infected individuals results in reductions in HIV viral titres, although the effect on HIV transmission remains unclear (Ward and Ronn, 2010).

The antiviral effects of IFNs have been extensively investigated in the context of both HSV-1 and -2 infection, with evidence that IFNs-α/β synergize with IFN-γ to inhibit virus replication (Domke *et al.*, 1985; Langenberg *et al.*, 1999; Sainz and Halford, 2002; Malmgaard and Paludan, 2003; Vollstedt *et al.*, 2004; Pierce *et al.*, 2005; Petrera and Coto, 2006; Svensson *et al.*, 2007; Carr *et al.*, 2009; Huang *et al.*, 2010b). Since HSV establishes a persistent latent infection, current therapies are aimed towards prevention of infection and control of recurrences. IFN-β treatment for viral ocular diseases – HSV-1 periocular skin lesions and Herpes zoster ophthalmicus – effectively shortens the course of these diseases by invoking viral clearance (Romano and Sadan, 1988). Recently, the use of a retroviral vector carrying IFN-γ has revealed that transgene expression significantly reduces the recurrence of HSV-1 reactivation in latently infected mice (Carr *et al.*, 2009). HSV-1 infection is lethal in IFN-α/β (IFNAR1) and IFN-γ receptor (IFNGR) double knockout mice (Luker *et al.*, 2003; Vollstedt *et al.*, 2004). By contrast, HSV-1 infection is not lethal in IFNAR1$^{-/-}$ mice, despite systemic spread to major organs including the lung, liver, and spleen (Luker *et al.*, 2003). In IFNGR$^{-/-}$ mice, HSV-1 pathogenesis is similar to that observed in wild-type mice (Luker *et al.*, 2003). The anti-HSV activities of IFN-β and IFN-γ inhibit latent infection of the trigeminal ganglia of mice (Sainz and Halford, 2002). Accordingly, with the emergence of acyclovir resistant strains of HSV, IFN treatment provides a potential alternative therapy, especially for immunocompromised individuals (Coen *et al.*, 1989; Huang *et al.*, 2010b).

IFN treatment for respiratory infections: rhinovirus, influenza virus and SARS-coronavirus (SARS-CoV)

Respiratory tract infections range from mild cases of the common cold – rhinovirus infection – to fatal cases of viral pneumonia. Respiratory viral infections are the most common cause of exacerbated asthma (Jackson, 2010). Recent global attention has been focused on the outbreak of SARS in 2003 and both avian (H5N1) and pandemic (H1N1) influenza viruses.

IFN treatment for rhinovirus infection

Studies have shown that when administered intranasally, IFN-β can reduce the symptoms of rhinovirus infection and reduce viral shedding, without incurring severe adverse events (Higgins *et al.*, 1986). In a double-blind, placebo-controlled study, IFN-β was administered 36 hours prior to and daily for 3 days post challenge with rhinovirus (Sperber *et al.*, 1988). Results indicated that, while neither high (12×10^6 IU/day) nor low (3×10^6 IU/day) dosing regimens of IFN-β were sufficient to prevent infection, only 25% of volunteers receiving high dose experienced colds, in comparison to 55% and 64% of volunteers receiving low dose or placebo, respectively (Sperber *et al.*, 1988). Comparing the efficacy of a four times daily low dose regimen of IFN-α for 4 days prior to challenge with rhinovirus, to a single high dose daily regimen for 5 days prior to challenge, both regimens by intranasal instillation, revealed that 78% of the volunteers receiving the multiple daily dosing were protected from virus infection and shedding and 100% of this group of volunteers experienced no cold symptoms. This multiple dosing regimen was superior to the single daily dose regimen (Hayden and Gwaltney, 1983). Similarly, a subsequent study evaluating efficacy of intranasal administration of IFN-a2b for seasonal prophylaxis of rhinovirus infection revealed a protective effect (Monto *et al.*, 1986). Unlike IFN-β and IFN-α, intranasal administration of IFN-γ treatment does not prevent rhinovirus infection or illness (Higgins *et al.*, 1988a). These results are in contrast to *in vitro* studies that have shown that IFN-γ treatment inhibited rhinovirus infection of human bronchial epithelial cells, specifically by down-regulating ICAM-1 expression (Sethi *et al.*, 1997; Whiteman and Spiteri, 2008). The chemical moiety, enviroxime (2-amino-1-[isopropyl sulphonyl]–6-benzimidazole phenyl ketone oxime), when administered via intranasal spray one day prior to and for five days following rhinovirus infection, leads to a reduction in virus-related symptoms (Phillpotts *et al.*, 1981). *In vitro*, both IFN-α and IFN-γ synergize with enviroxime to reduce viral infection (Ahmad and Tyrrell, 1986). Notably when volunteers were given IFN-α with or without additional enviroxime, this synergy was not reproduced in the volunteers (Higgins *et al.*, 1988b).

Different strains of rhinoviruses may circulate simultaneously in families, resulting in transmission among family members. In many cases of rhinovirus infection, children are symptomatic while adults are asymptomatic (Peltola *et al.*, 2008). In a double-blind study to evaluate IFN-α prophylaxis for protection from rhinovirus infection in a family setting, 120 adults of 46 families used IFN-α intranasal spray over the course of 6 months while another 109 adults of 49 families used placebo spray. In comparison to the placebo users, users of IFN-α encountered 33% fewer days with nasal symptoms and 41% fewer cases of respiratory illness, with specific protection against rhinovirus-related illness (Douglas *et al.*, 1986).

Similarly, daily intranasal dosing with IFN-α2a, for 5 days within 2 days of identification of an index case in a family, revealed that prophylactic treatment shortened the duration of the cold and reduced the severity of illness (Herzog *et al.*, 1986).

IFN treatment for influenza virus infection

In vitro studies have shown that influenza A virus (A/WSN/33) infection of mouse embryonic fibrocytes (MEFs) progresses more rapidly in IFNAR1$^{-/-}$ versus IFNGR$^{-/-}$ MEFs (Garcia-Sastre *et al.*, 1998; Goodman *et al.*, 2010). IFNAR1$^{-/-}$/IFNGR$^{-/-}$ MEFs and IFNAR1$^{-/-}$ MEFs had higher levels of viral protein synthesis and produced 100-fold more infectious virions than the IFNGR$^{-/-}$ and wild-type MEFs when infected with influenza A virus (Goodman *et al.*, 2010).

A number of different pre-clinical animal model studies have, likewise, indicated that IFN-α/β treatment may be beneficial for the clearance of influenza A viruses. Intranasal administration of human IFN-αB/D protected guinea pigs from infection with influenza A viruses (H1N1 and H5N1; Van Hoeven *et al.*, 2009). In ferrets, intranasal administration of IFN-α is protective against subtype H1N1, but not subtype H5N1 viruses (Kugel *et al.*, 2009). Moreover, the addition of Peg-IFN-α2b to the 2008–2009 seasonal human influenza virus vaccine (Vaxigrip®, Sanofi Pasteur Ltd.) prevented viral replication more effectively than vaccine alone, in ferrets (Fang *et al.*, 2010). IFN-β treatment of mice post-challenge with A/WSN/33 (H1N1) virus reduced lung viral titres and eosinophilia and enhanced type I immune responses (Yoo *et al.*, 2010).

Administration of mouse IFN-γ post-infection with influenza PR8 virus protects mice from infection, mediated by IFN-γ inducible NK cell activity (Weiss *et al.*, 2010). Notably, the mice are able to clear the virus in the early stages of infection, but succumb to the immunopathology associated with lung inflammation. IFN therapy did not affect lung viral titres, but increased NK cell proliferation in the bone marrow, spleen and lungs, concomitantly down-regulating T and NKT cell infiltration into the lungs at the inflammatory phase following infection. Co-administration of murine IFN-γ liposomes and a commercial subunit (HA and NA) vaccine has been tested in mice, to improve the immunogenicity of the vaccine (van Slooten *et al.*, 2001). Vaccination alone did not protect against subsequent intranasal viral challenge, yet the co-administration of IFN-γ liposomes enhanced immunogenicity and protected 83% of the mice from subsequent intranasal viral challenge (van Slooten *et al.*, 2001). Similar studies have demonstrated that IFN-α also functions as an adjuvant to improve immunogenicity of influenza vaccination (Proietti *et al.*, 2002; Bracci *et al.*, 2005).

However, IFN-α appears not to be an effective mucosal adjuvant for influenza vaccine in humans (Couch *et al.*, 2009). When healthy volunteers were given either trivalent, inactivated vaccine [2006–2007 formulation: A/New Caledonia/20/99 (H1N1), A/Wisconsin/67/205 (H3N2) and B/Malaysia/2506/2004-like] alone, or with 10^6 IU or 10^7 IU of IFN-α, no enhanced immunogenicity was detected in the IFN administered individuals (Couch *et al.*, 2009). Notably, these trials were conducted using considerably higher doses of IFN than those reported by Russian and Chinese groups in the 1980s, suggesting that dose–response trials are required to evaluate lower dose IFN efficacy. An influenza vaccine that includes IFN as an adjuvant is currently in clinical trials in the USA (http://clinicaltrials.gov/ct2/show/NCT00436046).

Administered intranasally in low doses, IFN was used in Eastern Europe and Russia as a method to prevent and treat influenza starting in the late 1960s (Soloviev, 1967). During the peak of the influenza epidemic in 1973, the number of sick cases reported was approximately 90,000 per day in Moscow. At the time, Russians used two types of live vaccines and exogenous IFN to treat and prevent the disease. During a 3-year period, several Soviet medical centres reported that human leucocyte IFN was effective as a prophylactic treatment for influenza. The patients were given 500 IU of leucocyte IFN intranasally three times a day for a 3 day period, followed by once daily for 2 days, resulting in a 60% decrease in influenza symptoms in IFN-treated patients, without adverse events (Jordan *et al.*, 1974).

In 1977, 83 volunteers were recruited to study the effects of prophylactic intranasal administration of IFN-α (Imanishi *et al.*, 1980). 5×10^3 IU/day for 4 months diminished the occurrence and intensity of influenza (H1N1, H3N2) virus-related respiratory disease. Fever was reported in only 6 of the 40 volunteers receiving IFN treatment, in contrast to 15 of the 43 in the placebo group. In addition, fever of >39°C was only reported in the placebo group, wherein twice the incidence of subjective symptoms including headache, fatigue, cough and myalgia were also reported. In a later study conducted in 1987, 16 volunteers were given 5×10^6 IU of IFN intranasally by spray gun, twice a day for 7 days, beginning 2 days prior to influenza virus (A/California/78) challenge (Treanor *et al.*, 1987). 3 of the 16 IFN-treated volunteers experienced influenza-related illness, while 13 of 16 shed virus, in comparison to the placebo group, in which 5 of 9 developed illness and 8 of 9 shed virus (Treanor *et al.*, 1987).

In the USA, over 85% of deaths from respiratory illness caused by influenza viruses are among the elderly, aged 65 years or older (Thompson *et al.*, 2003, 2004). Current seasonal vaccination strategies, however cost-effective, are unable to provide full protection to elderly individuals (Govaert *et al.*, 1994; McElhaney, 2005, 2008; Nichol *et al.*, 2007; Launay *et al.*, 2008). In a randomized double-blind placebo controlled study on the effects of sublingual IFN-α2b treatment administered just prior to influenza vaccination, in elderly individuals aged 75 or older, IFN-α did not provide any beneficial effects (Launay *et al.*, 2008). The subunit vaccine used in this study was composed of New Caledonia (H1N1) and New York (H3N2) strains of influenza A virus, and a Jiangsu strain of influenza B virus. Anti-New York (H3N2) HAI antibody titres and sIgA in saliva were lower in subjects given IFN-α2b, and the immune response against the other two strains present in the vaccine was unchanged. The results suggest that, at the doses tested, sublingual administration of IFN-α reduced the immune response to vaccination in these elderly individuals. Administration of high dose IFN (8×10^5 IU) by nasal spray also had little effect on influenza infection severity, viral shedding or seroconversion (Merigan *et al.*, 1973).

Children < 2 years of age represent another high risk group for serious influenza infection. In a 1982 placebo-controlled study of the prophylactic effects of human leucocyte IFN in children living in an orphanage, 10^4 IU per day of human leucocyte IFN administered by nasal drip did not prevent influenza-related illness during an outbreak of influenza A and B viruses (Isomura *et al.*, 1982). However, children who received IFN experienced milder symptoms than the placebo group. In another clinical study, Chinese physicians treated 140 children with IFN-α by oral and nasal spray twice a day for 4 days post infection during the influenza epidemics of 1983–1985. In comparison to 53 children who only received traditional Chinese medicines, children who received between 700–1600 IU of IFN experienced a faster recovery rate from fever (Dai *et al.*, 1987).

IFN treatment for SARS-CoV

Originating in southern China, SARS-CoV infection is the cause of severe acute respiratory syndrome (SARS), a potentially fatal respiratory illness, characterized by high fever (>38°C), myalgia, fatigue, dry cough, lymphopenia and atypical pneumonia (Zhao *et al.*, 2003). Replication of the SARS-CoV occurs mainly in the lower respiratory tract and results in diffuse alveolar damage. Between November 2002 and August 2003, 8422 cases of SARS were reported in 32 countries, resulting in 916 deaths (WHO). In the absence of controlled clinical studies to evaluate the efficacy of novel treatments for SARS, supportive care, corticosteroids and ribavirin treatment were the standards of care during the outbreak (Haagmans *et al.*, 2004).

During the 2003 outbreak in Toronto, a clinical study was conducted to evaluate the safety and efficacy of IFN-α treatment for SARS. In this preliminary study, use of IFN alfacon-1 plus corticosteroids was associated with reduced disease-associated impaired oxygen saturation, more rapid resolution of radiographic lung abnormalities and lower levels of severity of disease-linked creatine kinase (Loutfy *et al.*, 2003). Patients were administered subcutaneous IFN alfacon-1 for a total of 10 days, beginning with 9 µg/day for a minimum of 2 days and increased to 15 µg/day if no clinical response was observed.

Subsequent to the outbreak, a number of *in vitro* and *in vivo* studies have provided further evidence for the antiviral effectiveness of IFN treatment against the SARS-CoV (Cinatl *et al.*, 2004; Haagmans *et al.*, 2004; Sainz *et al.*, 2004; Morgenstern *et al.*, 2005; Theron *et al.*, 2005; Ward *et al.*, 2005). IFNs-α/β, but not IFN-γ, inhibit SARS-CoV replication *in vitro* when administered post-challenge (Cinatl *et al.*, 2004; Sainz *et al.*, 2004; Scagnolari *et al.*, 2004; Spiegel *et al.*, 2004; Stroher *et al.*, 2004; Zheng *et al.*, 2004). Prophylactic administration of Peg-IFN in cynomolgus macaques reduced viral replication and excretion, viral antigen presentation by type 1 pneumocytes, and pulmonary damage caused by subsequent challenge with the SARS-CoV (Haagmans *et al.*, 2004). When IFN alfacon-1 was administered post-infection, intermediate results were observed (Haagmans *et al.*, 2004). More recently, *in vitro* studies using human bronchial epithelial cells have confirmed that IFN alfacon-1 effectively inhibits SARS-CoV infection of human bronchial epithelial cells (Kumaki *et al.*, 2008, 2010).

Although a poor inhibitor of SARS-CoV replication *in vitro*, IFN-γ synergizes with either IFN-α or IFN-β to inhibit viral replication (Sainz *et al.*, 2004; Mossel *et al.*, 2006; Scagnolari *et al.*, 2007). In the earlier studies comparing different type I IFNs as inhibitors of SARS-CoV replication *in vitro*, IFN-β exhibited superior activity (Cinatl *et al.*, 2004; Scagnolari *et al.*, 2004). Moreover, IFN priming prior to SARS-CoV infection mitigated virus-mediated downregulation of IFN induction (Spiegel *et al.*, 2005; Kuri *et al.*, 2009).

Summary

Viewed altogether, the preceding review of the clinical uses of IFN-α(s) as a first-line treatment for acute and chronic virus infections confirm that these host-derived biological response modifiers are pleiotropic in their effects and will continue to have widespread application against existing and newly emerging virus infections.

References

Abe, H., Ochi, H., Maekawa, T., Hayes, C.N., Tsuge, M., Miki, D., Mitsui, F., Hiraga, N., Imamura, M., Takahashi, S., *et al.* (2010). Common variation of IL28 affects gamma-GTP levels and inflammation of the liver in chronically infected hepatitis C virus patients. J. Hepatol. *53*, 439–443.

Ahmad, A.L., and Tyrrell, D.A. (1986). Synergism between anti-rhinovirus antivirals: various human interferons and a number of synthetic compounds. Antiviral Res. *6*, 241–252.

Ahmed-Belkacem, A., Ahnou, N., Barbotte, L., Wychowski, C., Pallier, C., Brillet, R., Pohl, R.T., and Pawlotsky, J.M. (2010). Silibinin and related compounds are direct inhibitors of hepatitis C virus RNA-dependent RNA polymerase. Gastroenterol. *138*, 1112–1122.

Ank, N., West, H., Bartholdy, C., Eriksson, K., Thomsen, A.R., and Paludan, S.R. (2006a). Lambda interferon (IFN-lambda), a type III IFN, is induced by viruses and IFNs and displays potent antiviral activity against select virus infections *in vivo*. J. Virol. *80*, 4501–4509.

Ank, N., West, H., and Paludan, S.R. (2006b). IFN-lambda: novel antiviral cytokines. J. Interferon Cytokine Res. *26*, 373–379.

Antaki, N., Hermes, A., Hadad, M., Ftayeh, M., Antaki, F., Abdo, N., and Kebbewar, K. (2008). Efficacy of interferon plus ribavirin in the treatment of hepatitis C virus genotype 5. J. Viral Hepat. *15*, 383–386.

Antonucci, G., Longo, M.A., Angeletti, C., Vairo, F., Oliva, A., Comandini, U.V., Tocci, G., Boumis, E., Noto, P., Solmone, M.C., *et al.* (2007). The effect of age on response to therapy with peginterferon alpha plus ribavirin in a cohort of patients with chronic HCV hepatitis including subjects older than 65 yr. Am. J. Gastroenterol. *102*, 1383–1391.

Arase, Y., Suzuki, F., Akuta, N., Sezaki, H., Suzuki, Y., Kawamura, Y., Kobayashi, M., Hosaka, T., Yatsuji, H., Hirakawa, M., *et al.* (2010). Efficacy and safety of combination therapy of natural human interferon beta and ribavirin in chronic hepatitis C patients with genotype 2 and high virus load. Intern. Med. *49*, 965–970.

Bacon, B.R., Shiffman, M.L., Mendes, F., Ghalib, R., Hassanein, T., Morelli, G., Joshi, S., Rothstein, K., Kwo, P., and Gitlin, N. (2009). Retreating chronic hepatitis C with daily interferon alfacon-1/ribavirin after nonresponse to pegylated interferon/ribavirin: DIRECT results. Hepatology *49*, 1838–1846.

Bailly, F., Ahmed, S.N., Pradat, P., and Trepo, C. (2010). Management of nonresponsive hepatitis C. Expert Rev. Anti Infect. Ther. *8*, 379–395.

Bain, V.G., Kaita, K.D., Yoshida, E.M., Swain, M.G., Heathcote, E.J., Neumann, A.U., Fiscella, M., Yu, R., Osborn, B.L., Cronin, P.W., *et al.* (2006). A phase 2 study to evaluate the antiviral activity, safety, and pharmacokinetics of recombinant human albumin-interferon alfa fusion protein in genotype 1 chronic hepatitis C patients. J. Hepatol. *44*, 671–678.

Balan, V., Nelson, D.R., Sulkowski, M.S., Everson, G.T., Lambiase, L.R., Wiesner, R.H., Dickson, R.C., Garcia, A., Moore, P.A., Yu, R., *et al.* (2006a). Modulation of interferon-specific gene expression by albumin-interferon-alpha in interferon-alpha-experienced patients with chronic hepatitis C. Antivir. Ther. *11*, 901–908.

Balan, V., Nelson, D.R., Sulkowski, M.S., Everson, G.T., Lambiase, L.R., Wiesner, R.H., Dickson, R.C., Post, A.B., Redfield, R.R., Davis, G.L., *et al.* (2006b). A Phase I/II study evaluating escalating doses of recombinant human albumin–interferon-alpha fusion protein in chronic hepatitis C patients who have failed previous interferon-alpha-based therapy. Antivir. Ther. *11*, 35–45.

Barnes, E., Harcourt, G., Brown, D., Lucas, M., Phillips, R., Dusheiko, G., and Klenerman, P. (2002). The dynamics of T-lymphocyte responses during combination therapy for chronic hepatitis C virus infection. Hepatology *36*, 743–754.

Benhamou, Y., Afdhal, N.H., Nelson, D.R., Shiffman, M.L., Halliman, D.G., Heise, J., Chun, E., and Pockros, P.J. (2009). A phase III study of the safety and efficacy of viramidine versus ribavirin in treatment-naive patients with chronic hepatitis C: ViSER1 results. Hepatology *50*, 717–726.

Berenguer, J., Gonzalez-Garcia, J., Lopez-Aldeguer, J., Von-Wichmann, M.A., Quereda, C., Hernando, A., Sanz, J., Tural, C., Ortega, E., Mallolas, J., *et al.* (2009). Pegylated interferon-alpha-2a plus ribavirin versus pegylated interferon-alpha-2b plus ribavirin for the treatment of chronic hepatitis C in HIV-infected patients. J. Antimicrobial Chemother. *63*, 1256–1263.

Berman, K., and Kwo, P.Y. (2009). Boceprevir, an NS3 protease inhibitor of HCV. Clin. Liver Dis. *13*, 429–439.

Blatt, L.M., Davis, J.M., Klein, S.B., and Taylor, M.W. (1996). The biologic activity and molecular characterization of a novel synthetic interferon-alpha species, consensus interferon. J. Interferon Cytokine Res. *16*, 489–499.

Bracci, L., Canini, I., Puzelli, S., Sestili, P., Venditti, M., Spada, M., Donatelli, I., Belardelli, F., and Proietti, E. (2005). Type I IFN is a powerful mucosal adjuvant for a selective intranasal vaccination against influenza virus in mice and affects antigen capture at mucosal level. Vaccine 23, 2994–3004.

Brunetto, M.R., Moriconi, F., Bonino, F., Lau, G.K., Farci, P., Yurdaydin, C., Piratvisuth, T., Luo, K., Wang, Y., Hadziyannis, S., *et al.* (2009). Hepatitis B virus surface antigen levels: a guide to sustained response to peginterferon alfa-2a in HBeAg-negative chronic hepatitis B. Hepatology *49*, 1141–1150.

Bruno, S., Shiffman, M.L., Roberts, S.K., Gane, E.J., Messinger, D., Hadziyannis, S.J., and Marcellin, P. (2010). Efficacy and safety of peginterferon alfa-2a (40KD) plus ribavirin in hepatitis C patients with advanced fibrosis and cirrhosis. Hepatology *51*, 388–397.

Butt, A.S., Mumtaz, K., Aqeel, I., Shah, H.A., Hamid, S., and Jafri, W. (2009). Sustained virological response to pegylated interferon and ribavirin in patients with genotype 3 HCV cirrhosis. Trop. Gastroenterol. *30*, 207–212.

Caliceti, P. (2004). Pharmacokinetics of pegylated interferons: what is misleading? Dig. Liver Dis. *36* Suppl. *3*, S334–339.

Camerini, R., Ciancio, A., A., D.E.R., and Rizzetto, M. (2007). Studies of therapy with thymosin alpha1 in combination with pegylated interferon alpha2a and ribavirin in nonresponder patients with chronic hepatitis C. Ann. N. Y. Acad. Sci. *1112*, 368–374.

Carr, D.J., Austin, B.A., Halford, W.P., and Stuart, P.M. (2009). Delivery of Interferon-gamma by an adenovirus vector blocks herpes simplex virus Type 1 reactivation *in vitro* and *in vivo* independent of RNase L and double-stranded RNA-dependent protein kinase pathways. J. Neuroimmunol. *206*, 39–43.

Caruntu, F.A., Streinu-Cercel, A., Gheorghe, L.S., Grigorescu, M., Sporea, I., Stanciu, C., Andronescu, D., Voinea, F., Diculescu, M., Oproiu, A., *et al.* (2009). Efficacy and safety of peginterferon alpha-2a (40KD) in HBeAg-positive chronic hepatitis B patients. J. Gastrointestin. Liver Dis. *18*, 425–431.

Chan, H.L. Ren, H., Chow, W.C., Wee, T. (2007). Randomized trial of interferon beta-1a with or wihout ribavirin in Asian patients with chronic hepatitis C. Hepatology *46*, 315–323.

Chary, A., and Holodniy, M. (2010). Recent advances in hepatitis C virus treatment: review of HCV protease inhibitor clinical trials. Rev. Recent Clin. Trials *5*, 158–173.

Chatterji, U., Bobardt, M.D., Lim, P., and Gallay, P.A. (2010a). Cyclophilin A-independent recruitment of NS5A and NS5B into hepatitis C virus replication complexes. J. Gen. Virol. *91*, 1189–1193.

Chatterji, U., Lim, P., Bobardt, M.D., Wieland, S., Cordek, D.G., Vuagniaux, G., Chisari, F., Cameron, C.E., Targett-Adams, P., Parkinson, T., *et al.* (2010b). HCV resistance to cyclosporin A does not correlate with a resistance of the NS5A-cyclophilin A interaction to cyclophilin inhibitors. J. Hepatol. *53*, 50–56.

Ciancio, A., and Rizzetto, M. (2010). Thymalfasin in the treatment of hepatitis B and C. Ann. N. Y. Acad. Sci. *1194*, 141–146.

Cinatl, J., Jr., Michaelis, M., Scholz, M., and Doerr, H.W. (2004). Role of interferons in the treatment of severe acute respiratory syndrome. Expert Opin. Biol. Ther. *4*, 827–836.

Coen, D.M., Kosz-Vnenchak, M., Jacobson, J.G., Leib, D.A., Bogard, C.L., Schaffer, P.A., Tyler, K.L., and Knipe, D.M. (1989). Thymidine kinase-negative herpes simplex virus mutants establish latency in mouse trigeminal ganglia but do not reactivate. Proc. Natl. Acad Sci. U.S.A. *86*, 4736–4740.

Couch, R.B., Atmar, R.L., Cate, T.R., Quarles, J.M., Keitel, W.A., Arden, N.H., Wells, J., Nino, D., and Wyde, P.R. (2009). Contrasting effects of type I interferon as a mucosal adjuvant for influenza vaccine in mice and humans. Vaccine 27, 5344–5348.

Crabbe, R., Vuagniaux, G., Dumont, J.M., Nicolas-Metral, V., Marfurt, J., and Novaroli, L. (2009). An evaluation of the cyclophilin inhibitor Debio 025 and its potential as a treatment for chronic hepatitis C. Expert Opin. Investig. Drugs *18*, 211–220.

Dai, J.X., You, C.H., Qi, Z.T., Wang, X.M., Sun, P.Q., Bi, W.S., Qian, Y., Ding, R.L., Du, P., and He, Y. (1987). Children's respiratory viral diseases treated with interferon aerosol. Chin. Med. J. *100*, 162–166.

Dalgard, O., Bjoro, K., Ring-Larsen, H., Bjornsson, E., Holberg-Petersen, M., Skovlund, E., Reichard, O., Myrvang, B., Sundelof, B., Ritland, S., *et al.* (2008). Pegylated interferon alfa and ribavirin for 14 versus 24 weeks in patients with hepatitis C virus genotype 2 or 3 and rapid virological response. Hepatology *47*, 35–42.

Dalgard, O., Bjoro, K., Ring-Larsen, H., and Verbaan, H. (2010). In patients with HCV genotype 2 or 3 infection and RVR 14 weeks treatment is noninferior to 24 weeks. Pooled analysis of two Scandinavian trials. Eur. J. Gastroenterol. Hepatol. *22*, 552–556.

De Leede, L.G., Humphries, J.E., Bechet, A.C., Van Hoogdalem, E.J., Verrijk, R., and Spencer, D.G. (2008). Novel controlled-release Lemna-derived IFN-alpha2b (Locteron): pharmacokinetics, pharmacodynamics, and tolerability in a phase I clinical trial. J. Interferon Cytokine Res. *28*, 113–122.

Deschenes, M., Bain, V.G., Lee, S.S., Sherman, M., Cooper, C.L., Yoshida, E.M., Marotta, P.J., Krajden, M., Usaty, C., Balshaw, R., *et al.* (2010). Identifying HCV genotype 1 patients at risk of relapse. Eur. J. Gastroenterol. Hepatol. *22*, 546–551.

Dhillon, S., Kaker, A., Dosanjh, A., Japra, D., and Vanthiel, D.H. (2010). Irreversible pulmonary hypertension associated with the use of interferon alpha for chronic hepatitis C. Digestive Dis. Sciences *55*, 1785–1790.

Di Martino, V., Lunel, F., Cadranel, J.F., Hoang, C., Parlier, Y., Le Charpentier, Y., and Opolon, P. (1996). Long-term effects of interferon-alpha in five HIV-positive patients with chronic hepatitis B. J. Viral Hepat. *3*, 253–260.

Di Martino, V., Thevenot, T., Boyer, N., Cazals-Hatem, D., Degott, C., Valla, D., and Marcellin, P. (2002a). HIV coinfection does not compromise liver histological response to interferon therapy in patients with chronic hepatitis C. AIDS *16*, 441–445.

Di Martino, V., Thevenot, T., Colin, J.F., Boyer, N., Martinot, M., Degos, F., Coulaud, J.P., Vilde, J.L., Vachon, F., Degott, C., *et al.* (2002b). Influence of HIV infection on the response to interferon therapy and the long-term outcome of chronic hepatitis B. Gastroenterology *123*, 1812–1822.

van Dijkhuizen-Radersma, R., Metairie, S., Roosma, J.R., de Groot, K., and Bezemer, J.M. (2005). Controlled release of proteins from degradable poly(ether-ester) multiblock copolymers. J. Control Release *101*, 175–186.

Domke, I., Straub, P., Jacobsen, H., Kirchner, H., and Panet, A. (1985). Inhibition of replication of herpes simplex virus in mouse macrophages by interferons. J. Gen. Virol. *66 (Pt 10)*, 2231–2236.

Douglas, R.M., Moore, B.W., Miles, H.B., Davies, L.M., Graham, N.M., Ryan, P., Worswick, D.A., and Albrecht, J.K. (1986). Prophylactic efficacy of intranasal alpha 2-interferon against rhinovirus infections in the family setting. N. Engl. J. Med. *314*, 65–70.

Doyle, S.E., Schreckhise, H., Khuu-Duong, K., Henderson, K., Rosler, R., Storey, H., Yao, L., Liu, H., Barahmand-pour, F., Sivakumar, P., *et al.* (2006). Interleukin-29 uses a type 1 interferon-like program to promote antiviral responses in human hepatocytes. Hepatology *44*, 896–906.

Durantel, D. (2009). Celgosivir, an alpha-glucosidase I inhibitor for the potential treatment of HCV infection. Curr. Opin. Investig. Drugs *10*, 860–870.

Durantel, D., Alotte, C., and Zoulim, F. (2007). Glucosidase inhibitors as antiviral agents for hepatitis B and C. Curr. Opin. Investig. Drugs *8*, 125–129.

Dzyublyk, I., Yegorova, T., Moroz, L., Popovych, O., Zaytsev, I., Miroshnichenko, V., Kromminga, A., Wilkes, M.M., van Hoogdalem, E.J., and Humphries, J.E. (2010). Controlled release recombinant human interferon-alpha(2b) for treating patients with chronic hepatitis C genotype 1: a phase 2a clinical trial. J. Viral Hepat. [Epub ahead of print].

Elazar, M., Liu, M., McKenna, S.A., Liu, P., Gehrig, E.A., Puglisi, J.D., Rossignol, J.F., and Glenn, J.S. (2009). The anti-hepatitis C agent nitazoxanide induces phosphorylation of eukaryotic initiation factor 2alpha via protein kinase activated by double-stranded RNA activation. Gastroenterology *137*, 1827–1835.

Elefsiniotis, I.S., Vezali, E., Mihas, C., and Saroglou, G. (2009). Predictive value of complete and partial early virological response on sustained virological response rates of genotype-4 chronic hepatitis C patients treated with PEG-interferon plus ribavirin. Intervirology *52*, 247–251.

Fang, Y., Rowe, T., Leon, A.J., Banner, D., Danesh, A., Xu, L., Ran, L., Bosinger, S.E., Guan, Y., Chen, H., *et al.* (2010). Molecular characterization of *in vivo* adjuvant activity in influenza-vaccinated ferrets. J. Virol. *84*, 8369–8388.

Farnik, H., Lange, C.M., Sarrazin, C., Kronenberger, B., Zeuzem, S., and Herrmann, E. (2010). Meta-analysis shows extended therapy improves response of patients with chronic hepatitis C virus genotype 1 infection. Clin. Gastroenterol. Hepatol. *8*, 884–890.

Feld, J.J., Lutchman, G.A., Heller, T., Hara, K., Pfeiffer, J.K., Leff, R.D., Meek, C., Rivera, M., Ko, M., Koh, C., *et al.* (2010). Ribavirin improves early responses to peginterferon through improved interferon signaling. Gastroenterology *139*, 154–162 e154.

Ferenci, P., Scherzer, T.M., Kerschner, H., Rutter, K., Beinhardt, S., Hofer, H., Schoniger-Hekele, M., Holzmann, H., and Steindl-Munda, P. (2008). Silibinin is a potent antiviral agent in patients with chronic hepatitis C not responding to pegylated interferon/ribavirin therapy. Gastroenterology *135*, 1561–1567.

Flisiak, R., Dumont, J.M., and Crabbe, R. (2007). Cyclophilin inhibitors in hepatitis C viral infection. Expert Opin. Investig. Drugs *16*, 1345–1354.

Flisiak, R., Feinman, S.V., Jablkowski, M., Horban, A., Kryczka, W., Pawlowska, M., Heathcote, J.E., Mazzella, G., Vandelli, C., Nicolas-Metral, V., *et al.* (2009). The cyclophilin inhibitor Debio 025 combined with PEG IFNalpha2a significantly reduces viral load in treatment-naive hepatitis C patients. Hepatology *49*, 1460–1468.

Forestier, N., Reesink, H.W., Weegink, C.J., McNair, L., Kieffer, T.L., Chu, H.M., Purdy, S., Jansen, P.L., and Zeuzem, S. (2007). Antiviral activity of telaprevir (VX-950) and peginterferon alfa-2a in patients with hepatitis C. Hepatology *46*, 640–648.

Formann, E., Jessner, W., Bennett, L., and Ferenci, P. (2003). Twice-weekly administration of peginterferon-alpha-2b improves viral kinetics in patients with chronic hepatitis C genotype 1. J. Viral Hepat. *10*, 271–276.

Foster, G.R., Hezode, C., Bronowicki, J.-P., Carosi, G., Weiland, O., Verlinden, L., van Heeswijk, R., Van Baelen, B., Picchio, G., Beumont-Mauviel, M. (2010). Activity of telaprevir alone or in combination with peginterferon alfa-2a and ribavirin in treatment-naive genotype 2 and 3 hepatitis-C patients: final results of Study C209. 45th Annual Meeting of the European Association for the Study of the Liver (EASL 2010). Vienna, Austria, 14–18 April 2010.

Fried, M.W., Shiffman, M.L., Reddy, K.R., Smith, C., Marinos, G., Goncales, F.L., Jr., Haussinger, D., Diago, M., Carosi, G., Dhumeaux, D., *et al.* (2002). Peginterferon alfa-2a plus ribavirin for chronic hepatitis C virus infection. N. Engl. J. Med. *347*, 975–982.

Gaither, L.A., Borawski, J., Anderson, L.J., Balabanis, K.A., Devay, P., Joberty, G., Rau, C., Schirle, M., Bouwmeester, T., Mickanin, C., *et al.* (2010). Multiple cyclophilins involved in different cellular pathways mediate HCV replication. Virology *397*, 43–55.

Gale, M., Jr., Blakely, C.M., Kwieciszewski, B., Tan, S.L., Dossett, M., Tang, N.M., Korth, M.J., Polyak, S.J., Gretch, D.R., and Katze, M.G. (1998a). Control of PKR protein kinase by hepatitis C virus nonstructural 5A protein: molecular mechanisms of kinase regulation. Mol. Cell. Biol. *18*, 5208–5218.

Gale, M.J., Jr., Korth, M.J., and Katze, M.G. (1998b). Repression of the PKR protein kinase by the hepatitis C virus NS5A protein: a potential mechanism of interferon resistance. Clin. Diagn. Virol. *10*, 157–162.

Gale, M.J., Jr., Korth, M.J., Tang, N.M., Tan, S.L., Hopkins, D.A., Dever, T.E., Polyak, S.J., Gretch, D.R., and Katze, M.G. (1997). Evidence that hepatitis C virus resistance to interferon is mediated through repression of the PKR protein kinase by the nonstructural 5A protein. Virology *230*, 217–227.

Gao, B., Hong, F., and Radaeva, S. (2004). Host factors and failure of interferon-alpha treatment in hepatitis C virus. Hepatology *39*, 880–890.

Garcia-Sastre, A., Durbin, R.K., Zheng, H., Palese, P., Gertner, R., Levy, D.E., and Durbin, J.E. (1998). The role of interferon in influenza virus tissue tropism. J. Virol. *72*, 8550–8558.

Ge, D., Fellay, J., Thompson, A.J., Simon, J.S., Shianna, K.V., Urban, T.J., Heinzen, E.L., Qiu, P., Bertelsen, A.H., Muir, A.J., *et al.* (2009). Genetic variation in IL28B predicts hepatitis C treatment-induced viral clearance. Nature *461*, 399–401.

Gentile, I., Carleo, M.A., Borgia, F., Castaldo, G., and Borgia, G. (2010). The efficacy and safety of telaprevir – a new protease inhibitor against hepatitis C virus. Expert Opin. Investig. Drugs *19*, 151–159.

Gish, R.G., Arora, S., Rajender Reddy, K., Nelson, D.R., O'Brien, C., Xu, Y., and Murphy, B. (2007). Virological response and safety outcomes in therapy-nai ve patients treated for chronic hepatitis C with taribavirin or ribavirin in combination with pegylated interferon alfa-2a: a randomized, phase 2 study. J. Hepatol. *47*, 51–59.

Gonzalez, S.A., and Keeffe, E.B. (2009). Management of chronic hepatitis C treatment failures: role of consensus interferon. Biologics 3, 141–150.

Goodman, A.G., Zeng, H., Proll, S.C., Peng, X., Cilloniz, C., Carter, V.S., Korth, M.J., Tumpey, T.M., and Katze, M.G. (2010). The alpha/beta interferon receptor provides protection against influenza virus replication but is dispensable for inflammatory response signaling. J. Virol. *84*, 2027–2037.

Govaert, T.M., Thijs, C.T., Masurel, N., Sprenger, M.J., Dinant, G.J., and Knottnerus, J.A. (1994). The efficacy of influenza vaccination in elderly individuals. A randomized double-blind placebo-controlled trial. JAMA *272*, 1661–1665.

Grandini, E., Cannoletta, F., Scuteri, A., Fortini, C., Loggi, E., Cursaro, C., Riili, A., Di Donato, R., Gramenzi, A., Bernardi, M., *et al.* (2010). Immunological modifications during treatment with thymosin alpha1 plus antiviral therapy in chronic hepatitis C. Ann. N. Y. Acad. Sci. *1194*, 147–152.

Haagmans, B.L., Kuiken, T., Martina, B.E., Fouchier, R.A., Rimmelzwaan, G.F., van Amerongen, G., van Riel, D., de Jong, T., Itamura, S., Chan, K.H., *et al.* (2004). Pegylated interferon-alpha protects type 1 pneumocytes against SARS coronavirus infection in macaques. Nature Med. *10*, 290–293.

Habersetzer, F., Baumert, T.F., and Stoll-Keller, F. (2009). GI-5005, a yeast vector vaccine expressing an NS3-Core fusion protein for chronic HCV infection. Curr. Opin. Mol. Ther. *11*, 456–462.

Hadziyannis, S.J., Sette, H., Jr., Morgan, T.R., Balan, V., Diago, M., Marcellin, P., Ramadori, G., Bodenheimer, H., Jr., Bernstein, D., Rizzetto, M., *et al.* (2004). Peginterferon-alpha2a and ribavirin combination therapy in chronic hepatitis C: a randomized study of treatment duration and ribavirin dose. Ann. Intern. Med. *140*, 346–355.

Haller, A.A., Lauer, G.M., King, T.H., Kemmler, C., Fiolkoski, V., Lu, Y., Bellgrau, D., Rodell, T.C., Apelian, D., Franzusoff, A., *et al.* (2007). Whole recombinant yeast-based immunotherapy induces potent T cell responses targeting HCV NS3 and core proteins. Vaccine 25, 1452–1463.

Hawke, R.L., Schrieber, S.J., Soule, T.A., Wen, Z., Smith, P.C., Reddy, K.R., Wahed, A.S., Belle, S.H., Afdhal, N.H., Navarro, V.J., *et al.* (2010). Silymarin ascending multiple oral dosing phase I study in noncirrhotic patients with chronic hepatitis C. J. Clin. Pharmacol. *50*, 434–449.

Hayden, F.G., and Gwaltney, J.M., Jr. (1983). Intranasal interferon alpha 2 for prevention of rhinovirus infection and illness. J. Infect. Dis. *148*, 543–550.

Heathcote, E.J., Keeffe, E.B., Lee, S.S., Feinman, S.V., Tong, M.J., Reddy, K.R., Albert, D.G., Jr., Witt, K., and Blatt, L.M. (1998). Re-treatment of chronic hepatitis C with consensus interferon. Hepatology 27, 1136–1143.

Herzog, C., Berger, R., Fernex, M., Friesecke, K., Havas, L., Just, M., and Dubach, U.C. (1986). Intranasal interferon (rIFN-alpha A, Ro 22-8181) for contact prophylaxis against common cold: a randomized, double-blind and placebo-controlled field study. Antiviral Res. *6*, 171–176.

Hezode, C., Forestier, N., Dusheiko, G., Ferenci, P., Pol, S., Goeser, T., Bronowicki, J.P., Bourliere, M., Gharakhanian, S., Bengtsson, L., *et al.* (2009). Telaprevir and peginterferon with or without ribavirin for chronic HCV infection. N. Engl. J. Med. *360*, 1839–1850.

Higgins, P.G., al-Nakib, W., Barrow, G.I., and Tyrrell, D.A. (1988a). Recombinant human interferon-gamma as prophylaxis against rhinovirus colds in volunteers. J. Interferon Res. *8*, 591–596.

Higgins, P.G., Al-Nakib, W., Willman, J., and Tyrrell, D.A. (1986). Interferon-beta ser as prophylaxis against experimental rhinovirus infection in volunteers. J. Interferon Res. *6*, 153–159.

Higgins, P.G., Barrow, G.I., al-Nakib, W., Tyrrell, D.A., DeLong, D.C., and Lenox-Smith, I. (1988b). Failure to demonstrate synergy between interferon-alpha and a synthetic antiviral, enviroxime, in rhinovirus infections in volunteers. Antiviral Res. *10*, 141–149.

Hijikata, M., Ohta, Y., and Mishiro, S. (2000). Identification of a single nucleotide polymorphism in the MxA gene promoter (G/T at nt –88) correlated with the response of hepatitis C patients to interferon. Intervirology 43, 124–127.

Hiramatsu, N., Oze, T., Tsuda, N., Kurashige, N., Koga, K., Toyama, T., Yasumaru, M., Kanto, T., Takehara, T., Kasahara, A., *et al.* (2006). Should aged patients with chronic hepatitis C be treated with interferon and ribavirin combination therapy? Hepatol. Res. 35, 185–189.

Hiramatsu, N., Oze, T., Yakushijin, T., Inoue, Y., Igura, T., Mochizuki, K., Imanaka, K., Kaneko, A., Oshita, M., Hagiwara, H., *et al.* (2009). Ribavirin dose reduction raises relapse rate dose-dependently in genotype 1 patients with hepatitis C responding to pegylated interferon alpha-2b plus ribavirin. J. Viral Hepat. *16*, 586–594.

Honda, M., Sakai, A., Yamashita, T., Nakamoto, Y., Mizukoshi, E., Sakai, Y., Nakamura, M., Shirasaki, T., Horimoto, K., Tanaka, Y., *et al.* (2010). Hepatic Interferon-stimulated genes expression is associated with genetic variation in interleukin 28B and the outcome of interferon therapy for chronic hepatitis C. Gastroenterology *139*, 499–509.

Hoofnagle, J.H. (2002). Course and outcome of hepatitis C. Hepatology 36, S21–29.

Hoofnagle, J.H., and di Bisceglie, A.M. (1997). The treatment of chronic viral hepatitis. N. Engl. J. Med. *336*, 347–356.

Hopkins, S., Scorneaux, B., Huang, Z., Murray, M.G., Wring, S., Smitley, C., Harris, R., Erdmann, F., Fischer, G., and Ribeill, Y. (2010). SCY-635, a novel nonimmunosuppressive analog of cyclosporine that exhibits potent inhibition of hepatitis C virus RNA replication *in vitro*. Antimicrob. Agents Chemother. *54*, 660–672.

Huang, C.F., Yang, J.F., Dai, C.Y., Huang, J.F., Hou, N.J., Hsieh, M.Y., Lin, Z.Y., Chen, S.C., Wang, L.Y., Chang, W.Y., *et al.* (2010a). Efficacy and safety of pegylated interferon combined with ribavirin for the treatment of older patients with chronic hepatitis C. J. Infect. Dis. *201*, 751–759.

Huang, W.Y., Su, Y.H., Yao, H.W., Ling, P., Tung, Y.Y., Chen, S.H., and Wang, X. (2010b). Beta interferon plus gamma interferon efficiently reduces acyclovir-resistant herpes simplex virus infection in mice in a T-cell-independent manner. J. Gen. Virol. *91*, 591–598.

Ikeda, H., Suzuki, M., Okuse, C., Yamada, N., Okamoto, M., Kobayashi, M., Nagase, Y., Takahashi, H., Matsunaga, K., Matsumoto, N., *et al.* (2009). Short-term prolongation of pegylated interferon and ribavirin therapy for genotype 1b chronic hepatitis C patients with early viral response. Hepatol. Res. *39*, 753–759.

Ikeda, Y., Adachi, Y., Ishibashi, K., Miura, N., and Ohno, N. (2005). Activation of toll-like receptor-mediated NF-kappa beta by zymosan-derived water-soluble fraction: possible contribution of endotoxin-like substances. Immunopharmacol. Immunotoxicol. *27*, 285–298.

Imanishi, J., Karaki, T., Sasaki, O., Matsuo, A., Oishi, K., Pak, C.B., Kishida, T., Toda, S., and Nagata, H. (1980). The preventive effect of human interferon-alpha preparation on upper respiratory disease. J. Interferon Res. *1*, 169–178.

Inoue, K., Sekiyama, K., Yamada, M., Watanabe, T., Yasuda, H., and Yoshiba, M. (2003). Combined interferon alpha2b and cyclosporin A in the treatment of chronic hepatitis C: controlled trial. J. Gastroenterol. *38*, 567–572.

Inoue, Y., Hiramatsu, N., Oze, T., Yakushijin, T., Mochizuki, K., Hagiwara, H., Oshita, M., Mita, E., Fukui, H., Inada, M., *et al.* (2010). Factors affecting efficacy in patients with genotype 2 chronic hepatitis C treated by pegylated interferon alpha-2b and ribavirin: reducing drug doses has no impact on rapid and sustained virological responses. J. Viral Hepat. *17*, 336–344.

Isomura, S., Ichikawa, T., Miyazu, M., Naruse, H., Shibata, M., Imanishi, J., Matsuo, A., Kishida, T., and Karaki, T. (1982). The preventive effect of human interferon-alpha on influenza infection; modification of clinical manifestations of influenza in children in a closed community. Biken J. *25*, 131–137.

Jackson, D.J. (2010). The role of rhinovirus infections in the development of early childhood asthma. Curr. Opin. Allergy Clin. Immunol. *10*, 133–138.

Jain, M.K., Yuan, H.J., Adams-Huet, B., Reeck, A., Shelton, J., Attar, N., Zhang, S., Neumann, A.U., Carney, D.S., Gale, M., Jr., *et al.* (2009). Pegylated interferon and ribavirin promote early evolution of nonstructural 5A protein in individuals with hepatitis C who demonstrate a response to treatment. J. Infect. Dis. *200*, 866–876.

Johnson, L.F., and Lewis, D.A. (2008). The effect of genital tract infections on HIV-1 shedding in the genital tract: a systematic review and meta-analysis. Sex Transm. Dis. *35*, 946–959.

Jones, J., Shepherd, J., Baxter, L., Gospodarevskaya, E., Hartwell, D., Harris, P., and Price, A. (2009). Adefovir dipivoxil and pegylated interferon alpha for the treatment of chronic hepatitis B: an updated systematic review and economic evaluation. Health Technol. Assess. *13*, 1–172, iii.

Jordan, W.S., Jr., Dowdle, W.R., Easterday, B.C., Ennis, F.A., Gregg, B.M., Kilbourne, E.D., Seal, J.A., and Sloan, F.A. (1974). Influenza research in the Soviet Union--1974. J. Infect. Dis. *130*, 686–693.

Kamal, S.M., Fouly, A.E., Kamel, R.R., Hockenjos, B., Al Tawil, A., Khalifa, K.E., He, Q., Koziel, M.J., El Naggar, K.M., Rasenack, J., *et al.* (2006a). Peginterferon alfa-2b therapy in acute hepatitis C: impact of onset of therapy on sustained virologic response. Gastroenterology *130*, 632–638.

Kamal, S.M., Moustafa, K.N., Chen, J., Fehr, J., Abdel Moneim, A., Khalifa, K.E., El Gohary, L.A., Ramy, A.H., Madwar, M.A., Rasenack, J., *et al.* (2006b). Duration of peginterferon therapy in acute hepatitis C: a randomized trial. Hepatology *43*, 923–931.

Kaul, A., Stauffer, S., Berger, C., Pertel, T., Schmitt, J., Kallis, S., Zayas, M., Lohmann, V., Luban, J., and Bartenschlager, R. (2009). Essential role of cyclophilin A for hepatitis C virus replication and virus production and possible link to polyprotein cleavage kinetics. PLoS Pathogens *5*, e1000546.

Kearney, K.R., Thornton, J.J., and Navarro, V.J. (2008). Taribavirin for the treatment of chronic hepatitis C. Expert Opin. Pharmacother. *9*, 3243–3249.

Keating, G.M. (2009). Peginterferon-alpha-2a (40 kD): a review of its use in chronic hepatitis B. Drugs *69*, 2633–2660.

Keeffe, E.B., and Rossignol, J.F. (2009). Treatment of chronic viral hepatitis with nitazoxanide and second generation thiazolides. World J. Gastroenterol. *15*, 1805–1808.

Kieffer, T.L., Kwong, A.D., and Picchio, G.R. (2010). Viral resistance to specifically targeted antiviral therapies for hepatitis C (STAT-Cs). J. Antimicrob. Chemother. *65*, 202–212.

Kieffer, T.L., Sarrazin, C., Miller, J.S., Welker, M.W., Forestier, N., Reesink, H.W., Kwong, A.D., and Zeuzem, S. (2007). Telaprevir and pegylated interferon-alpha-2a inhibit wild-type and resistant genotype 1 hepatitis C virus replication in patients. Hepatology *46*, 631–639.

Kim, K.A., Lin, W., Tai, A.W., Shao, R.X., Weinberg, E., De Sa Borges, C.B., Bhan, A.K., Zheng, H., Kamegaya, Y., and Chung, R.T. (2009a). Hepatic SOCS3 expression is strongly associated with non-response to therapy and race in HCV and HCV/HIV infection. J. Hepatol. *50*, 705–711.

Kim, S.R., Imoto, S., Mita, K., Taniguchi, M., Sasase, N., Muramatsu, A., Kudo, M., Kitai, S., El-Shamy, A., Hotta, H., *et al.* (2009b). Pegylated interferon plus ribavirin combination therapy for chronic hepatitis C with high viral load of serum hepatitis C virus RNA, genotype 1b, discontinued on attaining sustained virological response at week 16 after onset of acute pancreatitis. Digestion *79*, 36–39.

Klein, S.B., Blatt, L.M., and Taylor, M.W. (1996). Cell surface binding characteristics correlate with consensus type I interferon enhanced activity. J. Interferon Cytokine Res. *16*, 1–6.

Ko, J.S., Choe, Y.H., Kim, E.J., Lee, E.H., Jang, J.J., and Seo, J.K. (2001). Interferon-alpha treatment of chronic hepatitis C in children with hemophilia. J. Ped. Gastro. Nutr. *32*, 41–44.

Korba, B.E., Elazar, M., Lui, P., Rossignol, J.F., and Glenn, J.S. (2008a). Potential for hepatitis C virus resistance to nitazoxanide or tizoxanide. Antimicrob. Agents Chemother. *52*, 4069–4071.

Korba, B.E., Montero, A.B., Farrar, K., Gaye, K., Mukerjee, S., Ayers, M.S., and Rossignol, J.F. (2008b). Nitazoxanide, tizoxanide and other thiazolides are potent inhibitors of hepatitis B virus and hepatitis C virus replication. Antiviral Res. *77*, 56–63.

Kotenko, S.V., Gallagher, G., Baurin, V.V., Lewis-Antes, A., Shen, M., Shah, N.K., Langer, J.A., Sheikh, F., Dickensheets, H., and Donnelly, R.P. (2003). IFN-lambdas mediate antiviral protection through a distinct class II cytokine receptor complex. Nat. Immunol. *4*, 69–77.

Kronenberger, B., and Zeuzem, S. (2008). Future treatment options for HCV: double, triple, what is the optimal combination? Best Pract. Res. Clin. Gastroenterol. *22*, 1123–1136.

Kugel, D., Kochs, G., Obojes, K., Roth, J., Kobinger, G.P., Kobasa, D., Haller, O., Staeheli, P., and von Messling, V. (2009). Intranasal administration of alpha interferon reduces seasonal influenza A virus morbidity in ferrets. J. Virol. *83*, 3843–3851.

Kumaki, Y., Day, C.W., Bailey, K.W., Wandersee, M.K., Wong, M.H., Madsen, J.R., Madsen, J.S., Nelson, N.M., Hoopes, J.D., Woolcott, J.D., *et al.* (2010). Induction of interferon-gamma-inducible protein 10 by SARS-CoV infection, interferon alfacon 1 and interferon inducer in human bronchial epithelial Calu-3 cells and BALB/c mice. Antivir. Agents Chemother. *20*, 169–177.

Kumaki, Y., Day, C.W., Wandersee, M.K., Schow, B.P., Madsen, J.S., Grant, D., Roth, J.P., Smee, D.F., Blatt, L.M., and Barnard, D.L. (2008). Interferon alfacon 1 inhibits SARS-CoV infection in human bronchial epithelial Calu-3 cells. Biochem. Biophys. Res. Commun. *371*, 110–113.

Kuri, T., Zhang, X., Habjan, M., Martinez-Sobrido, L., Garcia-Sastre, A., Yuan, Z., and Weber, F. (2009). Interferon priming enables cells to partially overturn the SARS coronavirus-induced block in innate immune activation. J. Gen. Virol. *90*, 2686–2694.

Kurokawa, M., Hiramatsu, N., Oze, T., Mochizuki, K., Yakushijin, T., Kurashige, N., Inoue, Y., Igura, T., Imanaka, K., Yamada, A., *et al.* (2009). Effect of interferon alpha-2b plus ribavirin therapy on incidence of hepatocellular carcinoma in patients with chronic hepatitis. Hepatol. Res. *39*, 432–438.

Kurosaki, M., Matsunaga, K., Hirayama, I., Tanaka, T., Sato, M., Yasui, Y., Tamaki, N., Hosokawa, T., Ueda, K., Tsuchiya, K., *et al.* (2010). A predictive model of response to peginterferon ribavirin in chronic hepatitis C using classification and regression tree analysis. Hepatol. Res. *40*, 251–260.

Landau, A., Batisse, D., Van Huyen, J.P., Piketty, C., Bloch, F., Pialoux, G., Belec, L., Petite, J.P., Weiss, L., and Kazatchkine, M.D. (2000). Efficacy and safety of combination therapy with interferon-alpha2b and ribavirin for chronic hepatitis C in HIV-infected patients. AIDS *14*, 839–844.

Landrieu, I., Hanoulle, X., Bonachera, F., Hamel, A., Sibille, N., Yin, Y., Wieruszeski, J.M., Horvath, D., Wei, Q., Vuagniaux, G., *et al.* (2010). Structural basis for the non-immunosuppressive character of the cyclosporin A analogue Debio 025. Biochem. *49*, 4679–4686.

Lange, C.M., Sarrazin, C., and Zeuzem, S. (2010). Review article: specifically targeted anti-viral therapy for hepatitis C – a new era in therapy. Aliment. Pharmacol. Ther. *32*, 14–28.

Langenberg, A.G., Corey, L., Ashley, R.L., Leong, W.P., and Straus, S.E. (1999). A prospective study of new infections with herpes simplex virus type 1 and type 2. Chiron HSV Vaccine Study Group. N. Engl. J. Med. *341*, 1432–1438.

Launay, O., Grabar, S., Bloch, F., Desaint, C., Jegou, D., Lallemand, C., Erickson, R., Lebon, P., and Tovey, M.G. (2008). Effect of sublingual administration of interferon-alpha on the immune response to influenza vaccination in institutionalized elderly individuals. Vaccine *26*, 4073–4079.

Lawitz, E., Rodriguez-Torres, M., Muir, A.J., Kieffer, T.L., McNair, L., Khunvichai, A., and McHutchison, J.G. (2008). Antiviral effects and safety of telaprevir, peginterferon alfa-2a, and ribavirin for 28 days in hepatitis C patients. J. Hepatol. *49*, 163–169.

Lee, D.Y., and Liu, Y. (2003). Molecular structure and stereochemistry of silybin A, silybin B, isosilybin A, and isosilybin B, Isolated from Silybum marianum (milk thistle). J. Nat. Prod. *66*, 1171–1174.

Lee, W.M., Reddy, K.R., Tong, M.J., Black, M., van Leeuwen, D.J., Hollinger, F.B., Mullen, K.D., Pimstone, N., Albert, D., and Gardner, S. (1998). Early hepatitis C virus-RNA responses predict interferon treatment outcomes in chronic hepatitis C. The Consensus Interferon Study Group. Hepatology *28*, 1411–1415.

Leemans, W.F., Flink, H.J., Janssen, H.L., Niesters, H.G., Schalm, S.W., and de Man, R.A. (2006). The effect of pegylated interferon-alpha on the treatment of lamivudine resistant chronic HBeAg positive hepatitis B virus infection. J. Hepatol. *44*, 507–511.

Legrand-Abravanel, F., Colson, P., Leguillou-Guillemette, H., Alric, L., Ravaux, I., Lunel-Fabiani, F., Bouviers-Alias, M., Trimoulet, P., Chaix, M.L., Hezode, C., *et al.* (2009). Influence of the HCV subtype on the virological response to pegylated interferon and ribavirin therapy. J. Med. Virol. *81*, 2029–2035.

Lin, R., Roach, E., Zimmerman, M., Strasser, S., and Farrell, G.C. (1995). Interferon alfa-2b for chronic hepatitis C: effects of dose increment and duration of treatment on response rates. Results of the first multicentre Australian trial. Australia Hepatitis C Study Group. J. Hepatol. *23*, 487–496.

Liu, C., Zhu, H., Subramanian, G.M., Moore, P.A., Xu, Y., and Nelson, D.R. (2007). Anti-hepatitis C virus activity of albinterferon alfa-2b in cell culture. Hepatol. Res. *37*, 941–947.

Liu, C.J., Chen, P.J., Lai, M.Y., Kao, J.H., Jeng, Y.M., and Chen, D.S. (2003). Ribavirin and interferon is effective for hepatitis C virus clearance in hepatitis B and C dually infected patients. Hepatology *37*, 568–576.

Loutfy, M.R., Blatt, L.M., Siminovitch, K.A., Ward, S., Wolff, B., Lho, H., Pham, D.H., Deif, H., LaMere, E.A., Chang, M., *et al.* (2003). Interferon alfacon-1 plus corticosteroids in severe acute respiratory syndrome: a preliminary study. JAMA *290*, 3222–3228.

Lukasiewicz, E., Gorfine, M., Freedman, L.S., Pawlotsky, J.M., Schalm, S.W., Ferrari, C., Zeuzem, S., and Neumann, A.U. (2010). Prediction of nonSVR to therapy with pegylated interferon-alpha2a and ribavirin in chronic hepatitis C genotype 1 patients after 4, 8 and 12 weeks of treatment. J. Viral Hepat. *17*, 345–351.

Luker, G.D., Prior, J.L., Song, J., Pica, C.M., and Leib, D.A. (2003). Bioluminescence imaging reveals systemic dissemination of herpes simplex virus type 1 in the absence of interferon receptors. J. Virol. *77*, 11082–11093.

Ma, S., Boerner, J.E., TiongYip, C., Weidmann, B., Ryder, N.S., Cooreman, M.P., and Lin, K. (2006). NIM811, a cyclophilin inhibitor, exhibits potent *in vitro* activity against hepatitis C virus alone or in combination with alpha interferon. Antimicrob. Agents Chemother. *50*, 2976–2982.

Maier, I., and Wu, G.Y. (2002). Hepatitis C and HIV co-infection: a review. World J. Gastroenterol. *8*, 577–579.

Malmgaard, L., and Paludan, S.R. (2003). Interferon (IFN)-alpha/beta, interleukin (IL)-12 and IL-18 coordinately induce production of IFN-gamma during infection with herpes simplex virus type 2. J. Gen. Virol. 84, 2497–2500.

Mangia, A., Santoro, R., Minerva, N., Ricci, G.L., Carretta, V., Persico, M., Vinelli, F., Scotto, G., Bacca, D., Annese, M., *et al.* (2005). Peginterferon alfa-2b and ribavirin for 12 vs. 24 weeks in HCV genotype 2 or 3. N. Engl. J. Med. 352, 2609–2617.

Manns, M.P., McHutchison, J.G., Gordon, S.C., Rustgi, V.K., Shiffman, M., Reindollar, R., Goodman, Z.D., Koury, K., Ling, M., and Albrecht, J.K. (2001). Peginterferon alfa-2b plus ribavirin compared with interferon alfa-2b plus ribavirin for initial treatment of chronic hepatitis C: a randomised trial. Lancet *358*, 958–965.

Marcellin, P., Horsmans, Y., Nevens, F., Grange, J.D., Bronowicki, J.P., Vetter, D., Purdy, S., Garg, V., Bengtsson, L., McNair, L., *et al.* (2007). Phase 2 study of the combination of merimepodib with peginterferon-alpha2b, and ribavirin in nonresponders to previous therapy for chronic hepatitis C. J. Hepatol. *47*, 476–483.

Matsuyama, N., Mishiro, S., Sugimoto, M., Furuichi, Y., Hashimoto, M., Hijikata, M., and Ohta, Y. (2003). The dinucleotide microsatellite polymorphism of the IFNAR1 gene promoter correlates with responsiveness of hepatitis C patients to interferon. Hepatol. Res. *25*, 221–225.

McElhaney, J.E. (2008). Influenza vaccination in the elderly: seeking new correlates of protection and improved vaccines. Aging Health *4*, 603–613.

McElhaney, J.E., Hooton, J.W., Hooton, N., and Bleackley, R.C. (2005). Comparison of single versus booster dose of influenza vaccination on humoral and cellular immune responses in older adults. Vaccine *23*, 3294–3300.

McHutchison, J.G., Everson, G.T., Gordon, S.C., Jacobson, I.M., Sulkowski, M., Kauffman, R., McNair, L., Alam, J., and Muir, A.J. (2009a). Telaprevir with peginterferon and ribavirin for chronic HCV genotype 1 infection. N. Engl. J. Med. *360*, 1827–1838.

McHutchison, J.G., Gordon, S.C., Schiff, E.R., Shiffman, M.L., Lee, W.M., Rustgi, V.K., Goodman, Z.D., Ling, M.H., Cort, S., and Albrecht, J.K. (1998). Interferon alfa-2b alone or in combination with ribavirin as initial treatment for chronic hepatitis C. Hepatitis Interventional Therapy Group. N. Engl. J. Med. *339*, 1485–1492.

McHutchison, J.G., Lawitz, E.J., Shiffman, M.L., Muir, A.J., Galler, G.W., McCone, J., Nyberg, L.M., Lee, W.M., Ghalib, R.H., Schiff, E.R., *et al.* (2009b). Peginterferon alfa-2b or alfa-2a with ribavirin for treatment of hepatitis C infection. N. Engl. J Med. *361*, 580–593.

McHutchison, J.G., Manns, M.P., Muir, A.J., Terrault, N.A., Jacobson, I.M., Afdhal, N.H., Heathcote, E.J., Zeuzem, S., Reesink, H.W., Garg, J., *et al.* (2010). Telaprevir for previously treated chronic HCV infection. N. Engl. J. Med. *362*, 1292–1303.

Mederacke, I., and Wedemeyer, H. (2009). Nitazoxanide for the treatment of chronic hepatitis C new opportunities but new challenges? Ann. Hepatol. *8*, 166–168.

Mederacke, I., Wedemeyer, H., and Manns, M.P. (2009). Boceprevir, an NS3 serine protease inhibitor of hepatitis C virus, for the treatment of HCV infection. Curr. Opin. Investig. Drugs *10*, 181–189.

Merigan, T.C., Reed, S.E., Hall, T.S., and Tyrrell, D.A. (1973). Inhibition of respiratory virus infection by locally applied interferon. Lancet *1*, 563–567.

Miller, D.M., Klucher, K.M., Freeman, J.A., Hausman, D.F., Fontana, D., and Williams, D.E. (2009). Interferon lambda as a potential new therapeutic for hepatitis C. Ann. NY Acad. Sci. *1182*, 80–87.

Mira, J.A., Gutierrez-Valencia, A., Gil Ide, L., Merino, D., Rivero, A., Rios-Villegas, M.J., Delgado, M., Gonzalez-Serrano, M., Collado, A., Torres-Tortosa, M., *et al.* (2009). Efficacy and safety of pegylated interferon plus ribavirin in HIV and hepatitis C virus-coinfected patients with advanced immunosuppression. Clin. Infect. Dis. *49*, e84–91.

Miyaaki, H., Ichikawa, T., Nakao, K., Matsuzaki, T., Muraoka, T., Honda, T., Takeshita, S., Shibata, H., Ozawa, E., Akiyama, M., *et al.* (2009). Predictive value of suppressor of cytokine signal 3 (SOCS3) in the outcome of interferon therapy in chronic hepatitis C. Hepatol. Res. *39*, 850–855.

Monto, A.S., Shope, T.C., Schwartz, S.A., Albrecht, J.K. (1986). Intranasal interferon-alpha 2b for seasonal prophylaxis of respiratory infection. J. Infect. Dis. *154*, 128–133.

Morello, J., Soriano, V., Barreiro, P., Medrano, J., Madejon, A., Gonzalez-Pardo, G., Jimenez-Nacher, I., Gonzalez-Lahoz, J., and Rodriguez-Novoa, S. (2010). Plasma ribavirin trough concentrations at week 4 predict hepatitis C virus (HCV) relapse in HIV-HCV-coinfected patients treated for chronic hepatitis C. Antimicrob. Agents Chemother. *54*, 1647–1649.

Moreno, C., Deltenre, P., Pawlotsky, J.M., Henrion, J., Adler, M., and Mathurin, P. (2010). Shortened treatment duration in treatment-naive genotype 1 HCV patients with rapid virological response: a meta-analysis. J. Hepatol. *52*, 25–31.

Morgenstern, B., Michaelis, M., Baer, P.C., Doerr, H.W., and Cinatl, J., Jr. (2005). Ribavirin and interferon-beta synergistically inhibit SARS-associated coronavirus replication in animal and human cell lines. Biochem. Biophys. Res. Commun. *326*, 905–908.

Mossel, E.C., Sainz, B., Jr., Garry, R.F., and Peters, C.J. (2006). Synergistic inhibition of SARS-coronavirus replication by type I and type II IFN. Adv. Exp. Med. Biol. *581*, 503–506.

Muir, A.J., Shiffman, M.L., Zaman, A., Yoffe, B., de la Torre, A., Flamm, S., Gordon, S.C., Marotta, P., Vierling, J.M., Carlos Lopez-Talavera, J., *et al.* (2010). Phase 1b study of pegylated interferon lambda 1 with or without ribavirin in patients with chronic genotype 1 hepatitis C virus infection. Hepatology *52*, 822–832.

Nakagawa, M., Sakamoto, N., Ueyama, M., Mogushi, K., Nagaie, S., Itsui, Y., Azuma, S., Kakinuma, S., Tanaka, H., Enomoto, N., *et al.* (2010). Mutations in the interferon sensitivity determining region and virological response to combination therapy with pegylated-interferon alpha 2b plus ribavirin in patients with chronic hepatitis C-1b infection. J. Gastroenterol. *45*, 656–665.

Negro, F. (2010). Adverse effects of drugs in the treatment of viral hepatitis. Best Pract. Res Clin. Gastroenterol. *24*, 183–192.

Neumann, A.U., Bain, V.G., Yoshida, E.M., Patel, K., Pulkstenis, E., and Subramanian, G.M. (2009a). Early prediction of sustained virological response at day 3 of treatment with albinterferon-alpha-2b in patients with genotype 2/3 chronic hepatitis C. Liver Int. *29*, 1350–1355.

Neumann, A.U., Pianko, S., Zeuzem, S., Yoshida, E.M., Benhamou, Y., Mishan, M., McHutchison, J.G., Pulkstenis, E., and Mani Subramanian, G. (2009b). Positive and negative prediction of sustained virologic response at weeks 2 and 4 of treatment with albinterferon alfa-2b or peginterferon alfa-2a in treatment-naive patients with genotype 1, chronic hepatitis C. J. Hepatol. *51*, 21–28.

Neumann, U.P., Biermer, M., Eurich, D., Neuhaus, P., and Berg, T. (2010). Successful prevention of hepatitis C virus (HCV) liver graft reinfection by silibinin monotherapy. J. Hepatol. *52*, 951–952.

Nichol, K.L., Nordin, J.D., Nelson, D.B., Mullooly, J.P., and Hak, E. (2007). Effectiveness of influenza vaccine in the community-dwelling elderly. N. Engl. J. Med. *357*, 1373–1381.

Nudo, C.G., Wong, P., Hilzenrat, N., and Deschenes, M. (2006). Elderly patients are at greater risk of cytopenia during antiviral therapy for hepatitis C. Can. J. Gastroenterol. *20*, 589–592.

Osborn, B.L., Olsen, H.S., Nardelli, B., Murray, J.H., Zhou, J.X., Garcia, A., Moody, G., Zaritskaya, L.S., and Sung, C. (2002). Pharmacokinetic and pharmacodynamic studies of a human serum albumin–interferon-alpha fusion protein in cynomolgus monkeys. J. Pharmacol. Exp. Ther. 303, 540–548.

Osinusi, A., Rasimas, J.J., Bishop, R., Proschan, M., McLaughlin, M., Murphy, A., Cortez, K.J., Polis, M.A., Masur, H., Rosenstein, D., *et al.* (2010). HIV/Hepatitis C virus-coinfected virologic responders to pegylated interferon and ribavirin therapy more frequently incur interferon-related adverse events than nonresponders do. J. AIDS 53, 357–363.

Pang, P.S., Planet, P.J., and Glenn, J.S. (2009). The evolution of the major hepatitis C genotypes correlates with clinical response to interferon therapy. PloS One *4*, e6579.

Pattullo, V., Ravindran, N.C., Mazzulli, T., Wong, D.K., and Heathcote, E.J. (2010). Pegylated interferon plus optimized weight-based ribavirin dosing negate the influence of weight and body mass index on early viral kinetics and sustained virological response in chronic hepatitis C. J. Viral Hepat. *17*, 834–838.

Pearlman, B.L., Ehleben, C., and Saifee, S. (2007). Treatment extension to 72 weeks of peginterferon and ribavirin in hepatitis c genotype 1-infected slow responders. Hepatology *46*, 1688–1694.

Pearlman, B.L., and Sjogren, M.H. (2010). Treatment options for HCV nonresponders and relapse patients. Gastroenterol. Hepatol. *6*, 1–12.

Peltola, V., Waris, M., Osterback, R., Susi, P., Ruuskanen, O., and Hyypia, T. (2008). Rhinovirus transmission within families with children: incidence of symptomatic and asymptomatic infections. J. Infect. Dis. *197*, 382–389.

Persico, M., Capasso, M., Russo, R., Persico, E., Croce, L., Tiribelli, C., and Iolascon, A. (2008). Elevated expression and polymorphisms of SOCS3 influence patient response to antiviral therapy in chronic hepatitis C. Gut 57, 507–515.

Petrera, E., and Coto, C.E. (2006). The synergistic effect of IFN-alpha and IFN-gamma against HSV-2 replication in Vero cells is not interfered by the plant antiviral 1-cinnamoyl-3, 11-dihydroxymeliacarpin. Virol. J. 3, 45.

Phillpotts, R.J., Jones, R.W., Delong, D.C., Reed, S.E., Wallace, J., and Tyrrell, D.A. (1981). The activity of enviroxime against rhinovirus infection in man. Lancet *1*, 1342–1344.

Piccolo, P., Lenci, I., Demelia, L., Bandiera, F., Piras, M.R., Antonucci, G., Nosotti, L., Mari, T., De Santis, A., Ponti, M.L., *et al.* (2009). A randomized controlled trial of pegylated interferon-alpha2a plus adefovir dipivoxil for hepatitis B e antigen-negative chronic hepatitis B. Antivir. Ther. *14*, 1165–1174.

Pierce, A.T., DeSalvo, J., Foster, T.P., Kosinski, A., Weller, S.K., and Halford, W.P. (2005). Beta interferon and gamma interferon synergize to block viral DNA and virion synthesis in herpes simplex virus-infected cells. J. Gen. Virol. *86*, 2421–2432.

Poordad, F.F., and Flamm, S.L. (2009). Virological relapse in chronic hepatitis C. Antivir. Ther. *14*, 303–313.

Posthouwer, D., Makris, M., Yee, T.T., Fischer, K., van Veen, J.J., Griffioen, A., van Erpecum, K.J., and Mauser-Bunschoten, E.P. (2007a). Progression to end-stage liver disease in patients with inherited bleeding disorders and hepatitis C: an international, multicenter cohort study. Blood *109*, 3667–3671.

Posthouwer, D., Yee, T.T., Makris, M., Fischer, K., Griffioen, A., Van Veen, J.J., and Mauser-Bunschoten, E.P. (2007b). Antiviral therapy for chronic hepatitis C in patients with inherited bleeding disorders: an international, multicenter cohort study. J. Thromb Haemost 5, 1624–1629.

Potthoff, A., Berg, T., and Wedemeyer, H. (2009). Late hepatitis B virus relapse in patients co-infected with hepatitis B virus and hepatitis C virus after antiviral treatment with pegylated interferon-a2b and ribavirin. Scand. J. Gastroenterol. *44*, 1487–1490.

Potthoff, A., Manns, M.P., and Wedemeyer, H. (2010). Treatment of HBV/HCV coinfection. Expert Opin. Pharmacother. *11*, 919–928.

Potthoff, A., Wedemeyer, H., Boecher, W.O., Berg, T., Zeuzem, S., Arnold, J., Spengler, U., Gruengreiff, K., Kaeser, T., Schuchmann, M., *et al.* (2008). The HEP-NET B/C co-infection trial: a prospective multicenter study to investigate the efficacy of pegylated interferon-alpha2b and ribavirin in patients with HBV/HCV co-infection. J. Hepatol. *49*, 688–694.

Poynard, T., Colombo, M., Bruix, J., Schiff, E., Terg, R., Flamm, S., Moreno-Otero, R., Carrilho, F., Schmidt, W., Berg, T., *et al.* (2009). Peginterferon alfa-2b and ribavirin: effective in patients with hepatitis C who failed interferon alfa/ribavirin therapy. Gastroenterology *136*, 1618–1628 e1612.

Poynard, T., Leroy, V., Cohard, M., Thevenot, T., Mathurin, P., Opolon, P., and Zarski, J.P. (1996). Meta-analysis of interferon randomized trials in the treatment of viral hepatitis C: effects of dose and duration. Hepatology *24*, 778–789.

Proietti, E., Bracci, L., Puzelli, S., Di Pucchio, T., Sestili, P., De Vincenzi, E., Venditti, M., Capone, I., Seif, I., De Maeyer, E., *et al.* (2002). Type I IFN as a natural adjuvant for a protective immune response: lessons from the influenza vaccine model. J. Immunol. *169*, 375–383.

Puoti, C., Barbarini, G., Picardi, A., Romano, M., Pellicelli, A., Barlattani, A., Mecenate, F., Guarisco, R., Costanza, O.M., Spilabotti, L., *et al.* (2010). Rapid virological response as a predictor of sustained response in HCV-infected patients with persistently normal alanine aminotransferase levels: a multicenter study. J. Viral Hepat. [Epub ahead of print].

Puyang, X., Poulin, D.L., Mathy, J.E., Anderson, L.J., Ma, S., Fang, Z., Zhu, S., Lin, K., Fujimoto, R., Compton, T., *et al.* (2010). Mechanism of resistance of hepatitis C virus replicons to structurally distinct cyclophilin inhibitors. Antimicrob. Agents Chemother. *54*, 1981–1987.

Reesink, H.W., Fanning, G.C., Farha, K.A., Weegink, C., Van Vliet, A., Van 't Klooster, G., Lenz, O., Aharchi, F., Marien, K., Van Remoortere, P., *et al.* (2010). Rapid HCV-RNA decline with once daily TMC435: a phase I study in healthy volunteers and hepatitis C patients. Gastroenterology *138*, 913–921.

Rehermann, B., and Nascimbeni, M. (2005). Immunology of hepatitis B virus and hepatitis C virus infection. Nat. Rev. Immunol. *5*, 215–229.

Riddell, L.A., Pinching, A.J., Hill, S., Ng, T.T., Arbe, E., Lapham, G.P., Ash, S., Hillman, R., Tchamouroff, S., Denning, D.W., *et al.* (2001). A phase III study of recombinant human interferon gamma to prevent opportunistic infections in advanced HIV disease. AIDS Res. Hum. Retroviruses *17*, 789–797.

Rodriguez-Torres, M., Sulkowski, M.S., Chung, R.T., Hamzeh, F.M., and Jensen, D.M. (2010). Factors associated with rapid and early virologic response to peginterferon alfa-2a/ribavirin treatment in HCV genotype 1 patients representative of the general chronic hepatitis C population. J. Viral Hepat. *17*, 139–147.

Romano, A., and Sadan, Y. (1988). Ten years of experience with human fibroblast interferon in treatment of viral ophthalmic infections. Metab. Pediatr. Syst. Ophthalmol. *11*, 43–46.

Rossignol, J.F., Elfert, A., El-Gohary, Y., and Keeffe, E.B. (2009a). Improved virologic response in chronic hepatitis C genotype 4 treated with nitazoxanide, peginterferon, and ribavirin. Gastroenterology *136*, 856–862.

Rossignol, J.F., Elfert, A., and Keeffe, E.B. (2009b). Treatment of chronic hepatitis C using a 4-week lead-in with nitazoxanide before peginterferon plus nitazoxanide. J. Clin. Gastroenterol. *44*, 504–509.

Rossignol, J.F., Kabil, S.M., El-Gohary, Y., Elfert, A., and Keeffe, E.B. (2008). Clinical trial: randomized, double-blind, placebo-controlled study of nitazoxanide monotherapy for the treatment of patients with chronic hepatitis C genotype 4. Aliment Pharmacol. Ther. *28*, 574–580.

Rustgi, V.K. (2009). Albinterferon alfa-2b, a novel fusion protein of human albumin and human interferon alfa-2b, for chronic hepatitis C. Curr. Med. Res. Opin. *25*, 991–1002.

Rustgi, V.K. (2010). Safety of small molecules in combination with interferon-based therapy for hepatitis C virus. Expert Opin. Drug. Saf. *9*, 883–892.

Rustgi, V.K., Lee, W.M., Lawitz, E., Gordon, S.C., Afdhal, N., Poordad, F., Bonkovsky, H.L., Bengtsson, L., Chandorkar, G., Harding, M., *et al.* (2009). Merimepodib, pegylated interferon, and ribavirin in genotype 1 chronic hepatitis C pegylated interferon and ribavirin nonresponders. Hepatology *50*, 1719–1726.

Sainz, B., Jr. and Halford, W.P. (2002). Alpha/Beta interferon and gamma interferon synergize to inhibit the replication of herpes simplex virus type 1. J. Virol. *76*, 11541–11550.

Sainz, B., Jr., Mossel, E.C., Peters, C.J., and Garry, R.F. (2004). Interferon-beta and interferon-gamma synergistically inhibit the replication of severe acute respiratory syndrome-associated coronavirus (SARS-CoV). Virology *329*, 11–17.

Sarrazin, C., Shiffman, M.L., Hadziyannis, S.J., Lin, A., Colucci, G., Ishida, H., and Zeuzem, S. (2010). Definition of rapid virologic response with a highly sensitive real-time PCR-based HCV RNA assay in peginterferon alfa-2a plus ribavirin response-guided therapy. J. Hepatol. *52*, 832–838.

Sauleda, S., Esteban, J.I., Altisent, C., Puig, L., Esteban, R., and Guardia, J. (2000). Treatment with interferon plus ribavirin in anti-HIV negative patients with congenital coagulation disorders and chronic hepatitis C. Thromb. Haemost. *83*, 807–810.

Sauleda, S., Juarez, A., Esteban, J.I., Altisent, C., Ruiz, I., Puig, L., Esteban, R., and Guardia, J. (2001). Interferon and ribavirin combination therapy for chronic hepatitis C in human immunodeficiency virus-infected patients with congenital coagulation disorders. Hepatology *34*, 1035–1040.

Scagnolari, C., Trombetti, S., Alberelli, A., Cicetti, S., Bellarosa, D., Longo, R., Spano, A., Riva, E., Clementi, M., and Antonelli, G. (2007). The synergistic interaction of interferon types I and II leads to marked reduction in severe acute respiratory syndrome-associated coronavirus replication and increase in the expression of mRNAs for interferon-induced proteins. Intervirology *50*, 156–160.

Scagnolari, C., Vicenzi, E., Bellomi, F., Stillitano, M.G., Pinna, D., Poli, G., Clementi, M., Dianzani, F., and Antonelli, G. (2004). Increased sensitivity of SARS-coronavirus to a combination of human type I and type II interferons. Antivir. Ther. *9*, 1003–1011.

Senturk, H., Baysal, B., Tahan, V., Zerdali, H., Ozaras, R., Tabak, F., Mert, A., Canbakan, B., Tabak, O., and Ozbay, G. (2010). Long-term effect of interferon therapy in patients with HBeAg positive chronic hepatitis B infection. Dig. Dis. Sci. [Epub ahead of print].

Sethi, S.K., Bianco, A., Allen, J.T., Knight, R.A., and Spiteri, M.A. (1997). Interferon-gamma (IFN-gamma) down-regulates the rhinovirus-induced expression of intercellular adhesion molecule-1 (ICAM-1) on human airway epithelial cells. Clin. Exp. Immunol. *110*, 362–369.

Sherman, K.E. (2010). Thymosin alpha 1 for treatment of hepatitis C virus: promise and proof. Ann. N.Y. Acad. Sci. *1194*, 136–140.

Shiffman, M.L. (2010). Treatment of hepatitis C in 2011: what can we expect? Curr. Gastroenterol. Rep. *12*, 70–75.

Shiffman, M.L., Suter, F., Bacon, B.R., Nelson, D., Harley, H., Sola, R., Shafran, S.D., Barange, K., Lin, A., Soman,

A., *et al.* (2007). Peginterferon alfa-2a and ribavirin for 16 or 24 weeks in HCV genotype 2 or 3. N. Engl. J. Med. *357*, 124–134.

Shoukry, N.H., Cawthon, A.G., and Walker, C.M. (2004). Cell-mediated immunity and the outcome of hepatitis C virus infection. Annu. Rev. Microbiol. *58*, 391–424.

Singal, A.G., Volk, M.L., Jensen, D., Di Bisceglie, A.M., and Schoenfeld, P.S. (2010a). A sustained viral response is associated with reduced liver-related morbidity and mortality in patients with hepatitis C virus. Clin. Gastroenterol. Hepatol. *8*, 280–288, 288, e281.

Singal, A.K., and Anand, B.S. (2009). Management of hepatitis C virus infection in HIV/HCV co-infected patients: clinical review. World J. Gastroenterol. *15*, 3713–3724.

Singal, A.K., and Anand, B.S. (2010). Tailoring treatment duration to 12 to 16 weeks in hepatitis C genotype 2 or 3 with rapid virologic response: systematic review and metaanalysis of randomized controlled trials. J. Clin. Gastroenterol. *44*, 583–587.

Singal, A.K., Singh, A., Jaganmohan, S., Guturu, P., Mummadi, R., Kuo, Y.F., and Sood, G.K. (2010b). Antiviral therapy reduces risk of hepatocellular carcinoma in patients with hepatitis C virus-related cirrhosis. Clin. Gastroenterol. Hepatol. *8*, 192–199.

Sjogren, M.H., Sjogren, R., Holtzmuller, K., Winston, B., Butterfield, B., Drake, S., Watts, A., Howard, R., and Smith, M. (2005). Interferon alfacon-1 and ribavirin versus interferon alpha-2b and ribavirin in the treatment of chronic hepatitis C. Dig. Dis. Sci. *50*, 727–732.

Sjogren, M.H., Sjogren, R., Jr., Lyons, M.F., Ryan, M., Santoro, J., Smith, C., Reddy, K.R., Bonkovsky, H., Huntley, B., and Faris-Young, S. (2007). Antiviral response of HCV genotype 1 to consensus interferon and ribavirin versus pegylated interferon and ribavirin. Dig. Dis. Sci. *52*, 1540–1547.

van Slooten, M.L., Hayon, I., Babai, I., Zakay-Rones, Z., Wagner, E., Storm, G., and Kedar, E. (2001). Immunoadjuvant activity of interferon-gamma-liposomes co-administered with influenza vaccines. Biochim. Biophys. Acta. *1531*, 99–110.

Sokal, E.M., Bourgois, A., Stephenne, X., Silveira, T., Porta, G., Gardovska, D., Fischler, B., and Kelly, D. (2010). Peginterferon alfa-2a plus ribavirin for chronic hepatitis C virus infection in children and adolescents. J. Hepatol. *52*, 827–831.

Soloviev, V.N. (1967). The Interferons (New York Academic Press, New York).

Sperber, S.J., Levine, P.A., Innes, D.J., Mills, S.E., and Hayden, F.G. (1988). Tolerance and efficacy of intranasal administration of recombinant beta serine interferon in healthy adults. J. Infect. Dis. *158*, 166–175.

Spiegel, M., Pichlmair, A., Martinez-Sobrido, L., Cros, J., Garcia-Sastre, A., Haller, O., and Weber, F. (2005). Inhibition of Beta interferon induction by severe acute respiratory syndrome coronavirus suggests a two-step model for activation of interferon regulatory factor 3. J. Virol. 79, 2079–2086.

Spiegel, M., Pichlmair, A., Muhlberger, E., Haller, O., and Weber, F. (2004). The antiviral effect of interferon-beta against SARS-coronavirus is not mediated by MxA protein. J. Clin. Virol. *30*, 211–213.

Straley, S.D., and Terrault, N.A. (2004). Chronic Hepatitis B. Curr. Treat Options Gastroenterol. 7, 477–489.

Stroher, U., DiCaro, A., Li, Y., Strong, J.E., Aoki, F., Plummer, F., Jones, S.M., and Feldmann, H. (2004). Severe acute respiratory syndrome-related coronavirus is inhibited by interferon-alpha. J. Infect. Dis. *189*, 1164–1167.

Stubbs, A.C., Martin, K.S., Coeshott, C., Skaates, S.V., Kuritzkes, D.R., Bellgrau, D., Franzusoff, A., Duke, R.C., and Wilson, C.C. (2001). Whole recombinant yeast vaccine activates dendritic cells and elicits protective cell-mediated immunity. Nat. Med. *7*, 625–629.

Subramanian, G.M., Fiscella, M., Lamouse-Smith, A., Zeuzem, S., and McHutchison, J.G. (2007). Albinterferon alpha-2b: a genetic fusion protein for the treatment of chronic hepatitis C. Nat. Biotechnol. *25*, 1411–1419.

Suppiah, V., Moldovan, M., Ahlenstiel, G., Berg, T., Weltman, M., Abate, M.L., Bassendine, M., Spengler, U., Dore, G.J., Powell, E., *et al.* (2009). IL28B is associated with response to chronic hepatitis C interferon-alpha and ribavirin therapy. Nat. Genet. *41*, 1100–1104.

Susser, S., Welsch, C., Wang, Y., Zettler, M., Domingues, F.S., Karey, U., Hughes, E., Ralston, R., Tong, X., Herrmann, E., *et al.* (2009). Characterization of resistance to the protease inhibitor boceprevir in hepatitis C virus-infected patients. Hepatology *50*, 1709–1718.

Suzuki, F., Akuta, N., Suzuki, Y., Sezaki, H., Yatsuji, H., Kawamura, Y., Hosaka, T., Kobayashi, M., Arase, Y., Ikeda, K., *et al.* (2009). Rapid loss of hepatitis C virus genotype 1b from serum in patients receiving a triple treatment with telaprevir (MP-424), pegylated interferon and ribavirin for 12 weeks. Hepatol. Res. *39*, 1056–1063.

Svensson, A., Bellner, L., Magnusson, M., and Eriksson, K. (2007). Role of IFN-alpha/beta signaling in the prevention of genital herpes virus type 2 infection. J. Reprod. Immunol. *74*, 114–123.

Tada, H., Nemoto, E., Shimauchi, H., Watanabe, T., Mikami, T., Matsumoto, T., Ohno, N., Tamura, H., Shibata, K., Akashi, S., *et al.* (2002). Saccharomyces cerevisiae- and Candida albicans-derived mannan induced production of tumor necrosis factor alpha by human monocytes in a CD14- and Toll-like receptor 4-dependent manner. Microbiol. Immunol. *46*, 503–512.

Takahashi, M., Yamada, G., Miyamoto, R., Doi, T., Endo, H., and Tsuji, T. (1993). Natural course of chronic hepatitis C. Am. J. Gastroenterol. *88*, 240–243.

Takkenberg, R.B., Weegink, C.J., Zaaijer, H.L., and Reesink, H.W. (2010). New developments in antiviral therapy for chronic hepatitis B. Vox Sang. *98*, 481–494.

Tanaka, Y., Nishida, N., Sugiyama, M., Kurosaki, M., Matsuura, K., Sakamoto, N., Nakagawa, M., Korenaga,

M., Hino, K., Hige, S., *et al.* (2009). Genome-wide association of IL28B with response to pegylated interferon-alpha and ribavirin therapy for chronic hepatitis C. Nat. Genet. *41*, 1105–1109.

Teoh, N.C., Farrell, G.C., and Chan, H.L. (2010). Individualisation of antiviral therapy for chronic hepatitis C. J. Gastroenterol. Hepatol. *25*, 1206–1216.

Theron, M., Huang, K.J., Chen, Y.W., Liu, C.C., and Lei, H.Y. (2005). A probable role for IFN-gamma in the development of a lung immunopathology in SARS. Cytokine *32*, 30–38.

Thompson, A., Patel, K., Tillman, H., and McHutchison, J.G. (2009). Directly acting antivirals for the treatment of patients with hepatitis C infection: a clinical development update addressing key future challenges. J. Hepatol. *50*, 184–194.

Thompson, A.J., Muir, A.J., Sulkowski, M.S., Ge, D., Fellay, J., Shianna, K.V., Urban, T., Afdhal, N.H., Jacobson, I.M., Esteban, R., *et al.* (2010). Interleukin-28B Polymorphism Improves Viral Kinetics and Is the Strongest Pretreatment Predictor of Sustained Virologic Response in Genotype 1 Hepatitis C Virus. Gastroenterology *139*, 120–129 e118.

Thompson, W.W., Shay, D.K., Weintraub, E., Brammer, L., Bridges, C.B., Cox, N.J., and Fukuda, K. (2004). Influenza-associated hospitalizations in the United States. JAMA *292*, 1333–1340.

Thompson, W.W., Shay, D.K., Weintraub, E., Brammer, L., Cox, N., Anderson, L.J., and Fukuda, K. (2003). Mortality associated with influenza and respiratory syncytial virus in the United States. JAMA *289*, 179–186.

Tong, M.J., Reddy, K.R., Lee, W.M., Pockros, P.J., Hoefs, J.C., Keeffe, E.B., Hollinger, F.B., Hathcote, E.J., White, H., Foust, R.T., *et al.* (1997). Treatment of chronic hepatitis C with consensus interferon: a multicenter, randomized, controlled trial. Consensus Interferon Study Group. Hepatology *26*, 747–754.

Toyoda, H., Kumada, T., Kiriyama, S., Tanikawa, M., Hisanaga, Y., Kanamori, A., Tada, T., Takagi, M., Hiramatsu, T., Hosokawa, T., *et al.* (2010). Transient reappearance of serum hepatitis C virus RNA observed by real-time PCR during antiviral therapy with peginterferon and ribavirin in patients with HCV genotype 1b. J. Clin. Virol. *47*, 258–262.

Treanor, J.J., Betts, R.F., Erb, S.M., Roth, F.K., and Dolin, R. (1987). Intranasally administered interferon as prophylaxis against experimentally induced influenza A virus infection in humans. J. Infect. Dis. *156*, 379–383.

Tsantrizos, Y.S. (2009). TMC-435, an NS3/4A protease inhibitor for the treatment of HCV infection. Curr. Opin. Investig. Drugs *10*, 871–881.

Van den Eynde, E., Tiraboschi, J.M., Tural, C., Sola, R., Mira, J.A., Podzamczer, D., Jou, A., Canete, N., Pineda, J.A., Pahissa, A., *et al.* (2010). Ability of treatment week 12 viral response to predict long-term outcome in genotype 1 hepatitis C virus/HIV coinfected patients. AIDS *24*, 975–982.

Van Hoeven, N., Belser, J.A., Szretter, K.J., Zeng, H., Staeheli, P., Swayne, D.E., Katz, J.M., and Tumpey, T.M. (2009). Pathogenesis of 1918 pandemic and H5N1 influenza virus infections in a guinea pig model: antiviral potential of exogenous alpha interferon to reduce virus shedding. J. Virol. *83*, 2851–2861.

Varghese, R., Al-Khaldi, J., Asker, H., Fadili, A.A., Al Ali, J., and Hassan, F.A. (2009). Treatment of chronic hepatitis C genotype 4 with peginterferon alpha-2a plus ribavirin. Hepatogastroenterology *56*, 218–222.

Villa, E., Lei, B., Taliani, G., Graziosi, A., Critelli, R., Luongo, M., Gennari, W., Bianchini, M., and Ferretti, I. (2009). Pretreatment with pegylated interferon prevents emergence of lamivudine mutants in lamivudine-naive patients: a pilot study. Antivir. Ther. *14*, 1081–1087.

Vollstedt, S., Arnold, S., Schwerdel, C., Franchini, M., Alber, G., Di Santo, J.P., Ackermann, M., and Suter, M. (2004). Interplay between alpha/beta and gamma interferons with B, T, and natural killer cells in the defense against herpes simplex virus type 1. J. Virol. *78*, 3846–3850.

Wagoner, J., Negash, A., Kane, O.J., Martinez, L.E., Nahmias, Y., Bourne, N., Owen, D.M., Grove, J., Brimacombe, C., McKeating, J.A., *et al.* (2010). Multiple effects of silymarin on the hepatitis C virus lifecycle. Hepatology *51*, 1912–1921.

Ward, H., and Ronn, M. (2010). Contribution of sexually transmitted infections to the sexual transmission of HIV. Curr. Opin. HIV AIDS *5*, 305–310.

Ward, S.E., Loutfy, M.R., Blatt, L.M., Siminovitch, K.A., Chen, J., Hinek, A., Wolff, B., Pham, D.H., Deif, H., LaMere, E.A., *et al.* (2005). Dynamic changes in clinical features and cytokine/chemokine responses in SARS patients treated with interferon alfacon-1 plus corticosteroids. Antivir. Ther. *10*, 263–275.

Watashi, K. (2010). Alisporivir, a cyclosporin derivative that selectively inhibits cyclophilin, for the treatment of HCV infection. Curr. Opin. Investig. Drugs *11*, 213–224.

Weiss, I.D., Wald, O., Wald, H., Beider, K., Abraham, M., Galun, E., Nagler, A., and Peled, A. (2010). IFN-gamma treatment at early stages of influenza virus infection protects mice from death in a NK cell-dependent manner. J. Interferon Cytokine Res. *30*, 439–449.

Whitby, K., Taylor, D., Patel, D., Ahmed, P., and Tyms, A.S. (2004). Action of celgosivir (6 O-butanoyl castanospermine) against the pestivirus BVDV: implications for the treatment of hepatitis C. Antivir. Chem. Chemother. *15*, 141–151.

Whiteman, S.C., and Spiteri, M.A. (2008). IFN-gamma regulation of ICAM-1 receptors in bronchial epithelial cells: soluble ICAM-1 release inhibits human rhinovirus infection. J. Inflamm. *5*, 8.

Wiegand, J., Buggisch, P., Boecher, W., Zeuzem, S., Gelbmann, C.M., Berg, T., Kauffmann, W., Kallinowski, B., Cornberg, M., Jaeckel, E., *et al.* (2006). Early monotherapy with pegylated interferon alpha-2b for acute hepatitis C infection: the HEP-NET acute-HCV-II study. Hepatology *43*, 250–256.

Wirth, S., Ribes-Koninckx, C., Calzado, M.A., Bortolotti, F., Zancan, L., Jara, P., Shelton, M., Kerkar, N., Galoppo,

M., Pedreira, A., *et al.* (2010). High sustained virologic response rates in children with chronic hepatitis C receiving peginterferon alfa-2b plus ribavirin. J. Hepatol. *52*, 501–507.

Wong, V.W., Wong, G.L., Yan, K.K., Chim, A.M., Chan, H.Y., Tse, C.H., Choi, P.C., Chan, A.W., Sung, J.J., and Chan, H.L. (2010). Durability of peginterferon alfa-2b treatment at 5 years in patients with hepatitis B e antigen-positive chronic hepatitis B. Hepatology *51*, 1945–1953.

Wormald, S., and Hilton, D.J. (2004). Inhibitors of cytokine signal transduction. J. Biol. Chem. *279*, 821–824.

Yada, M., Masumoto, A., Yamashita, N., Motomura, K., Koyanagi, T., and Sakamoto, S. (2010). Immediate virological response predicts the success of short-term peg-interferon monotherapy for chronic hepatitis C. World J. Gastroenterol. *16*, 1506–1511.

Yamazaki, M., Sato, A., Takeda, T., and Komatsu, M. (2010). Distinct clinical courses in type 1 diabetes mellitus induced by peg-interferon-alpha treatment for chronic hepatitis C. Int. Med. *49*, 403–407.

Yang, Y.F., Zhao, W., Zhong, Y.D., Xia, H.M., Shen, L., and Zhang, N. (2009). Interferon therapy in chronic hepatitis B reduces progression to cirrhosis and hepatocellular carcinoma: a meta-analysis. J. Viral Hepatol. *16*, 265–271.

Yoo, J.-K., Baker, D.P., Fish, E.N. (2010). Interferon-β modulates type I immunity during influenza virus infection. Antiviral. Res. *88*, 64–71.

Yu, M.L., Dai, C.Y., Huang, J.F., Hou, N.J., Lee, L.P., Hsieh, M.Y., Chiu, C.F., Lin, Z.Y., Chen, S.C., Wang, L.Y., *et al.* (2007). A randomised study of peginterferon and ribavirin for 16 versus 24 weeks in patients with genotype 2 chronic hepatitis C. Gut *56*, 553–559.

Zeuzem, S., and Poordad, F. (2010). Pegylated-interferon plus ribavirin therapy in the treatment of CHC: individualization of treatment duration according to on-treatment virologic response. Curr. Med. Res. Opin. *26*, 1733–1743.

Zhao, Z., Zhang, F., Xu, M., Huang, K., Zhong, W., Cai, W., Yin, Z., Huang, S., Deng, Z., Wei, M., *et al.* (2003). Description and clinical treatment of an early outbreak of severe acute respiratory syndrome (SARS) in Guangzhou, PR China. J. Med. Microbiol. *52*, 715–720.

Zheng, B., He, M.L., Wong, K.L., Lum, C.T., Poon, L.L., Peng, Y., Guan, Y., Lin, M.C., and Kung, H.F. (2004). Potent inhibition of SARS-associated coronavirus (SCOV) infection and replication by type I interferons (IFN-alpha/beta) but not by type II interferon (IFN-gamma). J. Interferon Cytokine Res. *24*, 388–390.

Zheng, Y., Zhao, L., Wu, T., Guo, S., Chen, Y., and Zhou, T. (2009). Efficacy of consensus interferon in treatment of HbeAg-positive chronic hepatitis B: a multicentre, randomized controlled trial. Virol. J. *6*, 99.

Zylberberg, H., Benhamou, Y., Lagneaux, J.L., Landau, A., Chaix, M.L., Fontaine, H., Bochet, M., Poynard, T., Katlama, C., Pialoux, G., *et al.* (2000). Safety and efficacy of interferon–ribavirin combination therapy in HCV-HIV coinfected subjects: an early report. Gut *47*, 694–697.

Index

T

V

W

Y

Z

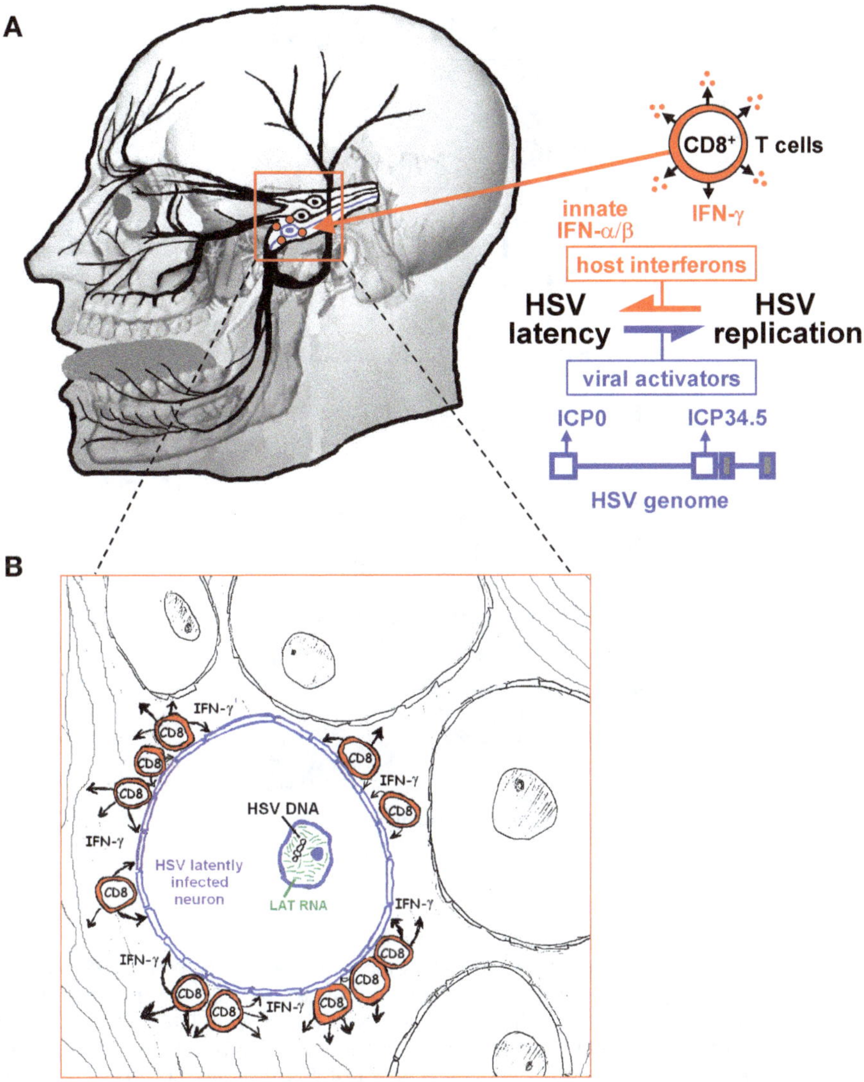

Plate 5.1 An equilibrium model of the latency-replication balance of HSV. (A) Schematic of a human trigeminal ganglion. The ophthalmic and maxillary branches collect nerve fibres from the eye and face, respectively. The lower, mandibular branch collects nerve fibres from the mouth and is the predominant site of latent HSV-1 infection in humans. Neuronal cell bodies collect sensory inputs from the epithelium and relay signals to the brain. A single HSV latently infected neuron is illustrated as a large blue cell surrounded by smaller, red CD8+ T-cells. In the schematic to the right, it is proposed that HSV infections of neurons are controlled by the opposing forces of host interferons (red) and viral activator proteins (blue). Innate interferons (α/β) and CD8+ T-cell-secreted interferon-γ are proposed to favour the establishment of latent HSV infections. Viral activator proteins ICP0 and ICP34.5 are proposed to favour viral spread by dismantling two interferon-inducible blocks to HSV replication. Adapted from the Netter Anatomy Illustration Collection© (Netter, 1997) with permission from Elsevier Inc. All rights reserved. (B) Schematic cross-section of a trigeminal ganglion showing a single, HSV latently infected neuron (blue cell) surrounded by CD8+ T-cells (red cells) that have infiltrated the tissue as part of the host response to viral infection. Green lines in the nucleus of the HSV latently infected neuron represent the HSV latency-associated transcript (LAT) RNAs that may be detected in the HSV latently infected ganglia of experimental animals and humans.

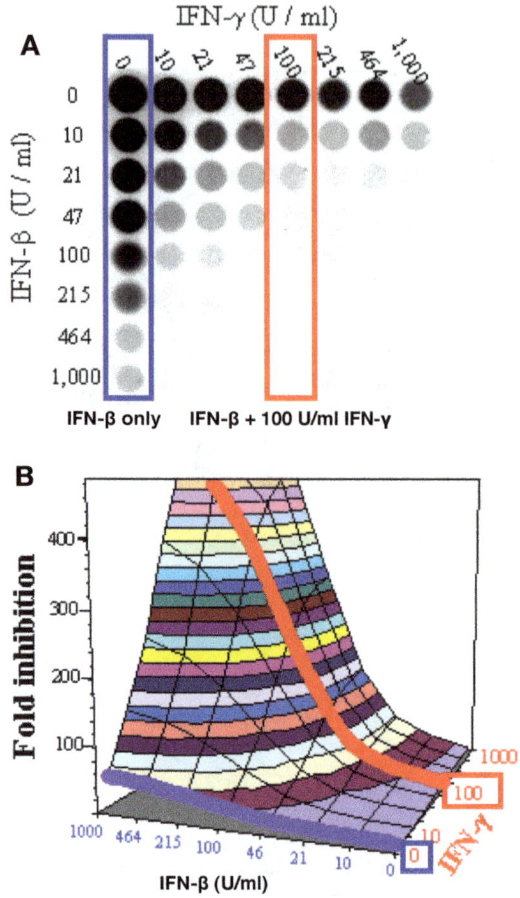

Plate 5.3 Interferon-γ multiplies the potency with which interferon-α/β inhibits HSV-1 replication. (A) Viral DNA harvested 30 hours p.i. from Vero cell cultures inoculated with 0.1 pfu per cell of HSV-1 and treated 16 hours earlier with 64 dose combinations of IFN-β and IFN-γ. Viral DNA yields were quantified by the relative abundance of *glycoprotein D* gene-specific oligonucleotide that hybridized to DNA harvested from HSV-1 infected cultures. The blue box denote the effect of IFN-β alone, compared with the red box which denotes the effect of IFN-β when combined with 100 U/ml of IFN-γ. (B) A mathematically modelled response surface that is based on the assumption that combinations of IFN-β and IFN-γ inhibit HSV-1 replication by an amount that is equal to the multiplied product of their individual inhibitory effects. Regression analysis validated that the modelled inhibitory effect (i.e. smoothed) did not significantly differ from the observed effect of combinations of IFN-β and IFN-γ ($P > 0{:}2$; two-tailed paired *t*-test). The thick blue line superimposed on the response surface denotes the inhibitory effect of IFN-β alone, compared with the thick red line which denotes the effect of IFN-β when combined with 100 U/ml of IFN-γ. Adapted from Halford, *et al.* (2005a) in accordance with the author reproduction policies of Elsevier, Inc.

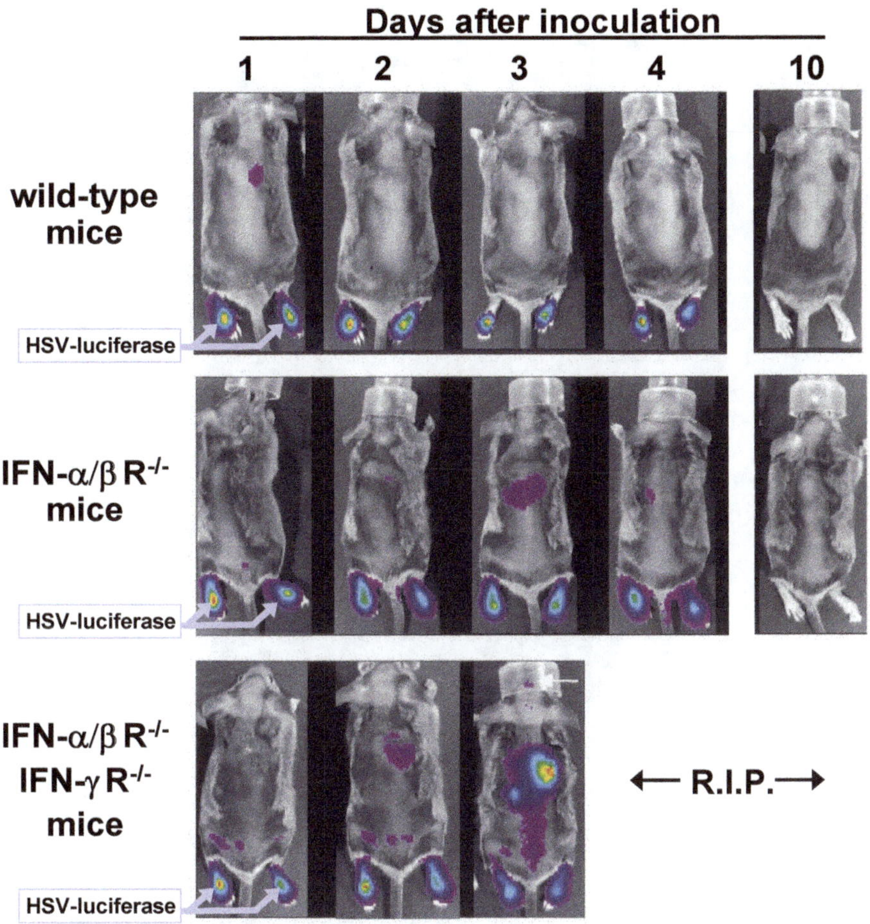

Plate 5.4 Uncontrolled spread of HSV-1 infection in mice lacking both interferon-α/β and interferon-γ signalling pathways. Comparison of the spread of a luciferase-expressing HSV-1 virus in a representative wild-type mouse, IFN-α/β receptor knockout mouse, or a mouse lacking both IFN-α/β receptors and IFN-β receptors. Mice were inoculated in the rear footpads with 2×10^6 pfu of a luciferase-expressing HSV-1 recombinant virus. Bioluminescence imaging was performed every day for 10 days; results of Days 1–4 and 10 are shown. All wild-type and IFN-α/β-receptor knockout mice survived until day 10 p.i.; 0% of double-knockout mice lacking IFN-α/β receptors and IFN-γ receptors survived beyond day 4 p.i. Adapted from Luker *et al.* (2003) with permission from the American Society for Microbiology.

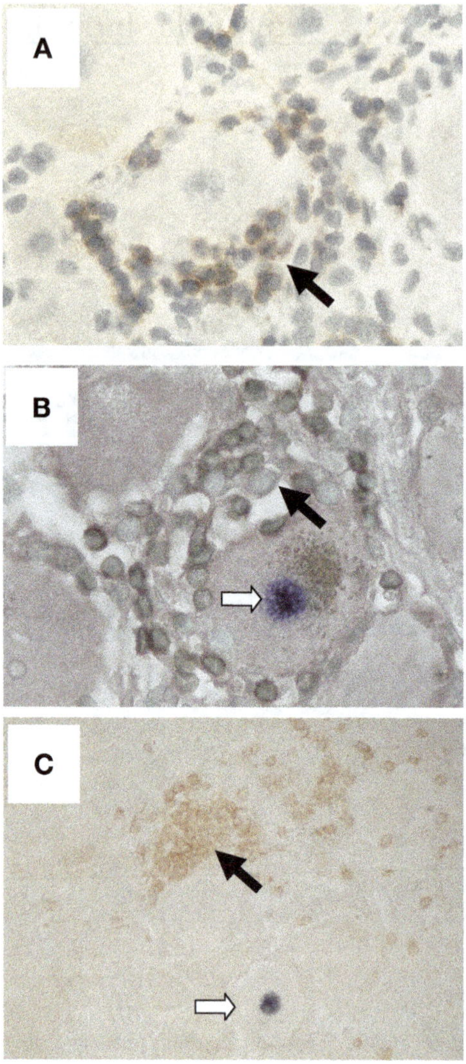

Plate 5.5 T-lymphocytes persist at sites of latent HSV infection in humans. Representative photomicrographs of a tissue section from a human trigeminal ganglion latently infected with HSV-1 stained with anti-CD3 antibodies, which recognize the cell surface CD3 molecule that is associated with the T-cell receptor of both CD4+ T-cells and CD8+ T-cells. (A) A single, large neuronal cell body encircled by an infiltrate of small, brown-stained CD3+ T-cells (denoted by black arrow). Reprinted from *Am. J. Pathol.* 2003, 163: 2179–2184 with permission from the American Society for Investigative Pathology. (B) *In situ* hybridization for HSV-1's LAT RNA introns highlights the nucleus of a single, human neuron that is latently infected with HSV-1 (denoted by white arrow), and immunohistochemical staining reveals the presence of small, brown-stained CD3+ T-cells in an adjacent mononuclear cell infiltrate. Adapted from Derfuss *et al.*, 2007, *Brain Pathol.* 17: 389–398 with permission from John Wiley and Sons. (C) An infiltrate of CD3+ T-cells (denoted by black arrow) drawn to a site several neuronal cell bodies away from an HSV-1 LAT+ neuron (denoted by white arrow). Original images were generously provided by Diethilde Theil.

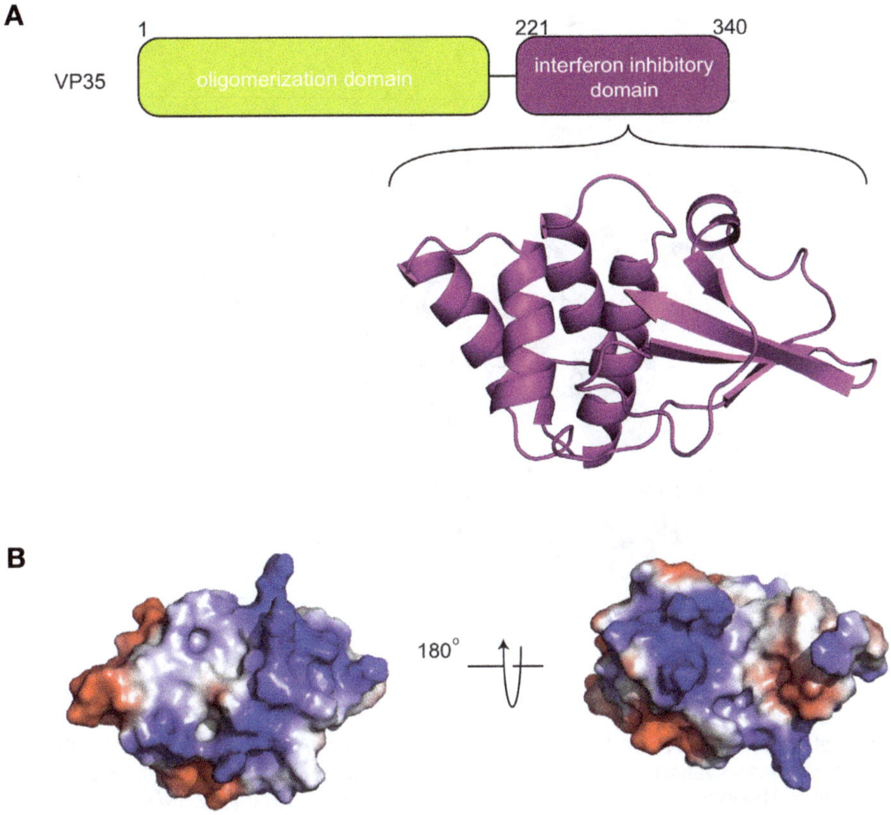

Plate 7.1 Domain organization and structure of the EBOV VP35 interferon inhibitory domain (IID). A. A diagram representing the organization of the EBOV VP35 protein is presented. A N-terminal domain contains an oligomerization function, while the C-terminal region from amino acids 221–340, comprises a dsRNA-binding domain required for maximum VP35 IFN-antagonist function. The C-terminal 'interferon inhibitory domain' has been crystallized, and the structure of this domain is shown. B. Two surface representations of the VP35 IID structure shown in A.

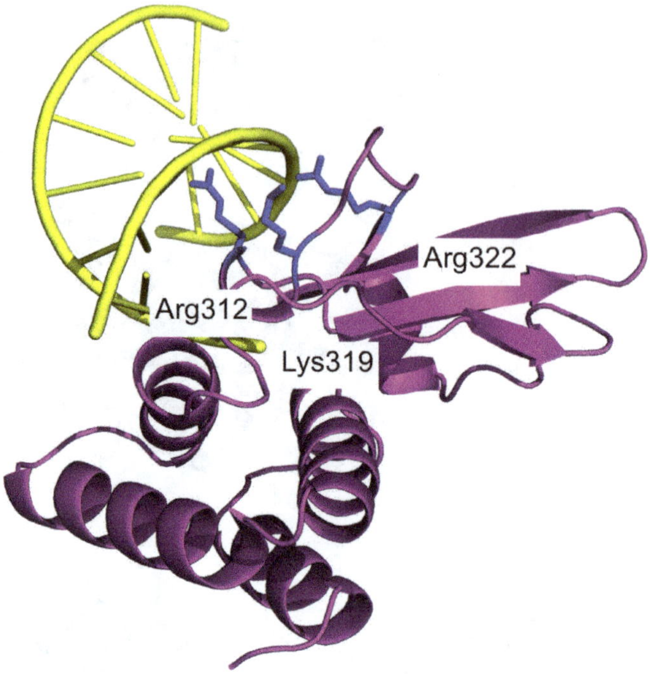

Plate 7.2 Location of Arg312, Lys319 and Arg322 in the VP35IID-dsRNA structure. The VP35 IID crystallized as a pair of IID dimers in association with an 8 basepair dsRNA. Shown is one VP35 IID monomer making contact with the dsRNA. The location of VP35 residues Arg312, Lys319 and Arg322 are shown. As detailed in the text, these residues are important for VP35-dsRNA binding and VP35 IFN-antagonist function. Mutation of these residues, in the context of an infectious EBOV, greatly attenuates the virus.

www.ingramcontent.com/pod-product-compliance
Lightning Source LLC
LaVergne TN
LVHW062227200926
848484LV00006B/1490
* 9 7 8 1 9 0 4 4 5 5 8 1 3 *